BARRON'S

AP®
Statistics

PREMIUM

WITH 9 PRACTICE TESTS

ELEVENTH EDITION

Martin Sternstein, Ph.D.

Professor Emeritus
Department of Mathematics
Ithaca College
Ithaca, New York

About the Author

Dr. Martin Sternstein, Professor Emeritus at Ithaca College, was honored by Princeton Review as one of the nation's "300 Best College Professors." He is a long-time College Board consultant and has been a Reader and Table Leader for the AP Statistics exam for many years. He has strong interests in national educational and social issues concerning equal access to math education for all. For two years, he was a Fulbright Professor in Liberia, West Africa, after which he developed a popular "Math in Africa" course, and he is the only mathematician to have given a presentation at the annual Conference on African Linguistics. He also taught the first U.S. course for college credit in chess theory.

Author's Note

Thanks to Annie Bernberg, editor at Kaplan, for her guidance. Thanks to my brother, Allan, my sister-in-law, Marilyn, my sons, Jonathan and Jeremy, my daughters-in-law, Asia and Cheryl, and my grandchildren, Jaiden, Jordan, Josiah, Luna, and Jayme, for their heartfelt love and support. Most of all, thanks are due to my wife, Faith, whose love, warm encouragement, and always calm and optimistic perspective on life provide a home environment in which deadlines can be met and goals easily achieved. My sincere appreciation goes to the participants who have attended my AP Statistics workshops for teaching me just as much as I have taught them, and special thanks for their many useful suggestions are due to the following exceptional teachers:

David Bock	Ithaca High School, Ithaca, NY (retired)
Paul Buckley	Gonzaga College High School, Washington, DC
Dawn Dentato	Somers High School, Lincolndale, NY
Jared Derksen	Rancho Cucamonga High School, Rancho Cucamonga, CA
Kristina Gazdik	Bishop Kelly High School, Boise, ID
Sarah Johnson	Grand Blanc High School, Grand Blanc, MI
Lee Kucera	University of California, Irvine, CA
Brendan Murphy	John Bapst Memorial High School, Bangor, ME
Kenn Pendleton	Montgomery College, Germantown, MD
Adam Riazi	Cabell Midland High School, Ona, WV
Roy St. Laurent	Northern Arizona University, Flagstaff, AZ
Daren Starnes	The Lawrenceville School, Lawrenceville, NJ
Josh Tabor	Canyon del Oro High School, Oro Valley, Arizona
Jane Viau	Frederick Douglass Academy, New York, NY
Dawn White	Silver Lake Regional High School, Kingston, MA

Ithaca College

Spring 2020

Martin Sternstein

© Copyright 2020, 2019, 2017, 2015, 2013, 2012, 2010, 2007 by Kaplan, Inc., d/b/a Barron's Educational Series
© Copyright 2004, 2000, 1998 by Kaplan, Inc., d/b/a Barron's Educational Series, under the title *How to Prepare for the AP Advanced Placement Exam in Statistics*.

Published by Kaplan, Inc.,
d/b/a Barron's Educational Series
750 Third Avenue
New York, NY 10017
www.barronseduc.com

ISBN: 978-1-5062-5892-8

10 9 8 7 6 5 4 3 2

Kaplan, Inc., d/b/a Barron's Educational Series print books are available at special quantity discounts to use for sales promotions, employee premiums, or educational purposes. For more information or to purchase books, please call the Simon & Schuster special sales department at 866-506-1949.

Contents

PART THREE: FINAL REVIEW

PART FOUR: PRACTICE TESTS

PART FIVE: APPENDICES

As you review the content in this book and work toward earning that **5** on your AP STATISTICS exam, here are five things that you **MUST** know:

Barron's **Essential**

1

Graders want to give you credit—help them! Make them understand *what* you are doing, *why* you are doing it, and *how* you are doing it. Don't make the reader guess at what you are doing.

- **Communication** is just as important as statistical knowledge!
- Be sure you understand **exactly what you are being asked to do or find or explain** and **approach each problem systematically**.
- Some problems look scary on first reading but are not overly difficult and are surprisingly straightforward. Questions that take you beyond the scope of the AP curriculum will be phrased in ways that you should be able to answer them based on what you have learned in your AP Statistics class.
- *Naked* or *bald answers* will receive little or **no** credit. You must show where answers come from.
- On the other hand, don't give more than one solution to the same problem—you will receive credit only for the weaker one.

2

Random sampling and random assignment are different ideas.

- Random sampling is the use of chance in selecting a sample from a population and is critical in being able to generalize from a sample to a population.
  - A *simple random sample* (SRS) is when every possible sample of a given size has the same chance of being selected.
  - A *stratified random sample* is when the population is divided into homogeneous units called strata, and random samples are chosen from each strata.
  - A *cluster sample* is when the population is divided into heterogeneous units called clusters, and a random sample of the clusters is chosen.
  - A *systematic sample* is when a random starting point is followed by choosing every *k*th member of a list.
- Random assignment in experiments is when subjects are randomly assigned to treatments and is critical in minimizing the effect of possible confounding variables.
  - This randomization evens out effects over which you have no control and allows for a valid comparison of treatments.
  - *Randomized block design* refers to when the randomization occurs only within groups of similar experimental units called blocks.

3

Distributions describe variability, and variability is the most fundamental concept in statistics. Understand the difference between:

- *population distribution* (variability in an entire population),
- *sample distribution* (variability within a particular sample), and
- *sampling distribution* (variability between samples).
- The larger the sample size, the more the **sample distribution** looks like the population distribution.
- Central Limit Theorem: the larger the sample size, the more the **sampling distribution** (probability distribution of the sample means) looks like a normal distribution.

4

Choosing the correct procedure and performing proper checks are critical.

- Categorical variables lead to proportions or chi-square procedures, while quantitative variables lead to means or linear regression.
- Decide whether there is a single population of interest or two populations being compared.
- Know the proper checks for each procedure and state them correctly. (Listing wrong conditions will lose points.)
- Verifying assumptions and conditions means more than simply listing them with little check marks—you must show work or give some reason to confirm verification.

5

Calculating the *P*-value is not the final step of a hypothesis test.

- There must be a *decision* to reject or fail to reject the null hypothesis.
- You must indicate how you interpret the *P*-value; that is, you need *linkage*. So, "Given that $P = 0.007$, I reject ..." isn't enough. You need something like, "Because $P = 0.007$ is less than 0.05, there is sufficient evidence to reject ..."
- The conclusion must refer to the population and be *in the context* of the problem.
- Reemphasizing from #1 above: **Communication** is just as important as statistical knowledge!

Introduction

In 1997, 7,667 students took the AP Statistics exam. As enrollment in AP Statistics classes increased at a higher rate than in any other AP class, over 220,000 took the exam in 2019. The number of students required to take statistics in college has surpassed the number of students required to take calculus. High schools across the country have recognized this trend and are developing and expanding their statistics offerings. The Common Core mathematics standards feature statistics and probability in a primary role throughout the high school curriculum. This Barron's book is intended both as a topical review during the year and as a final review in the weeks before the AP exam.

The contents of this book cover the topics recommended by the AP Statistics Development Committee. This review book follows the College Board's suggested organization into 9 units. Interspersed among these units are 37 quizzes (mini-AP exams with both multiple-choice and free-response questions), which should be used as progress checks. These quizzes are an opportunity to test yourself in what teachers call a "high rigor, low stress" setting. There are 6 full-length exams, 1 diagnostic and 5 practice, all with instructive, complete answers. This 11th edition of *Barron's AP Statistics* has been updated to adapt to the ongoing changes in topic emphasis and scoring guidelines of the AP exam.

On the AP Statistics exam, you will be furnished with a list of formulas (from (I) Descriptive Statistics, (II) Probability and Distributions, and (III) Inferential Statistics) and tables (including standard normal probabilities, t-distribution critical values, and χ^2 critical values). While you will be expected to bring a graphing calculator with statistics capabilities to the exam, it is not recommended leaving answers in terms of calculator syntax. Furthermore, many students have commented that calculator usage was less than they had anticipated. However, even though the calculator is simply a tool, to be used sparingly, as needed, you need to be proficient with this technology. You also must be comfortable with reading generic computer output.

The exam consists of two, equally weighted, parts: a 90-minute section with 40 multiple-choice questions and a 90-minute free-response section with five open-ended questions and one investigative task to complete. In grading, the two sections of the exam are given equal weight. Students have remarked that the first section involves "lots of reading," while the second section involves "lots of writing." The percentage of questions from each content area is approximately 25% data analysis, 15% experimental design, 25% probability, and 35% inference.

In the multiple-choice section, the questions are much more conceptual than computational, and thus use of the calculator is minimal. The score on the multiple-choice section is based on the number of correct answers, with no points deducted for incorrect answers. So don't leave any blank answers!

The multiple-choice section can be broken down as follows: Exploring One-Variable Data (15–23%), Exploring Two-Variable Data (5–7%), Collecting Data (12–15%), Probability,

Random Variables, and Probability Distributions (10–20%), Sampling Distributions (7–12%), Inference for Categorical Data: Proportions (12–15%), Inference for Quantitative Data: Means (10–18%), Inference for Categorical Data: Chi-Square (2–5%), and Inference for Quantitative Data: Slopes (2–5%).

In the free-response section, the first five open-ended questions can be broken down as follows:

- 1 multipart question with a primary focus on collecting data
- 1 multipart question with a primary focus on exploring data
- 1 multipart question with a primary focus on probability and sampling distributions
- 1 question with a primary focus on inference
- 1 question that combines 2 or more skill categories

The investigative task, the sixth question in the free-response section, assesses multiple skills and content in a nonroutine way.

In the free-response section, you must show all your work, and communication skills go hand in hand with statistical knowledge. You must indicate your methods clearly, as the problems will be graded on the correctness of the methods as well as on the accuracy of the results and explanation. That is, the free-response answers should address *why* a particular test was chosen, not just *how* the test is performed. Even if you use a calculator, such as the TI-84, Casio Prizm, or HP Prime, to perform a statistical test, formulas need to be understood. Choice of test in inference must include confirmation of underlying assumptions, and answers must be stated in context, not just as numbers.

Free-response questions are scored on a 0 to 4 scale with 1 point for a *minimal* response, 2 points for a *developing* response, 3 points for a *substantial* response, and 4 points for a *complete* response. Individual parts of these questions are scored as E for *essentially* correct, P for *partially* correct, and I for *incorrect*. Note that *essentially* correct does not mean *perfect*. Work is graded *holistically*—that is, a student's complete response is considered as a whole whenever scores do not fall precisely on an integer value on the 0 to 4 scale.

Each open-ended question counts 15% of the total free-response score and the investigative task counts 25% of the free-response score. The first open-ended question is typically the most straightforward, and after doing this one to build confidence, students might consider looking at the investigative task since it counts more.

Each completed AP exam will receive a grade based on a 5 point scale, with 5 the highest score and 1 the lowest score. Most colleges and universities accept a grade of 3 or better for credit or advanced placement or both. The cut scores on a recent exam together with the percent of students receiving each score on the 2019 exam are given in the following table.

AP Score	Total Points	Students (%)
5	73–100	14.5%
4	59–72	18.0%
3	44–58	26.7%
2	32–43	19.7%
1	0–31	21.1%

While a review book such as this can be extremely useful in helping you prepare for the AP exam (practice problems, practice more problems, and practice even more problems are the three strongest pieces of advice), nothing can substitute for a good high school teacher and a good textbook. This author personally recommends the following texts from among the many excellent books on the market: *Stats: Modeling the World* by Bock, Bullard, Velleman, and De Veaux; *The Practice of Statistics* by Starnes and Tabor; *Workshop Statistics: Discovery with Data* by Rossman and Chance; and *Statistics: Learning from Data* (AP Edition) by Peck and Olsen.

Other helpful sources of information are the College Board's websites: *https://apstudents. collegeboard.org/courses/ap-statistics* for students, and *https://apcentral.collegeboard.org/ courses/ap-statistics?course=ap-statistics* for teachers. When a teacher registers a class on the College Board site, much more information is at *https://myap.collegeboard.org.*

A good piece of advice is for you—from day one—to develop critical practices (like checking assumptions and conditions), to acquire strong technical skills, and to always write clear and thorough, yet to the point, interpretations and conclusions in context. Final answers to most problems should not be numbers but, rather, sentences explaining and analyzing numerical results. To help develop skills and insights to tackle AP free-response questions (which often choose contexts students haven't seen before), pick up newspapers and magazines and figure out how to apply what you are learning to better understand articles in print that reference numbers, graphs, and statistical studies.

While using this Barron's review book, you should study the text and illustrative examples carefully and try to complete the practice quiz problems before referring to the solution keys. Simply reading the detailed explanations to the answers without first striving to work through the questions on one's own is not the best approach. There is an old adage: *Mathematics is not a spectator sport!* Teachers clearly may use this book with a class in many profitable ways. Ideally, each individual unit review, together with practice quiz problems, should be assigned after the unit has been covered in class. The full-length practice tests should be reserved for final review shortly before the AP exam.

Although high-stakes assessments heighten anxiety and can actually hinder learning, research has shown that frequent testing, done right, can be an effective way to learn. With this in mind, use the practice quizzes at appropriate times throughout the school year. You will develop confidence and a deeper understanding of the material!

The Final Review part has several invaluable sections. These include:

- "Selecting an Appropriate Inference Procedure," which gives hints and insights into how to approach inference recognition, followed by two practice quizzes on naming the procedure you would use, defining parameters, listing conditions to be checked, and stating hypotheses if appropriate
- "Statistical Insights into Social Issues," which gives two quizzes of comprehensive review questions covering the whole AP Statistics curriculum with each set in a context that aims to give an appreciation of the power of statistics to give insights into some of society's most pressing issues
- "The Investigative Task," which helps you prepare for free-response Question 6, the last question on the exam, counting one-eighth of the whole exam; there are three illustrative examples followed by a quiz with seven practice investigative tasks
- "50 Misconceptions"
- "50 Common Errors on the AP Exam"
- "50 AP Exam Hints, Advice, and Reminders"

The AP Statistics course is one of the most important courses offered in the high school curriculum. If you work with your teacher and study hard, you will find this to be an enjoyable course, you will do well on the AP exam, and you will develop into a more thoughtful citizen of this world!

ANSWER SHEET
Diagnostic Test

1. Ⓐ Ⓑ Ⓒ Ⓓ Ⓔ
2. Ⓐ Ⓑ Ⓒ Ⓓ Ⓔ
3. Ⓐ Ⓑ Ⓒ Ⓓ Ⓔ
4. Ⓐ Ⓑ Ⓒ Ⓓ Ⓔ
5. Ⓐ Ⓑ Ⓒ Ⓓ Ⓔ
6. Ⓐ Ⓑ Ⓒ Ⓓ Ⓔ
7. Ⓐ Ⓑ Ⓒ Ⓓ Ⓔ
8. Ⓐ Ⓑ Ⓒ Ⓓ Ⓔ
9. Ⓐ Ⓑ Ⓒ Ⓓ Ⓔ
10. Ⓐ Ⓑ Ⓒ Ⓓ Ⓔ

11. Ⓐ Ⓑ Ⓒ Ⓓ Ⓔ
12. Ⓐ Ⓑ Ⓒ Ⓓ Ⓔ
13. Ⓐ Ⓑ Ⓒ Ⓓ Ⓔ
14. Ⓐ Ⓑ Ⓒ Ⓓ Ⓔ
15. Ⓐ Ⓑ Ⓒ Ⓓ Ⓔ
16. Ⓐ Ⓑ Ⓒ Ⓓ Ⓔ
17. Ⓐ Ⓑ Ⓒ Ⓓ Ⓔ
18. Ⓐ Ⓑ Ⓒ Ⓓ Ⓔ
19. Ⓐ Ⓑ Ⓒ Ⓓ Ⓔ
20. Ⓐ Ⓑ Ⓒ Ⓓ Ⓔ

21. Ⓐ Ⓑ Ⓒ Ⓓ Ⓔ
22. Ⓐ Ⓑ Ⓒ Ⓓ Ⓔ
23. Ⓐ Ⓑ Ⓒ Ⓓ Ⓔ
24. Ⓐ Ⓑ Ⓒ Ⓓ Ⓔ
25. Ⓐ Ⓑ Ⓒ Ⓓ Ⓔ
26. Ⓐ Ⓑ Ⓒ Ⓓ Ⓔ
27. Ⓐ Ⓑ Ⓒ Ⓓ Ⓔ
28. Ⓐ Ⓑ Ⓒ Ⓓ Ⓔ
29. Ⓐ Ⓑ Ⓒ Ⓓ Ⓔ
30. Ⓐ Ⓑ Ⓒ Ⓓ Ⓔ

31. Ⓐ Ⓑ Ⓒ Ⓓ Ⓔ
32. Ⓐ Ⓑ Ⓒ Ⓓ Ⓔ
33. Ⓐ Ⓑ Ⓒ Ⓓ Ⓔ
34. Ⓐ Ⓑ Ⓒ Ⓓ Ⓔ
35. Ⓐ Ⓑ Ⓒ Ⓓ Ⓔ
36. Ⓐ Ⓑ Ⓒ Ⓓ Ⓔ
37. Ⓐ Ⓑ Ⓒ Ⓓ Ⓔ
38. Ⓐ Ⓑ Ⓒ Ⓓ Ⓔ
39. Ⓐ Ⓑ Ⓒ Ⓓ Ⓔ
40. Ⓐ Ⓑ Ⓒ Ⓓ Ⓔ

Diagnostic Test

SECTION I

Questions 1–40

Spend 90 minutes on this part of the exam.

> **Directions:** The questions or incomplete statements that follow are each followed by five suggested answers or completions. Choose the response that best answers the question or completes the statement.

1. It is estimated that 56 percent of Americans have pets. However, favorite pet of choice differs by geographic location. A random sample of pet owners is cross-classified by geographic location and pet of choice. The results are summarized in the following segmented bar chart.

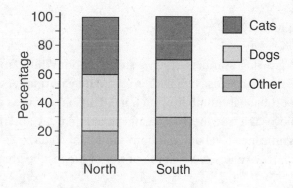

Which of the following is an *incorrect* conclusion?

(A) More pet owners in the South than in the North answered "Other."

(B) Twice as many pet owners in the North answered "Dogs" than answered "Other."

(C) The same number of pet owners in the South answered "Cats" as answered "Other."

(D) In both the North and South, the same proportion of pet owners answered "Dogs."

(E) A greater proportion of pet owners in the North than in the South answered "Cats."

2. Is there a linear relationship between calories and sodium content in beef hot dogs? A random sample of 20 beef hot dogs gives the following regression output:

```
Dependent variable is: Sodium

Predictor        Coef  SE Coef       T       P
Constant      -228.33    77.97   -2.93   0.009
Calories       4.0133   0.4922    8.15   0.000

S = 48.5799    R-Sq = 78.7%    R-Sq(adj) = 77.5%
```

Which of the following gives a 99% confidence interval for the slope of the regression line?

(A) $4.0133 \pm 2.861\left(\dfrac{0.4922}{\sqrt{20}}\right)$

(B) $4.0133 \pm (2.861)(0.4922)$

(C) $4.0133 \pm (2.878)(0.4922)$

(D) $4.0133 \pm 2.861\left(\dfrac{48.5799}{\sqrt{20}}\right)$

(E) $4.0133 \pm 2.878\left(\dfrac{48.5799}{\sqrt{20}}\right)$

3. In tossing a fair coin, which of the following sequences is more likely to appear?

(A) HHHHH
(B) HTHTHT
(C) HTHHTTH
(D) TTHTHHTH
(E) All are equally likely.

4. There have been growing numbers of news stories about White Americans calling the police on people of color whose behavior is completely normal. A criminologist hypothesizes that the mean number of such incidents across the country is 2 per day. A sociologist believes the true mean is greater than 2 per day and plans a hypothesis test at the 5% significance level on a random sample of 50 days. For which of the following possible true values of μ will the power of the test be greatest?

(A) 1.5
(B) 1.85
(C) 2.0
(D) 2.15
(E) 2.4

5. A simple random sample is defined by

   (A) the method of selection.
   (B) how representative the sample is of the population.
   (C) whether or not a random number generator is used.
   (D) the assignment of different numbers associated with the outcomes of some chance situation.
   (E) examination of the outcome.

6. Can shoe size be predicted from height? In a random sample of 50 teenagers, the standard deviation in heights was 8.7 cm, while the standard deviation in shoe size was 2.3. The least squares regression equation was:

   Predicted shoe size $= -33.6 + 0.25(Height\ in\ cm)$

   What was r, the correlation coefficient?

   (A) $\dfrac{(0.25)(8.7)}{2.3}$

   (B) $\dfrac{(0.25)(2.3)}{8.7}$

   (C) $\dfrac{2.3}{\left(\dfrac{8.7}{\sqrt{50}}\right)}$

   (D) $\dfrac{8.7}{\left(\dfrac{2.3}{\sqrt{50}}\right)}$

   (E) There is not enough information to calculate the correlation coefficient.

Questions 7–9 refer to the following situation:

   A researcher would like to show that a new oral diabetes medication she developed helps control blood sugar level better than insulin injection. She plans to run a hypothesis test at the 5% significance level.

7. What would be a Type I error?

   (A) The researcher concludes she has sufficient evidence that her new medication helps more than insulin injection, and her medication really is better than insulin injection.
   (B) The researcher concludes she has sufficient evidence that her new medication helps more than insulin injection, when in reality her medication is not better than insulin injection.
   (C) The researcher concludes she does not have sufficient evidence that her new medication helps more than insulin injection, and her medication really is not better than insulin injection.
   (D) The researcher concludes she does not have sufficient evidence that her new medication helps more than insulin injection, when in reality her medication is better than insulin injection.
   (E) The researcher concludes she has sufficient evidence that her new medication controls blood sugar level the same as insulin injection, and in reality there is a difference.

8. What would be a Type II error?

   (A) The researcher concludes she has sufficient evidence that her new medication helps more than insulin injection, and her medication really is better than insulin injection.
   (B) The researcher concludes she has sufficient evidence that her new medication helps more than insulin injection, when in reality her medication is not better than insulin injection.
   (C) The researcher concludes she does not have sufficient evidence that her new medication helps more than insulin injection, and her medication really is not better than insulin injection.
   (D) The researcher concludes she does not have sufficient evidence that her new medication helps more than insulin injection, when in reality her medication is better than insulin injection.
   (E) The researcher concludes she has sufficient evidence that her new medication controls blood sugar level the same as insulin injection, and in reality there is a difference.

9. The researcher thinks she can improve her chances by running five identical hypotheses tests, each using a different group of diabetic volunteers, hoping that at least one of the tests will show that her new oral diabetes medication helps control blood sugar level better than insulin injection. What is the probability of committing at least one Type I error?

   (A) 0.05
   (B) $5(0.05)(0.95)^4$
   (C) $1 - (0.95)^5$
   (D) $(0.95)^5$
   (E) 0.95

   _____

10. A financial analyst determines the yearly research and development investments for 50 blue chip companies. She notes that the distribution is distinctly not bell-shaped. If the 50 dollar amounts are converted to z-scores, what can be said about the standard deviation of the 50 z-scores?

   (A) It depends on the distribution of the raw scores.
   (B) It is less than the standard deviation of the raw scores.
   (C) It is greater than the standard deviation of the raw scores.
   (D) It is equal to the standard deviation of the raw scores.
   (E) It equals 1.

11. A coin is weighted so that the probability of heads is 0.6. The coin is tossed 20 times, and the number of heads is noted. This procedure is repeated a total of 200 times, and the number of heads is recorded each time. What kind of distribution has been simulated?

(A) The sampling distribution of the sample proportion with $n = 20$ and $p = 0.6$
(B) The sampling distribution of the sample proportion with $n = 200$ and $p = 0.6$
(C) The sampling distribution of the sample proportion with $\bar{x} = (20)(0.6)$ and $\sigma = \sqrt{20(0.6)(0.4)}$
(D) The binomial distribution with $n = 20$ and $p = 0.6$
(E) The binomial distribution with $n = 200$ and $p = 0.6$

12. A 100-question multiple-choice history exam is graded as number correct minus $\frac{1}{4}$ number incorrect, so scores can take values from -25 to $+100$. Suppose the standard deviation for one class's results is reported to be -3.14. What is the proper conclusion?

(A) More students received negative scores than positive scores.
(B) At least half the class received negative scores.
(C) Some students must have received negative scores.
(D) Some students must have received positive scores.
(E) An error was made in calculating the standard deviation.

13. Of the 423 seniors graduating this year from a city high school, 322 plan to go on to college. When the principal asks an AP student to calculate a 95% confidence interval for the proportion of this year's graduates who plan to go to college, the student says that this would be inappropriate. Why?

(A) The independence assumption may have been violated (students tend to do what their friends do).
(B) There is no evidence that the data come from a normal or nearly normal population (GPAs help determine college admission and may be skewed).
(C) Randomization was not used.
(D) There is a difference between a confidence interval and a hypothesis test with regard to the proportion of graduates planning on college.
(E) The population proportion is known, so a confidence interval has no meaning.

14. An AP Statistics student in a large high school plans to survey his fellow students with regard to their preference between using a laptop or using a tablet. Which of the following survey methods is unbiased?

(A) The student comes to school early and surveys the first 50 students who arrive.
(B) The student passes a survey card to every student with instructions to fill it out at home and drop the filled-out card in a box by the school entrance the next day.
(C) The student creates an online survey and asks everyone to respond.
(D) The student goes to all of the high school sports events for a week, hands out the survey, and waits for each student to fill it out and hand it back.
(E) None of the above sampling methods are unbiased.

15. In a random sample of 500 students, it was reported that test grades went up an average of at least 10 points for 70 percent of the students when usage of cell phones was banned during the school day. What was the degree of confidence if the margin of error was ± 2.5 percent?

   (A) $P(-0.025 < z < 0.025)$

   (B) $P(-0.675 < z < 0.725)$

   (C) $P\left(-\dfrac{0.025}{\sqrt{(0.5)(0.5)/500}} < z < \dfrac{0.025}{\sqrt{(0.5)(0.5)/500}}\right)$

   (D) $P\left(-\dfrac{0.025}{\sqrt{(0.7)(0.3)/500}} < z < \dfrac{0.025}{\sqrt{(0.7)(0.3)/500}}\right)$

   (E) $P\left(-\dfrac{0.675}{\sqrt{(0.7)(0.3)/500}} < z < \dfrac{0.725}{\sqrt{(0.7)(0.3)/500}}\right)$

16. In a random sample of 10 insects of a newly discovered species, an entomologist measures an average life expectancy of 17.3 days with a standard deviation of 2.3 days. Assuming all conditions for inference are met, what is a 95% confidence interval for the mean life expectancy for insects of this species?

   (A) $17.3 \pm 1.96\left(\dfrac{2.3}{\sqrt{9}}\right)$

   (B) $17.3 \pm 1.96\left(\dfrac{2.3}{\sqrt{10}}\right)$

   (C) $17.3 \pm 2.228\left(\dfrac{2.3}{\sqrt{9}}\right)$

   (D) $17.3 \pm 2.228\left(\dfrac{2.3}{\sqrt{10}}\right)$

   (E) $17.3 \pm 2.262\left(\dfrac{2.3}{\sqrt{10}}\right)$

17. A coin is weighted so that heads is twice as likely to occur as tails. The coin is flipped repeatedly until a tail occurs. Let X be the number of flips made. What is the most probable value for X?

   (A)   1
   (B)   2
   (C)   3
   (D)   4
   (E)   5

18. Suppose we are interested in determining whether or not a student's score on the AP Statistics exam is a reasonable predictor of the student's GPA in the first year of college. Which of the following is the best statistical test?

   (A) Two-sample t-test of population means
   (B) Linear regression t-test
   (C) Chi-square test of independence
   (D) Chi-square test of homogeneity
   (E) Chi-square test of goodness of fit

19. The following are the graphs of three normal curves and three cumulative distribution graphs:

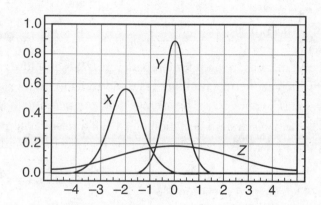

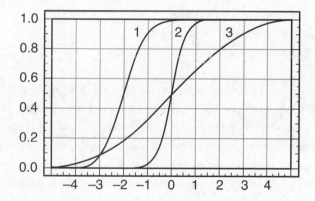

Which normal curve corresponds to which cumulative curve?

   (A) X and 1, Y and 2, Z and 3
   (B) X and 1, Y and 3, Z and 2
   (C) X and 2, Y and 1, Z and 3
   (D) X and 3, Y and 1, Z and 2
   (E) X and 3, Y and 2, Z and 1

20. A campus has 55% male and 45% female students. Suppose 30% of the male students pick basketball as their favorite sport compared to 20% of the females. If a randomly chosen student picks basketball as the student's favorite sport, what is the probability the student is male?

    (A) $\dfrac{0.30}{0.30 + 0.20}$

    (B) $\dfrac{0.55}{0.30 + 0.20}$

    (C) $\dfrac{0.30}{(0.55)(0.30) + (0.45)(0.20)}$

    (D) $\dfrac{0.55}{(0.55)(0.30) + (0.45)(0.20)}$

    (E) $\dfrac{(0.55)(0.30)}{(0.55)(0.30) + (0.45)(0.20)}$

21. The kelvin is a unit of measurement for temperature; 0 K is absolute zero, the temperature at which all thermal motion ceases. Conversion from Fahrenheit to Kelvin is given by $K = \dfrac{5}{9} \times (F - 32) + 273$. The average daily temperature in Monrovia, Liberia, is 78.35°F with a standard deviation of 6.3°F. If a scientist converts Monrovia daily temperatures to the Kelvin scale, what will be the new mean and standard deviation?

    (A) Mean, 25.75 K; standard deviation, 3.5 K
    (B) Mean, 231.75 K; standard deviation, 3.5 K
    (C) Mean, 298.75 K; standard deviation, 3.5 K
    (D) Mean, 298.75 K; standard deviation, 258.72 K
    (E) Mean, 298.75 K; standard deviation, 276.5 K

22. A cattle veterinarian is considering two experimental designs to compare two sources of bovine growth hormone, or BVH, to spur increased milk production in Guernsey cattle. Design 1 involves flipping a coin as each cow enters the stockade, and if *heads*, giving it BVH from bovine cadavers, and if *tails*, giving it BVH from engineered *E. coli*. Design 2 involves flipping a coin as each cow enters the stockade, and if *heads*, giving it BVH from bovine cadavers for a specified period of time and then switching to BVH from engineered *E. coli* for the same period of time, and if *tails*, the order is reversed. With both designs, daily milk production is noted. Which of the following is accurate?

    (A) Neither design uses randomization since there is no indication that cows will be randomly picked from the population of all Guernsey cattle.
    (B) Design 1 is a completely randomized design, while Design 2 is a block design.
    (C) Both designs use double-blinding, but neither uses a placebo.
    (D) In the second design, BVH from bovine cadavers and BVH from engineered *E. coli* are confounded.
    (E) One of the two designs is actually an observational study, while the other is an experiment.

23. The purpose of the linear regression *t*-test is

(A) to determine if there is a linear association between two numerical variables.
(B) to find a confidence interval for the slope of a regression line.
(C) to find the *y*-intercept of a regression line.
(D) to be able to calculate residuals.
(E) to be able to determine the consequences of Type I and Type II errors.

24. A fair die is tossed 12 times, and the number of 3's is noted. This is repeated 200 times. Which of the following distributions is the most likely to occur?

(A)
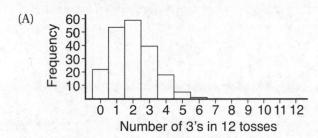

(B)
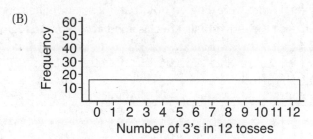

(C)
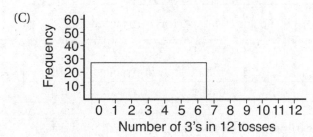

(D)
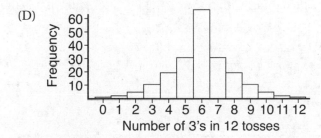

(E)
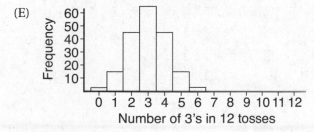

25. Which of the following is a true statement about sampling?

(A) If the sample is random, the size of the sample usually doesn't matter.
(B) If the sample is random, the size of the population usually doesn't matter.
(C) A sample of less than 1% of the population is too small for statistical inference.
(D) A sample of more than 10% of the population is too large for statistical inference.
(E) All of the above are true statements.

26. Suppose, in a study of mated pairs of soldier beetles, it is found that the measure of the elytron (hardened forewing) length is always 0.5 millimeters longer in the female. What is the correlation between elytron lengths of mated females and males?

(A) −1
(B) −0.5
(C) 0
(D) 0.5
(E) 1

27. A random sample of 100 individuals who were singled out at an international airport security checkpoint is reviewed, and the individuals are classified according to region of origin:

Region of origin	United States	Europe	Arabic	Asia, non-Arabic	Other
Number singled out	41	19	15	13	12

The proportion of travelers in each category who use this airport follows:

Region of origin	United States	Europe	Arabic	Asia, non-Arabic	Other
Proportion	0.64	0.12	0.08	0.09	0.07

We wish to test whether the distribution of people singled out is the same as the distribution of people who use the airport with regard to region of origin. What is the appropriate χ^2 statistic?

(A) $\dfrac{(41-64)^2}{64} + \dfrac{(19-12)^2}{12} + \dfrac{(15-8)^2}{8} + \dfrac{(13-9)^2}{9} + \dfrac{(12-7)^2}{7}$

(B) $\dfrac{(41-64)^2}{41} + \dfrac{(19-12)^2}{19} + \dfrac{(15-8)^2}{15} + \dfrac{(13-9)^2}{13} + \dfrac{(12-7)^2}{12}$

(C) $\dfrac{(0.41-0.64)^2}{0.64} + \dfrac{(0.19-0.12)^2}{0.12} + \dfrac{(0.15-0.08)^2}{0.08} + \dfrac{(0.13-0.09)^2}{0.09} + \dfrac{(0.12-0.07)^2}{0.07}$

(D) $\dfrac{(0.41-0.64)^2}{0.41} + \dfrac{(0.19-0.12)^2}{0.19} + \dfrac{(0.15-0.08)^2}{0.15} + \dfrac{(0.13-0.09)^2}{0.13} + \dfrac{(0.12-0.07)^2}{0.12}$

(E) $\dfrac{(41-64)^2}{20} + \dfrac{(19-12)^2}{20} + \dfrac{(15-8)^2}{20} + \dfrac{(13-9)^2}{20} + \dfrac{(12-7)^2}{20}$

28. The age distribution for a particular debilitating disease has a mean greater than the median. Which of the following graphs most likely illustrates this distribution?

(A)
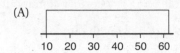

(B)
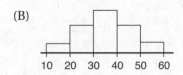

(C)
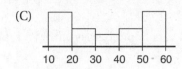

(D)
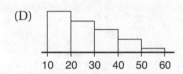

(E)
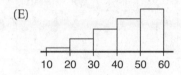

29. Suppose we have a random variable X where the probability associated with the value $\binom{10}{k} (0.38)^k (0.62)^{10-k}$ for $k = 0, \ldots, 10$.

What is the mean of X?

(A) 0.38
(B) 0.62
(C) 3.8
(D) 5.0
(E) 6.2

30. It is hypothesized that high school varsity pitchers throw fastballs at an average of 80 mph. A random sample of varsity pitchers is timed with radar guns resulting in a 95% confidence interval of (74.5, 80.5). Which of the following is a correct statement?

(A) There is a 95% chance that the mean fastball speed of all varsity pitchers is 80 mph.

(B) There is a 95% chance that the mean fastball speed of all varsity pitchers is 77.5 mph.

(C) Most of the interval is below 80, so there is evidence at the 5% significance level that the mean of all varsity pitchers is something other than 80 mph.

(D) The test $H_0: \mu = 80$, $H_a: \mu \neq 80$ is not significant at the 5% significance level, but it would be at the 1% level.

(E) It is likely that the true mean fastball speed of all varsity pitchers is within 3 mph of the sample mean fastball speed.

31. A recent study noted prices and battery lives of 10 top-selling tablet computers. The data follow:

	1	2	3	4	5	6	7	8	9	10
Cost	303	450	260	480	540	390	350	400	600	450
Battery life (hr)	8.5	10	7	11	10	9	8	9.5	11	9.5

The residual plot of the least squares model is

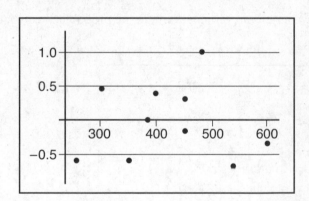

What is the model's predicted battery life for the tablet computer costing $480?

(A) 10 hr
(B) 10.5 hr
(C) 11 hr
(D) 11.5 hr
(E) 12 hr

32. Should college athletes be required to give their coaches their social media account usernames and passwords? A survey of student-athletes is to be taken. The statistician believes that Division I, II, and III players may differ in their views, so she selects a random sample of athletes from each division to survey. This is a

   (A) simple random sample.
   (B) stratified sample.
   (C) cluster sample.
   (D) systematic sample.
   (E) convenience sample.

33. In a random sample of 1500 college students, a pollster found 45% prefer a female president and 42% prefer a male president. To calculate a 95 percent confidence interval for the difference in the proportion of college students who prefer a female president over a male president, he uses $(0.45 - 0.42) \pm 1.96\sqrt{\frac{(0.45)(0.55)}{1500} + \frac{(0.42)(0.58)}{1500}}$. Were conditions for inference met?

   (A) Yes, because there was a random sample, 1500 is less than 10% of all college students, and 1500(0.45), 1500(0.55), 1500(0.42), and 1500(0.58) are all ≥ 10.
   (B) No, because there was no random assignment between female and male college students.
   (C) No, because 1500 is not greater than 10% of all college students.
   (D) No, because the independence assumption is violated.
   (E) No, because the random sample may not be truly representative of the population.

34. The population of the Greater Tokyo area is 34,400,000 and of Karachi is 17,200,000. A random sample of citizens is to be taken in each city, and 95% confidence intervals for the mean age in each city will be calculated. Assuming roughly equal sample standard deviations, to obtain the same margin of error for each confidence interval,

   (A) the sample sizes should be the same.
   (B) the sample in Greater Tokyo should be twice the size of the sample in Karachi.
   (C) the sample in Karachi should be twice the size of the sample in Greater Tokyo.
   (D) the sample in Greater Tokyo should be four times the size of the sample in Karachi.
   (E) the sample in Karachi should be four times the size of the sample in Greater Tokyo.

35. A truant officer determines the mean and standard deviation of the number of student absences for all school days during an academic year. Which of the following is the best description of the standard deviation?

   (A) Approximately the median difference between the number of students absent on individual days and the median number of absences on all days
   (B) Approximately the mean difference between the number of students absent on individual days and the mean number of absences on all days
   (C) The difference between the greatest number and the least number of absences among all days during the year
   (D) The difference between the greatest number and the mean number of absences among all days during the year
   (E) The difference between the greatest number and the least number of absences among the middle 50 percent of all the daily absences during the year

36. Which of the following is an *incorrect* statement?

(A) Statistics are random variables with their own probability distributions.
(B) The standard error does not depend on the size of the population.
(C) Bias means that, on average, our estimate of a parameter is different from the true value of the parameter.
(D) There are some statistics for which the sampling distribution is not approximately normal, no matter how large the sample size.
(E) The larger the sample size, the closer the sample distribution is to a normal distribution.

37. For male Air Force cadets, the recommended fitness level with regard to the number of push-ups is 34. In a test whether or not current classes of recruits can meet this standard, a t-test of H_0: $\mu = 34$ against H_a: $\mu < 34$ gives a P-value of 0.068. Using this data, among the following, which is the largest level of confidence for a two-sided confidence interval that does not contain 34?

(A) 85%
(B) 90%
(C) 92%
(D) 95%
(E) 96%

38. A particular car is tested for stopping distance in feet on wet pavement at 30 mph using tires with one tread design and then tires with another tread design. For each set of tires, the test is repeated 30 times, and the following parallel boxplots give a comparison of the resulting five-number summaries.

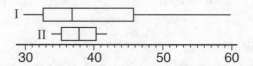

Which of the following is a reasonable conclusion?

(A) Distribution I is skewed right, while distribution II is bell-shaped.
(B) Distribution I is skewed left, while distribution II is a normal distribution.
(C) The mean of distribution I is greater than the mean of distribution II.
(D) The range of distribution I is approximately $46 - 33 = 13$.
(E) The upper 50% of the values in distribution I are all greater than the lower 50% of the values in distribution II.

39. In American roulette, there are 18 red pockets, 18 black pockets, and two green pockets (labeled 0 and 00). The ball is equally likely to land in any of the 38 pockets. What is the probability that a player ends up with a positive outcome, that is, makes money, after 50 equal bets on "red" (that is, for each of 50 spins of the wheel, the player wins or loses the specified identical dollar bet depending on whether or not the ball lands in a red or nonred pocket, respectively)?

(A) 0.105
(B) 0.212
(C) 0.303
(D) 0.408
(E) 0.500

40. Do middle school and high school students have different views on what makes someone popular? Random samples of 100 middle school and 100 high school students yield the following counts with regard to three choices: lots of money, good at sports, and good looks:

	Money	Sports	Looks
Middle school	22	48	30
High school	36	24	40

A chi-square test of homogeneity yields which of the following test statistics?

(A) $\dfrac{(22-29)^2}{22}+\dfrac{(48-36)^2}{48}+\dfrac{(30-35)^2}{30}+\dfrac{(36-29)^2}{36}+\dfrac{(24-36)^2}{24}+\dfrac{(40-35)^2}{40}$

(B) $\dfrac{(22-29)^2}{29}+\dfrac{(48-36)^2}{36}+\dfrac{(30-35)^2}{35}+\dfrac{(36-29)^2}{29}+\dfrac{(24-36)^2}{36}+\dfrac{(40-35)^2}{35}$

(C) $\left(\dfrac{22}{29}\right)^2+\left(\dfrac{48}{36}\right)^2+\left(\dfrac{30}{35}\right)^2+\left(\dfrac{36}{29}\right)^2+\left(\dfrac{24}{36}\right)+\left(\dfrac{40}{35}\right)^2$

(D) $\dfrac{(22)(29)}{58}+\dfrac{(48)(36)}{72}+\dfrac{(30)(35)}{70}+\dfrac{(36)(29)}{58}+\dfrac{(24)(36)}{72}+\dfrac{(40)(35)}{70}$

(E) $\sqrt{(22-29)^2+(48-36)^2+(30-35)^2+(36-29)^2+(24-36)^2+(40-35)^2}$

STOP

If there is still time remaining, you may check your work on this section.

SECTION II

Part A

QUESTIONS 1–5

Spend about 65 minutes on this part of the exam.
Percentage of Section II grade—75

> **Directions:** You must show all work and indicate the methods you use. You will be graded on the correctness of your methods and on the accuracy of your results and explanations.

1. A horticulturist plans a study on the use of compost tea for plant disease management. She obtains 16 identical beds, each containing a random selection of five minipink rose plants. She plans to use two different composting times (two and five days), two different compost preparations (aerobic and anaerobic), and two different spraying techniques (with and without adjuvants). Midway into the growing season she will check all plants for rose powdery mildew disease.
    (a) List the complete set of treatments.
    (b) Describe a completely randomized design for the treatments above.
    (c) Explain the advantage of using only minipink roses in this experiment.
    (d) Explain a disadvantage of using only minipink roses in this experiment.

2. A top-100, 7.0-rated tennis pro wishes to compare a new Wilson N1 racket against his current model. He strings the new racket with the same Luxilon strings at 60 pounds tension that he uses on his old racket. From past testing, he knows that the average forehand cross court volley with his old racket is 82 miles per hour (mph). On an indoor court, using a ball machine set at 70 mph, the same speed he had his old racket tested against, he takes 47 swings with the new racket. An associate with a speed gun records an average of 83.5 mph with a standard deviation of 3.4 mph. Assuming that the 47 swings represent a random sample of his swings, is there statistical evidence that his speed with the new racket is an improvement over the old? Justify your answer.

3. On the social media platform Snapchat, users who contact each other once per day develop a Snapstreak. The Snapstreaks of students at a large suburban high school have an approximately normal distribution with mean 26.4 and standard deviation 8.2.

(a) What is the probability that a given Snapstreak is over 30?
(b) In a random sample of five independent Snapstreaks of students at this school, what is the probability that a majority are over 30?
(c) What is the probability that the mean of the five independent Snapstreaks is over 30?

4. Two hundred fifty randomly chosen people raised in the United States were asked their expression for "soft drink" and the geographic region where they were raised. The results are summarized in the following two-way table.

	Geographic region within the U. S.			
Expression	Northeast	Midwest	West	South
Coke	10	5	5	30
Soda	50	20	40	5
Pop	20	50	10	5

(a) Of the people calling soft drinks "soda," what proportion were from the Northeast?
(b) Of the people from the West, what proportion called soft drinks "pop"?
(c) The mosaic plot below displays the distribution of soft drink expressions given by people from different geographic locations. Describe what this plot reveals about the association between these two variables for the 250 people in the study.

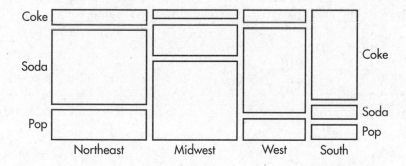

5. Researchers want to study whether a capuchin monkey named Rafiki can correctly predict (better than guesswork) whether the Dow Jones Industrial Average (DJIA) will go up or down on any given day. In a random sample of 10 days, Rafiki correctly predicted the rise or fall of the DJIA 7 of the 10 days (by choosing to eat a banana from a box with an "up" arrow or a box with a "down" arrow).

(a) What are the null and alternative hypotheses?

(b) If you conduct a simulation to investigate whether the observed result provides strong evidence that Rafiki can correctly predict the rise or fall of the DJIA, what would you use for the probability of success, for the sample size, and for the number of samples?

(c) Which of the following graphs could reasonably have come from the simulation?

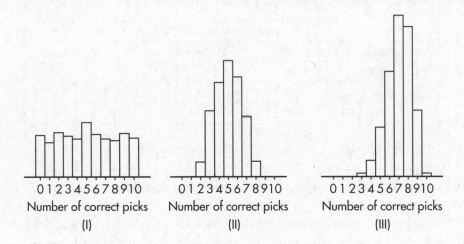

(d) Which of 6.5, 0.65, 0.15, and 0.015 is closest to the *P*-value? Interpret it in context.

(e) Make a conclusion based on the simulation and *P*-value.

SECTION II

Part B

QUESTION 6

Spend about 25 minutes on this part of the exam.
Percentage of Section II grade—25

6. A college counselor is interested in whether or not the number of hours a student studies per week has a statistically significant linear association to the student's GPA. She takes a random sample of 25 students and, for each, records weekly study hours versus GPA on an anonymous survey. The resulting scatterplot along with regression output for a linear model is shown below.

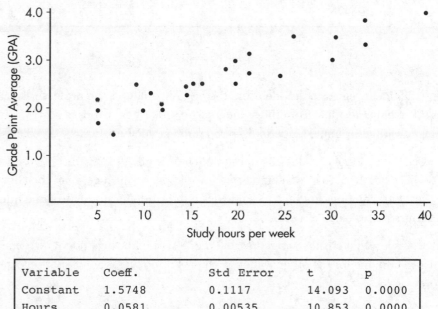

Variable	Coeff.		Std Error	t	p
Constant	1.5748		0.1117	14.093	0.0000
Hours	0.0581		0.00535	10.853	0.0000
S = 0.2620	R-Sq = 0.8366	R-Sq(adj) = 0.8295			

(a) Interpret the *y*-intercept in context.
(b) Interpret the slope in context.
(c) What proportion of the variation in GPAs is not accounted for by the linear regression model?
(d) Assuming all conditions for inference are met, find a 95% confidence interval for the slope, and interpret this in context.

A math professor further analyzes the data and notes that 10 of the students took AP Statistics in high school. The modified scatterplot showing this additional information is shown below.

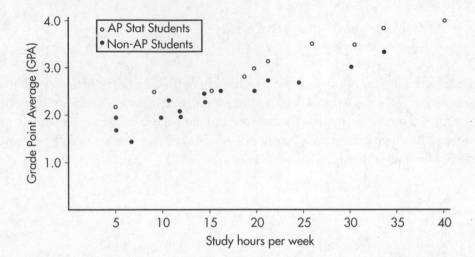

Using the linear regression model from the original analysis, the professor calculates that the average residual from the students who took AP Statistics is 0.2414, while the average residual from the students who did not take AP Statistics is −0.1609.

(e) Use the residual calculations to estimate how much greater the GPA for a student who takes AP Statistics in high school would be, on average, than the GPA for a student who does not take AP Statistics if the students study the same number of hours per week.

The professor then creates two regression models, one for the students who take AP Statistics and one for the other students. The resulting regression equations are shown below.

Linear Fit for Students who take AP Statistics in High School:

   Predicted GPA = 1.9427 + 0.0524(Hours)

Linear Fit for Students who do not take AP Statistics in High School:

   Predicted GPA = 1.5396 + 0.0502(Hours)

(f) A 95% confidence interval for the true difference in the two slopes is 0.0022 ± 0.0134. Based on this interval, is there a significant difference in the two slopes? Explain.

(g) A 95% confidence interval for the true difference in the two y-intercepts is 0.4031 ± 0.2767. Based on this interval, is there a significant difference in the two y-intercepts? Explain.

If there is still time remaining, you may check your work on this section.

Answer Key

SECTION I

1. **A**	9. **C**	17. **A**	25. **B**	33. **D**
2. **C**	10. **E**	18. **B**	26. **E**	34. **A**
3. **A**	11. **D**	19. **A**	27. **A**	35. **B**
4. **E**	12. **E**	20. **E**	28. **D**	36. **E**
5. **A**	13. **E**	21. **C**	29. **C**	37. **A**
6. **A**	14. **E**	22. **B**	30. **E**	38. **C**
7. **B**	15. **D**	23. **A**	31. **A**	39. **C**
8. **D**	16. **E**	24. **A**	32. **B**	40. **B**

Answers and Explanations

SECTION I

1. **(A)** Without knowing the actual number of the sample pet owners from each location, only proportions, not numbers of pet owners, can be compared between locations.

2. **(C)** The critical t-values with $df = 20 - 2 = 18$ and 0.005 in each tail are $\pm$ invT(0.995,18) $= \pm 2.878$. Thus, we have $b \pm t^* \times SE(b) = 4.0133 \pm (2.878)(0.4922)$.

3. **(A)** The shortest sequence has a greater probability than any longer sequence.

4. **(E)** Power, the probability of rejecting a false null hypothesis, will be the greatest for true values furthest from the hypothesized value, in the direction of the alternative hypothesis. Here, $H_a: \mu > 2.0$, and 2.4 is furthest from 2.0 among the value choices which are greater than 2.0.

5. **(A)** A simple random sample may or may not be representative of the population. It is a method of selection in which every possible sample of the desired size has an equal chance of being selected.

6. **(A)** A formula relating the given statistics is $b = r\dfrac{S_y}{S_x}$, which in this case gives $0.25 = r\dfrac{2.3}{8.7}$ and thus $r = \dfrac{(0.25)(8.7)}{2.3}$.

7. **(B)** The null hypothesis is that the new medication is no better than insulin injection, while the alternative hypothesis is that the new medication is better. A Type I error means a mistaken rejection of a true null hypothesis.

8. **(D)** A Type II error means a mistaken failure to reject a false null hypothesis. A false null hypothesis here means that her medication really is better than insulin injection, and failure to realize this means she does not have sufficient evidence that it is better.

9. **(C)** Running a hypothesis test at the 5% significance level means that the probability of committing a Type I error is 0.05. Then the probability of not committing a Type I error is 0.95. Assuming the tests are independent, the probability of not committing a Type I error on any of the five tests is $(0.95)^5$, and the probability of at least one Type I error is $1 - (0.95)^5$.

10. **(E)** No matter what the distribution of raw scores, the set of z-scores always has mean 0 and standard deviation 1.

11. **(D)** There are two possible outcomes (heads and tails), with the probability of heads always 0.6 (independent of what happened on the previous toss), and we are interested in the number of heads in 20 tosses. Thus, this is a binomial model with $n = 20$ and $p = 0.6$. Repeating this over and over (in this case 200 times) simulates the resulting binomial distribution.

12. **(E)** The standard deviation can never be negative.

13. **(E)** When a complete census is taken (all 423 seniors were in the study), the population proportion is known and a confidence interval has no meaning.

14. **(E)** The method described in (A) is a convenience sample, (B) and (C) are voluntary response surveys, and (D) suffers from undercoverage bias.

15. **(D)** $SE(\hat{p}) = \sqrt{(0.7)(0.3)/500}$. Then $z\, SE(\hat{p}) = 0.025$ gives critical z-scores of $\pm\dfrac{0.025}{SE(\hat{p})}$. The degree of confidence is the probability between these two values.

16. **(E)** With $df = n - 1 = 10 - 1 = 9$ and 95% confidence, the critical t-values are $\pm\texttt{invT(0.975, 9)} = \pm 2.262$. $SE(\bar{x}) = \dfrac{s}{\sqrt{n}} = \dfrac{2.3}{\sqrt{10}}$. The confidence interval is $\bar{x} \pm t * SE(\bar{x}) = 17.3 \pm 2.262\left(\dfrac{2.3}{\sqrt{10}}\right)$.

17. **(A)** $P(X = 1) = \frac{1}{3}$, $P(X = 2) = \left(\frac{2}{3}\right)\left(\frac{1}{3}\right)$, $P(X = 3) = \left(\frac{2}{3}\right)^2\left(\frac{1}{3}\right)$, $P(X = 4) = \left(\frac{2}{3}\right)^3\left(\frac{1}{3}\right)$, $P(X = 5) = \left(\frac{2}{3}\right)^4\left(\frac{1}{3}\right)$, ..., and we see that this distribution, like all geometric distributions, has the greatest probability at $X = 1$.

18. **(B)** The chi-square tests all involve counts, and comparing means doesn't make sense in this context. We can plot (AP Stat score, GPA) for a random sample of students, look for a pattern in the scatterplot, and perform a linear regression t-test.

19. **(A)** The median of X is -2, and this is also true of distribution 1 (note that a horizontal line from 0.5 strikes curve 1 above -2 on the x-axis). Y has a smaller standard deviation than Z (tighter clustering around the mean), so Y must correspond to distribution 2, which shows almost all values are between -1 and 1.

20. **(E)**

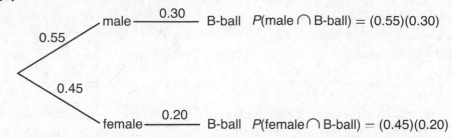

$$P(\text{B-ball}) = (0.55)(0.30) + (0.45)(0.20)$$

$$P(\text{male} \mid \text{B-ball}) = \frac{P(\text{male} \cap \text{B-ball})}{P(\text{B-ball})} = \frac{(0.55)(0.30)}{(0.55)(0.30) + (0.45)(0.20)}$$

21. **(C)** Adding the same constant to every value in a set adds the same constant to the mean but leaves the standard deviation unchanged. Multiplying every value in a set by the same constant multiplies the mean and standard deviation by that constant. So the new mean is $\frac{5}{9} \times (78.35 - 32) + 273 = 298.75$, and the new standard deviation is $\frac{5}{9} \times 6.3 = 3.5$.

22. **(B)** Design 2 is an example of a matched pairs design, a special case of a block design; here, each subject is compared to itself with respect to the two treatments. Both designs definitely use randomization with regard to assignment of treatments, but since they do not use randomization in selecting subjects from the general population, care must be taken in generalizing any conclusions. It's not clear whether or not the researchers who do the observations and measurements know which treatment individual cows are receiving, so there is no way to conclude if there is or is not blinding. The two sources of BVH are different treatments, and so they are not being confounded. In both designs, treatments are randomly applied, so neither is an observational study.

23. **(A)** The linear regression t-test generally has null hypothesis $H_0: \beta = 0$ that there is no linear relationship; if the P-value is small enough, there is evidence of a linear association; that is, there is evidence that $\beta \neq 0$.

24. **(A)** We have a binomial distribution with $n = 12$, and $p = \frac{1}{6}$. With mean $= np = 12\left(\frac{1}{6}\right) = 2$, choice (A) is the only reasonable choice.

25. **(B)** The size of the sample always matters: the larger the sample, the greater the power of statistical tests. One percent of a large population is large. Larger samples are better, but if the sample is greater than 10% of the population, the best statistical techniques are not those covered in the AP curriculum. While not in the AP curriculum, if the population is small, its size may matter, as shown by the finite population correction factor.

26. **(E)** The points on the scatterplot all fall on the straight line:

$$\textit{Female length} = \textit{Male length} + 0.5$$

27. **(A)** $\chi^2 = \sum \dfrac{(\textit{observed} - \textit{expected})^2}{\textit{expected}}$. Expected values are found by multiplying the proportions times the sample size of 100.

28. **(D)** The distributions in (A), (B), and (C) appear roughly symmetric, so the mean and median will be roughly the same. The distribution in (D) is skewed to the right, so the mean will most likely be greater than the median, while the distribution in (E) is skewed to the left, so the mean will most likely be less than the median.

29. **(C)** This is a binomial with $n = 10$ and $p = 0.38$, so the mean is $np = 10(0.38) = 3.8$.

30. **(E)** Answers (A), (B), and (C) are common misconceptions. Since the 95% confidence interval contains 80, a two-sided test would not be significant at the 5% significance level or lower. The interval can be expressed as 77.5 ± 3—that is, we are 95% confident that the true mean fastball speed is within 3 mph of 77.5 mph.

31. **(A)** *Residual = Observed − Predicted*, so $1.0 = 11 - Predicted$ and *Predicted* = 10.

32. **(B)** Stratified sampling is when the population is divided into homogeneous groups (the three divisions in this example), and a random sample of individuals is chosen from each group.

33. **(D)** The two-sample z-interval for a difference in proportions is inappropriate here, because the two proportions do not come from *independent* samples.

34. **(A)** The margin of error, $\pm t^* \dfrac{s}{\sqrt{n}}$, depends on n, the sample size, not the population size.

35. **(B)** The standard deviation is an approximate average measure of how far individual values of a set differ from the mean of the set.

36. **(E)** The larger the sample size, the closer the sample distribution is to the population distribution. The central limit theorem roughly says that if multiple samples of size n are drawn randomly and independently from a population, the histogram of the means of those samples will be approximately normal. Statistics have probability distributions called sampling distributions. The standard error is based on the spread of the sample and on the sample size. The central limit theorem does not apply to all statistics as it does to sample means. Many sampling distributions are not normal; for example, the sampling distribution of the sample max is not a normal distribution. An estimator of a parameter is unbiased if we have a method that, through repeated samples, is on average the same value as the parameter.

37. **(A)** With 0.068 in a tail, the confidence interval with 34 at one end would have a confidence level of $1 - 2(0.068) = 0.864$, so anything higher than 86.4% confidence will contain 34.

38. **(C)** From a boxplot there is no way of telling if a distribution is bell-shaped (very different distributions can have the same five-number summary). Distribution I appears strongly skewed right, and so its mean is probably much greater than its median, while distribution II appears roughly symmetric, and so its mean is probably close to its median. The interquartile range, not the range, in distribution I is 13.

39. **(C)** To make money, there must be more wins than losses, so with 50 plays, we need to calculate $P(X > 25)$. We have a binomial distribution with $n = 50$ and probability of success $p = \frac{18}{38}$. On a calculator such as the TI-84 we find $P(X > 25) = 1 - P(X \leq 25) = 1 - \texttt{binomcdf(50, 18/38, 25)} = 0.303$. (Or use Bcd on the Casio Prizm or BINOMIAL_CDF on the HP Prime.)

40. **(B)** $\chi^2 = \sum \dfrac{(observed - expected)^2}{expected}$ and cell calculations [expected value of a cell equals (row total)(column total)/(table total)] or χ^2-test on a calculator such as the TI-84, Casio Prizm, or HP Prime will yield expected cells of 29, 36, 35, 29, 36, and 35.

SECTION II: PART A

1. (a) There are $2 \times 2 \times 2 = 8$ different treatments:
    Two-day, aerobic, with adjuvant
    Two-day, aerobic, without adjuvant
    Two-day, anaerobic, with adjuvant
    Two-day, anaerobic, without adjuvant
    Five-day, aerobic, with adjuvant
    Five-day, aerobic, without adjuvant
    Five-day, anaerobic, with adjuvant
    Five-day, anaerobic, without adjuvant

    (b) We must randomly assign the treatment combinations to the beds. (Roses have already been randomly assigned to the beds.) With 8 treatments and 16 beds, each treatment should be assigned to 2 beds. For example, give each bed a random number between 1 and 16 (use a computer to generate 16 random integers between 1 and 16 *without replacement*), and then assign the first treatment in the above list to the beds with the numbers 1 and 2, assign the second treatment in the above list to the beds with the numbers 3 and 4, and so on.

    (c) Using only minipink roses in this experiment gives reduced variability and increases the likelihood of determining differences among the treatments.

    (d) Using only minipink roses in this experiment limits the scope and makes it difficult to generalize the results to other species of roses.

Section 1 is essentially correct for correctly listing all eight treatment combinations and is incorrect otherwise.

Section 2 is essentially correct if each treatment combination is randomly assigned to two beds of roses. Section 2 is partially correct if each treatment is randomly assigned to two beds but the method is unclear or if a method of randomization is correctly described but the method may not assure that each treatment is assigned to *two* beds.

Section 3 is essentially correct for 1) noting *reduced variability* in (c), 2) explaining that this increases the likelihood of determining differences among the treatments, 3) noting *limited scope* in (d), and 4) explaining that this makes generalization to other species difficult. Section 3 is partially correct for only two or three of these four components.

4	**Complete Answer**	All three sections essentially correct.
3	**Substantial Answer**	Two sections essentially correct and one section partially correct.
2	**Developing Answer**	Two sections essentially correct OR one section essentially correct and one or two sections partially correct OR all three sections partially correct.
1	**Minimal Answer**	One section essentially correct OR two sections partially correct.

2. Section 1:

> *Hypotheses:* H_0: $\mu = 82$ and H_a: $\mu > 82$, where μ is the mean speed in the population of all swings with the new racquet
>
> *Procedure:* One-sample t-test for a population mean

Section 2:

> *Conditions:* Random sample (given), $n = 47$ is less than 10% of all possible swings with the new racket, and $n = 47 \geq 30$ is sufficiently large for the CLT to apply.
>
> *Mechanics:* Calculator software (such as T-Test on the TI-84 or 1-Sample t-Test on the Casio Prizm) gives $t = 3.0246$ and $P = 0.00203$.

Section 3:

> *Conclusion in context with linkage to the* P-*value:* With this small a P-value, $0.00203 < 0.05$, there is sufficient evidence to reject H_0; that is, there is sufficient evidence that his mean speed with the new racket is an improvement over the old.

SCORING

Section 1 is essentially correct for correct hypotheses and the test correctly identified. Section 1 is partially correct if only one of these two elements is correct.

Section 2 is essentially correct if 1) the randomization assumption is checked, 2) the sample size condition is checked, 3) the t-statistic is correctly reported, and 4) the P-value is correctly reported. Section 2 is partially correct if only two or three of these four elements are correct.

Section 3 is essentially correct if the correct conclusion is given in context and the conclusion is linked to the P-value. Section 3 is partially correct if the correct conclusion is given in context but there is no linkage to the P-value OR if the conclusion is correct with linkage to the P-value, but context is missing.

4	**Complete Answer**	All three sections essentially correct.
3	**Substantial Answer**	Two sections essentially correct and one section partially correct.
2	**Developing Answer**	Two sections essentially correct OR one section essentially correct and one or two sections partially correct OR all three sections partially correct.
1	**Minimal Answer**	One section essentially correct OR two sections partially correct.

TIP

Full credit would also be given in 3(a) for: Normal, $\mu = 26.4$, $\sigma = 8.2$, $P(X > 30) = 0.330$.

TIP

Full credit would also be given in 3(b) for: Binomial, $n = 5$, $p = 0.3303$, $P(x \geq 3) = 0.2054$.

TIP

Full credit would also be given in 3(c) for: Normal, $\mu_{\bar{x}} = 26.4$, $\sigma_{\bar{x}} = \dfrac{8.2}{\sqrt{5}} = 3.667$, $P(\bar{x} > 30) = 0.1631$.

3. (a) Normal distribution with $\mu = 26.4$ and $\sigma = 8.2$.

$$P(x > 30) = P\left(z > \frac{30 - 26.4}{8.2}\right) = P(z > 0.4390) = 0.3303$$

(b) $B(n = 5, p = 0.3303)$, and $1 - \text{binomcdf}(5, 0.3303, 2) = 0.2054$

[or $P(\text{at least 3 out of 5 are} > 30) = 10(0.3303)^3(0.6697)^2 + 5(0.3303)^4(0.6697) + (0.3303)^5 = 0.2054]$

(c) The distribution of $\bar{x}$ is normal with mean $\mu_{\bar{x}} = 26.4$ and standard deviation $\sigma_{\bar{x}} = \dfrac{8.2}{\sqrt{5}} = 3.667$. So, $P(\bar{x} > 30) = P\left(z > \dfrac{30 - 26.4}{3.667}\right) = P(z > 0.9817) = 0.1631$.

SCORING

Part (a) is essentially correct if the distribution is named (normal), the parameters are given (mean and standard deviation), the cutoff with direction is indicated (> 30), and the correct probability is calculated. Simply writing normalcdf(30,∞,26.4,8.2) = 0.3303 is a partially correct response; however, writing normalcdf(lower bound = 30, upper bound = ∞, mean = 26.4, standard deviation = 8.2) = 0.3303 is essentially correct.

Part (b) is essentially correct if the correct probability is calculated and the derivation is clear. Part (b) is partially correct for indicating a binomial with $n = 5$ and $p =$ answer from (a) but calculating incorrectly. Simply writing $1 - $ binomcdf(5,0.3303,2) = 0.2054 without designating the parameters is also a partially correct response.

Part (c) is essentially correct for specifying a normal distribution and both $\mu_{\bar{x}}$ and $\sigma_{\bar{x}}$ and for correctly calculating the probability. Part (c) is partially correct for specifying a normal distribution and both $\mu_{\bar{x}}$ and $\sigma_{\bar{x}}$ but incorrectly calculating the probability, or for failing to specify a normal distribution and both $\mu_{\bar{x}}$ and $\sigma_{\bar{x}}$ but correctly calculating the probability.

4   **Complete Answer**   All three parts essentially correct.

3   **Substantial Answer**   Two parts essentially correct and one part partially correct.

2   **Developing Answer**   Two parts essentially correct OR one part essentially correct and one or two parts partially correct OR all three parts partially correct.

1   **Minimal Answer**   One part essentially correct OR two parts partially correct.

4. (a) $\dfrac{50}{50+20+40+5} = \dfrac{50}{115}$

   (b) $\dfrac{10}{5+40+10} = \dfrac{10}{55}$

   (c) Based upon the mosaic plot, there is an association between the geographic location where someone is raised and what they call a soft drink. In this study, most people from the Northeast called soft drinks "soda," most people from the Midwest called soft drinks "pop," most people from the West called soft drinks "soda," and most people from the South called soft drinks "Coke."

SCORING

Part (a) is essentially correct for the expression $\dfrac{50}{115}$ and partially correct if only 0.435 without showing where this came from.

Part (b) is essentially correct for the expression $\dfrac{10}{55}$ and partially correct if only 0.182 without showing where this came from.

Part (c) is essentially correct for correctly noting the observed preferences for each geographic location and partially correct for noting only two or three of the preferences.

4 Complete Answer   All three parts essentially correct.

3 Substantial Answer   Two parts essentially correct and one part partially correct.

2 Developing Answer   Two parts essentially correct OR one part essentially correct and one or two parts partially correct OR all three parts partially correct.

1 Minimal Answer   One part essentially correct OR two parts partially correct.

5. (a) H_0: Rafiki can do no better than guesswork in predicting winners, H_0: $p = 0.5$.
H_a: Rafiki can correctly predict winners better than guesswork, H_a: $p > 0.5$.

(b) $p = 0.5$, $n = 10$, and a large number for number of samples.

(c) Graph (II). The simulation should be centered at the value based on the null hypothesis probability, not the observed value.

(d) Looking at areas under (II), the correct graph, among the given choices, 7 or greater correct picks seems to have a probability closest to 0.15. If the null hypothesis were true, that is, if Rafiki can do no better than guesswork, the estimated probability of Rafiki correctly predicting 7 or more out of 10 rises or falls of the DJIA is about 0.15.

(e) With this large a P-value, $0.15 > 0.05$, there is not sufficient evidence to reject H_0; that is, there is not sufficient evidence that Rafiki can pick the rise and fall of the DJIA better than guesswork.

SCORING

Part (a) is essentially correct for the correct hypotheses either in words or formulas and is incorrect otherwise.

Parts (b) and (c) together are essentially correct for $p = 0.5$, $n = 10$, any large number for number of samples, and picking Graph (II), and are partially correct for two or three of these four answers correct.

Part (d) is essentially correct for correctly picking 0.15 and giving a correct interpretation of the P-value and is partially correct for correctly picking 0.15 but giving a weak interpretation of the P-value.

Part (e) is essentially correct for a correct conclusion in context with linkage to the simulated P-value and is partially correct if only missing linkage or only missing context.

Count partially correct answers as one-half an essentially correct answer.

4	**Complete Answer**	Four essentially correct answers.
3	**Substantial Answer**	Three essentially correct answers.
2	**Developing Answer**	Two essentially correct answers.
1	**Minimal Answer**	One essentially correct answer.

Use a holistic approach to decide a score totaling between two numbers.

SECTION II: PART B

6. (a) **Think:** Remember, this has to be referred to as a predicted value or an average value when $x = 0$, where x in this problem is weekly study hours.

   Answer: The y-intercept, 1.5748, refers to a student who studies 0 hours per week. Thus, a student who doesn't study any hours at all is predicted to have a GPA of 1.5748.

   (b) **Think:** Again, you have to refer to a predicted or an average increase in the y-value for each unit increase in the x-value.

   Answer: The slope is 0.0581. This means that for each additional hour of study per week, the predicted GPA increases by 0.0581. In other words, this model predicts that the average GPA increases by 0.0581 for each additional hour of study per week.

   (c) **Think:** The coefficient of determination, R-Sq, gives the proportion of the variation in the y-variable that *is* accounted for by the linear regression model. So the proportion *not* accounted for by the regression model must be $1 - $ R-Sq.

   Answer: $1 - 0.8366 = 0.1634$

   (d) **Think:** This is the easiest confidence interval to find because the computer printout gives the standard error of the sample slope as well as the sample slope. Remember that for linear regression inference, $df = n - 2$.

   Answer: With $df = 25 - 2 = 23$ and 0.025 in each tail, the critical t-values are $\pm\texttt{invT(0.975, 23)} = 2.0687$. The 95% confidence interval of the true slope is $b \pm ts_b = 0.0581 \pm 2.0687(0.00535) = 0.0581 \pm 0.0111$. We are 95% confident that for every additional study hour per week, the average increase in GPA is between 0.0470 and 0.0692.

   (e) **Think:** The residual averages suggest that the regression line tends to underestimate the GPA of students who take AP Statistics in high school by 0.2414 and to overestimate the GPA of students who do not take AP Statistics in high school by 0.1609. The difference between these two residual averages is what we are looking for!

   Answer: We calculate $0.2414 - (-0.1609) = 0.4023$. Thus, for two students, one of whom had taken AP Statistics in high school, who study the same number of hours weekly, the student who had taken AP Statistics in high school would be estimated to have a GPA 0.4023 higher than the student who had not taken AP Statistics in high school.

   (f) **Think:** If the interval is entirely above 0 or entirely below 0, there is statistical significance, but if 0 is in the interval, there is not. We also note that a 95 percent confidence level is equivalent to a 5 percent significance level.

   Answer: 0.0022 ± 0.0134 gives the interval $(-0.0112, 0.0156)$ which does contain 0. Thus, no, the confidence interval does not indicate a significant difference (at the 5 percent significance level) between the two slopes (one slope from each of the two regression models, one for the students who take AP Statistics and one for the students do not take AP Statistics).

(g) **Think:** Just as above, this depends on whether or not 0 is in the interval.

Answer: 0.4031 ± 0.2767 gives the interval $(0.1264, 0.6798)$, which is entirely positive and does not contain 0. Thus, yes, the confidence interval does indicate a significant difference (at the 5 percent significance level) between the two *y*-intercepts (one *y*-intercept from each of the two regression models, one for the students who take AP Statistics and one for the students do not take AP Statistics).

SCORING

Section 1 is essentially correct for correct interpretations of the *y*-intercept in (a) and slope in context in (b) and a correct calculation in (c). Section 1 is partially correct if two of these three parts are correct.

Section 2 is essentially correct in (d) for a correct confidence interval together with a correct conclusion in context and is partially correct for one of the following: an incorrect *df* is used or just missing context or just missing no-deterministic language.

Section 3 is essentially correct in (e) for correctly calculating the difference, including noting which is greater, and using nondeterministic language ("predicted"), and is partially correct if the magnitude of the calculated difference is correct but which is greater is not noted or if nondeterministic language isn't used.

Section 4 is essentially correct when correct answers with explanations are given in both (f) and (g), and is partially correct for correct conclusions with weak explanations or for a completely correct statement for only one of (f) and (g).

Count partially correct answers as one-half an essentially correct answer.

4   Complete Answer   Four essentially correct answers.

3   Substantial Answer   Three essentially correct answers.

2   Developing Answer   Two essentially correct answers.

1   Minimal Answer   One essentially correct answer.

Use a holistic approach to decide a score totaling between two numbers.

AP SCORE FOR THE DIAGNOSTIC TEST

Multiple-Choice section (40 questions)

Number correct × 1.25 = _____

Free-Response section (5 open-ended questions plus an investigative task)

Question 1 _____ × 1.875 = _____
out of 4

Question 2 _____ × 1.875 = _____
out of 4

Question 3 _____ × 1.875 = _____
out of 4

Question 4 _____ × 1.875 = _____
out of 4

Question 5 _____ × 1.875 = _____
out of 4

Question 6 _____ × 3.125 = _____
out of 4

Total points from Multiple-Choice and Free-Response sections = _____

Conversion chart based on a recent AP exam:

Total Points	AP Score
73–100	5
59–72	4
44–58	3
32–43	2
0–31	1

STUDY GUIDE FOR THE DIAGNOSTIC TEST
MULTIPLE-CHOICE QUESTIONS

Circle all of the questions you answered incorrectly, and focus your review on the unit or units that contain the most circled questions.

Unit 1: Exploring One-Variable Data
Questions 1, 10, 12, 19, 21, 28, 35, 38

Unit 2: Exploring Two-Variable Data
Questions 6, 26, 31

Unit 3: Collecting Data
Questions 5, 13, 14, 22, 25, 32

Unit 4: Probability, Random Variables, and Probability Distributions
Questions 3, 9, 11, 17, 20, 24, 29, 39

Unit 5: Sampling Distributions
Questions 19, 24, 36

Unit 6: Inference for Categorical Data: Proportions
Questions 7, 8, 9, 13, 15, 33

Unit 7: Inference for Quantitative Data: Means
Questions 4, 7, 8, 9, 16, 30, 34, 37

Unit 8: Inference for Categorical Data: Chi-Square
Questions 27, 40

Unit 9: Inference for Quantitative Data: Slopes
Questions 2, 18, 23

Part Two: Units Review

Exploring One-Variable Data

1

(15–23% AP EXAM WEIGHTING)

→ **CATEGORICAL VARIABLES**

→ **REPRESENTING A QUANTITATIVE VARIABLE WITH TABLES AND GRAPHS**

→ **DESCRIBING THE DISTRIBUTION OF A QUANTITATIVE VARIABLE**

→ **QUIZ 1**

→ **QUIZ 2**

→ **SUMMARY STATISTICS FOR A QUANTITATIVE VARIABLE**

→ **GRAPHICAL REPRESENTATIONS OF SUMMARY STATISTICS**

→ **COMPARING DISTRIBUTIONS OF A QUANTITATIVE VARIABLE**

→ **QUIZ 3**

→ **QUIZ 4**

→ **THE NORMAL DISTRIBUTION**

→ **QUIZ 5**

In this unit, you will learn how to distinguish between two variable types and how to describe and compare distributions using a variety of visual and calculational approaches. You will begin to understand how to use statistics to investigate specific claims. Finally, you will be introduced to the normal distribution model. Patterns, uncertainty, and variability are key ideas that will appear and reappear throughout the AP Statistics curriculum. This unit and the next unit primarily involve descriptive statistics (summarizing and describing data), whereas later units will involve inferential statistics (using samples to generalize or infer something about a population).

UNIT LEARNING OBJECTIVES

- To be able to create and interpret a bar graph for categorical data given as a frequency or relative frequency table.
- To be able to calculate summary statistics for univariate (one-variable) quantitative data including center (mean and median), variability (range, interquartile range, variance, and standard deviation), and the five-number summary (minimum, Q_1, median, Q_3, and maximum).
- To be able to describe univariate quantitative data including shape (such as bell-shaped, symmetric, right or left skewed, uniform, unimodal, or bimodal), center, spread (variability), and unusual features (such as gaps, clusters, and outliers), and always remembering to do so in context (including units).

- To be able to create displays for quantitative data including dotplots, stemplots, histograms, and boxplots.
- To be able to read and interpret specific information from the above displays and from cumulative frequency graphs.
- To be able to compare two distributions, being sure to use comparative terminology rather than simply making two separate lists, and again not forgetting context.

CATEGORICAL VARIABLES

A categorical (or qualitative) variable takes on values that are category names or group labels. The values can be organized into frequency tables or relative frequency tables or can be represented graphically by displays such as bar graphs and dotplots.

➥ **EXAMPLE 1.1**

NOTE

Frequency tables give the number of cases falling into each category, while relative frequency tables give the proportion of cases falling into each category.

In a survey taken during the first week of January 2020, 1,100 parents wanted to keep the school year to the current 180 days, 300 wanted to shorten it to 160 days, 500 wanted to extend it to 200 days, and 100 expressed no opinion. (Noting that there were 2,000 parents surveyed, percentages can be calculated.)

A table showing frequencies and relative frequencies:

Desired School Length	Number of Parents	Percent of Parents
180 days	1,100	55%
160 days	300	15%
200 days	500	25%
No opinion	100	5%

Bar graphs showing frequencies and relative frequencies:

TIP

Graphs must have appropriate labeling and scaling, or they will lose credit!

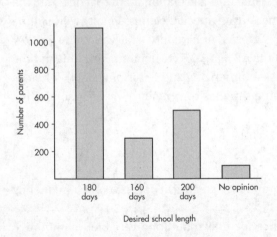
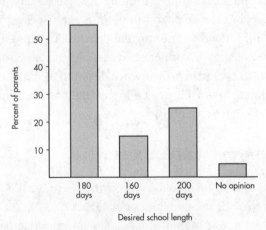

When asked to choose their favorite hip-hop artist, 8 students chose Eminem, 5 picked 2Pac, 6 picked Notorius B.I.G., 3 picked Nas, and 3 picked Jay-Z. These data can be displayed in the following dotplot and pie chart.

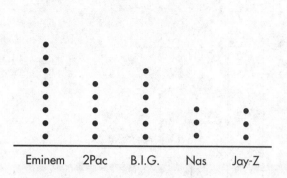
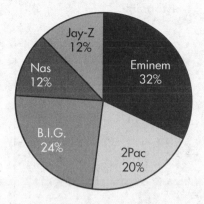

REPRESENTING A QUANTITATIVE VARIABLE WITH TABLES AND GRAPHS

A quantitative variable takes on numerical values for a measured or counted quantity. The values can be organized into frequency tables or relative frequency tables or can be represented graphically by displays such as dotplots, histograms, stemplots, cumulative relative frequency plots, or boxplots.

➡ **EXAMPLE 1.3**

The *dotplot* below shows the lengths of stay (in days) for all patients admitted to a rural hospital during the first week of January 2020.

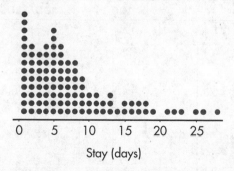

Histograms, useful for large data sets involving quantitative variables, show counts or percents falling either at certain values or between certain values. While the AP Statistics exam does not stress the construction of histograms, there are often questions on interpreting given histograms.

NOTE

To construct a histogram using the TI-84, go to STAT → EDIT and put the data in a list, then turn a STAT PLOT on, choose the histogram icon under Type, specify the list where the data are, and use ZoomStat or adjust the WINDOW (Xscl determines the width of the bin).

Suppose there are 2,200 seniors in a city's six high schools. Four hundred of the seniors are taking no AP classes, 500 are taking one, 900 are taking two, 300 are taking three, and 100 are taking four. These data can be displayed in the following histogram:

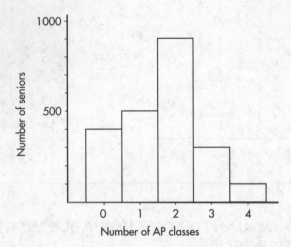

Sometimes, instead of labeling the vertical axis with frequencies, it is more convenient or more meaningful to use *relative frequencies*, that is, frequencies divided by the total number in the population.

Number of AP classes	Frequency	Relative frequency
0	400	400/2200 = 0.18
1	500	500/2200 = 0.23
2	900	900/2200 = 0.41
3	300	300/2200 = 0.14
4	100	100/2200 = 0.05

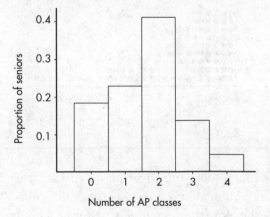

Note that the shape of the histogram is the same whether the vertical axis is labeled with frequencies or with relative frequencies. Sometimes we show both frequencies and relative frequencies on the same graph.

➡ EXAMPLE 1.5

Consider the following histogram of the numbers of pairs of shoes owned by 2,000 women.

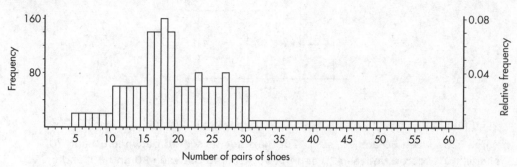

What can we learn from this histogram? For example, none of the women had fewer than 5 or more than 60 pairs of shoes. One hundred sixty of the women had 18 pairs of shoes. Twenty women had 5 pairs of shoes. Half the total area is less than or equal to 19, so half the women have 19 or fewer pairs of shoes. Fifteen percent of the area is more than 30, so 15 percent of the women have more than 30 pairs of shoes. Five percent of the area is more than 50, so 5 percent of the women have more than 50 pairs of shoes.

➡ EXAMPLE 1.6

Consider the following histogram of exam scores, where the vertical axis has not been labeled.

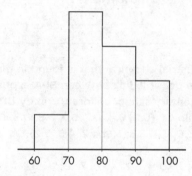

What can we learn from this histogram?

Answer: It is impossible to determine the actual frequencies; that is, we have no idea if there were 25 students, 100 students, or any particular number of students who took the exam. However, we can determine the relative frequencies by noting the fraction of the total area that is over any interval.

(continued)

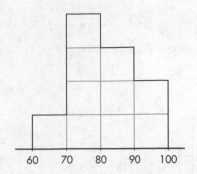

We can divide the area into ten equal portions and then note that $\frac{1}{10}$, or 10%, of the area is between 60 and 70, so 10% of the students scored between 60 and 70. Similarly, 40% scored between 70 and 80, 30% scored between 80 and 90, and 20% scored between 90 and 100.

Although it is usually not possible to divide histograms so nicely into ten equal areas, the principle of relative frequencies corresponding to relative areas still applies. Also note how this example shows the number of exam scores falling *between* certain values, whereas the previous two examples showed the number of AP classes taken and number of shoes owned for *each* value.

Although a histogram may show how many scores fall into each grouping or interval, the exact values of individual scores are lost. An alternative pictorial display, called a *stemplot* (also called a stem and leaf display), retains this individual information and is useful for giving a quick overview of a distribution, displaying the relative density and shape of the data. A stemplot contains two columns separated by a vertical line. The left column contains the stems, and the right column contains the leaves.

➡ **EXAMPLE 1.7**

Bisphenol A (BPA) is an industrial chemical that is found in many hard plastic bottles. Recent studies have shown a possible link between BPA exposure and childhood obesity. In one study of 27 elementary school children, urinary BPA levels in nanograms/milliliter (ng/mL) were as follows: {0.2, 0.4, 0.7, 0.7, 0.8, 0.8, 0.9, 1.0, 1.0, 1.3, 1.4, 1.4, 1.4, 1.7, 1.9, 2.1, 2.4, 2.5, 2.8, 2.8, 3.0, 3.3, 3.3, 3.8, 4.2, 4.5, 5.2}.

0	2 4 7 7 8 8 9
1	0 0 3 4 4 4 7 9
2	1 4 5 8 8
3	0 3 3 8
4	2 5
5	2

Urinary BPA level
(0|2 means 0.2 ng/mL)

Note: Those with urine BPA level of 2 ng/mL or higher had more than twice the risk of being overweight.

➡ **EXAMPLE 1.8**

How many nonstop push-ups can a teenager between the ages of 15 and 18 do? In one study in a mixed-gender high school gym class, the numbers of push-ups were {2, 5, 7, 10, 12, 12, 14, 16, 16, 18, 19, 20, 21, 29, 32, 34, 35, 37, 37, 38, 39, 39, 42, 44, 50}.

```
0 | 2 5 7
1 | 0 2 2 4 6 6 8 9
2 | 0 1 9
3 | 2 4 5 7 7 8 9 9
4 | 2 4
5 | 0
```
Number of push-ups
(5|0 means 50 push-ups)

Sometimes we sum frequencies and show the result visually in a cumulative relative frequency plot (also known as an *ogive*).

➡ **EXAMPLE 1.9**

The following graph shows 2019 school enrollment in the United States by age.

2019 School Enrollment in the U.S. by Age

TIP

More generally, a cumulative graph can show either the number or proportion of a data set less than or equal to a given number.

What can we learn from this cumulative relative frequency plot? For example, going up to the graph from age 5, we see that 0.15, or 15%, of school enrollment is below age 5. Going over to the graph from 0.5 on the vertical axis, we see that 50% of the school enrollment is below and 50% is above a middle age of 11. Going up from age 29, we see that 0.95, or 95%, of the enrollment is below age 29, and thus 5% is above age 29. Going over from 0.25 and 0.75 on the vertical axis, we see that the middle 50% of school enrollment is between ages 6 and 7 at the lower end and age 16 at the upper end.

DESCRIBING THE DISTRIBUTION OF A QUANTITATIVE VARIABLE

TIP

Center and spread should always be described together.

Looking at a graphical display, we see that two important aspects of the overall pattern are

1. the *center*, which separates the values (or area under the curve in the case of a histogram) roughly in half, and
2. the *spread*, that is, the scope of the values from smallest to largest.

In the histogram of Example 1.4, the center is 2 AP classes while the spread is from 0 to 4 AP classes.

In the histogram of Example 1.5, the center is about 19 and the spread is from 5 to 60; in the histogram of Example 1.6, the center is about 80 and the spread is from about 60 to about 100.

NOTE

In a histogram, often there is no way to know the exact minimum or maximum value, only intervals in which these values fall.

In the stemplot of Example 1.7, the center is 1.7 (middle of the 27 values) and the spread is from 0.2 to 5.2; in the stemplot of Example 1.8, the center is 21 (middle of the 25 values) and the spread is from 2 to 50.

Other important aspects of the overall pattern are

1. *clusters*, which show natural subgroups into which the values fall (for example, the salaries of teachers in Ithaca, NY, fall into three overlapping clusters: one for public school teachers, a higher one for Ithaca College professors, and an even higher one for Cornell University professors).
2. *gaps*, which show holes where no values fall (for example, the Office of the Dean sends letters to students being put on the honor roll and to those being put on academic warning for low grades; thus the GPA distribution of students receiving letters from the Dean has a huge middle gap).

➡ EXAMPLE 1.10

Hodgkin's lymphoma is a cancer of the lymphatic system, the system that drains excess fluid from the blood and protects against infection. Consider the following histogram:

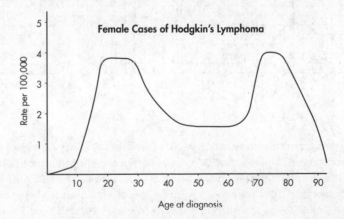

Simply saying that the average age at diagnosis for female cases is around 50 clearly misses something. The distribution of ages at diagnosis for female cases of Hodgkin's lymphoma is bimodal with two distinct clusters, centered at 25 and 75.

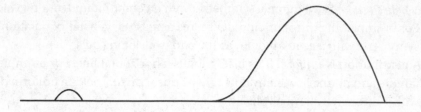

TIP

Pay attention to outliers! They deserve special consideration.

Extreme values, called outliers, are found in many distributions. Sometimes they are the result of errors in measurements and deserve scrutiny; however, outliers can also be the result of natural chance variation. Outliers may occur on one side or both sides of a distribution.

Some distributions have one or more major peaks, called modes. (The values with the peaks above them are the modes.) With exactly one or two such peaks, the distribution is said to be unimodal or bimodal, respectively. You should always look at the big picture and decide whether or not two (or more) phenomena are affecting the histogram.

NOTE

There is a formal mathematical definition of an outlier as a value at least a certain distance away from Q_1 or Q_3. You will learn how to calculate this distance later in this unit.

➡ **EXAMPLE 1.11**

The histogram below shows employee computer usage (number accessing the Internet) at given times at a company's main office.

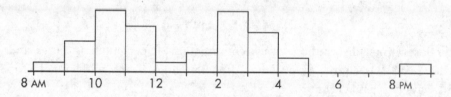

TIP

Some distributions have many little ups (and downs), which should not be confused with modes. Every little bump in the data is not a mode!

Note that this is a *bimodal* distribution. Computer usage at this company appears heaviest at midmorning and midafternoon, with a dip in usage during the noon lunch hour. There is an evening outlier possibly indicating employees returning after dinner (or perhaps a hacker using company computing power for free!).

Note that, as illustrated above, it is usually instructive to look for reasons behind outliers and modes.

Distributions come in an endless variety of shapes; however, certain common patterns are worth special mention:

1. A *symmetric* distribution is one in which the two halves are mirror images of each other. For example, the weights of all people in some organizations fall into symmetric distributions with two mirror-image bumps, one for men's weights and one for women's weights.
2. A distribution is *skewed to the right* if it spreads far and thinly toward the higher values. For example, ages of nonagenarians (people in their 90s) is a distribution with sharply decreasing numbers as one moves from 90-year-olds to 99-year-olds. *tail to the right*
3. A distribution is *skewed to the left* if it spreads far and thinly toward the lower values. For example, track meet times usually show a distribution bunched at a higher end with a few low values. *tail to the left*

4. A *bell-shaped* distribution is symmetric with a center mound and two sloping tails. For example, the distribution of IQ scores across the general population is roughly symmetric with a center mound at 100 and two sloping tails.

5. A distribution is *uniform* if its histogram is a horizontal line. For example, tossing a fair die and noting how many dots (pips) appear on top yields a uniform distribution with 1 through 6 all equally likely.

TIP

In the real world, distributions are rarely perfectly symmetric, perfectly bell-shaped, or perfectly uniform, so we usually say "roughly" or "approximately" symmetric, bell-shaped, or uniform.

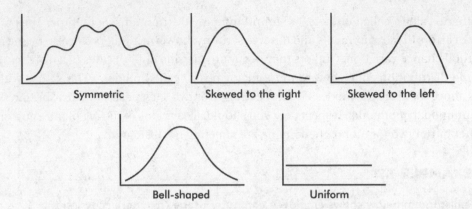

Symmetric   Skewed to the right   Skewed to the left

Bell-shaped   Uniform

IMPORTANT

When describing a distribution, always comment on Shape, Outliers, Center, and Spread (SOCS). Or, alternatively, Center, Unusual values, Shape, and Spread (CUSS). And always describe in context.

Even when a basic shape is noted, it is important also to note if some of the data deviate from this shape.

A distribution skewed to the left has a cumulative frequency plot that rises slowly at first and then steeply later, while a distribution skewed to the right has a cumulative frequency plot that rises steeply at first and then slowly later.

➡ EXAMPLE 1.12

Consider the essay-grading policies of three teachers: Abrams, who gives very low scores; Brown, who gives equal numbers of low and high scores; and Connors, who gives very high scores. Histograms of the grades (*with 1 the lowest score and 4 the highest score*) are as follows:

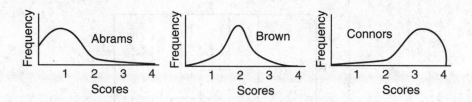

These translate into the following cumulative frequency plots:

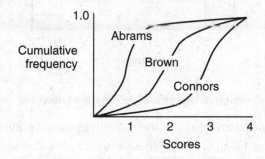

Multiple-Choice Questions

> **Directions:** The questions that follow are each followed by five suggested answers. Choose the response that best answers the question.

1. The times, in minutes, taken by 155 high school students to complete a level of the video game *Angry Birds* are summarized in the histogram below.

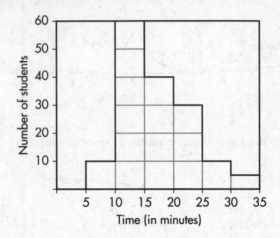

Based on the histogram, which of the following must be true?

(A) The minimum time taken by any of these students was 5 minutes.

(B) The maximum time taken by any of these students was 35 minutes.

(C) If the times are arranged in order, the middle time would be between 10 and 15 minutes.

(D) If the times are arranged in order, the middle time would be between 15 and 20 minutes.

(E) The same number of students took less than 15 minutes as took over 20 minutes.

2. Which of the following is a true statement?

(A) Stemplots are useful for both quantitative and categorical data sets.

(B) Stemplots are equally useful for small and very large data sets.

(C) Stemplots can show symmetry, gaps, clusters, and outliers.

(D) Stemplots may or may not show individual values.

(E) Stems may be skipped if there is no data value for a particular stem.

Questions 3–5 refer to the following five histograms:

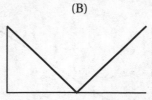 (A)   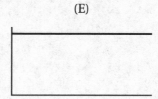 (B)   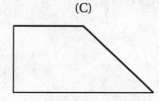 (C)

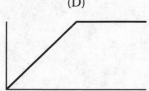 (D)   (E)

3. To which of the above five histograms does the following cumulative relative frequency plot correspond?

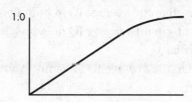

(A) A
(B) B
(C) C
(D) D
(E) E

4. To which of the above five histograms does the following cumulative relative frequency plot correspond?

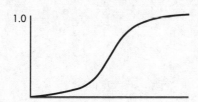

(A) A
(B) B
(C) C
(D) D
(E) E

5. To which of the above five histograms does the following cumulative relative frequency plot correspond?

(A) A
(B) B
(C) C
(D) D
(E) E

6. Allan famously quipped that when he moved from Pennsylvania to California, the average IQ dropped in both states. What would have had to been true for this to happen?

(A) Allan's IQ was greater than the average IQ in both Pennsylvania and California.
(B) Allan's IQ was greater than the average IQ in Pennsylvania and less than the average IQ in California.
(C) Allan's IQ was less than the average IQ in Pennsylvania and greater than the average IQ in California.
(D) Allan's IQ was less than the average IQ in both Pennsylvania and California.
(E) There is no way for Allan's statement to be true.

7. A car manufacturer claims that the cars it sells have an average fuel efficiency of 35 mpg. A consumer group believes that the true figure is lower. The consumer group obtains a random sample of 5 of the company's cars and determines an average of 32 mpg. In response to the consumer group's findings, the company runs a simulation by randomly picking 5 cars 160 times from a fleet of cars of known 35 mpg efficiency and calculating the resulting averages to show that 32 mpg was possible. The company makes the following dotplot.

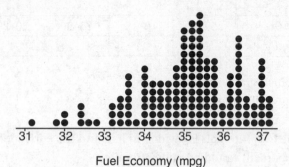

Fuel Economy (mpg)

Does the company's argument seem reasonable given this dotplot?

(A) Yes, because the dotplot shows that anything is possible in the real world.

(B) Yes, because the dotplot shows that the average was not 35 for every 5-car set that the company randomly picked.

(C) Yes, because the dotplot shows that 32 mpg is a possible average of 5 cars from a fleet of cars of known 35 mpg efficiency.

(D) No, because the dotplot shows that anything is possible in the real world.

(E) No, because the possibility of picking 5 cars with an average of 32 mpg or lower from a fleet of cars of known 35 mpg efficiency is very small, only 4 out of 160, or 0.025.

Free-Response Questions

1. The miles run by 25 randomly selected runners during a given week are given below.

Miles run	5–10	10–15	15–20	20–25
Frequency	3	11	7	4

   (a) Construct a histogram.
   (b) What percent of the runners ran under 15 miles?
   (c) Construct a cumulative relative frequency plot.
   (d) If each of the four intervals is divided in half and the corresponding numbers given, what percent would be under 15 miles?

2. The dotplot below shows the numbers of goals scored by the 20 teams playing in a city's high school soccer games on a particular day.

Goals scored by each team

   (a) Describe the distribution.
   (b) One superstar scored six goals, but his team still lost. What are all possible final scores for that game? Explain.
   (c) Is it possible that all the teams scoring exactly two goals won their games? Explain.

3. The winning percentages for a major league baseball team over the past 22 years are shown in the following stemplot:

```
46 | 0 8
47 | 1 4 7 9
48 | 5 8 8 9
49 | 3 4 7
50 |
51 |
52 | 5 8
53 | 2 5 6
54 | 4 8 9
55 | 6
```

(55 | 6   means 55.6%)

(a) Interpret the lowest value.
(b) Describe the distribution.
(c) Give a reason that one might argue that the team is more likely to lose a given game than win it.
(d) Give a reason that one might argue that the team is more likely to win a given game than lose it.

The answers for this quiz can be found in the Appendix on page 576.

QUIZ 2

Multiple-Choice Questions

> **Directions:** The questions that follow are each followed by five suggested answers. Choose the response that best answers the question.

1. The stemplot below shows ages of CEOs of a select group of corporations.

$$
\begin{array}{c|l}
2 & 2\ 8 \\
3 & 3\ 6 \\
4 & 2\ 7 \\
5 & 0\ 2\ 5\ 8 \\
6 & 3\ 3\ 7\ 9 \\
7 & 4\ 8 \\
8 & 3 \\
9 & 0 \\
\end{array}
$$

(9 | 0   means an age of 90)

Which of the following is *not* a correct statement about this distribution?

(A) The distribution is roughly bell-shaped.
(B) The distribution is skewed left and right.
(C) The center is around 60.
(D) The spread is from 22 to 90.
(E) There are no outliers.

2. Which of the following is an *incorrect* statement?

(A) In histograms, relative areas correspond to relative frequencies.
(B) In histograms, frequencies can be determined from relative heights.
(C) Symmetric histograms may have multiple peaks.
(D) Two students working with the same set of data may come up with histograms that look different.
(E) Displaying outliers may be more problematic when using histograms than when using stemplots.

Questions 3–5 refer to the following five cumulative relative frequency plots:

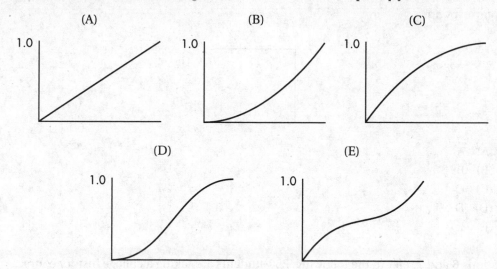

3. To which of the above cumulative relative frequency plots does the following histogram correspond?

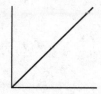

   (A) A
   (B) B
   (C) C
   (D) D
   (E) E

4. To which of the above cumulative relative frequency plots does the following histogram correspond?

   (A) A
   (B) B
   (C) C
   (D) D
   (E) E

5. To which of the above cumulative relative frequency plots does the following histogram correspond?

(A) A
(B) B
(C) C
(D) D
(E) E

Questions 6 and 7 refer to the following: 250 students are taking a college history course. There is one class of size 150 and four classes of size 25.

6. What is the average class size among the five classes?

(A) 35
(B) 50
(C) 100
(D) 125
(E) 137.5

7. Among the 250 students, what is the average size, per student, of their history class?

(A) 35
(B) 50
(C) 100
(D) 125
(E) 137.5

Free-Response Questions

1. A college basketball team keeps records of career average points per game of players playing at least 75% of team games during their college careers. The cumulative relative frequency plot below summarizes statistics of players graduating over the past 10 years.

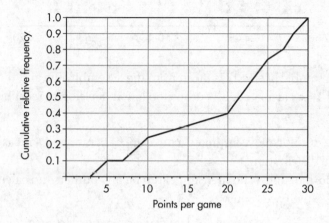

(a) Interpret the point (20, 0.4) in context.
(b) Interpret the intersection of the plot with the horizontal axis in context.
(c) Interpret the horizontal section of the plot from 5 to 7 points per game in context.
(d) The players with the top 10% of the career average points per game will be listed on a plaque. What is the cutoff score for being included on the plaque?
(e) What proportion of the players averaged between 10 and 20 points per game?

2. A company engineer creates a diagnostic measurement, $W = \dfrac{\text{Max} + \text{Min}}{2}$, which should be at least 24.10 in a sample of size 12 if certain machinery is operating correctly. To explore this diagnostic measurement, the machine is perfectly calibrated. Then 100 random samples of size 12 of the product are taken from the assembly line. For each of these 100 samples, the diagnostic measurement W is calculated and shown plotted below.

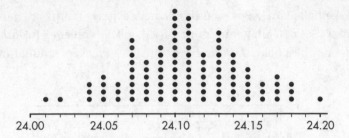

Each day, one sample of size 12 is taken from the assembly line and the diagnostic measurement W is calculated. If W drops too low, a decision to recalibrate the machinery is made.

(a) From the dotplot above, estimate a measure of center and a measure of variability for the distribution.

(b) For the dotplot above, do there appear to be any outliers (no calculations required)? Justify your answer.

One day the random sample is {24.2, 24.84, 25.05, 23.43, 23.9, 25.01, 23.01, 24.5, 24.23, 23.76, 24.69, 23.21}.

(c) Based on the dotplot above, does the engineer have sufficient evidence to conclude that recalibration is necessary? Justify your answer.

3. In a random sample of 32 newlyweds, the numbers of months they dated before becoming engaged are displayed in the following stemplot.

```
12 | 0 1 1 3 9
13 | 0 1 3 4 4 7 7 9
14 | 2 2 5 6 8 8 8 9
15 | 3 3 7 7
16 | 5 6
17 | 0 7
18 | 4
19 | 2
20 |
21 |
22 |
23 | 1
```

23 | 1   means 23.1 months

(a) Describe the distribution of the data in this stemplot.

(b) Describe the shape of the cumulative frequency plot of these data. (You are not asked to draw this plot.)

The answers for this quiz can be found in the Appendix on page 577.

SUMMARY STATISTICS FOR A QUANTITATIVE VARIABLE

Given a raw set of data, often we can detect no overall pattern. Perhaps some values occur more frequently, a few extreme values may stand out, and the range of values is usually apparent. The presentation of data, including summarizations and descriptions, and involving such concepts as representative or average values, measures of dispersion, positions of various values, and the shape of a distribution, falls under the broad topic of *descriptive statistics*. This aspect of statistics is in contrast to *inferential statistics*, the process of drawing inferences from limited data, a subject discussed in later units.

The word *average* is used in phrases common to everyday conversation. People speak of bowling and batting averages or the average life expectancy of a battery or a human being. Actually the word *average* is derived from the French *avarie*, which refers to the money that shippers contributed to help compensate for losses suffered by other shippers whose cargo did not arrive safely (i.e., the losses were shared, with everyone contributing an average amount). In common usage, *average* has come to mean a representative score or a typical value or the center of a distribution. Mathematically, there are a variety of ways to define the average of a set of data. In practice, the *mean* is the most useful such measure with regard to statistical inference. However, beware of a headline with the word *average*; the writer has probably chosen the method that emphasizes the point he or she wishes to make.

In the following paragraphs, we consider the two primary ways of denoting the "center" of a distribution:

1. The *median,* which is the middle number of a set of numbers arranged in numerical order.
2. The *mean,* which is found by summing items in a set and dividing by the number of items.

➡ EXAMPLE 1.13

Consider the following set of home run distances (in feet) to center field in 13 ballparks: {387, 400, 400, 410, 410, 410, 414, 415, 420, 420, 421, 457, 461}. What is the average?

Answer: The median is 414 (there are six values below 414 and six values above), while the mean is

$$\frac{387 + 400 + 400 + 410 + 410 + \ldots + 457 + 461}{13} = 417.3 \text{ feet}$$

The word *median* is derived from the Latin *medius,* which means "middle." The values under consideration are arranged in ascending or descending order. If there is an odd number of values, the median is the middle one. If there is an even number of values, the median is found by adding the two middle values and dividing by 2. Thus the median of a set has the same number of elements above it as below it.

The median is not affected by exactly how large the larger values are or by exactly how small the smaller values are. Thus it is a particularly useful measurement when the extreme values, *outliers,* are in some way suspicious or when you want to diminish their effect. For example, if ten mice try to solve a maze, and nine succeed in less than 15 minutes while one is still trying after 24 hours, the most representative value is the median (not the mean,

REMEMBER

Don't forget to put the data in order before finding the median.

which is over 2 hours). Similarly, if the salaries of four executives are each between $240,000 and $245,000 while a fifth is paid less than $20,000, again the most representative value is the median (the mean is under $200,000). It is often said that the median is "resistant" to extreme values.

In certain situations the median offers the most economical and quickest way to calculate an average. For example, suppose 10,000 lightbulbs of a particular brand are installed in a factory. An average life expectancy for the bulbs can most easily be found by noting how much time passes before exactly one-half of them have to be replaced. The median is also useful in certain kinds of medical research. For example, to compare the relative strengths of different poisons, a scientist notes what dosage of each poison will result in the death of exactly one-half the test animals. If one of the animals proves especially susceptible to a particular poison, the median lethal dose is not affected.

While the median is often useful in descriptive statistics, the *mean,* or more accurately, the *arithmetic mean*, is most important for statistical inference and analysis. Also, for the layperson, the "average" is usually understood to be the mean.

The mean of a *whole population* (the complete set of items of interest) is often denoted by the Greek letter μ (mu), while the mean of a *sample* (a part of a population) is often denoted by $\bar{x}$. For example, the mean value of the set of all houses in the United States might be $\mu = \$156,400$, while the mean value of 100 randomly chosen houses might be $\bar{x} = \$152,100$ or perhaps $\bar{x} = \$163,800$ or even $\bar{x} = \$224,000$.

In statistics we learn how to estimate a population mean from a sample mean. Throughout this book, the word *sample* often implies a *simple random sample* (SRS), that is, a sample selected in such a way that every possible sample of the desired size has an equal chance of being included. (It is also true that each element of the population has an equal chance of being included.) In the real world, this process of random selection is often very difficult to achieve, and so we proceed, with caution, as long as we have good reason to believe that our sample is representative of the population.

NOTE

Methods of collecting data, including the SRS, will be fully discussed in Unit 3.

Mathematically, the mean $= \dfrac{\sum x}{n}$, where Σx represents the sum of all the elements of the set under consideration and n is the actual number of elements. Σ is the uppercase Greek letter sigma.

➥ EXAMPLE 1.14

Suppose that the numbers of unnecessary procedures recommended by five doctors in a one-month period are given by the set {2, 2, 8, 20, 33}. Note that the median is 8 and the mean is $\dfrac{2+2+8+20+33}{5} = 13$. If it is discovered that the fifth doctor recommended an *additional* 25 unnecessary procedures, how will the median and mean be affected?

Answer: The set is now {2, 2, 8, 20, 58}. The median is still 8; however, the mean changes to $\dfrac{2+2+8+20+58}{5} = 18$.

This example illustrates how the mean, unlike the median, is sensitive to a change in any value.

➡ EXAMPLE 1.15

Suppose the salaries of six employees are $3,000, $7,000, $15,000, $22,000, $23,000, and $38,000, respectively.

a. What is the mean salary?

Answer: $\dfrac{3,000 + 7,000 + 15,000 + 22,000 + 23,000 + 38,000}{6} = \$18,000$

b. What will the new mean salary be if everyone receives a $3000 increase?

Answer: $\dfrac{6,000 + 10,000 + 18,000 + 25,000 + 26,000 + 41,000}{6} = \$21,000$

Note that $18,000 + $3,000 = $21,000.

c. What if instead everyone receives a 10% raise?

Answer: $\dfrac{3,300 + 7,700 + 16,500 + 24,200 + 25,300 + 41,800}{6} = \$19,800$

Note that 110% of $18,000 is $19,800.

Example 1.15 illustrates how adding the same constant to each value increases the mean by a like amount. Similarly, multiplying each value by the same constant multiplies the mean by a like amount.

NOTE

The same is true for medians: adding the same constant to each value increases the median by a like amount, and multiplying each value by the same constant multiplies the median by a like amount.

In describing a set of numbers, not only is it useful to designate an average value but also it is important to be able to indicate the *variability* or the *dispersion* of the measurements. An explosion engineer in mining operations aims for small variability—it would not be good for the 30-minute fuses actually to have a range of 10–50 minutes before detonation. On the other hand, a teacher interested in distinguishing better students from poorer students aims to design exams with large variability in results—it would not be helpful if all the students scored exactly the same. The players on two basketball teams may have the same average height, but this observation doesn't tell the whole story. If the dispersions are quite different, one team may have a 7-foot player, whereas the other has no one over 6 feet tall. Two Mediterranean holiday cruises may advertise the same average age for their passengers. One, however, may have passengers only between 20 and 25 years old, while the other has only middle-aged parents in their forties together with their children under age 10.

There are four primary ways of describing variability, or dispersion:

1. The *range*, which is the difference between the largest and smallest values
2. The *interquartile range*, IQR, which is the difference between the largest and smallest values after removing the lower and upper quarters (i.e., IQR is the range of the middle 50%); that is, $IQR = Q_3 - Q_1 = $ 75th percentile minus 25th percentile
3. The *variance*, which is determined by averaging the squared differences of all the values from the mean
4. The *standard deviation*, which is the square root of the variance

TIP

Variability is the single most fundamental concept in statistics and is the key to understanding statistics.

NOTE

The first quartile, denoted by Q_1, is the median of the lower half of the data set. That is, about 25% of the numbers in the data set lie below Q_1 and about 75% lie above Q_1. Similarly, the third quartile, denoted by Q_3, is the median of the upper half of the data set. Thus, about 75% of the numbers in the data set lie below Q_3 and about 25% lie above Q_3.

NOTE

Correct terminology is important in mathematics and essential in statistics. On the exam, words like *range* should be used only with their proper statistical meaning.

The monthly rainfall in Monrovia, Liberia, where May through October is the rainy season and November through April is the dry season, is as follows:

Month:	Jan	Feb	Mar	Apr	May	June	July	Aug	Sept	Oct	Nov	Dec
Rain (in.):	1	2	4	6	18	37	31	16	28	24	9	4

The mean is $\dfrac{1+2+4+6+18+37+31+16+28+24+9+4}{12} = 15$ inches

What are the measures of variability?

Answer: *Range:* The maximum is 37 inches (June), and the minimum is 1 inch (January). Thus the range is $37 - 1 = 36$ inches of rain.

Interquartile range: Removing the lower and upper quarters leaves 4, 6, 9, 16, 18, and 24. Thus the interquartile range is $24 - 4 = 20$. [The interquartile range is sometimes calculated as follows: The median of the lower half is $Q_1 = \dfrac{4+4}{2} = 4$, the median of the upper half is $Q_3 = \dfrac{24+28}{2} = 26$, and the interquartile range is $Q_3 - Q_1 = 22$. When there is a large number of values in the set, the two methods give the same answer.]

Variance:

$$\frac{(15-1)^2 + (15-2)^2 + \ldots + (15-4)^2}{12} =$$

$$\frac{14^2 + 13^2 + 11^2 + 9^2 + 3^2 + 22^2 + 16^2 + 1^2 + 13^2 + 9^2 + 6^2 + 11^2}{12} = 143.7$$

Standard deviation: $\sqrt{143.7} = 11.9$ inches

The simplest, most easily calculated measure of variability is the *range*. The difference between the largest and smallest values can be noted quickly, and the range gives some impression of the dispersion. However, it is entirely dependent on the two extreme values and is insensitive to the ones in the middle.

One use of the range is to evaluate samples with very few items. For example, some quality control techniques involve taking periodic small samples and basing further action on the range found in several such samples.

Finding the *interquartile range* is one method of removing the influence of extreme values on the range. It is calculated by arranging the data in numerical order, removing the upper and lower quarters of the values, and noting the range of the remaining values. That is, it is the range of the middle 50% of the values.

The numerical rule for designating outliers is to calculate 1.5 times the interquartile range (IQR) and then call a value an outlier if it is more than $1.5 \times$ IQR below the first quartile or $1.5 \times$ IQR above the third quartile.

NOTE

The description to the right is the most commonly accepted mathematical definition of *outliers* and should be used whenever you are asked to determine if a value qualifies as an outlier.

➡️ EXAMPLE 1.17

Suppose that the starting salaries (in thousands of dollars) for college graduates who took AP Statistics in high school and at least one additional statistics course in college have the following characteristics: the smallest value is 18.8, 10% of the values are below 25.6, 25% are below 41.1, the median is 59.3, 60% are below 84.3, 75% are below 101.9, 90% are below 118.0, and the top value is 201.7.

a. What is the range?

   Answer: The range is 201.7 − 18.8 = 182.9 (thousand dollars) = $182,900.

b. What is the interquartile range?

   Answer: The interquartile range, that is, the range of the middle 50% of the values, is $Q_3 - Q_1$ = 101.9 − 41.1 = 60.8 (thousand dollars) = $60,800.

c. Using the numerical rule that determines if a value is an outlier, should either the smallest or largest value be called an outlier?

   Answer: 1.5 × IQR = 1.5 × 60.8 = 91.2. If a value is more than 91.2 below the first quartile, 41.1, or more than 91.2 above the third quartile, 101.9, it will be called an outlier. Because the lowest value, 18.8, is not less than 60.8 − 91.2 = −30.4, the smallest value is not considered an outlier by the given numerical rule. Because the largest value, 201.7, is greater than 101.9 + 91.2 = 193.1, the largest value is considered an outlier by the given numerical rule.

Dispersion is often the result of various chance happenings. For example, consider the motion of microscopic particles suspended in a liquid. The unpredictable motion of any particle is the result of many small movements in various directions caused by random bumps from other particles. If we average the total displacements of all the particles from their starting points, the result will not increase in direct proportion to time. If, however, we average the *squares* of the total displacements of all the particles, this result will increase in direct proportion to time.

The same holds true for the movement of paramecia. Their seemingly random motions as seen under a microscope can be described by the observation that the average of the squares of the displacements from their starting points is directly proportional to time. Also, consider Ping-Pong balls dropped straight down from a high tower and subjected to chance buffeting in the air. We can measure the deviations from a center spot on the ground to the spots where the balls actually strike. As the height of the tower is increased, the average of the squared deviations increases proportionately.

In a wide variety of cases we are in effect trying to measure dispersion from the mean due to a multitude of chance effects. The proper tool in these cases is the average of the squared deviations from the mean; it is called the *variance* and is denoted by σ^2 (σ is the lowercase Greek letter sigma):

$$\sigma^2 = \frac{\Sigma(x - \mu)^2}{n}$$

For circumstances specified later in the units on inference, the variance of a sample, denoted by s^2, is calculated as

$$s^2 = \frac{\Sigma(x - \bar{x})^2}{n - 1}$$

The points per game (PPG) during the 2014–2015 season of the Golden State Warriors players were {23.8, 21.7, 11.7, 10.1, 7.8, 6.3, 5.9, 7.9, 10.4, 7.1, 4.3, 4.4, 4.1, 0.9, 1.3}. What was the variance?

Answer: The variance can be quickly found on any calculator that has a simple statistical package, or it can be found as follows:

$$\mu = \frac{23.8 + 21.7 + 11.7 + 10.1 + 7.8 + 6.3 + 5.9 + 7.9 + 10.4 + 7.1 + 4.3 + 4.4 + 4.1 + 0.9 + 1.3}{15} = 8.51$$

$$\sigma^2 = \frac{(23.8 - 8.51)^2 + (21.7 - 8.51)^2 + (11.7 - 8.51)^2 + \ldots + (1.3 - 8.51)^2}{15} = 40.17$$

Suppose we wish to pick a representative value for the variability of a certain population. The preceding discussions indicate that a natural choice is the value whose square is the average of the squared deviations from the mean. Thus we are led to consider the square root of the variance. This value is called the *standard deviation* (SD), is denoted by σ, and is calculated on your calculator or as follows:

$$\sigma = \sqrt{\frac{\Sigma(x - \mu)^2}{n}}$$

Similarly, the standard deviation of a sample is denoted by s and is calculated on your calculator or as follows:

$$s = \sqrt{\frac{\Sigma(x - \bar{x})^2}{n - 1}}$$

While variance is measured in square units, standard deviation is measured in the same units as are the data.

For the various x-values, the deviations $x - \bar{x}$ are called *residuals*, and s is a "typical value" for the residuals. While s is not the average of the residuals (the average of the residuals is always 0), s does give a measure of the spread of the x-values around the sample mean.

➥ **EXAMPLE 1.19**

The number of calories in 12-ounce servings of five popular beers are {95, 152, 188, 205, 131}. Using the TI-84, 1-Var Stats gives:

$$1 - \text{Var Stats}$$
$$\bar{x} = 154.2$$
$$\Sigma x = 771$$
$$\Sigma x^2 = 126659$$
$$Sx = 44.076$$
$$\sigma x = 39.423$$

Since these data represent a sample of beers, the standard deviation is 44.076.

⇒ EXAMPLE 1.20

Consider the following dotplots of student scores on an AP exam of four AP Calculus classes, each with 10 students. Without making any calculations, arrange the four classes in order from smallest SD to largest SD.

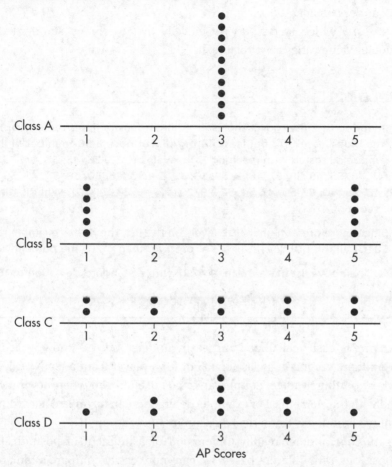

AP Scores

Answer: While all four classes have a mean score of 3, variability is the issue, as it is unique for each class. The scores in Class A have no variability at all, so the SD is smallest for Class A. The scores in Class B are all as far as possible away from the mean, so the SD is greatest for Class B. Note that Class D has more scores near the middle, while Class C has more scores further from the middle. So the SD for Class D is less than the SD for Class C. The correct ordering of the classes from smallest SD to largest SD is: A-D-C-B.

We have seen ways of choosing a value to represent the center of a distribution. We also need to be able to talk about the *position* of other values. In some situations, such as wine tasting, simple rankings are of interest. Other cases, for example, evaluating college applications, may involve positioning according to percentile rankings. There are also situations in which position can be specified by making use of measurements of both central tendency and variability.

There are three important, recognized procedures for designating position:

1. *Simple ranking*, which involves arranging the elements in some order and noting where in that order a particular value falls;
2. *Percentile ranking*, which indicates what percentage of all values fall at or below the value under consideration;
3. The *z-score*, which states very specifically by how many standard deviations a particular value varies from the mean.

NOTE

A negative z-score indicates that a value is less than the mean.

➡ **EXAMPLE 1.21**

It is recommended that the "good cholesterol," high-density lipoprotein (HDL), be present in the blood at levels of at least 40 mg/dl. Suppose a 50-member high school football team are all tested with resulting HDL levels of {53, 26, 45, 33, 64, 29, 73, 29, 21, 58, 70, 41, 48, 55, 55, 39, 57, 48, 9, 59, 56, 39, 68, 50, 65, 30, 38, 54, 49, 35, 56, 70, 43, 86, 52, 40, 28, 40, 67, 50, 47, 54, 59, 29, 29, 42, 45, 37, 51, 40}. What is the position of the HDL score of 41?

Answer: Since there are 31 higher HDL levels on the list, the 41 has a simple ranking of 32 out of 50. Of all the HDL levels, 19 are at or lower than 41, so the percentile ranking is $\frac{19}{50} = 38\%$. The above list has a mean of 47.22 with a standard deviation of 15.05, so the HDL score of 41 has a z-score of $\frac{41 - 47.22}{15.05} = -0.413$.

Simple ranking is easily calculated and easily understood. We know what it means for someone to graduate second in a class of 435 or for a player from a team of size 30 to have the seventh-best batting average. Simple ranking is useful even when no numerical values are associated with the elements. For example, detergents can be ranked according to relative cleansing ability without any numerical measurements of strength.

Percentile ranking, another readily understood measurement of position, is helpful in comparing positions with different bases. We can more easily compare a rank of 176 out of 704 with a rank of 187 out of 935 by noting that the first has a rank of 75% and the second a rank of 80%. Percentile rank is also useful when the exact population size is not known or is irrelevant. For example, it is more meaningful to say that Jennifer scored in the 90th percentile on a national exam rather than trying to determine her exact ranking among some large number of test takers.

The *quartiles*, Q_1 and Q_3, lie one-quarter and three-quarters of the way up a list, respectively. Their percentile ranks are 25% and 75%, respectively. The interquartile range defined earlier can also be defined as $Q_3 - Q_1$. The *deciles* lie one-tenth and nine-tenths of the way up a list, respectively, and have percentile ranks of 10% and 90%.

The *z-score* is a measure of position that takes into account both the center and the dispersion of the distribution. More specifically, the *z*-score of a value tells how many standard deviations the value is from the mean. Mathematically, $x - \mu$ gives the raw distance from μ to x; dividing by σ converts this to number of standard deviations. Thus $z = \frac{x - \mu}{\sigma}$, where x is the raw score, μ is the mean, and σ is the standard deviation. If the score x is greater than the mean μ, z is positive; if x is less than μ, z is negative.

Given a z-score, we can reverse the procedure and find the corresponding raw score. Solving for x gives $x = \mu + z\sigma$.

➥ EXAMPLE 1.22

In 2019, because of hurricanes, refinery outages, and Middle East tensions, the average (mean) price of gasoline in one northeast city hit \$3.80 per gallon with a standard deviation of \$0.05. Then \$3.90 has a z-score of $\frac{3.90 - 3.80}{0.05} = +2$, while \$3.65 has a z-score of $\frac{3.65 - 3.80}{0.05} = -3$. Alternatively, a z-score of $+2.2$ corresponds to a raw score of $3.80 + 2.2(0.05) = 3.80 + 0.11 = 3.91$, while a z-score of -1.6 corresponds to $3.80 - 1.6(0.05) = 3.72$.

It is often useful to portray integer z-scores and the corresponding raw scores as follows:

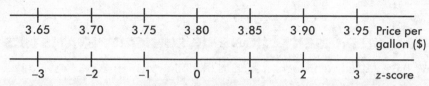

One purpose of z-scores is to act as a common "measuring stick" with which to compare values from different populations. For example, a high school senior might have a higher GPA than a high school sophomore, but the senior's GPA might be less far above the average GPA of seniors than the sophomore's GPA is above the average GPA of sophomores. We can make a comparison using z-scores.

Changing units, for example, from dollars to rubles or from miles to kilometers, is common in a world that seems to become smaller all the time. It is instructive to note how measures of center and spread are affected by such changes.

Adding the same constant to every value increases the mean and median by that same constant; however, the distances between the increased values stay the same, and so the range and standard deviation are unchanged. Multiplying every value by the same constant multiplies the mean, median, range, and standard deviation by that same constant.

> **NOTE**
>
> One consequence is that no matter what the distribution of raw scores, the set of z-scores always has mean 0 and standard deviation 1.

➥ EXAMPLE 1.23

A set of experimental measurements of the freezing point of an unknown liquid yield a mean of 25.32 degrees Celsius with a standard deviation of 1.47 degrees Celsius. If all the measurements are converted to the Kelvin scale, what are the new mean and standard deviation? (Kelvins are equivalent to degrees Celsius plus 273.16.)

Answer: The new mean is $25.32 + 273.16 = 298.48$ kelvins. The standard deviation, however, remains numerically the same, 1.47 kelvins. Graphically, you should picture the whole distribution moving over by the constant 273.16; the mean moves, but the standard deviation (which measures spread) doesn't change.

➡ **EXAMPLE 1.24**

Measurements of the sizes of farms in an upstate New York county yield a mean of 59.2 hectares with a standard deviation of 11.2 hectares. If all the measurements are converted from hectares (metric system) to acres (one acre was originally the area a yoke of oxen could plow in one day), what are the new mean and standard deviation? (One hectare is equivalent to 2.471 acres.)

Answer: The new mean is $2.471 \times 59.2 = 146.3$ acres with a standard deviation of $2.471 \times 11.2 = 27.7$ acres. Graphically, multiplying each value by the constant 2.471 both moves and spreads out the distribution.

GRAPHICAL REPRESENTATIONS OF SUMMARY STATISTICS

Suppose we have a detailed histogram such as the following:

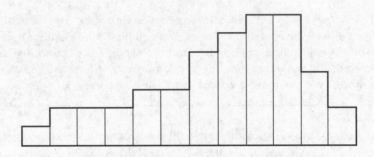

Our measures of central tendency fit naturally into such a diagram.

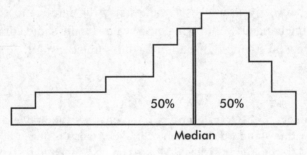

The *median* divides a distribution in half, so it is represented by a line that divides the area of the histogram in half.

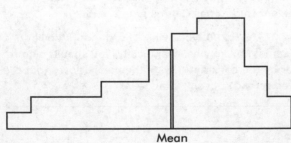

The *mean* is affected by the spacing of all the values. Therefore, if the histogram is considered to be a solid region, the mean corresponds to a line passing through the center of gravity, or balance point.

The above distribution, spread thinly far to the low side, is said to be *skewed to the left*. Note that in this case the mean is usually less than the median. Similarly, a distribution spread far to the high side is *skewed to the right*, and its mean is usually greater than its median.

➡ EXAMPLE 1.25

Suppose that the faculty salaries at a college have a median of $82,500 and a mean of $88,700. What does this indicate about the shape of the distribution of the salaries?

Answer: The median is less than the mean, and so the salaries are probably skewed to the right. There are a few highly paid professors, with the bulk of the faculty at the lower end of the pay scale.

It should be noted that the above principle is a useful, but not hard-and-fast, rule.

➡ EXAMPLE 1.26

The set given by the dotplot below is skewed to the right; however, its median (3) is greater than its mean (2.97).

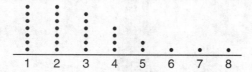

We have seen that relative frequencies are represented by relative areas, and so labeling the vertical axis is not crucial. If we know the standard deviation, the horizontal axis can be labeled in terms of z-scores. In fact, if we are given the percentile rankings of various z-scores, we can construct a histogram.

➡ EXAMPLE 1.27

Suppose we are asked to construct a histogram from these data:

z-score:	−2	−1	0	1	2
Percentile ranking:	0	20	60	70	100

We note that the entire area is less than z-score +2 and greater than z-score −2. Also, 20% of the area is between z-scores −2 and −1, 40% is between −1 and 0, 10% is between 0 and 1, and 30% is between 1 and 2. Thus the histogram is as follows:

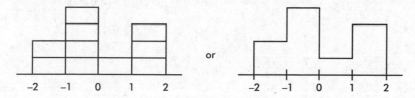

(continued)

Now suppose we are given four in-between *z*-scores as well:

z-score	Percentile Ranking
2.0	100
1.5	80
1.0	70
0.5	65
0.0	60
−0.5	30
−1.0	20
−1.5	5
−2.0	0

Then we have:

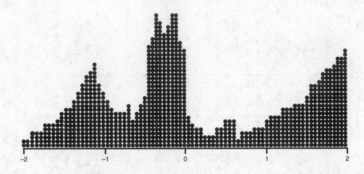

With 1,000 *z*-scores perhaps the histogram would look like:

The height at any point is meaningless; what is important is relative areas. For example, in the final diagram above, what percentage of the area is between *z*-scores of +1 and +2?

Answer: Still 30%.

What percent is to the left of 0?

Answer: Still 60%.

TIP

The IQR is the *length* of the box, not the box itself. So, the median is in the box, or is between Q_1 and Q_3, but is not *in* the IQR.

A *boxplot* (also called a *box and whisker display*) is a visual representation of dispersion that shows the smallest value, the largest value, the middle (median), the middle of the bottom half of the set (Q_1), and the middle of the top half of the set (Q_3).

➡ EXAMPLE 1.28 _____

After an AP Statistics teacher hears that every one of her 27 students received a 3 or higher on the AP exam, she treats the class to a game of bowling. The individual student bowling scores are 210, 130, 150, 140, 150, 210, 150, 125, 85, 200, 70, 150, 75, 90, 150, 115, 120, 125, 160, 140, 100, 95, 100, 215, 130, 160, and 200. Their students note that the greatest score is 215, the smallest is 70, the middle is 140, the middle of the top half is 160, and the middle of the bottom half is 100. A boxplot of these five numbers is

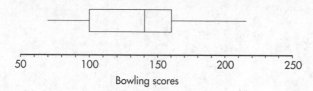

Bowling scores

Note that the display consists of two "boxes" together with two "whiskers"—hence the alternative name. The boxes show the spread of the two middle quarters; the whiskers show the spread of the two outer quarters. This relatively simple display conveys information not immediately available from histograms or stem and leaf displays.

Putting the above data into a list, for example, L1, on the TI-84, not only gives the five-number summary

```
1-Var Stats
minX=70
Q1=100
Med=140
Q3=160
MaxX=215
```

but also gives the boxplot itself using STAT PLOT, choosing the boxplot from among the six type choices, and then using ZoomStat or in WINDOW letting Xmin=0 and Xmax=225.

On the TI-Nspire the data can be put in a list (here called *index*), and then a simultaneous multiple view is possible.

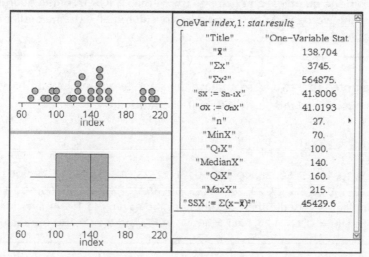

When a distribution is strongly skewed, or when it has pronounced outliers, drawing a boxplot with its five-number summary including median, quartiles, and extremes, gives a more useful description than calculating a mean and a standard deviation.

Values more than $1.5 \times$ IQR (1.5 times the interquartile range) outside the two boxes are plotted separately as outliers. (The TI-84, Casio Prizm, and HP Prime have this boxplot option.)

➡ EXAMPLE 1.29

Inputting the lengths of words in a selection of Shakespeare's plays results in a calculator output of

```
1-Var Stats
minX=1
Q₁=3
Med=4
Q₃=5
maxX=12
```

NOTE

In formula form, outliers are any value less than $Q_1 - 1.5(IQR)$ or greater than $Q_3 + 1.5(IQR)$.

Outliers consist of any word lengths less than $Q_1 - 1.5(IQR) = 3 - 1.5(5 - 3) = 0$ or greater than $Q_3 + 1.5(IQR) = 5 + 1.5(5 - 3) = 8$. A boxplot indicating outliers, together with a histogram (on the TI-84 up to three different graphs can be shown simultaneously) is

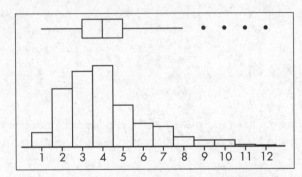

NOTE

When outliers are present, the whiskers are drawn to the smallest and largest values that are not outliers; they are *not* drawn to $Q_1 - 1.5$ (IQR) and $Q_3 + 1.5$ (IQR) unless these are actual data values.

Note: Some computer output shows *two* levels of outliers—mild (between 1.5 IQR and 3 IQR from the quartiles) and extreme (more than 3 IQR from the quartiles). In this example, the word length of 12 would be considered an extreme outlier since it is greater than $5 + 3(5 - 3) = 11$.

It should be noted that two sets can have the same five-number summary and thus the same boxplots but have dramatically different distributions.

NOTE

Remember that concepts like skewness and symmetry are generally very difficult to conclude from just a boxplot.

➡ EXAMPLE 1.30

Let $A = \{0, 5, 10, 15, 25, 30, 35, 40, 45, 50, 71, 72, 73, 74, 75, 76, 77, 78, 100\}$ and $B = \{0, 22, 23, 24, 25, 26, 27, 28, 29, 50, 55, 60, 65, 70, 75, 85, 90, 95, 100\}$. Simple inspection indicates very different distributions; however, the TI-84 gives identical boxplots with Min = 0, $Q_1 = 25$, Med = 50, $Q_3 = 75$, and Max = 100 for each.

COMPARING DISTRIBUTIONS OF A QUANTITATIVE VARIABLE

➡ **EXAMPLE 1.31**_____

The numbers of wins for the 30 NBA teams at the end of the 2018–2019 season is shown in the following *back-to-back stemplot*.

Eastern Conference		Western Conference
9 7	1	9
9 2	2	
9 9 2	3	3 3 3 6 7 9
9 8 2 2 1	4	8 8 9
8 1	5	0 3 3 4 7
0	6	

7 | 1 | 9 represents Eastern and Western Conference teams with 17 and 19 wins, respectively.

> **NOTE**
>
> The leaves radiate out from a common stem.

> **NOTE**
>
> Although both sets in this example are of the same size, in general you do not have to have the same number of data points on each side of the plot.

When comparing shape, center, and spread, we have:

Shape: The distribution of wins in the Eastern Conference (EC) is roughly bell-shaped, while the distribution of wins in the Western Conference (WC) is roughly uniform with a low outlier.

Center: Counting values (8th out of 15) gives medians of $m_{EC} = 41$ and $m_{WC} = 49$. Thus, the WC distribution of wins has the greater center.

Spread: The range of the EC distribution of wins is $60 - 17 = 43$, while the range of the WC distribution of wins is $57 - 19 = 38$. Thus, the EC distribution of wins has the greater spread.

> **IMPORTANT**
>
> When asked for a comparison, don't forget to address <u>s</u>hape, <u>o</u>utliers (<u>u</u>nusual values), <u>c</u>enter, and <u>s</u>pread (SOCS or CUSS) and to refer to context. You must use *comparative* words—that is, you must state which center and which spread is larger (or if they are approximately the same). Simply making two separate lists is not enough and will be penalized.

➡ **EXAMPLE 1.32**

The following are parallel boxplots showing the daily price fluctuations of a certain common stock over the course of 5 years. What trends do the boxplots show?

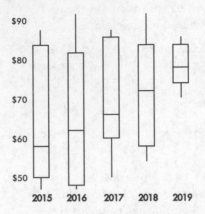

TIP

Don't forget to label and provide a scale for all graphs!

The parallel boxplots show that from year to year the median daily stock price has steadily risen 20 points from about $58 to about $78, the thlrd quartile value has been roughly stable at about $84, the yearly low has never decreased from that of the previous year, and the interquartile range has never increased from one year to the next.

➡ **EXAMPLE 1.33**

The graph below compares cumulative frequency plotted against age for the U.S. population in 1860 and in 1980.

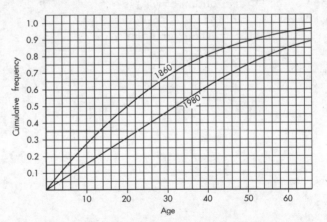

How do the medians and interquartile ranges compare?

Answer: Looking across from 0.5 on the vertical axis, we see that in 1860 half the population was under the age of 20, while in 1980 all the way up to age 32 must be included to encompass half the population. Looking across from 0.25 and 0.75 on the vertical axis, we see that for 1860, $Q_1 = 9$ and $Q_3 = 35$ and so the interquartile range is $35 - 9 = 26$ years, while for 1980, $Q_1 = 16$ and $Q_3 = 50$ and so the interquartile range is $50 - 16 = 34$ years. Thus, both the median and the interquartile range were greater in 1980 than in 1860.

QUIZ 3

Multiple-Choice Questions

> **Directions:** The questions that follow are each followed by five suggested answers. Choose the response that best answers the question.

1. The graph below shows household income in Laguna Woods, California.

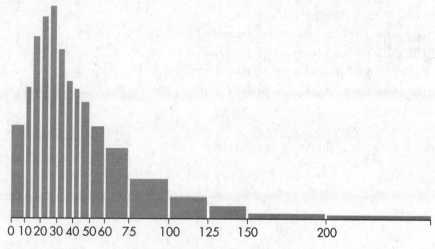

Household income distribution ($1000)

What can be said about the ratio $\dfrac{\text{Mean household income}}{\text{Median household income}}$?

(A) Approximately zero

(B) Less than one, but definitely above zero

(C) Approximately one

(D) Greater than one

(E) Cannot be answered without knowing the standard deviation

2. Students in an algebra class were timed in seconds while solving a series of mathematical brainteasers. One student's time had a standardized score of $z = 2.40$. If the times are all changed to minutes, what will then be the student's standardized score?

(A) $z = 0.04$

(B) $z = 0.4$

(C) $z = 1.80$

(D) $z = 2.40$

(E) The new standardized score cannot be determined without knowing the class mean.

3. Dieticians are concerned about sugar consumption in teenagers' diets (a 12-ounce can of soft drink typically has 10 teaspoons of sugar). In a random sample of 55 students, the number of teaspoons of sugar consumed for each student on a randomly selected day is tabulated. Summary statistics are noted below:

Min = 10   Max = 60   First quartile = 25   Third quartile = 38

Median = 31   Mean = 31.4 $n = 55$ $s = 11.6$

Which of the following is a true statement?

(A) None of the values are outliers.
(B) The value 10 is an outlier, and there can be no others.
(C) The value 60 is an outlier, and there can be no others.
(D) Both 10 and 60 are outliers, and there may be others.
(E) The value 60 is an outlier, and there may be others at the high end of the data set.

Questions 4–5 refer to the following five boxplots:

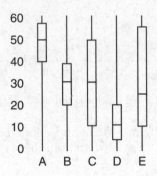

4. To which of the above boxplots does the following histogram correspond?

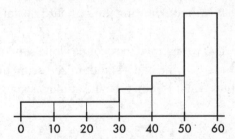

(A) A
(B) B
(C) C
(D) D
(E) E

5. To which of the above boxplots does the following histogram correspond?

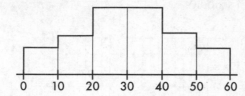

(A) A
(B) B
(C) C
(D) D
(E) E

6. Suppose the average score on a national test is 500 with a standard deviation of 100. If each score is increased by 25, what are the new mean and standard deviation?

(A) Mean = 500 and SD = 100
(B) Mean = 500 and SD = 125
(C) Mean = 525 and SD = 100
(D) Mean = 525 and SD = 105
(E) Mean = 525 and SD = 125

7. If quartiles $Q_1 = 20$ and $Q_3 = 30$, which of the following must be true?

I. The median is 25.
II. The mean is between 20 and 30.
III. The standard deviation is at most 10.

(A) I only
(B) II only
(C) III only
(D) All must be true.
(E) None must be true.

8. The graph below shows cumulative proportions plotted against grade point averages for a large public high school.

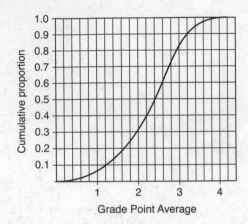

What are the median grade point average and the IQR?

(A)   Median = 0.8, IQR = 1.8
(B)   Median = 2.0, IQR = 2.8
(C)   Median = 2.4, IQR = 1.0
(D)   Median = 2.5, IQR = 1.0
(E)   Median = 2.6, IQR = 1.8

Free-Response Questions

Directions: You must show all work and indicate the methods you use. You will be graded on the correctness of your methods and on the accuracy of your final answers.

1. Victims spend from 5 to 5840 hours repairing the damage caused by identity theft with a mean of 330 hours and a standard deviation of 245 hours.

    (a) What would be the mean, range, standard deviation, and variance for hours spent repairing the damage caused by identity theft if each of the victims spent an additional 10 hours?

    (b) What would be the mean, range, standard deviation, and variance for hours spent repairing the damage caused by identity theft if each of the victims' hours spent increased by 10%?

2. Are women better than men at multitasking? Suppose in one study of multitasking a random sample of 200 female and 200 male high school students were assigned several tasks at the same time, such as solving simple mathematics problems, reading maps, and answering simple questions while talking on a telephone. Total times taken to complete all the tasks are given in the histograms below.

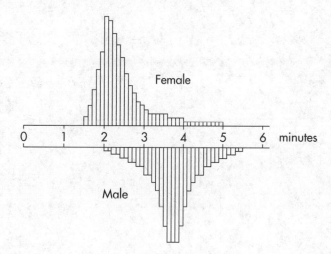

Write a few sentences comparing the distributions of times to complete all tasks by females and by males.

3. In independent random samples of 20 men and 20 women, the number of minutes spent on grooming on a given day were:

    Men: 27, 32, 82, 36, 43, 75, 45, 16, 23, 48, 51, 57, 60, 64, 39, 40, 69, 72, 54, 57

    Women: 49, 50, 35, 69, 75, 35, 49, 54, 98, 58, 22, 34, 60, 38, 47, 65, 79, 38, 42, 87

    Using back-to-back stemplots, compare the two distributions.

4. To analyze the social media behavior differences between boys and girls, Mrs. V's FDA high school AP Statistics class was asked to count the number of text messages that they sent over a three-day weekend. The following table summarizes the data:

	Values under Q_1	Q_1	Median	Q_3	Values over Q_3
Females	15, 43, 100	130	175	358	450, 573, 1098
Males	3, 59	72	183	273	293, 337

(a) Construct parallel boxplots of this set of data.

(b) Do the data indicate that females or males had the greater mean number of texts? Explain.

The answers for this quiz can be found in the Appendix on page 579.

QUIZ 4

Multiple-Choice Questions

> **Directions:** The questions that follow are each followed by five suggested answers. Choose the response that best answers the question.

Questions 1–2 refer to the following five histograms:

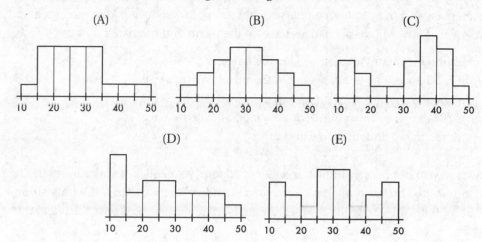

1. To which of the above histograms does the following boxplot correspond?

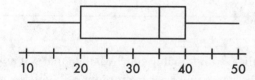

- (A) A
- (B) B
- (C) C
- (D) D
- (E) E

2. To which of the above histograms does the following boxplot correspond?

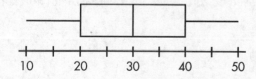

- (A) A
- (B) B
- (C) C
- (D) D
- (E) E

3. Below is a boxplot of CO_2 levels (in grams per kilometer) for a sampling of year 2018 vehicles.

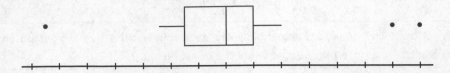

Suppose follow-up testing determines that the low outlier should be 10 grams per kilometer less and the two high outliers should each be 5 grams per kilometer greater. What effect, if any, will these changes have on the mean and median CO_2 levels?

(A) Both the mean and median will be unchanged.
(B) The median will be unchanged, but the mean will increase.
(C) The median will be unchanged, but the mean will decrease.
(D) The mean will be unchanged, but the median will increase.
(E) Both the mean and median will change.

4. Below is a boxplot of yearly tuition and fees of all four-year colleges and universities in a western state. The low outlier is from a private university that gives full scholarships to all accepted students, while the high outlier is from a private college catering to the very rich.

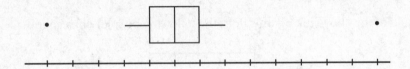

Removing both outliers will probably effect what changes, if any, on the mean and median costs for this state's four-year institutions of higher learning?

(A) Both the mean and the median will be unchanged.
(B) The median will be unchanged, but the mean will increase.
(C) The median will be unchanged, but the mean will decrease.
(D) The mean will be unchanged, but the median will increase.
(E) Both the mean and median will change.

5. Suppose the average score on a national test is 500 with a standard deviation of 100. If each score is increased by 25%, what are the new mean and standard deviation, respectively?

(A) 500 and 100
(B) 525 and 100
(C) 625 and 100
(D) 625 and 105
(E) 625 and 125

6. A teacher is teaching two AP Statistics classes. On the final exam, the 20 students in the first class averaged 92, while the 25 students in the second class averaged only 83. If the teacher combines the classes, what will the average final exam score be?

(A) 87
(B) 87.5
(C) 88
(D) None of the above.
(E) More information is needed to make this calculation.

7. Suppose 10% of a data set lies between 40 and 60. If 5 is first added to each value in the set and then each result is doubled, which of the following is true?

(A) 10% of the resulting data will lie between 85 and 125.
(B) 10% of the resulting data will lie between 90 and 130.
(C) 15% of the resulting data will lie between 80 and 120.
(D) 20% of the resulting data will lie between 45 and 65.
(E) 30% of the resulting data will lie between 85 and 125.

8. The following boxplots were constructed from SAT math scores of boys and girls at a high school:

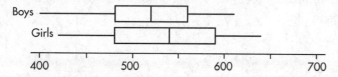

Which of the following is a possible boxplot for the combined scores of all the students?

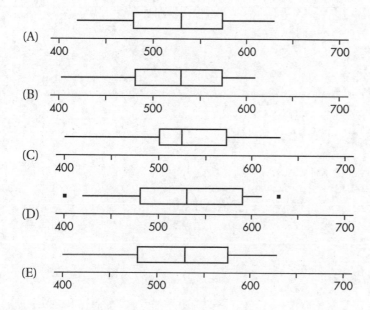

Free-Response Questions

1. Suppose a distribution has mean 300 and standard deviation 25. If the z-score of Q_1 is -0.7 and the z-score of Q_3 is 0.7, what values would be considered outliers?

2. The speeds (in mph) of runners in one marathon was summarized in the following table.

n	Mean	StDev	Min	Q_1	Med	Q_3	Max
135	8.9	1.2	6.1	7.9	9.5	10.8	12.6

Note that 1 mile $= 1.6$ kilometers and 1 kilometer $= 0.625$ miles.

(a) What was the median speed in km/hr?

(b) Suppose it turns out that due to an equipment malfunction, all 135 speeds were overreported by 0.2 mph. What was the true IQR in km/hr?

(c) If 14 km/hr has a percentile ranking of 65%, what speed in mph will be at the 65th percentile?

3. Below are two population pyramids from the U.S. Census Bureau.

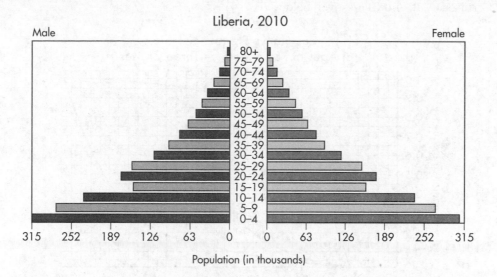

Liberia, 2010

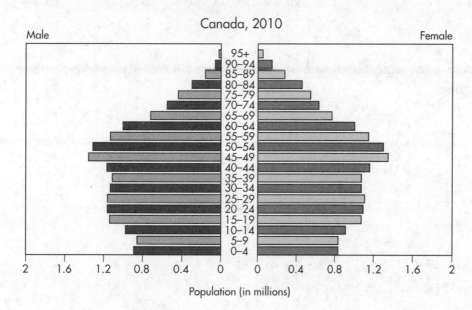

Canada, 2010

(a) The approximate median age of the Liberian population falls in which of these intervals: 0–4, 15–19, 30–34, 40–44? Explain.

(b) Explain why it is impossible to calculate the mean age of either population.

(c) Which country has more children younger than 10 years of age? Explain.

(d) Does the population pyramid indicate that Canadian men or Canadian women live longer? Explain.

(e) In 2010, Liberia had recently come out of a civil war with the extensive use of child soldiers. How is this visible in the population pyramid?

4. Cumulative frequency graphs of the ages of people on three different Caribbean cruises (A, B, and C) are given below:

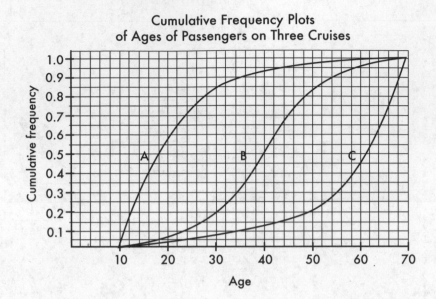

Cumulative Frequency Plots
of Ages of Passengers on Three Cruises

Write a few sentences comparing the distributions of ages of people on the three cruises.

The answers for this quiz can be found in the Appendix on page 580.

THE NORMAL DISTRIBUTION

One bell-shaped distribution, called the normal distribution, is valuable in describing various natural phenomena. However, we will see that the real importance of the normal distribution in statistics is that it can be used to describe the results of many sampling procedures.

The normal distribution curve is bell-shaped and symmetric and has an infinite base. Long, flat-looking tails cover many values but only a small proportion of the area. The flat appearance of the tails is deceptive. Actually, far out in the tails, the curve drops proportionately at an ever-increasing rate. In other words, when two intervals of equal length are compared, the one closer to the center may experience a greater numerical drop, but the one further out in the tail experiences a greater drop when measured as a proportion of the height at the beginning of the interval.

The mean here is the same as the median and is located at the center. We want a unit of measurement that applies equally well to any normal distribution, and we choose a unit that arises naturally out of the curve's shape. There is a point on each side where the slope is steepest. These two points are called *points of inflection*, and the distance from the mean to either point is precisely equal to one standard deviation. Thus, it is convenient to measure distances under the normal curve in terms of *z*-scores (recall from earlier this unit that *z*-scores are fractions or multiples of standard deviations from the mean).

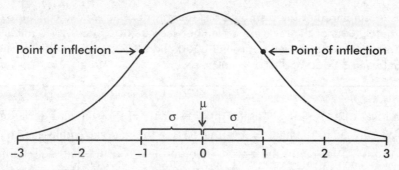

Unlike most other distributions, any normal distribution is completely determined by its mean μ and standard deviation σ.

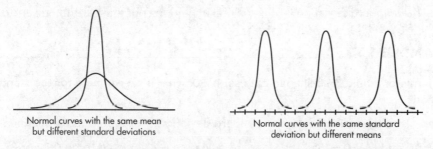

Normal curves with the same mean but different standard deviations

Normal curves with the same standard deviation but different means

The *empirical rule* (also called the *68-95-99.7 rule*) applies specifically to normal distributions. In this case, about 68% of the values lie within 1 standard deviation of the mean, about 95% of the values lie within 2 standard deviations of the mean, and more than 99% of the values lie within 3 standard deviations of the mean.

In the following figures, the horizontal axes show z-scores:

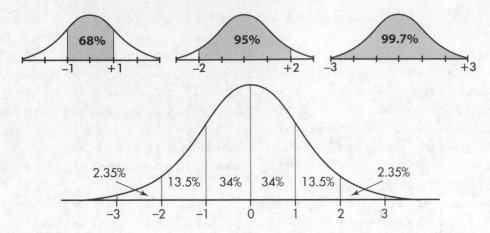

➡ **EXAMPLE 1.34** _____

Suppose that taxicabs in New York City are driven an average of 75,000 miles per year with a standard deviation of 12,000 miles. What information does the empirical rule give us?

Answer: Assuming that the distribution is roughly normal, we can conclude that approximately 68% of the taxis are driven between 63,000 and 87,000 miles per year, approximately 95% are driven between 51,000 and 99,000 miles, and virtually all are driven between 39,000 and 111,000 miles.

The empirical rule also gives a useful quick estimate of the standard deviation in terms of the range. We can see in the figure above that 95% of the data fall within a span of 4 standard deviations (from -2 to $+2$ on the z-score line) and 99.7% of the data fall within 6 standard deviations (from -3 to $+3$ on the z-score line). It is therefore reasonable to conclude that for these data the standard deviation is roughly between one-fourth and one-sixth of the range of a finite set. Since we can find the range of a finite set almost immediately, the empirical rule technique for estimating the standard deviation is often helpful in pointing out gross arithmetic errors.

➡ **EXAMPLE 1.35** _____

If the range of a finite bell-shaped data set is 60, what is an estimate for the standard deviation?

Answer: By the empirical rule, the standard deviation is expected to be between $\left(\frac{1}{6}\right)60 = 10$ and $\left(\frac{1}{4}\right)60 = 15$. If the standard deviation is calculated to be 0.32 or 87, there is probably an arithmetic error; a calculation of 12, however, is reasonable.

However, it must be stressed that the above use of the range is not intended to provide an accurate value for the standard deviation. It is simply a tool for pointing out unreasonable answers rather than, for example, blindly accepting computer outputs.

Don't be confused into thinking that z-scores mean normality. A z-score can be calculated whenever the mean and standard deviation are known, no matter what the distribution. Only in the very special case of nearly normal distributions can one move easily from z-scores to percentiles and probabilities.

QUIZ 5

Multiple-Choice Questions

> **Directions:** The questions that follow are each followed by five suggested answers. Choose the response that best answers the question.

1. Consider the following two normal curves.

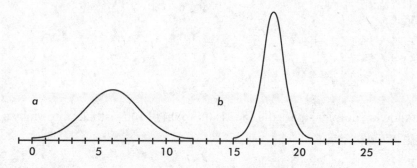

Which has the smaller mean, and which has the smaller standard deviation?

(A) Smaller mean, *a*; smaller standard deviation, *a*

(B) Smaller mean, *a*; smaller standard deviation, *b*

(C) Smaller mean, *b*; smaller standard deviation, *a*

(D) Smaller mean, *b*; smaller standard deviation, *b*

(E) Smaller mean, *a*; same standard deviation

Questions 2–4 refer to the following. Suppose an average life expectancy in a particular country is 74.3 years with a standard deviation of 9.7 years. Assuming a roughly normal distribution, consider the following graph of the distribution.

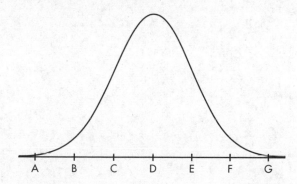

2. Point E on this normal curve corresponds to

(A) 54.9 years.

(B) 64.6 years.

(C) 74.3 years.

(D) 84.0 years.

(E) 93.7 years.

3. About what percent of the years are between 54.9 and 93.7 (points B and F)?

   (A)   17%
   (B)   34%
   (C)   68%
   (D)   95%
   (E)   99.7%

4. About what percent of the years are less than 64.6 (point C)?

   (A)   17%
   (B)   34%
   (C)   68%
   (D)   95%
   (E)   99.7%

5. The following dotplot shows the speeds (in mph) of 100 fastballs thrown by a major league pitcher.

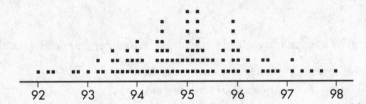

   Which of the following is the best estimate of the standard deviation of these speeds?

   (A)   0.5 mph
   (B)   1.1 mph
   (C)   1.6 mph
   (D)   2.2 mph
   (E)   6.0 mph

Free-Response Questions

> **Directions:** You must show all work and indicate the methods you use. You will be graded on the correctness of your methods and on the accuracy of your final answers.

1. It is estimated that 23% of American teen spending goes toward food. Assume that yearly amounts spent on food can be approximated by a normal model with a mean of $600 and a standard deviation of $120.

    (a) A teen spending $720 would be at what percentile?
    (b) What percent of American teens spend between $360 and $840 per year on food?
    (c) 99.7% of American teens spend between what two dollar amounts on food per year?

2. A high school guidance counselor tallied the number of colleges that seniors applied to in Fall 2019 and created the following descriptive statistics for the data set.

Variable	n	Mean	Median	StDev	Min	Max	Q_1	Q_3
Colleges	50	6.4	5	4.80	0	15	2	10

    Do the numbers of colleges applied to by this school's seniors in Fall 2019 seem to roughly follow a normal model?

The answers for this quiz can be found in the Appendix on page 582.

- The four keys to describing a distribution are shape, center, spread, and unusual features.
- Note clusters, gaps, modes, and outliers.
- Always provide context.
- Look for reasons behind any unusual features.
- A few common shapes which arise are symmetric, skewed to the right, skewed to the left, bell-shaped, and uniform distributions.
- Categorical variables are summarized by counts (frequency) and proportions (relative frequency) and visually displayed by dotplots and bar graphs.
- For quantitative data, dotplots, histograms, cumulative relative frequency plots (ogives), stemplots, and boxplots give useful displays.
- In a histogram, relative area corresponds to relative frequency.
- The two principal measurements of the center of a distribution are the mean and the median.
- The principal measurements of the spread of a distribution are the range (maximum value minus minimum value), the interquartile range (IQR $= Q_3 - Q_1$), the variance, and the standard deviation.
- Adding the same constant to every value in a set adds the same constant to the mean and median but leaves all the above measures of spread unchanged.
- Multiplying every value in a set by the same constant multiplies the mean, median, range, IQR, and standard deviation by that constant.
- The mean, range, variance, and standard deviation are sensitive to extreme values, while the median and interquartile range are not.
- The principal measurements of position are simple ranking, percentile ranking, and the z-score (which measures the number of standard deviations from the mean).
- No matter what the distribution of raw scores, the set of z-scores always has mean 0 and standard deviation 1.
- The empirical rule (the 68-95-99.7 rule) applies specifically to normal distributions.
- In skewed left data, the mean is usually less than the median, while in skewed right data, the mean is usually greater than the median.
- Boxplots show the five-number summary: the minimum value, the first quartile (Q_1), the median, the third quartile (Q_3), and the maximum value; they indicate outliers as distinct points.
- Two sets can have the same five-number summary and thus the same boxplots but have dramatically different distributions.
- Outliers are any values less than $Q_1 - 1.5(\text{IQR})$ or greater than $Q_3 + 1.5(\text{IQR})$.
- When asked for a comparison of two data sets, in addition to comparing shape and unusual features, you must state which center and which spread is greater (two separate calculations are not a comparison).

Exploring Two-Variable Data 2

(5–7% AP EXAM WEIGHTING)

In this unit, you will work with two-way tables, together with graphs and calculations, to explore relationships in two-variable categorical data sets. To understand association between two quantitative variables, you will analyze scatterplots and generic computer output, assess correlation, interpret slopes and intercepts in context, investigate residuals, and consider nonlinearity.

UNIT LEARNING OBJECTIVES

- To be able to calculate frequencies and conditional frequencies from a two-way table.
- To be able to interpret side-by-side bar charts, segmented bar charts, and mosaic plots.
- To be able to describe bivariate data from a scatterplot using the terminology direction (positive or negative), form (linear or nonlinear), strength (such as weak, moderate, or strong), and any unusual points, and always in context.
- To be able to find the equation of a least squares regression line from generic computer regression output.
- To be able to calculate the slope and y-intercept of the least squares regression line using relevant summary statistics (means, standard deviations, and correlation).
- To be able to interpret the slope and y-intercept of a linear regression line in context, using nondeterministic language.
- To be able to calculate and interpret a residual.
- To be able to roughly estimate a correlation coefficient from a scatterplot.
- To be able to interpret the coefficient of determination.
- To be able to identify outliers, influential points, and points of high leverage in a scatterplot.

TWO CATEGORICAL VARIABLES

Qualitative data often encompass two categorical variables that may or may not have a dependent relationship. These data can be displayed in a *two-way table* (also called a *contingency table*).

➥ EXAMPLE 2.1

The Cuteness Factor: A Japanese study had volunteers look at pictures of cute baby animals, adult animals, or tasty-looking foods, after which they tested their focus in solving puzzles.

		Level of focus		
		Low	Medium	High
Pictures viewed	Baby animals	5	20	40
	Adult animals	30	40	15
	Tasty foods	55	35	10

Pictures viewed is the *row variable*, whereas level of focus is the *column variable*. One method of analyzing these data involves calculating the totals for each row and each column.

		Level of focus			
		Low	Medium	High	*Total*
Pictures viewed	Baby animals	5	20	40	*65*
	Adult animals	30	40	15	*85*
	Tasty foods	55	35	10	*100*
	Total	*90*	*95*	*65*	*250*

These totals are placed in the right and bottom margins of the table and thus are called *marginal frequencies* (or *marginal totals*). These marginal frequencies can then be put in the form of proportions or percentages. The *marginal distribution* of the level of focus is

Low: $\dfrac{90}{250} = 0.36 = 36\%$

Medium: $\dfrac{95}{250} = 0.38 = 38\%$

High: $\dfrac{65}{250} = 0.26 = 26\%$

NOTE

The *marginal relative frequencies* are the row and column totals divided by the total for the entire table.

This distribution can be displayed in a bar graph as follows:

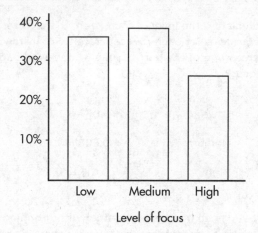

Similarly, we can determine the marginal distribution for the pictures viewed:

Baby animals: $\dfrac{65}{250} = 0.26 = 26\%$

Adult animals: $\dfrac{85}{250} = 0.34 = 34\%$

Tasty foods: $\dfrac{100}{250} = 0.40 = 40\%$

The representative bar graph is

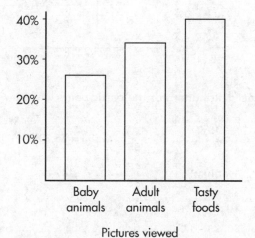

The marginal distributions described and calculated above do not describe or measure the relationship between the two categorical variables. For this we must consider the information in the body of the table, not just the sums in the margins.

We are interested in predicting the level of focus from the pictures viewed, and so we look at *conditional frequencies* for each row separately. For instance, in Example 2.1, what proportion or percentage of the participants who viewed baby animals then had each of the levels of focus?

Low: $\dfrac{5}{65} = 0.077 = 7.7\%$

Medium: $\dfrac{20}{65} = 0.308 = 30.8\%$

High: $\dfrac{40}{65} = 0.615 = 61.5\%$

This *conditional distribution* can be displayed either with groupings of bars or by a *segmented bar chart* where each segment has a length corresponding to its relative frequency:

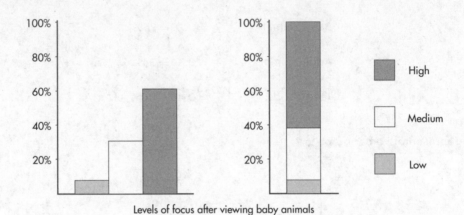

Levels of focus after viewing baby animals

Similarly, the conditional distribution for the participants who viewed adult animals are

Low: $\dfrac{30}{85} = 0.353 = 35.3\%$

Medium: $\dfrac{40}{85} = 0.471 = 47.1\%$

High: $\dfrac{15}{85} = 0.176 = 17.6\%$

For the participants who viewed the tasty foods, we have:

Low: $\dfrac{55}{100} = 0.55 = 55\%$

Medium: $\dfrac{35}{100} = 0.35 = 35\%$

High: $\dfrac{10}{100} = 0.10 = 10\%$

Both of the following bar charts give good visual representations of the data:

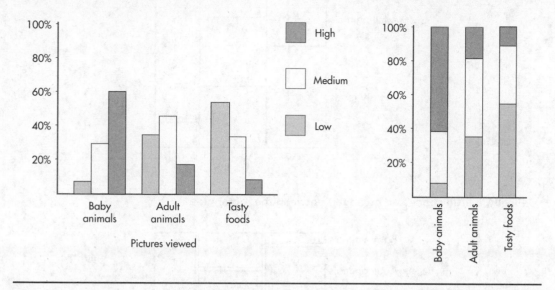

➡️ **EXAMPLE 2.3**_____

Mosaic plots are another graphical display that show the cell frequencies in a two-way table. They look like segmented bar graphs in which the width of each rectangle is proportional to the number of individuals in the corresponding group. In the above example, we can start with vertical sections based on level of focus. The area of each rectangle is proportional to the frequency of that group.

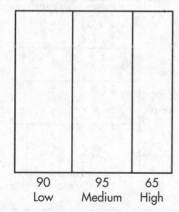

Then, each vertical section is split into horizontal sections based on picture viewed, again with areas of rectangles proportional to frequencies of groups.

(continued)

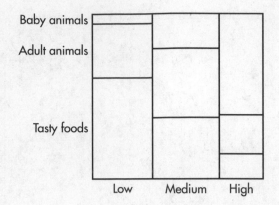

Finally, a little space can be added between the rectangles.

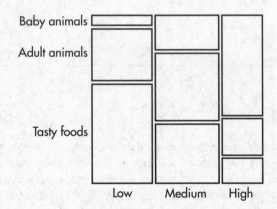

Alternatively, we could have first started with horizontal sections based on picture viewed.

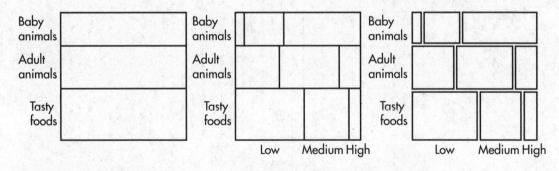

EXAMPLE 2.4

A study was made to compare year in high school with preference for vanilla or chocolate ice cream with the following results:

	Vanilla	Chocolate
Freshman	20	10
Sophomore	24	12
Junior	18	9
Senior	22	11

What are the conditional frequencies for each class?

Freshmen: $\frac{20}{30} = 0.667$ prefer vanilla and $\frac{10}{30} = 0.333$ prefer chocolate.

Sophomores: $\frac{24}{36} = 0.667$ prefer vanilla and $\frac{12}{36} = 0.333$ prefer chocolate.

Juniors: $\frac{18}{27} = 0.667$ prefer vanilla and $\frac{9}{27} = 0.333$ prefer chocolate.

Seniors: $\frac{22}{33} = 0.667$ prefer vanilla and $\frac{11}{33} = 0.333$ prefer chocolate.

In such a case, where all the conditional frequency distributions are identical, we say that the two variables show *perfect independence.* (However, it should be noted that even if the two variables are completely independent, the chance is very slim that a resulting two-way table will show perfect independence.)

EXAMPLE 2.5

Suppose you need heart surgery and are trying to decide between two surgeons, Dr. Fixit and Dr. Patch. You find out that each operated 250 times last year with the following results:

	Dr. F	Dr. P
Died	60	50
Survived	190	200

Which surgeon would you choose? Among Dr. Fixit's 250 patients, 190 survived, for a survival rate of $\frac{190}{250} = 0.76$ or 76%, while among Dr. Patch's 250 patients, 200 survived, for a survival rate of $\frac{200}{250} = 0.80$ or 80%. Your choice seems clear.

(continued)

However, everything may not be so clear-cut. Suppose that on further investigation you determine that the surgeons operated on patients who were in either good or poor condition with the following results:

	Good condition			Poor condition	
	Dr. F	Dr. P		Dr. F	Dr. P
Died	8	17	Died	52	33
Survived	60	120	Survived	130	80

Note that adding corresponding boxes from these two tables gives the original table above.

How do the surgeons compare when operating on patients in good health? In poor health?

Dr. Fixit's 68 patients in good condition have a survival rate of $\frac{60}{68} = 0.882$, or 88.2%, while Dr. Patch's 137 patients in good condition have a survival rate of $\frac{120}{137} = 0.876$, or 87.6%. Similarly, we note that Dr. Fixit's 182 patients in poor condition have a survival rate of $\frac{130}{182} = 0.714$ or 71.4%, while Dr. Patch's 113 patients in poor condition have a survival rate of $\frac{80}{113} = 0.708$ or 70.8%.

Thus, Dr. Fixit does better with patients in good condition (88.2% versus Dr. Patch's 87.6%) and also does better with patients in poor condition (71.4% versus Dr. Patch's 70.8%). However, Dr. Fixit has a lower overall patient survival rate (76% versus Dr. Patch's 80%)! How can this be?

This problem is an example *of Simpson's paradox*, where a conjecture can be reversed when more than one group is combined to form a single group. The effect of another variable is masked when the groups are combined. In this particular example, closer scrutiny reveals that Dr. Fixit operates on many more patients in poor condition than Dr. Patch, and these patients in poor condition are precisely the ones with lower survival rates. Thus, even though Dr. Fixit does better with all patients, his overall rating is lower. Our original table hid the effect of the variable related to the condition of the patients.

QUIZ 6

Multiple-Choice Questions

> **Directions:** The questions that follow are each followed by five suggested answers. Choose the response that best answers the question.

Questions 1–5 are based on the following: Students, teachers, and administrators were asked which of three teacher characteristics (challenging, enthusiastic, strict) they considered most important for a successful classroom experience. Five hundred people in a high school community were surveyed with the following results:

	Challenging	Enthusiastic	Strict
Student	50	150	50
Teacher	125	50	25
Administrator	15	10	25

1. What percentage of those surveyed were students?

   (A) 10%
   (B) 20%
   (C) 30%
   (D) 40%
   (E) 50%

2. What percentage of those surveyed picked challenging as most important and were teachers?

   (A) 25%
   (B) 38%
   (C) 40%
   (D) 62.5%
   (E) 65.8%

3. What percentage of administrators picked strict as most important?

   (A) 5%
   (B) 10%
   (C) 20%
   (D) 25%
   (E) 50%

4. What percentage of those picking enthusiastic as most important were students?

(A) 30%

(B) 42%

(C) 50%

(D) 60%

(E) 71.4%

5. Which group of people were most likely to pick strict as most important?

(A) Student

(B) Teacher

(C) Administrator

(D) Teacher and administrator, equally

(E) Student, teacher, and administrator, equally

6. The mosaic plot below displays the distribution of humanities and science students at a particular college and whether or not they live on or off campus.

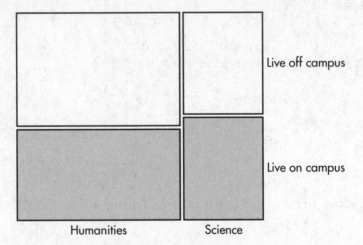

Based on this plot, which of the following statements is NOT true?

(A) The number of humanities students who live on campus is greater than the number of science students who live on campus.

(B) The proportion of humanities students who live on campus is less than the proportion of science students who live on campus.

(C) The number of humanities students at this college is greater than the number of science students at this college.

(D) The number of students who live on campus is less than the number living off campus.

(E) Among the students living on campus, the proportion who study humanities is less than the proportion who study science.

7. For what value of *n* does the following table show perfect independence?

20	50
30	*n*

(A) 10

(B) 40

(C) 60

(D) 75

(E) 100

Free-Response Questions

Directions: You must show all work and indicate the methods you use. You will be graded on the correctness of your methods and on the accuracy of your final answers.

1. The following table gives the numbers (in thousands) of officers and enlisted personnel by military branch in the U.S. armed forces.

	Army	Navy	Marine Corps	Air Force
Officers	88	52	20	65
Enlisted	452	276	178	258

(a) Calculate the percentage

   i. of military personnel who are enlisted.
   ii. of military personnel who are not Marine Corps officers.
   iii. of officers who are in the Navy.

(b) Construct a graphical display showing the association between career path (officer vs. enlisted) and military branch.

(c) Summarize what the graphical display illustrates about the association between career path (officer vs. enlisted) and military branch.

2. A guaranteed, or basic, income is a controversial idea. Some believe that a basic income would increase people's laziness and thus lower productivity, while others believe that guaranteeing a basic income would improve people's mental and physical health and thus increase productivity. In a random sample of 750 adults, people identified as being liberal or conservative and indicated their support for or opposition to a guaranteed income. The following two-way table summarizes the data.

	Liberal	Conservative	Total
Support	273	92	365
Opposition	122	193	315
No opinion	25	45	70
Total	420	330	750

(a) Of the people identifying as liberal, what proportion support a guaranteed income?

(b) Of the people who oppose a guaranteed income, what percentage identify as conservative?

(c) The mosaic plot displays the distribution of support for a guaranteed income by identification as liberal or conservative. Describe what this graph reveals about the association between these two variables for the 750 adults in this survey.

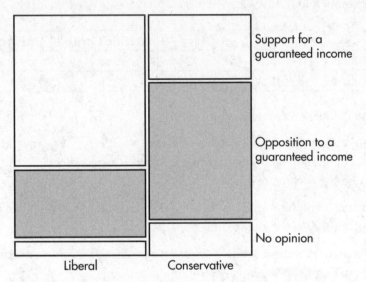

3. In 2005, the racial composition and mean incomes of a company's employees were 150 White, 50 Black, and 50 Hispanic (non-white), with mean incomes of $37,000, $22,000, and $23,000, respectively. Ten years later, in 2015, the data were 90 White, 60 Black, and 100 Hispanic, with mean incomes of $40,000, $23,000, and $25,000, respectively.

(a) What was the mean income trend for each racial group between 2005 and 2015?

(b) What was the mean income trend for the whole company between 2005 and 2015?

(c) If you worked in the company's HR office, how would you explain the company's income trend to a prospective employee?

The answers for this quiz can be found in the Appendix on page 583.

QUIZ 7

Multiple-Choice Questions

Directions: The questions that follow are each followed by five suggested answers. Choose the response that best answers the question.

Questions 1–5 are based on the following: A study of music preferences in three geographic locations resulted in the following segmented bar chart:

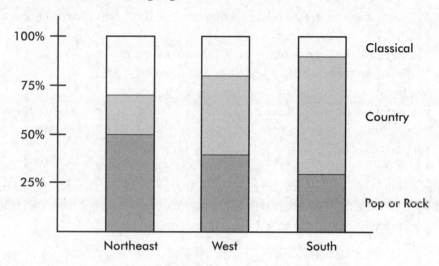

1. What percentage of those surveyed from the Northeast prefer country music?

   (A) 20%

   (B) 30%

   (C) 40%

   (D) 50%

   (E) 70%

2. Which of the following is greatest?

   (A) The percentage of those from the Northeast who prefer classical

   (B) The percentage of those from the West who prefer country

   (C) The percentage of those from the South who prefer pop or rock

   (D) The above are all equal.

   (E) It is impossible to determine the answer without knowing the actual numbers of people involved.

3. Which of the following is greatest?

   (A) The number of people in the Northeast who prefer pop or rock

   (B) The number of people in the West who prefer classical

   (C) The number of people in the South who prefer country

   (D) The above are all equal.

   (E) It is impossible to determine the answer without knowing the actual numbers of people involved.

4. All three bars have a height of 100%. Which of the following is true?

(A) This is a coincidence.
(B) This happened because each bar shows a complete distribution.
(C) This happened because there are three bars each divided into three segments.
(D) This happened because of the nature of musical patterns.
(E) None of the above is true.

5. Based on the given segmented bar chart, does there seem to be a relationship between geographic location and music preference?

(A) Yes, because the corresponding segments of the three bars have different lengths.
(B) Yes, because the heights of the three bars are identical.
(C) Yes, because there are three segments and three bars.
(D) No, because the heights of the three bars are identical.
(E) No, because summing the corresponding segments for classical, summing the corresponding segments for country, and summing the corresponding segments for pop or rock all give approximately the same total.

6. High school students were surveyed as to whether or not their GPA was 3.0 or higher and what was the most severe punishment they had ever received from their parents. The results are displayed in the following mosaic plot.

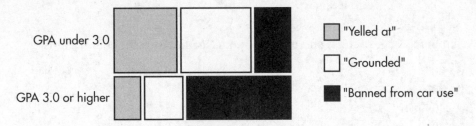

Based on this plot, which of the following is a true statement?

(A) The number of students with GPAs under 3.0 who were banned from car use is greater than the number of students with GPAs 3.0 or higher who were banned from car use.
(B) More students had a GPA 3.0 or higher and were either yelled at or grounded than had a GPA under 3.0 and were yelled at.
(C) Of the students who were grounded, a greater proportion had GPAs 3.0 or higher than GPAs under 3.0.
(D) More students had a GPA under 3.0 than a GPA 3.0 or higher.
(E) More students were yelled at than were grounded.

7. A company employs both men and women in its secretarial and executive positions. In reports filed with the government, the company shows that the percentage of female employees who receive raises is higher than the percentage of male employees who receive raises. A government investigator claims that the percentage of male secretaries who receive raises is higher than the percentage of female secretaries who receive raises and that the percentage of male executives who receive raises is higher than the percentage of female executives who receive raises. Is this possible?

(A) No, either the company report is wrong or the investigator's claim is wrong.
(B) No, if the company report is correct, either a greater percentage of female secretaries than of male secretaries receive raises or a greater percentage of female executives than of male executives receive raises.
(C) No, if the investigator is correct, by summation of the corresponding numbers, the total percentage of male employees who receive raises would have to be greater than the total percentage of female employees who receive raises.
(D) All of the above are true.
(E) It is possible for both the company report to be true and the investigator's claim to be correct.

Free-Response Questions

> **Directions:** You must show all work and indicate the methods you use. You will be graded on the correctness of your methods and on the accuracy of your final answers.

1. In 1973, a question arose concerning possible gender discrimination in admissions to the graduate school at the University of California at Berkeley. The graduate school analyzed admissions to the largest graduate programs, and below are partial results of that study:

Program	Men Accepted	Men Rejected	Women Accepted	Women Rejected
A	511	314	89	19
E	53	138	95	298

(a) Find the percentage of men and the percentage of women accepted by each of the two programs. Comment on any pattern or bias you see.

(b) Find the percentage of men and the percentage of women accepted overall by these two programs. Does this appear to contradict the results from part (a)?

(c) If you worked in the Graduate Admissions Office, what would you say to an inquiring reporter who is investigating gender bias in graduate admissions?

2. For one study of racial disparities in traffic stops, 225 random records for speeding pullovers at night (when it is difficult to distinguish race at a distance) are analyzed. The following two-way table summarizes one aspect of the data.

	White	Black	Latino	Total
Ticket	36	30	27	93
Warning	84	30	18	132
Total	120	60	45	225

(a) Of the speeding pullovers who were given a ticket, what proportion were Latino?

(b) Of the speeding pullovers who were White, what percent were given a warning?

(c) The mosaic plot displays the distribution of tickets versus warnings given by race. Describe what this graph reveals about the association between these two variables for the 225 drivers pulled over for speeding.

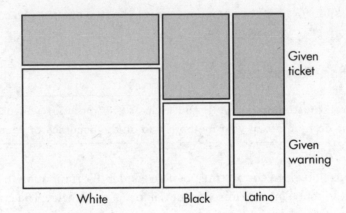

3. Three tennis players were comparatively ranked by tournament directors voting on their play on hardcourt, clay, and grass:

	Hardcourt	Clay	Grass
Player A	3	2	1
Player B	2	1	2
Player C	1	3	3

Are these differences significant, or could the players have equal ability on all surfaces and these particular rankings be due to chance? If the players have equal ability on all surfaces, the expected average ranking for each would be 2.

Consider the following test statistic: $Q = 3\left[\left(\bar{R}_A - 2\right)^2 + \left(\bar{R}_B - 2\right)^2 + \left(\bar{R}_C - 2\right)^2\right]$

where $\bar{R}_A$, $\bar{R}_B$, and $\bar{R}_C$ are the average rankings for the three players.

(a) Calculate the test statistic Q for the above data.

One hundred simulated values of the test statistic were calculated assuming the players have equal ability on all surfaces. The results are shown in the dotplot below.

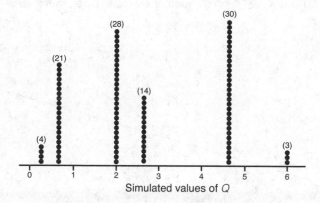

Simulated values of Q

(b) Use these simulated values and the test statistic to determine if directors' rankings provide evidence of a significant difference in the players' abilities.

The answers for this quiz can be found in the Appendix on page 585.

TWO QUANTITATIVE VARIABLES

Many important applications of statistics involve examining whether two or more quantitative (numerical) variables are related to one another. For example, is there a relationship between the smoking histories of pregnant women and the birth weights of their children? Between SAT scores and success in college? Between amount of fertilizer used and amount of crop harvested?

Two questions immediately arise. First, how can the strength of an apparent relationship be measured? Second, how can an observed relationship be put into functional terms? For example, a real estate broker might not only wish to determine whether a relationship exists between the prime rate and the number of new homes sold in a month, but might also find useful an expression with which to predict the number of home sales given a particular value of the prime rate.

A graphical display, called a scatterplot, gives an immediate visual impression of a possible relationship between two variables, while a numerical measurement, called the *correlation coefficient* (or simply the *correlation*), is often used as a quantitative value of the strength of a linear relationship. In either case, evidence of a relationship is not evidence of causation.

Suppose a relationship is perceived between two quantitative variables, X and Y. We are interested in the strength and direction of this relationship and in any deviation from the basic pattern of this relationship. We graph the pairs (x, y). In this topic, we examine whether the relationship, as illustrated by the scatterplot, can be reasonably explained in terms of a linear function, that is, one whose graph is a straight line.

 TIP

Recognize explanatory (*x*) and response (*y*) variables in context.

➡ EXAMPLE 2.6_____

Comic books heroes and villains can be compared with regard to various attributes.
The scatterplot below looks at speed (measured on a 20-point scale) versus strength
(measured on a 50-point scale) for 17 such characters. Does there appear to be a linear
association?

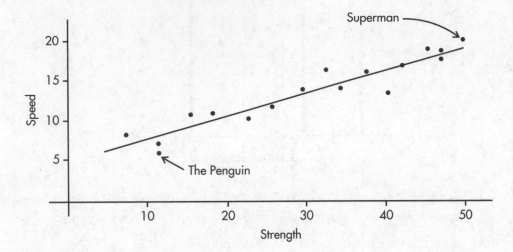

NOTE

There are various
so-called *lines
of best fit*. We
will be working
with one called
the *least squares
regression line*.

We need to know what the term *best-fitting straight line* means and how we can find
this line. Furthermore, we want to be able to gauge whether the relationship between
the variables is strong enough so that finding and making use of this straight line is
meaningful.

_____

When larger values of one variable are associated with larger values of a second variable,
the variables are called *positively associated*. When larger values of one are associated with
smaller values of the other, the variables are called *negatively associated*. The strength of the
association is gauged by how close the plotted points are to a straight line.

NOTE

No matter
whether the
variables are
positively or
negatively
associated, for
any two points,
anything is
possible, and
either point
could be higher
or lower than the
other.

➡ EXAMPLE 2.7_____

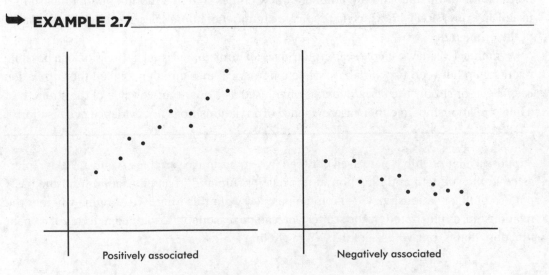

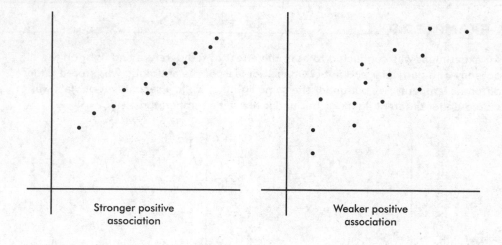

Stronger positive
association

Weaker positive
association

Sometimes different dots in a scatterplot are labeled with different symbols or different colors to show a categorical variable. The resulting labeled scatterplot might distinguish, for example, between men and women, or between stocks and bonds, and so on.

➡ **EXAMPLE 2.8**

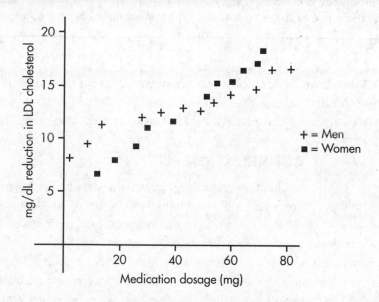

The above diagram is a labeled scatterplot distinguishing men with plus signs and women with square dots to show a categorical variable, gender. The plot indicates that higher doses of the medication lead to greater reductions in LDL (bad cholesterol) for both men and women. However, lower doses of the medication help men more than women in lowering LDL, while higher doses help women more than men.

When analyzing the overall pattern in a scatterplot, it is also important to note *clusters* and *outliers*.

➥ **EXAMPLE 2.9**

An experiment was conducted to note the effect of temperature and light on the potency of a particular antibiotic. One set of vials of the antibiotic was stored under different temperatures but under the same lighting, while a second set of vials was stored under different lightings but under the same temperature.

TIP

Note when the data fall in distinct groups.

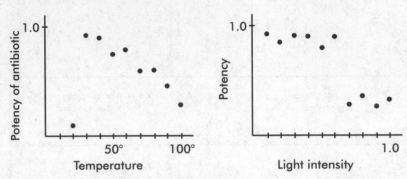

In the first scatterplot, note the roughly linear pattern with one outlier far outside this pattern. A possible explanation is that the antibiotic is more potent at lower temperatures but only down to a certain temperature, at which it drastically loses potency.

In the second scatterplot, note the two clusters. It appears that below a certain light intensity the potency is one value, while above that intensity it is another value. In each cluster, there seems to be no association between intensity and potency.

As seen from all the above, to describe a scatterplot you must consider *form* (linear or nonlinear), *direction* (positive or negative), *strength* (weak, moderate, or strong), and *unusual features* (such as outliers and clusters). As usual, all answers must also mention *context*.

IMPORTANT

Correlation does not imply causation!

CORRELATION

Although a scatter diagram usually gives an intuitive visual indication when a linear relationship is strong, in most cases it is quite difficult to visually judge the specific strength of a relationship. For this reason there is a mathematical measure called *correlation* (or the *correlation coefficient*). Important as correlation is, we always need to keep in mind that significant correlation does not necessarily indicate *causation*. For example, there is a very strong correlation between ice cream sales and frequency of forest fires! (As ice cream sales rise, in general so does temperature, and with higher temperatures, there are more forest fires.) Furthermore, correlation measures the strength of only a *linear* relationship.

Correlation, designated by *r*, has the formula

NOTE

The division in the formula also gives that the correlation is unit-free.

$$r = \frac{1}{n-1} \sum \left(\frac{x_i - \bar{x}}{s_x} \right)\left(\frac{y_i - \bar{y}}{s_y} \right)$$

TIP

With standardized data (*z*-scores), *r* is the slope of the regression line.

in terms of the means and standard deviations of the two sets. We note that the formula is actually the sum of the products of the corresponding *z*-scores divided by 1 less than the sample size. However, you should be able to calculate correlation quickly using the statistical package on your calculator. (Examining the formula helps you understand where correlation is coming from, but you will NOT have to use the formula to calculate *r*.)

Note from the formula that correlation does not distinguish between which variable is called *x* and which is called *y*. The formula is also based on standardized scores (*z*-scores), and so changing units does not change the correlation. Finally, since means and standard deviations can be strongly influenced by outliers, correlation is also strongly affected by extreme values.

The value of *r* always falls between -1 and $+1$, with -1 indicating perfect negative correlation and $+1$ indicating perfect positive correlation. It should be stressed that a correlation at or near zero doesn't mean there isn't a relationship between the variables; there may still be a strong *nonlinear* relationship. Additionally, a correlation close to -1 or $+1$ does not necessarily mean that a linear model is the *most* appropriate model.

➡ **EXAMPLE 2.10**

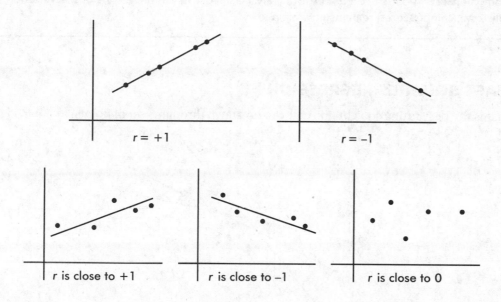

Very roughly, for the purpose of the AP exam, one might say that a correlation with absolute value above 0.8 is "high," from 0.5 to 0.8 is "moderate," and below 0.5 is "low." However, in the real world, everything depends on context. For example, a doctor might feel that an indicator for the severity of cancer spread that has a correlation of 0.9 is not good enough if the result determines major differences in treatment. On the other hand, a financial advisor who discovers a variable with a correlation of 0.1 with stock performance might feel that this is enough to recommend major investments.

It can be shown that r^2, called the *coefficient of determination*, is the ratio of the variance of the predicted values $\hat{y}$ to the variance of the observed values y. That is, there is a partition of the *y*-variance, and r^2 is the proportion of this variance that is predictable from a knowledge of *x*. We can say that r^2 gives the percentage of variation in the response variable, *y*, that is explained by the *variation* in the explanatory variable, *x*. Or we can say that r^2 gives the percentage of variation in *y* that is explained by the linear regression model between *x* and *y*. In any case, always interpret r^2 in context of the problem. Remember when calculating *r* from r^2 that *r* may be positive or negative.

While the correlation *r* is given as a decimal between -1.0 and 1.0, the coefficient of determination r^2 is usually given as a percentage. An r^2 of 100% is a perfect fit, with all the variation in *y* explained by variation in *x*. How large a value of r^2 is desirable depends on the application under consideration. While scientific experiments often aim for an r^2 in the 90%

NOTE

Alternatively, r^2 is 1 minus the proportion of unexplained variance:
$$r^2 = 1 - \frac{\Sigma(y_i - \hat{y}_i)^2}{\Sigma(y_i - \bar{y})^2}.$$

or above range, observational studies with r^2 of 10% to 20% might be considered informative. Note that while a correlation, r, of 0.6 is twice a correlation, r, of 0.3, the corresponding r^2 of 36% is *four* times the corresponding r^2 of 9%.

➡ EXAMPLE 2.11

The correlation between Total Points Scored and Total Yards Gained for the 2019 season among a set of college football teams is $r = 0.84$. What information is given by the coefficient of determination?

Answer: $r^2 = (0.84)^2 = 0.7056$. Thus, 70.56% of the variation in Total Points Scored can be accounted for by (or predicted by or explained by) the linear relationship between Total Points Scored and Total Yards Gained. The other 29.44% of the variation in Total Points Scored remains unexplained.

LEAST SQUARES REGRESSION

What is the best-fitting straight line that can be drawn through a set of points?

TIP

If a scatterplot indicates a nonlinear relationship, don't try to force a straight-line fit.

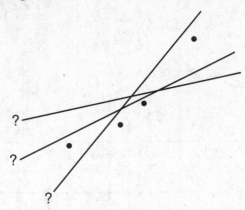

On the basis of our experience with measuring variances, by *best-fitting straight line* we mean the straight line that minimizes the sum of the squares of the vertical differences between the observed values and the values predicted by the line.

TIP

A variable with a hat $(\hat{y})$ is a predicted version of the variable.

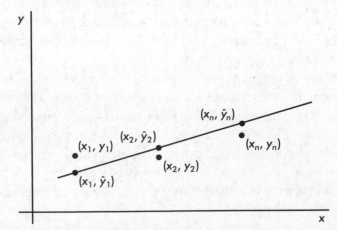

That is, in the above figure, we wish to minimize

$$\left(y_1 - \hat{y}_1\right)^2 + \left(y_2 - \hat{y}_2\right)^2 + \ldots + \left(y_n - \hat{y}_n\right)^2$$

It is reasonable, intuitive, and correct that the best-fitting line will pass through $(\bar{x}, \bar{y})$, where $\bar{x}$ and $\bar{y}$ are the means of the variables X and Y. Then, from the basic expression for a line with a given slope through a given point, we have:

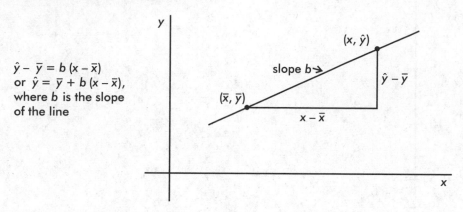

$\hat{y} - \bar{y} = b(x - \bar{x})$
or $\hat{y} = \bar{y} + b(x - \bar{x})$,
where b is the slope
of the line

NOTE

This equation is typically expressed as $\hat{y} = a + bx$.

The slope b can be determined from the formula

$$b = r\frac{s_y}{s_x}$$

where r is the correlation and s_x and s_y are the standard deviations of the two sets. That is, each standard deviation change in x results in a change of r standard deviations in $\hat{y}$. If you graph z-scores for the y-variable against z-scores for the x-variable, the slope of the regression line is precisely r, and in fact, the linear equation becomes $z_y = rz_x$.

This best-fitting straight line, that is, the line that minimizes the sum of the squares of the differences between the observed values and the values predicted by the line, is called the *least squares regression line*, or simply the *regression line*. It can be calculated directly by entering the two data sets and using the statistics package on your calculator.

NOTE

The slope is the amount that the *predicted y*-value changes for every unit increase in *x*. Similarly, the *y*-intercept is the *predicted y*-value for *x* = 0.

TIP

Just because we can calculate a regression line doesn't mean it is useful.

➡ **EXAMPLE 2.12**_____

A sociologist conducts a survey of 15 teens. The number of "close friends" and the number of times Facebook is checked every evening are noted for each student. Letting X and Y represent the number of close friends and the number of Facebook checks, respectively, we have:

X: 25 23 30 25 20 33 18 21 22 30 26 26 27 29 20
Y: 10 11 14 12   8 18   9 10 10 15 11 15 12 14 11

Find the equation of the best-fitting straight line for the data.

(continued)

Answer: Plotting the 15 points (25, 10), (23, 11), . . . , (20, 11) gives an intuitive visual impression of the relationship:

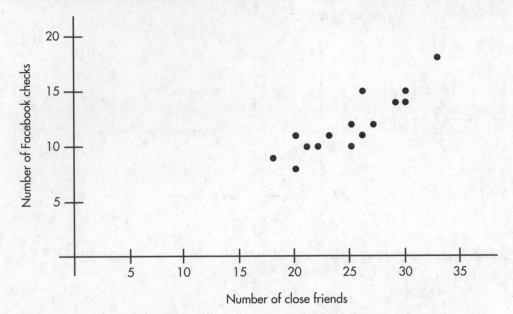

This scatterplot indicates the existence of a relationship that appears to be *linear*; that is, the points lie roughly on a straight line. Furthermore, the linear relationship is *positive*; that is, as one variable increases, generally so does the other (the straight line slopes upward).

By using a calculator, we find the correlation to be $r = 0.8836$, the coefficient of determination to be $r^2 = 0.78$ (indicating that 78% of the variation in the number of Facebook checks is accounted for by the variation in the number of close friends), and the regression line to be

$$\hat{y} = a + bx = -1.73 + 0.5492x$$

We also write: Predicted Facebook checks = $-1.73 + 0.5492$(Close friends)

$$\overset{\frown}{\text{Facebook checks}} = -1.73 + 0.5492(\text{Close friends})$$

Adding this to our scatterplot yields:

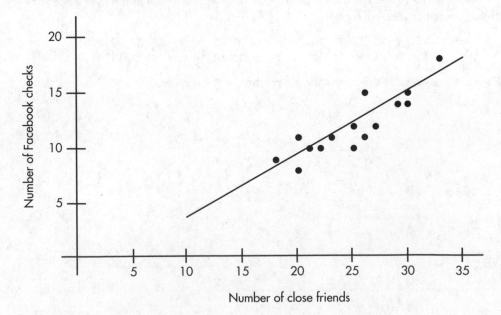

Thus, for example, we might predict that students with 24 best friends will average 0.5492(24) 1.73 = 11.45 Facebook checks per evening. We also note that each additional best friend seems to lead to an average 0.5492 more Facebook checks.

⇨ EXAMPLE 2.13

The following are advertising expenditures and total sales for six detergent products:

Advertising ($1000) ($x$): 2.3 5.7 4.8 7.3 5.9 6.2

Total sales ($1000) ($y$): 77 105 96 118 102 95

Predict the total sales if $5000 is spent on advertising, and interpret the slope of the regression line. What if $100,000 is spent on advertising?

Answer: With your calculator, the equation of the regression line is found to be:

$$\hat{y} = 59.683 + 7.295x$$

It is also worthwhile to replace the x and y with more appropriately named variables, resulting, for example, in:

$$\widehat{\text{Sales}} = 59.683 + 7.295 \, (\text{Advcost})$$

TIP

Use meaningful variable names in your regression equations.

The regression line predicts that if $5000 is spent on advertising, the resulting total sales will be 59.683 + 7.295(5) = 96.158 thousands of dollars ($96,158).

The slope of the regression line indicates that every extra $1000 spent on advertising will result in an average of $7295 in added sales.

If $100,000 is spent on advertising, we calculate 59.683 + 7.295(100) = 789.183 thousands of dollars ($789,183). How much confidence should we have in this answer? Not much! We are trying to use the regression line to predict a value far outside the other data values. This procedure is called *extrapolation* and must be used with great care.

TIP

Be careful about extrapolation beyond the observed *x*-values.

It should be noted that when we use the regression line to predict a y-value for a given x-value, we are actually predicting the *mean y-value* for that given x-value. For any given x-value, there are many possible y-values, and we are predicting their mean. So, if $5000 is spent many times on advertising, various resulting total sales figures may result, but their predicted average is $96,158.

The TI-Nspire gives:

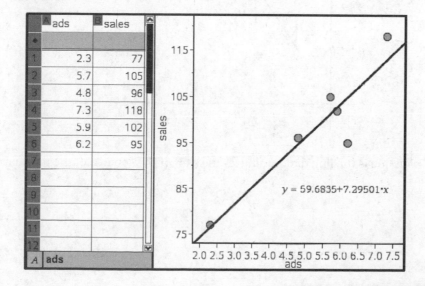

➡ EXAMPLE 2.14

A random sample of 30 U.S. farm regions surveyed during the summer of 2020 produced the following statistics:

Average temperature (°F) during growing season: $\bar{x} = 81$, $s_x = 3$
Average corn yield per acre (bushels): $\bar{y} = 131$, $s_y = 5$
Correlation $r = 0.32$

Based on this study, what is the mean predicted corn yield for a region where the average growing season temperature is 76.5°F?

Answer: $\frac{76.5 - 81}{3} = -1.5$, so 76.5 is −1.5 standard deviations away from (1.5 standard deviations *below*) the average temperature reported in the study. With a correlation of $r = 0.32$, the predicted corn yield is 0.32(−1.5) = −0.48 standard deviations from (0.48 standard deviations *below*) the average corn yield, or 131 − 0.48(5) = 128.6 bushels per acre.

Alternatively, we could have found the linear regression equation relating these variables: slope $= r\frac{s_y}{s_x} = 0.32\left(\frac{5}{3}\right) \approx 0.533$, intercept $\approx 131 - 81(0.533) \approx 87.8$, and thus $\widehat{\text{Yield}} = 0.533(\text{Temp}) + 87.8$. Then 0.533(76.5) + 87.8 ≈ 128.6.

RESIDUALS

CAREFUL!

Order is important: *residual* equals *observed* minus *predicted*.

The difference between an observed and a predicted value is called the *residual*. When the regression line is graphed on the scatterplot, the residual of a point is the vertical distance the point is from the regression line.

NOTE

When the data point is above the regression line, the residual is positive; a data point below the line gives a negative residual.

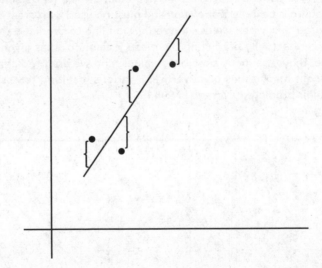

The regression line is the line that minimizes the sum of the squares of the residuals.

→ EXAMPLE 2.15

We calculate the predicted values from the regression line in Example 2.13 and subtract from the observed values to obtain the residuals:

x	2.3	5.7	4.8	7.3	5.9	6.2
y	77	105	96	118	102	95
$\hat{y}$	76.5	101.3	94.7	112.9	102.7	104.9
$y - \hat{y}$	0.5	3.7	1.3	5.1	−0.7	−9.9

Note that the sum of the residuals is

$$0.5 + 3.7 + 1.3 + 5.1 - 0.7 - 9.9 = 0.0$$

The above equation is true in general; that is, *the sum and thus the mean of the residuals is always zero.*

The notation for residuals is $\hat{e}_i = y_i - \hat{y}_i$ and so $\sum_{i=1}^{n} \hat{e}_i = 0$. The standard deviation of the residuals, s_e, gives a measure of how the points are spread around the regression line.

Plotting the residuals gives further information. In particular, a residual plot with a definite pattern is an indication that a nonlinear model will show a better fit to the data than the straight regression line. In addition to whether or not the residuals are randomly distributed, one should look at the balance between positive and negative residuals and also at the size of the residuals in comparison to the associated x-values.

The residuals can be plotted against either the x-values or the $\hat{y}$-values. (Because $\hat{y}$ is a linear transformation of x, the plots are identical except for scale and a left-right reversal when the slope is negative.)

It is also important to understand that a linear model may be appropriate, but weak, with a low correlation. Alternatively, a linear model may not be the best model (as evidenced by the residual plot), but it still might be a very good model with high r^2.

> **NOTE**
>
> You will not have to do so on the exam, but the standard deviation of the residuals can be calculated by
>
> $$s_e = \sqrt{\frac{\sum e_i^2}{n-2}}$$
> $$= \sqrt{\frac{\sum (y_i - \hat{y}_i)^2}{n-2}}.$$

→ EXAMPLE 2.16

What do each of the following residual plots indicate about the linear model?

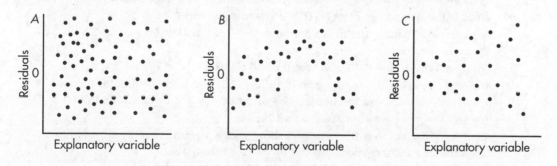

Answer: In **A**, there is no apparent pattern in the residual plot, indicating that a linear model is appropriate.

In **B**, the residual plot shows a strong pattern, indicating that a nonlinear model will be a better fit than a straight-line model.

In **C**, the residual plot shows "fanning," indicating that the linear model gives a stronger fit for smaller x-values than for larger x-values.

The ability to interpret computer output is important not only to do well on the AP Statistics exam, but also to understand statistical reports in the business and scientific world.

➡ EXAMPLE 2.17

Miles per gallon versus speed for a new model automobile is fitted with a least squares regression line. The graph of the residuals and some computer output for the regression are as follows:

Regression Analysis: MPG Versus Speed

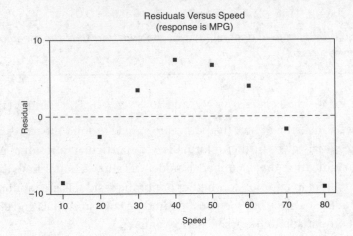

```
The regression equation is MPG = 38.9 − 0.218 (Speed)

Predictor        Coef      SE Coef           T          P
Constant       38.929        5.651        6.89      0.000
Speed         −0.2179        0.1119       −1.95      0.099

S = 7.252   R−Sq = 38.7%  R−Sq(adj) = 28.5%
```

a. Interpret the slope of the regression line in context.
    Answer: The slope of the regression line is −0.2179, indicating that, on average, the MPG drops by 0.2179 for every increase of one mile per hour in speed.

b. What is the mean predicted MPG at a speed of 30 mph?
    Answer: At 30 mph, the mean predicted MPG is −0.2179(30) + 38.929, or about 32.4 MPG.

c. What was the actual MPG at a speed of 30 mph?
    Answer: The residual for 30 mph is about +3.5, and since residual = actual − predicted, we estimate the actual MPG to be 32.4 + 3.5, or about 36 MPG.

d. Is a line the most appropriate model? Explain.
    Answer: The fact that the residuals show such a strong curved pattern indicates that a nonlinear model would be more appropriate.

e. What does "S = 7.252" refer to?
    Answer: The standard deviation of the residuals, $s_e = 7.252$, is a "typical value" of the residuals and gives a measure of how the points are spread around the regression line.

➡ EXAMPLE 2.18 _____

The number of children playing Little League baseball in Ithaca, New York, during the years 2010–2018 is fitted with a least squares regression line. The graph of the residuals and some computer output for their regression are as follows:

Regression Analysis: Number of Players Versus Years Since 2010

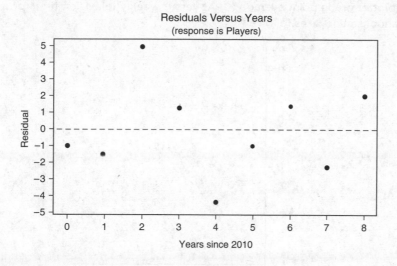

Residuals Versus Years
(response is Players)

Predictor	Coef	SE Coef	T	P
Constant	123.800	1.798	68.84	0.000
Years	12.6333	0.3778	33.44	0.000

S = 2.926   R–Sq = 99.4%   R–Sq(adj) = 99.3%

a. Does it appear that a line is an appropriate model for the data? Explain.

Answer: Yes. R–Sq = 99.4% is large, and the residual plot shows no pattern. Thus, a linear model is appropriate.

b. What is the equation of the regression line (in context)?

Answer: Predicted # of players = 123.8 + 12.6(years since 2010)

c. Interpret the slope of the regression line in the context of the problem.

Answer: The slope of the regression line is 12.6, indicating that the predicted number of children playing Little League baseball in Ithaca increased by an average of 12.6 players per year during the 2010–2018 time period.

d. Interpret the y-intercept of the regression line in the context of the problem.

Answer: The y-intercept, 123.8, refers to the year 2010. Thus, the number of Little League players in Ithaca in 2010 was predicted to be around 124.

e. What is the predicted number of players in 2012?

Answer: For 2012, $x = 2$, so the predicted number of players is 12.6(2) + 123.8 = 149.

f. What was the actual number of players in 2012?

Answer: The residual for 2012 ($x = 2$) from the residual plot is +5, so actual minus predicted equals 5, and thus the actual number of players in 2012 must have been 5 + 149 = 154.

g. What years, if any, did the number of players decrease from the previous year? Explain.

Answer: The number would decrease if one residual were more than 12.6 greater than the next residual. This never happens, so the number of players never decreased.

TIP

Simply using a calculator to find a regression line is not enough; you must understand it (for example, be able to interpret the slope and intercepts in context).

OUTLIERS, INFLUENTIAL POINTS, AND LEVERAGE

In a scatterplot, *regression outliers* are indicated by points falling far away from the overall pattern. That is, a point is an outlier if its residual is an outlier in the set of residuals.

➡ EXAMPLE 2.19

A scatterplot of grade point average (GPA) versus weekly television time for a group of high school seniors is as follows:

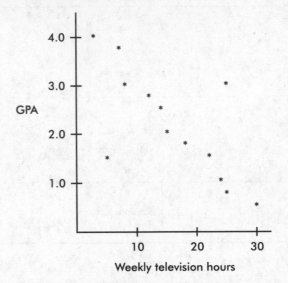

By direct observation of the scatterplot, we note that there are two outliers: one person who watches 5 hours of television weekly yet has only a 1.5 GPA, and another person who watches 25 hours weekly yet has a 3.0 GPA. Note also that while the value of 30 weekly hours of television may be considered an outlier for the television hours variable and the 0.5 GPA may be considered an outlier for the GPA variable, the point (30, 0.5) is *not* an outlier in the regression context because it does not fall off the straight-line pattern.

Scores whose removal would sharply change the regression line are called *influential scores*. Sometimes this description is restricted to points with extreme x-values. An influential score may have a small residual but still have a greater effect on the regression line than scores with possibly larger residuals but average x-values.

Consider the following scatterplot of six points and the regression line:

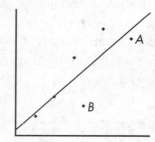

The heavy line in the scatterplot on the left below shows what happens when point *A* is removed, and the heavy line in the scatterplot on the right below shows what happens when point *B* is removed.

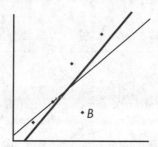 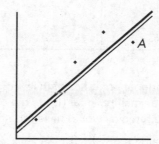

Note that the regression line is greatly affected by the removal of point *A* but not by the removal of point *B*. Thus, point *A* is an *influential score*, while point *B* is not. This is true in spite of the fact that point *A* is closer to the original regression line than point *B*.

There are many ways in which a point can be influential. Removal of the point may markedly change the slope, the *y*-intercept, or the correlation. In the AP class, we are primarily interested in changes to the slope, but this often shows up in the correlation too.

➥ **EXAMPLE 2.21**

Consider the effect of outliers in the *x*, the *y*, and both *x* and *y* directions on *b*, the slope of the regression line, and on *r*, the correlation:

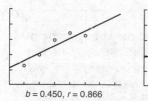

$b = 0.450, r = 0.866$

outlier in *x* direction

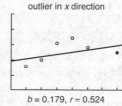

$b = 0.179, r = 0.524$

outlier in *y* direction

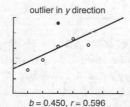

$b = 0.450, r = 0.596$

outlier in both *x* and *y* directions

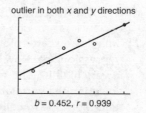

$b = 0.452, r = 0.939$

NOTE

The only "regression outlier" is found in the third graph where the highlighted point has a residual far greater than any other residual.

Note how an outlier in the *x* direction can dramatically affect both the slope and the correlation. Also note how an outlier in both the *x* and *y* directions, if maintaining the same pattern, will result in roughly the same slope but a much increased correlation (in absolute value).

A point is said to have *high leverage* if its *x*-value is far from the mean of the *x*-values. Such a point has the strong potential to change the regression line. If it happens to line up with the pattern of the other points, its inclusion might not influence the equation of the regression line, but it could well strengthen the correlation and r^2, the coefficient of determination. However, if the point does not line up, its inclusion can dramatically change the regression line; in this case, it is also an influential point. The only clear way of telling whether or not a point with high leverage is an influential point or not is to calculate the linear model twice, both with and without the point in question.

➡ **EXAMPLE 2.22**

Consider the four scatterplots below, each with a cluster of points and one additional point separated from the cluster.

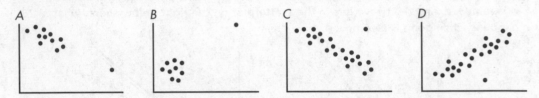

In *A*, the additional point has high leverage (its *x*-value is much greater than the mean *x*-value), has a small residual (it fits the overall pattern), and does not appear to be influential (its removal would have little effect on the regression equation).

In *B*, the additional point has high leverage (its *x*-value is much greater than the mean *x*-value), probably has a small residual (the regression line would pass close to it), and is very influential (removing it would dramatically change the slope of the regression line to close to 0).

In *C*, the additional point has some leverage (its *x*-value is greater than the mean *x*-value but not very much greater), has a large residual compared to other residuals (so it's a regression outlier), and is somewhat influential (its removal would change the slope to more negative).

In *D*, the additional point has no leverage (its *x*-value appears to be close to the mean *x*-value), has a large residual compared to other residuals (so it's a regression outlier), and is not influential (its removal would increase the *y*-intercept very slightly and would have very little if any effect on the slope).

➡ **EXAMPLE 2.23**

Imagine a scatterplot of (GPA, IQ) for 150 high school students and the corresponding regression line. Now add one student from the chess club who has an average GPA but a very high IQ. She may raise the average IQ a little, shifting the line up marginally and lowering the correlation slightly. But because of her average GPA, she has no leverage, and the slope remains the same. She's an outlier in the (GPA, IQ) relationship but not influential.

MORE ON REGRESSION

The regression equation $\hat{y} = a + bx = a + \left(r\dfrac{s_y}{s_x}\right)x$ has important implications.

1. If the correlation $r = +1$, then $\hat{y} = a + \left(\dfrac{s_y}{s_x}\right)x$, and for each standard deviation s_x increase in x, the *predicted* y-value increases by s_y.

2. If, for example, $r = +0.4$, then $\hat{y} = a + 0.4\left(\dfrac{s_y}{s_x}\right)x = a + \left(\dfrac{0.4s_y}{s_x}\right)x$, and for each standard deviation s_x increase in x, the *predicted* y-value increases by $0.4\,s_y$.

➡ **EXAMPLE 2.24** _____

Suppose x = attendance at a movie theater, y = number of boxes of popcorn sold, and we are given that there is a roughly linear relationship between x and y. Suppose further we are given the summary statistics: $\bar{x} = 250$, $s_x = 30$, $\bar{y} = 160$, $s_y = 20$, and $r = 0.8$.

  a. When attendance is 250, what is the predicted number of boxes of popcorn sold?
      Answer: The least squares regression line passes through $(\bar{x}, \bar{y})$, so the predicted number of boxes of popcorn sold is 160.

  b. When attendance is 295, what is the predicted number of boxes of popcorn sold?
      Answer: 295 is $\dfrac{295 - 250}{30} = 1.5$ standard deviations above $\bar{x} = 250$, so the predicted number of boxes of popcorn sold is $(0.8)1.5 = 1.2$ standard deviations above $\bar{y}$, or $160 + (1.2)20 = 184$ boxes.

3. The regression equation for predicting x from y has the slope $r\dfrac{s_x}{s_y}$.

➡ **EXAMPLE 2.25** _____

Use the same attendance and popcorn summary statistics from Example 2.24 above.
  a. When 160 boxes of popcorn are sold, what is the predicted attendance?
      Answer: The least squares regression line passes through $(\bar{y}, \bar{x})$, so the predicted attendance is 250.

  b. When 184 boxes of popcorn are sold, what is the predicted attendance?
      Answer: 184 is $\dfrac{184 - 160}{20} = 1.2$ standard deviations above $\bar{y} = 160$, so the predicted attendance is $(0.8)1.2 = 0.96$ standard deviations above $\bar{x}$, or $250 + (0.96)30 = 278.8$ people.

4. It is important to understand that while y versus x and x versus y have the same correlation r, the line we get from predicting y from x is *not* the equivalent line we get from predicting x from y, just solved for the other variable. For example, the slopes are not reciprocals as they were in algebra when we "solved for the other variable." We note that $r\dfrac{s_y}{s_x}$ is not the reciprocal of $r\dfrac{s_x}{s_y}$. For example, we see above that while an attendance of 295 gave a prediction of 184 boxes of popcorn, starting with 184 boxes of popcorn gave a predicted attendance of 278.8, not 295.

TRANSFORMATIONS TO ACHIEVE LINEARITY

Often a straight-line pattern is not the best model for depicting a relationship between two variables. A clear indication of this problem is when the scatterplot shows a distinctive curved pattern. Another indication is when the residuals show a distinctive pattern rather than a random scattering. In such a case, the nonlinear model can sometimes be revealed by transforming one or both of the variables and then noting a linear relationship. Useful transformations often result from using the *log* or *ln* buttons on your calculator to create new variables.

➡ **EXAMPLE 2.26**

Consider the following years and corresponding populations:

Year, x:	1980	1990	2000	2010	2020
Population (1000s), y:	44	65	101	150	230

The linear model is $\hat{y} = -9022 + 4.57x$ with $r^2 = 94.3\%$.

So, 94.3% of the variability in population is accounted for by the linear model. However, the scatterplot and residual plot indicate that a nonlinear relationship would be an even stronger model.

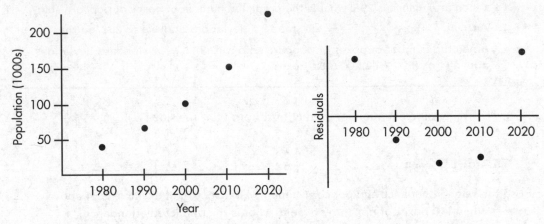

Linear fits to (x, log y) and (log x, log y) result in the following two residual plots:

TIP

After transformation of data, either increased randomness in the residual plot or an increase in r^2, both offer evidence that the least squares regression line for the transformed data is a more appropriate model than the regression line for the untransformed data.

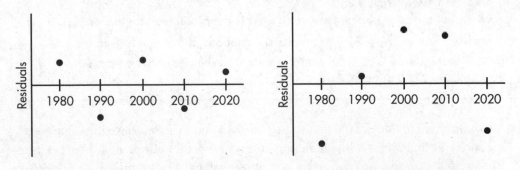

$\widehat{\log y} = -34.0 + 0.018x$ with $r^2 = 99.9\%$ $\widehat{\log y} = 1.56 + 0.0037(\log x)$ with $r^2 = 94.5\%$

Two of the three residual plots have distinct curved patterns. The residual plot coming from log y versus x illustrates a more random pattern in the residual plot as well as the greatest r^2. The best model is thus $\widehat{\log y} = -34.0 + 0.018x$. In context we have: $\widehat{\log(\text{Pop})} = -34.0 + 0.018(\text{Year})$. So, for example, the population predicted for the year 2025 would be calculated $\widehat{\log(\text{Pop})} = -34.0 + 0.018(2025) = 2.45$, and so $\widehat{\text{Pop}} = 10^{2.45} = 282$ thousand or 282,000.

There are many useful transformations. For example:

Log y as a linear function of x, $\log y = a + bx$, re-expresses as an *exponential*:

$$y = 10^{a+bx}, \text{ or } y = c10^{bx}, \text{ where } c = 10^a$$

Log y as a linear function of log x, $\log y = a + b \log x$, re-expresses as a *power*:

$$y = 10^{a+b\log x}, \text{ or } y = cx^b, \text{where } c = 10^a$$

$\sqrt{y}$ as a linear function of x, $\sqrt{y} = a + bx$, re-expresses as a *quadratic*:

$$y = (a + bx)^2$$

$\frac{1}{y}$ as a linear function of x, $\frac{1}{y} = a + bx$, re-expresses as a *reciprocal*:

$$y = \frac{1}{a + bx}$$

y as a linear function of log x, $y = a + b \log x$, is a *logarithmic* function.

> **NOTE**
>
> These very brief reexpressions are shown here for completeness and background, but you will not be responsible for performing these transformations. You do need to be able to recognize the need for a transformation, justify its appropriateness (residuals plot), and use the model to make predictions.

QUIZ 8

Multiple-Choice Questions

Directions: The questions that follow are each followed by five suggested answers. Choose the response that best answers the question.

1. A rural college is considering constructing a windmill to generate electricity but is concerned over noise levels. A study is performed measuring noise levels (in decibels) at various distances (in feet) from the campus library, and a least squares regression line is calculated with a correlation of 0.74. Which of the following is a proper and most informative conclusion for an observation with a negative residual?

   (A) The measured noise level is 0.74 times the predicted noise level.
   (B) The predicted noise level is 0.74 times the measured noise level.
   (C) The measured noise level is greater than the predicted noise level.
   (D) The predicted noise level is greater than the measured noise level.
   (E) The slope of the regression line at that point must also be negative.

2. A study of department chairperson ratings and student ratings of the performance of high school statistics teachers reports a correlation of $r = 1.15$ between the two ratings. From this information we can conclude that

   (A) chairpersons and students tend to agree on who is a good teacher.
   (B) chairpersons and students tend to disagree on who is a good teacher.
   (C) there is little relationship between chairperson and student ratings of teachers.
   (D) there is strong association between chairperson and student ratings of teachers, but it would be incorrect to infer causation.
   (E) a mistake in arithmetic has been made.

Questions 3–5 refer to the following:

The relationship between winning game proportions when facing the sun and when the sun is on one's back is analyzed for a random sample of 10 professional tennis players. The computer printout for regression is below:

Predictor	Coef	SE Coef	T	P
Constant	0.05590	0.02368	2.36	0.046
Facing	0.92003	0.03902	23.58	0.000

S = 0.0242922   R-Sq = 98.6%   R-Sq(adj) = 98.4%

3. What is the equation of the regression line, where "facing" and "back" are the winning game proportions when facing the sun and with back to the sun, respectively?

   (A) $\widehat{facing} = 0.056 + 0.920\,(back)$
   (B) $\widehat{back} = 0.056 + 0.920\,(facing)$
   (C) $\widehat{facing} = 0.920 + 0.056\,(back)$
   (D) $\widehat{back} = 0.920 + 0.056\,(facing)$
   (E) $\widehat{facing} = 0.024 + 0.039\,(back)$

4. What is the correlation?

   (A) −0.986
   (B) −0.984
   (C) 0.984
   (D) 0.986
   (E) 0.993

5. For one player, the winning game proportions were 0.55 and 0.59 for "facing" and "back," respectively. What was the associated residual?

   (A) −0.0488
   (B) −0.028
   (C) 0.028
   (D) 0.0488
   (E) 0.3608

6. Data are obtained for a group of college freshmen examining their SAT scores (Math + Evidence-Based Reading and Writing) from their senior year of high school and their GPAs during their first year of college. The resulting regression equation is

   $$\widehat{GPA} = 0.55 + 0.00161(\text{SAT score}) \quad \text{with} \quad r = 0.632$$

   What percentage of the variation in GPAs can be accounted for by looking at the linear relationship between GPAs and SAT scores?

   (A) 0.161%
   (B) 16.1%
   (C) 39.9%
   (D) 63.2%
   (E) This value cannot be computed from the information given.

7. Suppose the correlation between two variables is $r = 0.23$. What will the new correlation be if 0.14 is added to all values of the x-variable, every value of the y-variable is doubled, and the two variables are interchanged?

   (A) 0.74
   (B) 0.37
   (C) 0.23
   (D) −0.23
   (E) −0.74

8. Consider the following scatterplot and regression analysis of 15 data points:

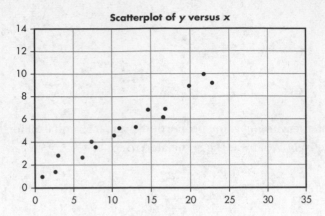

Slope $b = 0.377$

Coefficient of determination $r^2 = 95.0\%$

Standard deviation of the residuals $s = 0.640$

With the addition of a data point at (35, 14), which one of the following choices gives the most likely new regression statistics?

(A) Slope $b = 0.377$
    Coefficient of determination $r^2 = 97.0\%$
    Standard deviation of the residuals $s = 0.663$

(B) Slope $b = 0.377$
    Coefficient of determination $r^2 = 97.0\%$
    Standard deviation of the residuals $s = 0.617$

(C) Slope $b = 0.377$
    Coefficient of determination $r^2 = 93.0\%$
    Standard deviation of the residuals $s = 0.640$

(D) Slope $b = 0.377$
    Coefficient of determination $r^2 = 93.0\%$
    Standard deviation of the residuals $s = 0.663$

(E) Slope $b = 0.397$
    Coefficient of determination $r^2 = 95.0\%$
    Standard deviation of the residuals $s = 0.640$

9. A study of the number of highway deaths due to failure to wear seat belts in a year versus percentage seat belt usage in that year in 300 national regions shows a strong negative linear association. The least squares regression equation is: Predicted number of deaths $= 11{,}100 - 305.1$(Belt usage). What does "negative" mean in this context?

(A) If no one wore seat belts in a given region, it is predicted that there will be 11,100 highway deaths due to failure to wear seat belts.

(B) The correlation, r, between the number of highway deaths due to failure to wear seat belts and the percentage of seat belt usage is negative.

(C) If a given region has a lower percentage of seat belt usage than a second region, the given region will have a higher number of highway deaths due to failure to wear seat belts than the second region.

(D) Regions with a higher percentage of seat belt usage tend to have lower numbers of highway deaths due to failure to wear seat belts.

(E) If a region has a one percent gain in seat belt usage, then it will have a reduction of 305 highway deaths due to failure to wear seat belts, on average.

10. A simple random sample of 35 world-ranked chess players provides the following statistics:

Number of hours of study per day: $\bar{x} = 6.2$, $s_x = 1.3$
Yearly winnings: $\bar{y} = \$208{,}000$, $s_y = \$42{,}000$
Correlation $r = 0.15$

Based on these data, what is the resulting linear regression equation?

(A) $\widehat{\text{Winning}} = 178{,}000 + 4850\,(\text{Hours})$

(B) $\widehat{\text{Winning}} = 169{,}000 + 6300\,(\text{Hours})$

(C) $\widehat{\text{Winning}} = 14{,}550 + 31{,}200\,(\text{Hours})$

(D) $\widehat{\text{Winning}} = 7750 + 32{,}300\,(\text{Hours})$

(E) $\widehat{\text{Winning}} = -52{,}400 + 42{,}000\,(\text{Hours})$

11. Consider the set of points $\{(2, 5), (3, 7), (4, 9), (5, 12), (10, n)\}$. What should n be so that the correlation between the x- and y-values is 1?

(A) 21
(B) 24
(C) 25
(D) A value different from any of the above.
(E) No value for n can make $r = 1$.

12. The scatterplot below has one point labeled *X*. Does this point have high leverage, a large residual, both, or neither?

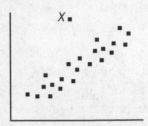

(A) High leverage and a large residual

(B) High leverage and a small residual

(C) Low leverage and a large residual

(D) Low leverage and a small residual

(E) This cannot be answered without calculating the regression line.

Free-Response Questions

Directions: You must show all work and indicate the methods you use. You will be graded on the correctness of your methods and on the accuracy of your final answers.

1. The following scatterplot shows the pulse rate drop (in beats per minute) plotted against the amount of medication (in grams) of an experimental drug being field-tested in several hospitals.

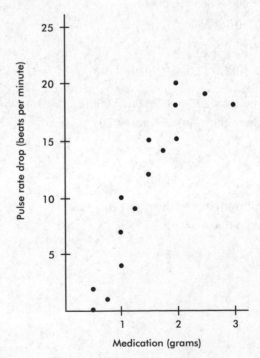

A computer printout showing the results of fitting a straight line to the data by the method of least squares gives:

```
PulseRateDrop = -1.68 + 8.5 Grams
R-sq = 81.9%
```

(a) Find the correlation coefficient for the relationship between pulse rate drop and grams of medication.

(b) What is the slope of the regression line, and what does it signify?

(c) Predict the pulse rate drop for a patient given 2.25 grams of medication.

(d) A patient given 5 grams of medication has his pulse rate drop to zero. Does this invalidate the regression equation? Explain.

(e) How will the size of the correlation coefficient change if the 3-gram result is removed from the data set? Explain.

(f) How will the size of the slope of the least squares regression line change if the 3-gram result is removed from the data set? Explain.

2. Facebook showed a remarkable rate of growth during its early years. A scatterplot of MAUs (monthly active users, in millions) during the company's first nine years is as follows:

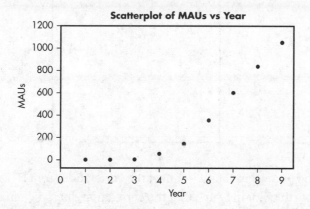

(a) Describe the relationship between MAUs and year.

Transforming using logarithms and then fitting a linear regression model gives the following computer output and residual plot:

Regression Analysis: log(MAUs) versus log(Year)

```
The regression equation is
log(MAUs) = - 0.196 + 3.39 log(Year)

Predictor       Coef    SE Coef        T       P
Constant      -0.1963    0.1292     -1.52    0.172
log(Year)      3.3886    0.1888     17.95    0.000
S = 0.166815    R-Sq = 97.9%    R-Sq(adj) = 97.6%
```

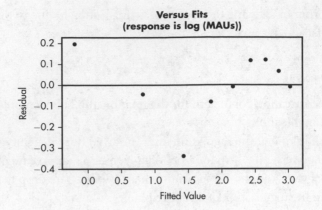

Versus Fits
(response is log (MAUs))

(b)   Interpret the value of r^2.

(c)   Is it appropriate to use a linear regression model on the transformed variables?

3. An outlier can have a striking effect on the correlation r. Consider the following three scatterplots:

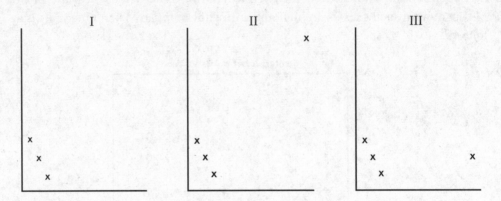

(a)   How do the outliers in II and III affect the correlation from I?

(b)   Insert a fourth point in I such that the correlation doesn't change.

4. A continuing political debate concerns how to address the large student debt carried by college graduates. Graph (I) is a scatterplot showing the amount of student debt of 25 randomly selected college graduates and their stress level as self-reported on a 1–100 scale.

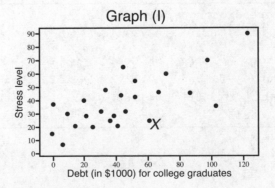

Graph (I)

Computer output from linear regression gives:

Predictor	Coef	SE Coef
Constant	19.376	4.782
Debt	0.42640	0.08906

(a) Describe the association between stress level and debt for this sample of 25 college graduates.

(b) The point labeled X represents a student with $60,000 debt and a stress level of 25. Calculate and interpret the residual of that point.

Two other possible explanatory variables with regard to stress level are income and health as measured on a 3 to 8 HRQoL (health-related quality of life) index. Graph (II) is a scatterplot of health plotted with the corresponding residuals from the regression of stress on debt. Graph (III) is a scatterplot of income plotted with the corresponding residuals from the regression of stress on debt.

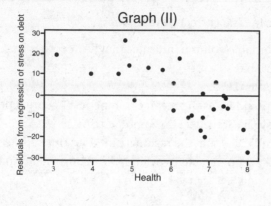

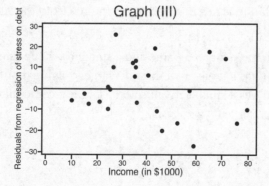

(c) Which variable, health or income, in addition to debt should be used to improve the prediction of stress level? Explain.

The answers for this quiz can be found in the Appendix on page 587.

QUIZ 9

Multiple-Choice Questions

> **Directions:** The questions that follow are each followed by five suggested answers. Choose the response that best answers the question.

1. A study collects data on average combined SAT scores (Math + Evidence-Based Reading and Writing) and percentage of students who took the exam at 100 randomly selected high schools. The following is part of the computer printout for regression:

Variable	Coefficient	s.e. of coeff	t-ratio	prob
Constant	1176.32	12.65	92.99	≤ 0.0001
SAT	−2.84276	0.2461	−11.55	≤ 0.0001

   R-squared = 76.5%   R-squared (adj) = 76.1%

   Which of the following is a correct conclusion?

   (A) "SAT" in the variable column indicates that SAT score is the dependent (response) variable.
   (B) The correlation is 0.875.
   (C) The y-intercept indicates the mean combined SAT score if percent of students taking the exam has no effect on combined SAT scores.
   (D) The r^2 value indicates that the residual plot does not show a strong pattern.
   (E) Schools with lower percentages of students taking the exam tend to have higher average combined SAT scores.

2. Suppose the correlation is negative. Given two points from the scatterplot, which of the following is possible?

   I. The first point has a larger x-value and a smaller y-value than the second point.
   II. The first point has a larger x-value and a larger y-value than the second point.
   III. The first point has a smaller x-value and a larger y-value than the second point.

   (A) I only
   (B) II only
   (C) III only
   (D) I and III only
   (E) I, II, and III

3. Suppose the regression line for a set of data, $\hat{y} = 3x + b$, passes through the point (2, 5). If $\bar{x}$ and $\bar{y}$ are the sample means of the x- and y-values, respectively, then $\bar{y} =$

   (A) $\bar{x}$.
   (B) $\bar{x} - 2$.
   (C) $\bar{x} + 5$.
   (D) $3\bar{x}$.
   (E) $3\bar{x} - 1$.

4. Which of the following statements about correlation r is true?

    (A) A correlation of 0.2 means that 20% of the points are highly correlated.
    (B) Perfect correlation, that is, when the points lie exactly on a straight line, results in $r = 0$.
    (C) Correlation is not affected by which variable is called x and which is called y.
    (D) Correlation is not affected by extreme values.
    (E) A correlation of 0.75 indicates a relationship that is 3 times as linear as one for which the correlation is only 0.25.

5. Which of the following statements about residuals from the least squares line are true?

    I. The mean of the residuals is always zero.
    II. The regression line for a residual plot is a horizontal line.
    III. A definite pattern in the residual plot is an indication that a nonlinear model will show a better fit to the data than the straight regression line.

    (A) I and II only
    (B) I and III only
    (C) II and III only
    (D) I, II, and III
    (E) None of the above gives the complete set of true responses.

6. In a study of winning percentage in home games versus average home attendance for professional baseball teams, the resulting regression line is

$$\text{Predicted winning percentage} = 44 + 0.0003(\text{Attendance})$$

What is the residual if a team has a winning percentage of 55% with an average attendance of 34,000?

    (A) −11.0
    (B) −0.8
    (C) 0.8
    (D) 11.0
    (E) 23.0

7. Suppose the correlation between two variables is −0.57. If each of the y-scores is multiplied by −1, which of the following is true about the new scatterplot?

    (A) It slopes up to the right, and the correlation is −0.57.
    (B) It slopes up to the right, and the correlation is +0.57.
    (C) It slopes down to the right, and the correlation is −0.57.
    (D) It slopes down to the right, and the correlation is +0.57.
    (E) None of the above is true.

8. A study of selling prices of homes in a southern California community (in $1000) versus size of the homes (in 1000s of square feet) shows a moderate positive linear association. The least squares regression equation is: Predicted selling price = 35.3 + 214.1(Size). What does "linear" mean in this context?

   (A) The points in the scatterplot line up in a straight line.
   (B) There is no distinct pattern in the residual plot.
   (C) The coefficient of determination, r^2, is large (close to 1).
   (D) As home size increases by 1000 square feet, the selling price tends to change by a constant amount, on average.
   (E) Each increase of 1000 square feet in home size gives an increase of 214.1($1000) in selling price.

9. Does lower sun exposure, as measured by greater distance from the equator, result in lower incidence of skin cancer? A study of skin cancer mortality versus latitude of 100 northern hemisphere locales shows a strong negative linear association. What does "strong" mean in this context?

   (A) More sun exposure causes greater numbers of skin cancer deaths.
   (B) More sun exposure is associated with greater numbers of skin cancer deaths.
   (C) A locale's incidence of skin cancer deaths has a linear association with its latitude.
   (D) A least squares model predicts that the greater the latitude, the lower the average incidence of skin cancer.
   (E) The actual incidence of skin cancer at a given latitude will be very close to what is predicted by a least squares model.

10. Consider n pairs of numbers. Suppose $\bar{x} = 2$, $s_x = 3$, $\bar{y} = 4$, and $s_y = 5$. Of the following, which could be the least squares line?

   (A) $\hat{y} = -2 + x$
   (B) $\hat{y} = 2x$
   (C) $\hat{y} = -2 + 3x$
   (D) $\hat{y} = \dfrac{5}{3} - x$
   (E) $\hat{y} = 6 - x$

11. Consider the three points (2, 11), (3, 17), and (4, 29). Given any straight line, we can calculate the sum of the squares of the three vertical distances from these points to the line. What is the smallest possible value this sum can be?

   (A) 6
   (B) 9
   (C) 29
   (D) 57
   (E) None of these values

12. The scatterplot below has one point labeled *X*. Does this point have high leverage, a large residual, both, or neither?

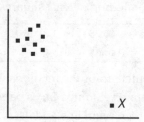

(A) High leverage and a large residual

(B) High leverage and a small residual

(C) Low leverage and a large residual

(D) Low leverage and a small residual

(E) This cannot be answered without calculating the regression line.

Free-Response Questions

Directions: You must show all work and indicate the methods you use. You will be graded on the correctness of your methods and on the accuracy of your final answers.

1. The following scatterplot shows the grades for research papers for a sociology professor's class plotted against the lengths of the papers (in pages).

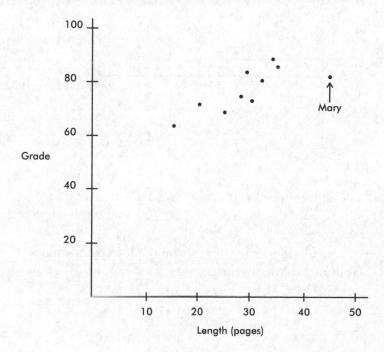

Mary turned in her paper late and was told by the professor that her grade would have been higher if she had turned it in on time. A computer printout fitting a straight line to the data (not including Mary's score) by the method of least squares gives:

```
Grade = 46.51 + 1.106 Length
R-sq = 74.6%
```

(a) Find the correlation coefficient for the relationship between grade and length of paper based on these data (excluding Mary's paper).

(b) What is the slope of the regression line and what does it signify?

(c) How will the correlation coefficient change if Mary's paper is included? Explain your answer.

(d) How will the slope of the regression line change if Mary's paper is included? Explain your answer.

2. Data show a trend in winning long jump distances for an international competition over the years 1972–1992. With jumps recorded in inches and dates in years since 1900, a least squares regression line is fit to the data. The computer output and a graph of the residuals are as follows:

```
R squared = 92.1%
Variable        Coefficient     SE of Coeff     t-ratio     Prob
Constant        256.576         11.59           22.1        0.0001
Year            0.95893         0.141           6.81        0.0024
```

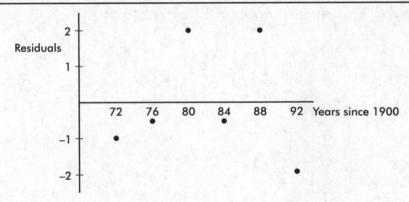

(a) Does a line appear to be an appropriate model? Explain.

(b) What is the slope of the least squares line? Give an interpretation of the slope.

(c) What is the correlation?

(d) What is the predicted winning distance for the 1980 competition?

(e) What was the actual winning distance in 1980?

3. A scatterplot of the number of accidents per day on a particular interstate highway during a 30-day month is as follows:

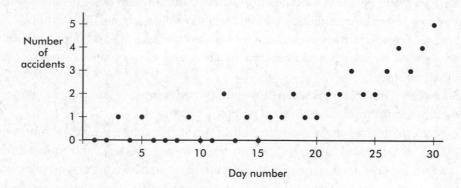

(a) Draw a histogram of the frequencies of the number of accidents.
(b) Draw a boxplot of the number of accidents.
(c) Name a feature apparent in the scatterplot but not in the histogram or boxplot.
(d) Name a feature clearly shown by the histogram and boxplot but not as obvious in the scatterplot.

4. In a study of how exam scores are influenced by hours of sleep the night before the exam, a random sample of 13 students was outfitted with wristbands that tracked sleep time. The scatterplot of the resulting data follows.

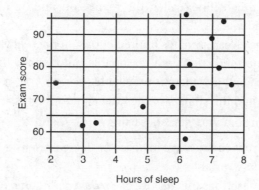

(a) Describe the association between exam score and hours of sleep for these 13 students.
(b) Describe the association between exam score and hours of sleep for the 8 students who received at least 6 hours of sleep the night before the exam.
(c) Explain how the appropriateness of using linear regression is critically influenced by one student's data.

The answers for this quiz can be found in the Appendix on page 588.

SUMMARY

- Two-way tables are useful in showing relationships between two categorical variables.
- The row and column totals lead to calculations of the marginal distributions.
- Focusing on single rows or columns leads to calculations of conditional distributions.
- Segmented bar charts and mosaic plots are useful visual tools to show conditional distributions.
- Simpson's paradox occurs when the results from a combined grouping seem to contradict the results from the individual groups.
- A scatterplot gives an immediate indication of the shape (linear or not), strength (weak, moderate, or strong), direction (positive or negative), and unusual features (such as outliers or clusters) of a possible relationship between two variables.
- If the relationship appears roughly linear, the correlation coefficient, r, and especially the coefficient of determination, r^2, are useful measurements.
- The value of r is always between -1 and $+1$, with positive values indicating positive association and negative values indicating negative association; values close to -1 or $+1$ indicate a stronger linear association than values close to 0, which indicate a weaker linear association.
- Evidence of an association is not evidence of a cause-and-effect relationship!
- Correlation is not affected by which variable is called x and which y or by changing units.
- Correlation can be strongly affected by extreme values.
- The differences between the observed and predicted values are called residuals.
- The best-fitting straight line, called the regression line, minimizes the sum of the squares of the residuals.
- The regression line gives estimates or predictions, not actual values.
- The regression line predictions are reasonable only within the domain of the data; extrapolation outside the domain is unreliable.
- The y-intercept a and the slope b of the least squares regression line, $\hat{y} = a + bx$, can be calculated from summary statistics using $b = r\dfrac{s_y}{s_x}$ and $a = \bar{y} - b\bar{x}$.
- The slope b describes the *average* increase or decrease in the y-variable for each one unit increase in the x-variable.
- The y-intercept a describes the *average* or *predicted* outcome of y if $x = 0$.
- For the linear regression model, the mean of the residuals is always 0.
- A definite pattern in the residual plot indicates that a nonlinear model may fit the data better than the straight regression line.
- The coefficient of determination, r^2, gives the percentage of variation in y that is accounted for by the least squares regression line. Alternatively, $1 - r^2$ gives how much of the variability in y is *unaccountable* by the regression line.
- Outliers are points whose residuals in absolute value are large when compared to those of other residuals.
- Influential points are points whose removal would sharply change the regression line.
- High leverage points are those whose x-values are far from the mean of the x-values.
- Nonlinear models can sometimes be studied by transforming one or both variables and then noting a linear relationship.
- It is very important to be able to interpret generic computer output.

Collecting Data

3

(12–15% AP EXAM WEIGHTING)

- → **RETROSPECTIVE VERSUS PROSPECTIVE OBSERVATIONAL STUDIES**
- → **BIAS**
- → **SAMPLING METHODS**
- → **SAMPLING VARIABILITY**
- → **QUIZ 10**
- → **QUIZ 11**
- → **EXPERIMENTAL VERSUS OBSERVATIONAL STUDIES**
- → **THE LANGUAGE OF EXPERIMENTS**
- → **REPLICATION AND GENERALIZABILITY OF RESULTS**
- → **INFERENCE AND EXPERIMENTS**
- → **QUIZ 12**
- → **QUIZ 13**

In this unit, you will learn principles of sampling and experimental design, you will see how bias can result, and you will understand how to control for confounding. You will also be able to distinguish between retrospective versus prospective studies, observational versus experimental studies, random sampling versus random assignment, stratified versus cluster sampling, stratification versus blocking, experimental units versus treatment groups, blocks versus treatment groups, and completely randomized design versus randomized block design. Then, finally, you will begin to understand the meaning of statistical significance and what kind of conclusions can result from well-designed experiments.

UNIT LEARNING OBJECTIVES

- To be able to distinguish between a sample and a population and between a statistic and a parameter.
- To be able to explain whether an observational study is retrospective or prospective.
- To be able to distinguish between a convenience sample, a voluntary response sample, and a random sample.
- To be able to identify, describe, and implement different sampling methods including a simple random sample, a cluster sample, a stratified sample, and a systematic sample.
- To be able to describe the benefits and drawbacks of various sampling methods.
- To be able to explain a particular bias and how it is likely to result in the sample underestimating or overestimating a population value.

- To be able to explain why a particular study is observational or an experiment.
- To be able to design an experiment using one or more of the following: different levels of an explanatory variable, a placebo, blinding or double-blinding, blocking, and matched pairs design.
- To be able to explain why or why not a cause-and-effect conclusion is justified.
- To understand the difference between, and how to perform, a random sampling and a random assignment.
- To be able to explain in context why a particular variable might lead to confounding.
- To be able to explain why or why not the results of an experiment can be generalized.

TIP

You should always be aware to what population you can reasonably generalize the results of your study.

In the real world, time and cost considerations usually make it impossible to analyze an entire population. Do companies like Apple question every potential consumer before designing a new smartphone? Does a television producer check every household's viewing preferences before deciding whether a pilot program will be continued? By studying statistics, we learn how to estimate a population characteristic (called a population *parameter*) by considering a sample measurement (called a sample *statistic*). Later in this review book we will see how to estimate population proportions, means, and slopes by looking at sample proportions, means, and slopes.

To derive conclusions about the larger population, we need to be confident that the sample we have chosen represents the population fairly. Analyzing the data with computers is often easier than gathering the data, but the frequently quoted "garbage in, garbage out" applies here. Nothing can help if the data are collected poorly. Unfortunately, many of the statistics that we are bombarded with in newspapers, on the radio, and on television are based on poorly designed data collection procedures.

NOTE

MYTH: A random sample will *always* be representative of the population.

FACT: A random sample, *simply by chance*, might turn out not to be very representative.

There are two important principles in picking a representative sample in order to be able to generalize our findings to the whole population:

1. The use of *randomization* in sample selection increases the chance that the resulting sample exhibits all the known and unknown features of the population.
2. The *sample size*, not the fraction of the population surveyed, is key.

There are two important principles in designing an experiment in order to be able to make a cause-and-effect conclusion:

1. The use of *randomization* in assignment to treatment groups increases the chance that treatment groups are as similar as possible other than which treatment each group receives.
2. The size of treatment groups, that is, the number of subjects in each treatment group, is key.

TIP

If there is an important difference among the subjects, grouping similar subjects into *blocks* and randomizing treatments within the blocks becomes a third principle.

RETROSPECTIVE VERSUS PROSPECTIVE OBSERVATIONAL STUDIES

Observational studies aim to gather information about a population without disturbing the population. *Retrospective studies* look backward, examining existing data, while *prospective studies* watch for outcomes, tracking individuals into the future.

⇒ EXAMPLE 3.1 _____

Retrospective studies of the 2014–2016 Ebola epidemic in West Africa have looked at the timing, numbers, and locations of reported cases and have tried to understand how the outbreak spread. There is now much better understanding of transmission through contact with bodily fluids of infected people. Several prospective studies involve ongoing surveillance to see how experience and tools to rapidly identify cases will now limit future epidemics.

Advantages and Disadvantages

Retrospective studies tend to be smaller scale, quicker to complete, and less expensive. In cases such as addressing diseases with low incidence, the study begins right from the start with people who have already been identified. However, researchers have much less control, usually having to rely on past record keeping of others. Furthermore, the existing data will often have been gathered for other purposes than the topic of interest. Then there is the problem of subjects' inaccurate memories and possible biases.

Prospective studies usually have greater accuracy in data collection and are less susceptible to subject recall error. Researchers do their own record keeping and can monitor exactly what variables they are interested in. However, these studies can be very expensive and time consuming, as they often follow large numbers of subjects for a long time.

BIAS

The one thing that most quickly invalidates a sample and makes obtaining useful information and drawing meaningful conclusions impossible is *bias*. **A sampling method is biased if in some critical way it consistently results in samples that do not represent the population.** This typically leads to certain responses being repeatedly favored over others.

Voluntary response surveys are based on individuals who choose to participate, typically give too much emphasis to people with strong opinions, and undersample people who don't care much about a topic. For example, radio call-in programs about controversial topics such as gun control, abortion, and school segregation do not produce meaningful data on what proportion of the population favor or oppose related issues. Online surveys posted to websites are a common source of voluntary response bias. Although a voluntary response survey may be easy and inexpensive to conduct, the sample is likely to be composed of strongly opinionated people, especially those with negative opinions on a subject.

Convenience surveys, like interviews at shopping malls, are based on choosing individuals who are easy to reach. These surveys tend to produce data highly unrepresentative of the entire population. For example, door-to-door household surveys miss people who do not happen to be at home and typically miss groups such as college students, prison inmates, and the homeless. Although convenience samples may allow a researcher to conduct a survey quickly and inexpensively, generalizing from the sample to a population is almost impossible.

There are many other sources of sampling bias. Although it is not necessary to know every possible kind of bias by name, it is important to recognize that the sample designs are flawed, to be able to explain why they are flawed, and to understand the *direction* of the bias, that is, whether it will likely cause an under- or an overestimation of a proportion or a mean.

NOTE

The way data are obtained is crucial. A large sample size cannot make up for a poor survey design or for faulty data collection techniques.

TIP

Think about potential biases before collecting data!

TIP

Unless specifically asked, it's probably better not to try to *name* these other sources of bias, because a wrong name will be penalized. Simply explain the flaw and the *direction* of bias.

Undercoverage bias happens when there is inadequate representation, and thus some groups in the population are left out of the process of choosing the sample. For example, telephone surveys ignore those who don't have telephones. *Response bias* occurs when the question itself can lead to misleading results because people don't want to be perceived as having unpopular or unsavory views or don't want to admit to having committed crimes. *Nonresponse bias*, where there are low response rates, occurs when individuals chosen for the sample can't be contacted or refuse to participate, and it is often unclear which part of the population is responding. *Quota sampling bias*, where interviewers are given free choice in picking people in the (problematic, if not impossible) attempt to pick representatively with no randomization, is a recipe for disaster. *Question wording bias* can occur when nonneutral or poorly worded questions lead to very unrepresentative responses or even when the order in which questions are asked makes a difference.

SAMPLING METHODS

From politics to business, from science to sociology, and from sports to fighting crime, statisticians use available data to understand and make predictions about entire populations.

Collecting data from every individual in a population, called a *census*, might seem to provide complete information, but unless the population is small, this method would be prohibitively expensive and time-consuming. Furthermore, it could be ridiculous, as, for example, tasting every single pizza a company makes would leave nothing left to sell. Finally, if a questionnaire is poorly worded, answers, even from an entire population, might be meaningless.

Given that it is almost always impractical, if not impossible, to examine a whole population, we must settle for examining a sample. Of primary importance is how to choose a representative sample.

How can we increase our chance of choosing a representative sample? One technique is to write the name of each member of the population on a card, mix the cards thoroughly in a large box, and pull out a specified number of cards. This method gives everyone in the population an equal chance of being selected as part of the sample. Unfortunately, this method is usually too time-consuming, and bias might still creep in if the mixing is not thorough. A *simple random sample* (SRS), **one in which every possible sample of the desired size has an equal chance of being selected**, can be more easily obtained by assigning a number to everyone in the population and using a random digit table or having a computer generate random numbers to indicate choices.

➡ EXAMPLE 3.2

Suppose 80 students are taking an AP Statistics course and the teacher wants to pick a sample of 10 students randomly to try out a practice exam. She first assigns the students numbers 01, 02, 03, . . . , 80. While reading off two digits at a time from a random digit table, she ignores any numbers over 80, ignores 00, and ignores repeats, stopping when she has a set of 10. If the table began 75425 56573 90420 48642 27537 61036 15074 84675, she would choose the students numbered 75, 42, 55, 65, 73, 04, 27, 53, 76, and 10. Note that 90 and 86 are ignored because they are over 80. Note that the second and third occurrences of 42 are ignored because they are repeats.

→ EXAMPLE 3.3

An alternative solution to Example 3.2 is to make use of a random number generator on a computer. Assign the students numbers 1, 2, 3, . . ., 80. Use a computer to generate 10 random integers between 1 and 80 *without replacement*, that is, throw out repeats. The sample consists of the students with assigned numbers corresponding to the 10 unique computer-generated numbers.

Advantages of simple random sampling include the following:

- The simplicity of simple random sampling makes it relatively easy to interpret data collected.

- This method requires minimal advance knowledge of the population other than knowing the complete sampling frame.

- Simple random sampling allows us to make generalizations (i.e., statistical inferences) from the sample to the population.

> **IMPORTANT**
>
> To receive full credit, there are 3 steps:
>
> 1. assigning numbers,
> 2. using a computer random number generator to generate distinct numbers in a given range, and
> 3. linking selected numbers with corresponding individuals selected for the sample.

Disadvantages of simple random sampling include the following:

- The need for a list of all potential subjects can be a formidable task.

- Although simple random sampling is a straightforward procedure to understand, it can be difficult to execute, especially if the population is large. For example, how would you obtain a simple random sample of students from the population of all high school students in the United States?

- This method may require you to contact nonrespondents repeatedly, which can be very time-consuming.

- This method may leave out groups you want to be sure are represented.

NOTE

The list of all individuals from which the sample is drawn is called the *sampling frame*.

There are other sampling methods available. Each has its own set of advantages and disadvantages.

Stratified sampling involves dividing the population into homogeneous groups called *strata*, then picking random samples from each of the strata, and finally combining these individual samples into what is called a *stratified random sample*. For example, we can stratify by age, gender, income level, or race; pick a sample of people from each stratum; and combine to form the final sample. For instance, a political pollster might want to be sure to include respondents from various minority groups. Stratifying by religion or race and including people from each group would ensure this diversity in the final sample.

Advantages of stratified sampling include the following:

- Samples taken within a stratum have reduced variability, which means the resulting estimates are more precise than when using other sampling methods.
- Important differences among groups can become more apparent.

Disadvantages of stratified sampling include the following:

- Like an SRS, this method might be difficult to implement with large populations.
- Forcing subdivisions when none really exist is meaningless.

NOTE

All individuals in a given stratum have a characteristic in common.

TIP

You should be ready to explain why you think that stratification should be used in a given problem.

NOTE

It's beyond this course, but stratification reduces the standard error of estimates.

We could further do *proportional sampling*, where the sizes of the random samples from each stratum depend on the proportion of the total population represented by the stratum.

Cluster sampling involves dividing the population into heterogeneous groups called *clusters* and then picking everyone in a random selection of one or more of the clusters. For example, to survey high school seniors, we could randomly pick several senior class homerooms in which to conduct our study and sample all students in those selected homerooms.

Advantages of cluster sampling include the following:

- Clusters are often chosen for ease, convenience, and quickness.
- With limited fixed funds, cluster sampling usually allows for a larger sample size than do other sampling methods.

Disadvantages of cluster sampling include the following:

- With a given sample size, cluster sampling usually provides less precision than either an SRS or a stratified sample provides.
- If the population doesn't have natural clusters and the designated clusters are not representative of the population, selection could easily result in a biased sample.

The differences between strata and clusters include the following:

- While all strata are represented in the sample, only a subset of clusters are in the sample.
- Stratified sampling is accurate when each stratum consists of *homogeneous* elements, while cluster sampling is accurate when each cluster consists of *heterogeneous* elements.

Systematic sampling is a relatively simple and quick method. It involves listing the population in some order (for example, alphabetically), choosing a random point to start, and then picking every tenth (or hundredth, or thousandth, or *k*th) person from the list. This gives a reasonable sample as long as the original order of the list is not in any way related to the variables under consideration.

An advantage of systematic sampling is that if similar members in the list are grouped together, we can end up with a kind of stratified sample, only more easily implemented.

A disadvantage of systematic sampling is that if the list happens to have a periodic structure similar to the period of the sampling, a very biased sample could result.

➡ EXAMPLE 3.4

Suppose a sample of 100 high school students from a Chicago school of size 5000 is to be chosen to determine their views on whether they think the Cubs will win another World Series this century. One method would be to have each student write his or her name on a card, put the cards into a box, and have the principal reach in and pull out 100 cards. However, questions could arise regarding how well the cards are mixed. For example, how might the outcome be affected if all students in one PE class toss their names in at the same time so that their cards are clumped together? Another method would be to assign each student a number from 0001 to 5000 and then use a random digit table, picking out four digits at a time, ignoring repeats, 0000, and numbers over 5000, until a unique set of 100 numbers are picked. Then choose the students corresponding to the selected 100 numbers (simple random sampling). What are alternative procedures?

Answer: From a list of the students, the surveyor could choose a random starting name and then simply note every 50th name (systematic sampling). Since students in each class have certain characteristics in common, the surveyor could use a random selection method to pick 25 students from each of the separate lists of freshmen, sophomores, juniors, and seniors (stratified sampling). If each homeroom has a random mix of 20 students from across all grade levels, the surveyor could randomly pick five homerooms with the sample consisting of all the students in these five rooms (cluster sampling).

NOTE

Understand that these other procedures are still proper random samples even if they are not simple random samples.

Sometimes we randomly pick a single sample, cross-categorize on two variables, and then test to see if there is an association between the variables. However, sometimes we instead randomly pick a sample from each of two or more populations and then compare the distribution of some variable among the different populations. This second method is called *sampling for homogeneity*.

NOTE

We'll revisit this sampling method later when developing chi-square inference, but this choice of sampling method is really a design issue and worth recognizing at this time.

SAMPLING VARIABILITY

No matter how well designed and well conducted a survey is, it still gives a sample *statistic* as an estimate for a population *parameter*. Different samples give different sample statistics, all of which are estimates for the same population parameter. So *sampling variability*, also called *sampling error*, is naturally present. This variability can be described using probability; that is, we can say how likely we are to have a certain size error. Generally, the associated probabilities are smaller when the sample size is larger. However, **whenever a sample is taken, sampling variability (sampling error) will be present**. This is not an error that someone is committing but is simply sample-to-sample variation.

One shouldn't confuse **accuracy** (centered at the right place) with **precision** (low variability).

TIP

Sampling variability is to be expected and can be quantified, while bias is to be avoided.

➡️ **EXAMPLE 3.5**

Suppose we are trying to estimate the mean age of high school teachers and have four methods of choosing samples. We choose 10 samples using each method. Later we find the true mean, $\mu = 42$. Plots of the results of each sampling method are given below.

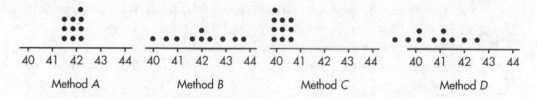

Method A exhibits high accuracy and high precision. Method B exhibits high accuracy and low precision. Method C exhibits low accuracy and high precision. Method D exhibits low accuracy and low precision.

NOTE

Low accuracy (not centered at the right place) indicates probable bias in the selection method.

NOTE

Sampling variability is discussed in more detail in Unit 5: Sampling Distributions.

It needs to be emphasized that *bias* is a question of sampling methodology, not a characteristic of an individual sample. It's possible for a biased methodology to produce a sample that characterizes the population well, but that's just sampling variability resulting in a lucky break. With bias, samples will collectively misrepresent the population, but that does not mean that all of them will do so. With an unbiased methodology, samples will collectively represent the population; however, because of sampling variability, any given sample still can be a "bad" sample. In this case, it would be wrong to say that the sample is biased, but rather it's simply an example of sampling variability.

QUIZ 10

Multiple-Choice Questions

> **Directions:** The questions that follow are each followed by five suggested answers. Choose the response that best answers the question.

1. A school psychologist wants to investigate student depression. She selects, at random, samples of 30 freshmen, 30 sophomores, 30 juniors, and 30 seniors to interview. Which of the following best describes the principal's sampling plan?

   (A) A convenience sample
   (B) A simple random sample
   (C) A stratified random sample
   (D) A cluster sample
   (E) A systematic sample

2. Which of the following is a true statement?

   (A) If bias is present in a sampling procedure, it can be overcome by dramatically increasing the sample size.
   (B) There is no such thing as a "bad sample."
   (C) Sampling techniques that use probability techniques effectively eliminate bias.
   (D) Convenience samples often lead to undercoverage bias.
   (E) Voluntary response samples often underrepresent people with strong opinions.

3. Which of the following is a true statement about sampling error?

   (A) Sampling error can be eliminated only if a survey is both extremely well designed and extremely well conducted.
   (B) Sampling error reflects natural variation between samples, is always present, and can be described using probability.
   (C) Sampling error is generally larger when the sample size is larger.
   (D) Sampling error implies an error, possibly very small, but still an error on the part of the surveyor.
   (E) Sampling error is higher when bias is present.

4. Many colleges are moving toward coed dormitories. A Residential Life director plans to sample student attitudes toward this arrangement. He randomly selects three of the 12 dorms and sends a questionnaire to all residents living in them. Which of the following best describes the director's sampling plan?

   (A) A convenience sample
   (B) A simple random sample
   (C) A stratified random sample
   (D) A cluster sample
   (E) A systematic sample

5. Ann Landers, who wrote a daily advice column appearing in newspapers across the country, once asked her readers, "If you had it to do over again, would you have children?" Of the more than 10,000 readers who responded, 70% said no. What does this show?

   (A) Voluntary response bias makes the survey meaningless.
   (B) No meaningful conclusion is possible without knowing something more about the characteristics of her readers.
   (C) The survey would have been more meaningful if she had picked a random sample of the 10,000 readers who responded.
   (D) The survey would have been more meaningful if she had used a control group.
   (E) This was a legitimate sample, randomly drawn from her readers and of sufficient size to allow the conclusion that most of her readers who are parents would have second thoughts about having children.

6. Each of the 30 MLB teams has 25 active roster players. A sample of 60 players is to be chosen as follows. Each team will be asked to place 25 cards with its players' names into a hat and randomly draw out two names. The two names from each team will be combined to make up the sample. Will this method result in a simple random sample of the 750 baseball players?

   (A) Yes, because each player has the same chance of being selected.
   (B) Yes, because each team is equally represented.
   (C) Yes, because this is an example of stratified sampling, which is a special case of simple random sampling.
   (D) No, because the teams are not chosen randomly.
   (E) No, because not each group of 60 players has the same chance of being selected.

Free-Response Questions

Directions: You must show all work and indicate the methods you use. You will be graded on the correctness of your methods and on the accuracy of your final answers.

1. A student is interested in estimating the average length of words in a 600-page textbook and plans the following three-stage sampling procedure:

   I.   After noting that each of the 12 chapters has a different author, the student decides to obtain a sample of words from each chapter.
   II. Each chapter is approximately 50 pages long. The student uses a random number generator to pick three pages from each chapter.
   III. On each chosen page, the student notes the length of every tenth word.

      (a) The first stage above represents what kind of sampling procedure? Give an advantage in using it in this context.
      (b) The second stage above represents what kind of sampling procedure? Give an advantage in using it in this context.
      (c) The third stage above represents what kind of sampling procedure? Give a *disadvantage* to using it dependent upon an author's writing style.

2. Two versions of a Pew Research poll were conducted:

Version I: "Is it more important for the U.S. president to focus on domestic policy or foreign policy?"

Version II: "Is it more important for the U.S. president to focus on domestic policy or the war on terrorism?"

The results of one wording were 33% choosing domestic policy, while with the other wording 52% chose domestic policy.

Which version probably resulted in which result? Explain.

3.

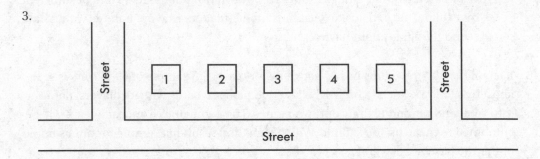

You are supposed to interview the residents of two of the above five houses.

(a) How would you choose which houses to interview in order to obtain an SRS?
(b) You plan to visit the homes at 9:00 a.m. If someone isn't home, explain the reasons for and against substituting another house.
(c) Are there any differences you might expect to find among the residents based on the above sketch, and noting this, what other sampling procedure might you choose to use?

4. An emotionally charged topic in government these days is immigration reform. Suppose a congresswoman wishes to survey her constituents concerning their opinions on whether the federal government should expand the Dream Act (a process for some undocumented immigrants to obtain permanent residency). Discuss possible sources of bias with regard to the following four options: (1) conducting a survey via random telephone dialing into her district, (2) sending out a mailing using a registered voter list, (3) having a pollster interview everyone who walks past her downtown office, and (4) broadcasting a radio appeal urging interested citizens in her district to call in their opinions to her office.

The answers for this quiz can be found in the Appendix on page 591.

QUIZ 11

Multiple-Choice Questions

> **Directions:** The questions that follow are each followed by five suggested answers. Choose the response that best answers the question.

1. To evaluate their catering service with regard to culinary excellence, airline executives plan to pick a random passenger to start with and then survey every tenth passenger departing from a Beijing to San Francisco flight. Which of the following best describes the executives' sampling plan?

   (A) A convenience sample
   (B) A simple random sample
   (C) A stratified random sample
   (D) A cluster sample
   (E) A systematic sample

2. School uniforms are being adopted by U.S. public schools in increasing numbers. Two possible wordings for a question on whether or not students should have to wear school uniforms are as follows:

   I. Many educators believe in creating a level playing field to reduce socioeconomic disparities. Do you believe that students should have to wear school uniforms?

   II. Many sociologists believe that students have a right to express their individuality. Do you believe that students should have to wear school uniforms?

   One of these questions showed that 18% of the population favors school uniforms, while the other question showed that 23% of the population favors school uniforms. Which question probably produced which result and why?

   (A) The first question probably showed 23% of the population favors school uniforms, and the second question probably showed 18% because of the lack of randomization in the choice of pro-uniform and anti-uniform arguments as evidenced by the wording of the questions.
   (B) The first question probably showed 18% and the second question probably showed 23% because of stratification in the wording of the questions.
   (C) The first question probably showed 23% and the second question probably showed 18% because of the lack of a neutral cluster in the sample.
   (D) The first question probably showed 18% and the second question probably showed 23% because of response bias due to the wording of the questions.
   (E) The first question probably showed 23% and the second question probably showed 18% because of response bias due to the wording of the questions.

3. In a study of successes and failures in adopting Common Core standards, a random sample of high school principals will be selected from each of the 50 states. Selected individuals will be asked a series of evaluative questions. Why is stratification used here?

    (A) To minimize response bias
    (B) To minimize nonresponse bias
    (C) To minimize voluntary response bias
    (D) Because each state is roughly representative of the U.S. population as a whole
    (E) To obtain higher statistical precision because variability of responses within a state is likely less than variability of responses found in the overall population

4. To find out the average occupancy size of student-rented apartments, a researcher picks a simple random sample of 100 such apartments. Even after one follow-up visit, the interviewer is unable to make contact with anyone in 27 of these apartments. Concerned about nonresponse bias, the researcher chooses another simple random sample and instructs the interviewer to continue this procedure until contact is made with someone in a total of 100 apartments. The average occupancy size in the final 100-apartment sample is 2.78. Is this estimate probably too low or too high?

    (A) Too low, because of undercoverage bias
    (B) Too low, because convenience samples overestimate average results
    (C) Too high, because of undercoverage bias
    (D) Too high, because convenience samples overestimate average results
    (E) Too high, because voluntary response samples overestimate average results

5. To survey the opinions of bleacher fans at Wrigley Field, the home stadium of the Cubs baseball team, a surveyor plans to select every one-hundredth fan entering the bleachers one afternoon. Will this result in a simple random sample of Cub fans who sit in the bleachers?

    (A) Yes, because each bleacher fan has the same chance of being selected.
    (B) Yes, but only if there is a single entrance to the bleachers.
    (C) Yes, because the 99 out of 100 bleacher fans who are not selected will form a control group.
    (D) Yes, because this is an example of systematic sampling, which is a special case of simple random sampling.
    (E) No, because not every sample of the intended size has an equal chance of being selected.

6. A researcher plans a study to examine the depth of belief in God among the adult population. He obtains a simple random sample of 100 adults as they leave church one Sunday morning. All but one of them agree to participate in the survey. Which of the following is a true statement?

(A) Proper use of chance as evidenced by the simple random sample makes this a well-designed survey.
(B) The high response rate makes this a well-designed survey.
(C) Selection bias makes this a poorly designed survey.
(D) The validity of this survey depends on whether or not the adults attending this church are representative of all churches.
(E) The validity of this survey depends upon whether or not similar numbers of those surveyed are male and female.

Free-Response Questions

> **Directions:** You must show all work and indicate the methods you use. You will be graded on the correctness of your methods and on the accuracy of your final answers.

1. A cable company plans to survey potential customers in a small city currently served by satellite dishes. Two sampling methods are being considered. Method A is to randomly select a sample of 25 city blocks and survey every family living on those blocks. Method B is to randomly select a sample of families from each of the five natural neighborhoods making up the city.

(a) What is the statistical name for the sampling technique used in Method A, and what is a possible reason for using it rather than an SRS?
(b) What is the statistical name for the sampling technique used in Method B, and what is a possible reason for using it rather than an SRS?

2. A questionnaire is being designed to determine whether most people are or are not in favor of sending U.S. troops to participate in U.N. peacekeeping operations. Give two examples of poorly worded questions, one biased toward each response.

3. Pelicans are considered an endangered species, and there is a need to protect them from becoming extinct. Unfortunately, the U.S. Fish & Wildlife Service is unable to protect all pelican colonies; however, it plans to protect those colonies identified as most important. One criterion in determining the importance of a geographical region is the density of nesting sites within it.

Given an aerial view of the nests in a geographical region used for pelican nesting, the service plans to divide the rectangular picture into 150 equal-size plots as below:

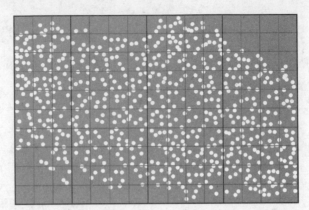

The service will randomly pick 15 plots (10% of the total), count the nests in each of those plots, sum and multiply by 10 to estimate the total number of nests in the region, and then calculate density. Two methods for counting the number of nests are under consideration.

Method I: Randomly pick one of the above rows and use all the 15 plots in that row.

Method II: Randomly pick one of the plots in each of the 15 columns to obtain the 15 plots.

(a) What sampling procedure is each method striving for, and how can each method be implemented?

(b) Which of these two methods is preferable here? Explain.

4. For each of the following survey designs, identify the problem and the effect it would have on the estimate of interest.

(a) Should spending be increased for the athletic program? A school official decides to sample spectators randomly at the next home basketball game.

(b) Do people believe that eating organic will lower the risk of cancer? A health magazine sends a questionnaire to all their subscribers.

(c) A supermarket quality control inspector opens boxes of strawberries. If no mold is seen among the visible strawberries, he accepts the box.

(d) Are the patients following their diet instructions? A family physician asks each of her patients during their office visits.

(e) Parents are asked to write to a newspaper with a "yes" or "no" response to whether or not they are happy with the education their children are receiving in the local public schools.

The answers for this quiz can be found in the Appendix on page 593.

EXPERIMENTS VERSUS OBSERVATIONAL STUDIES

In an experiment, we impose some change or treatment and measure the response. In an observational study, we simply observe and measure something that has taken place or is taking place while trying not to cause any changes by our presence. Experiments often suggest causal relationships, while observational studies can show only the existence of associations.

➡ **EXAMPLE 3.6** _____

A study is to be designed to determine whether a particular commercial review course actually helps raise SAT scores among students attending a particular high school. How could an observational study be performed? An experiment? Which is more appropriate here?

Answer: An observational study would interview students at the school who have taken the SAT exam, asking whether or not they took the review course, and then the SAT results of those who have and have not taken the review course would be compared.

An experiment performed on students at the school who are planning to take the SAT exam would use chance to randomly assign some students to take the review course while others to not take the course and would then compare the SAT results of those who have and have not taken the review course.

The experimental approach is more appropriate here. With the observational study, there could be many explanations for any SAT score difference noted between those who took the course and those who didn't. For example, students who took the course might be the more serious high school students. Higher score results might be due to their more serious outlook and have nothing to do with taking the course. The experiment tries to control for confounding variables by using random assignment to make the groups taking and not taking the course as similar as possible except for whether or not the individuals take the course.

_____

➡ **EXAMPLE 3.7** _____

A study is to be designed to examine the GPAs of students who study at least 3 hours every evening and those who don't. Which is more appropriate, an observational study or an experiment?

Answer: Requiring some students to cut their studying to under 3 hours every evening would be unethical. The proper procedure here is an observational study, having students anonymously fill out questionnaires asking about their study habits and GPA.

_____

Several primary principles deal with the proper planning and conducting of experiments. First, possible confounding variables are controlled by making conditions (other than the treatments) similar for all treatment groups. Second, chance should be used in the random assignment of subjects to treatment groups. Third, natural variation in outcomes can be lessened by using more subjects.

NOTE

Observational studies are generally less expensive and less time-consuming than experiments. Observational studies are often conducted first and suggest experiments to be performed next.

TIP

It is important to identify the *response variable* (SAT scores in this example) right away. This is critical in answering any experimental design question.

NOTE

Anytime we observe a relationship between two variables, there's always the possibility that some third variable also relates to the two variables and is responsible for "confounding" the apparent relationship between the two original variables. This important term will be more fully explained in the next section on the language of experiments.

NOTE

Recognizing when a study is an experiment versus when it is an observational study is a critical life skill!

TIP

You should always be asking whether or not you can reasonably draw cause-and-effect connections between explanatory and response variables.

TIP

If asked to list all the treatments (explanatory variables), don't create ones that aren't there!

NOTE

When there is more than one factor, the number of treatments is the product of the number of levels for each factor.

THE LANGUAGE OF EXPERIMENTS

An experiment is performed on objects called *experimental units*. If the units are people, they are called *subjects*. Experiments involve *explanatory variables*, called *factors*, that are believed to have an effect on *response variables*. A group is treated with some *level* of the explanatory variable, and the outcome of the response variable is measured.

➡ EXAMPLE 3.8

In an experiment to test exercise and blood pressure reduction, volunteers are randomly assigned to participate in either 0, 1, or 2 hours a day of exercise, 5 days a week, for 6 months. What are the explanatory and response variables, and what are the levels?

Answer: The explanatory variable, hours of exercise, is being implemented at three levels: 0, 1, and 2 hours a day. The response variable is not specified but could be the measured change in either systolic or diastolic blood pressure readings after 6 months.

Suppose the volunteers were further randomly assigned to follow either the DASH (Dietary Approaches to Stop Hypertension) or the TLC (Therapeutic Lifestyle Changes) diet for the 6 months. There would then be two factors, one with three levels and one with two levels, and a total of six treatments (DASH diet with 0 hours daily exercise, DASH with 1 hour exercise, DASH with 2 hours exercise, TLC diet with 0 hours daily exercise, TLC with 1 hour exercise, and TLC with 2 hours exercise).

An explanatory variable and another variable are *confounded* when their effects on a response variable cannot be distinguished. For example, in an experiment to determine if different fertilizers cause differences in plant growth, a horticulturist might decide to have many test plots using one or the other of the fertilizers, with equal numbers of sunny and shady plots for each fertilizer, so that fertilizer and sun are not confounded. However, if two fertilizers require different amounts of watering, it might be difficult to determine if the difference in fertilizers or the difference in watering is the real cause of observed differences in plant growth. In this case, fertilizer and water would be confounded.

➡ EXAMPLE 3.9

Many are concerned that vaping produces the so-called gateway effect. Studies have shown that young people who vape are more likely to go on to smoke cigarettes than young people who do not vape.

  (a) What are the explanatory and response variables?

  (b) Are these observational studies or experiments?

Some researchers suggest that a confounding variable might be that many people feel that smoking in general is "cool."

  (c) Explain what this means to be a confounding variable in the above context.

  (d) Explain how this confounding variable might lead to a higher or lower proportion of cigarette smokers among those who used vaping as young adults than among those who did not use vaping as young adults.

Answer:

(a) The explanatory variable is whether or not a young adult vapes, and the response variable is whether or not the person becomes a cigarette smoker later in life.

(b) These are observational studies. It would have been highly unethical to randomly select young adults and instruct them to vape as part of an experiment to see if they go on to cigarette smoking.

(c) People who think smoking is "cool" are more likely to vape as young adults and are more likely to smoke cigarettes later in life. So, it may be that it's not vaping that leads to cigarette smoking, but rather that thinking smoking is "cool" leads both to vaping as a young adult and to smoking cigarettes later in life.

(d) This confounding variable would lead to a higher proportion of cigarette smokers who vaped as young adults than who did not vape as young adults.

> **IMPORTANT**
>
> It is not enough simply to say that two variables are confounded. You must be able to explain the possible incorrect conclusion in context!

In an experiment, there is often a *control group* to determine if the treatment of interest has an effect. There are several types of control groups. A control group can be a collection of experimental units not given any treatment, or given the current treatment, or given a treatment with an inactive substance (a placebo). When a control group receives the current treatment or a placebo, these count as treatments if asked to enumerate the treatments.

Random assignment to the various treatment groups, including to the control group, can help reduce the problem posed by confounding variables. Random assignment to treatments, called *completely randomized design*, is critical in good experimental design. This is especially true if the subjects are not randomly selected, as is the case in most medical and drug experiments.

> **NOTE**
> The goal of random assignment is to create treatment groups that are similar in all respects except for the treatment imposed.

> **NOTE**
> Random assignment *minimizes*, but can't *eliminate*, the effect of possible confounding variables.

➡ **EXAMPLE 3.10**

Sixty patients, ages 5 to 12, all with common warts are enrolled in a study to determine if application of duct tape is as effective as cryotherapy in the treatment of warts. Subjects will receive either cryotherapy (liquid nitrogen applied to each wart for 10 seconds every 2 weeks) for 6 treatments or duct tape occlusion (applied directly to the wart) for 2 months. Describe a completely randomized design.

Answer: Assign each patient a number from 1 to 60. Use a random integer generator on a calculator to pick integers between 1 and 60, throwing away repeats, until 30 unique integers have been selected. (Or use a random number table, reading off two digits at a time, ignoring repeats, 00, and numbers over 60, until 30 unique numbers have been selected.) The 30 patients corresponding to the 30 selected integers will be given the cryotherapy treatment. (A third design would be to put the 60 names on identical slips of paper, put the slips in a hat, mix them well, and then pick out 30 slips, without replacement, with the corresponding names given cryotherapy.) The remaining 30 patients will receive the duct tape treatment. At the end of the treatment schedules, compare the proportion of each group that had complete resolution of the warts being studied.

> **NOTE**
> Random assignment allows us to make a valid comparison of treatments.

It is a fact that many people respond to any kind of perceived treatment. This is called the *placebo effect*. For example, when given a sugar pill after surgery but told that it is a strong pain reliever, many people feel immediate relief from their pain. *Blinding* occurs when the subjects don't know which of the different treatments (such as placebos) they are receiving. *Double-blinding* is when neither the subjects nor the *response evaluators* know who is receiving which treatment.

➡ **EXAMPLE 3.11** _____

There is a pressure point on the wrist that some doctors believe can be used to help control the nausea experienced following certain medical procedures. The idea is to place a band containing a small marble firmly on a patient's wrist so that the marble is located directly over the pressure point. Describe how a double-blind experiment might be run on 50 postoperative patients.

Answer: Assign each patient a number from 1 to 50. Use a random integer generator on a calculator to pick integers between 1 and 50, ignoring repeats, until 25 unique integers have been selected. (Or from a random number table read off two digits at a time, throwing away repeats, 00, and numbers over 50, until 25 unique numbers have been selected.) Put wristbands with marbles over the pressure point on the patients with these assigned numbers. (A third experimental design would be to put the 50 names on identical slips of paper, put the slips in a hat, mix them well, and then pick out 25 slips, without replacement, with the corresponding names given wristbands with marbles over the pressure point.) Put wristbands with marbles on the remaining patients also, but *not* over the pressure point. Have a researcher check by telephone with all 50 patients at designated time intervals to determine the degree of nausea being experienced. Neither the patients nor the researcher on the telephone should know which patients have the marbles over the correct pressure point.

A *matched pairs design* (also called a *paired comparison design*) is when two treatments are compared based on the responses of paired subjects, one of whom receives one treatment while the other receives the second treatment. Often the paired subjects are really single subjects who are given both treatments, one at a time in random order.

➡ **EXAMPLE 3.12** _____

Does seeing pictures of accidents caused by drunk drivers influence one's opinion on penalties for drunk drivers? How could a comparison test be designed for 100 subjects?

Answer: The subjects could be asked questions about drunk driving penalties before and then again after seeing the pictures, and any change in answers noted. This would be a poor design because there is no control group, there is no use of randomization, and subjects might well change their answers because they realize that that is what is expected of them after seeing the pictures.

A better design is to use randomization to split the subjects into two groups, half of whom simply answer the questions while the other half first see the pictures and then answer the questions. For example, put the 100 names on identical slips of paper, put the slips in a hat, mix them well, and then pick out 50 slips (without replacement) with the corresponding names designating the subjects who see the pictures.

Another possibility is to use a group of 50 biological twins as subjects. One of each set of twins is randomly picked (e.g., based on choosing an odd or even digit from a random number table) to answer the questions without seeing the pictures, while the other first sees the pictures and then answers the questions. The answers could be compared from each set of twins. This is a paired comparison test that might help minimize confounding variables due to family environment, heredity, and so on.

TIP

Use proper terminology! The language of experiments is different from the language of sampling. Don't mix up *blocking* and *stratification*.

Just as stratification is sampling that first divides the population into representative groups called *strata*, *blocking* in experimental design first divides the subjects into representative groups called *blocks*. Subjects within *each* block are randomly assigned treatments. One can think of blocking as running parallel experiments before combining the results. This technique helps control certain possible confounding variables by bringing them directly into the picture and helps make conclusions more specific. The paired comparison (matched pairs) design is a special case of blocking in which each pair can be considered a block of size 2.

➡ EXAMPLE 3.13

There is a rising trend for star college athletes to turn professional without finishing their degrees. A study is performed to assess whether reading an article about professional salaries has an impact on such decisions. Randomization can be used to split the subjects into two groups and those in one group given the article before answering questions. How can a block design be incorporated into the design of this experiment?

Answer: The subjects can be split into two blocks, underclass and upperclass, before using randomization to assign some to read the article before questioning. With this design, the impact of the salary article on freshmen and sophomores can be distinguished from the impact on juniors and seniors.

TIP

Treatments are not randomly assigned to blocks, and blocks are not formed at random. This experimental design uses randomization *within* blocks.

Similarly, blocking can be used to separately analyze men and women, those with high GPAs and those with low GPAs, those in different sports, those with different majors, and so on.

Completely Randomized Design

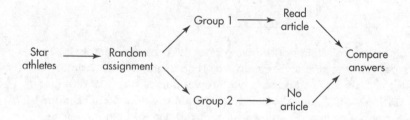

NOTE

When the results are combined, this estimates the effect, separate from blocks, of the independent variable of interest.

Block Design

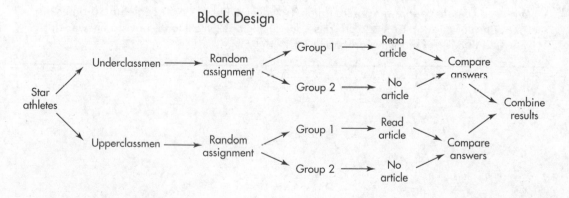

REPLICATION AND GENERALIZABILITY OF RESULTS

TIP

Replication refers to having more than one experimental unit in each treatment group, not multiple trials of the same experiment.

When differences are observed in a comparison test, the researcher must decide whether these differences are *statistically significant* or whether they can be explained by natural variation. One important consideration is the size of the sample: the larger the sample, the more significant the observation. This is the principle of *replication*. In other words, the treatment should be repeated on a sufficient number of subjects so that real response differences are more apparent.

Randomization, in the form of random assignment, is critical to minimize the effect of confounding variables. However, in order to generalize experimental results to a larger population (as we try to do in sample surveys), it would also be necessary that the group of subjects used in the experiment be randomly selected from the population. For example, it is hard to generalize from the effect a television commercial has on students at a private midwestern high school to the effect the same commercial has on retired senior citizens in Florida. If the experiment was performed on those high school students, it can be generalized only to the population of all students at private midwestern high schools.

TIP

Most experiments are not performed on random samples from a population. So, don't use *random sample* terminology when you mean *random assignment*.

INFERENCE AND EXPERIMENTS

What kind of conclusions result from well-designed experiments? Can we conclude that observed differences in the treatment groups are statistically significant; that is, are the differences so large as to be unlikely to have occurred by chance? How can we decide whether the differences are large enough?

If the treatments make no difference, just by chance there probably will be some variation. In later units, you will learn mathematically how to decide if something is different enough to be considered statistically significant. However, for now, you should understand that something is statistically significant if the probability of it happening just by chance is so small that you're convinced there must be another explanation.

➡ **EXAMPLE 3.14** _____

For a fair pair of dice, the probability of throwing "7 or 11" is approximately 22.2 percent. We would expect for the sum to come up 7 or 11 an average of 22.2 times for every 100 tosses if the dice are fair. Suppose we throw a pair of dice 100 times, and 7 or 11 comes up 26 times. This would appear to be within the bounds of random fluctuations, and we would have no reason to suspect the dice of being unfair. However, if 7 or 11 came up 55 times, that would raise our suspicions. While it would still be possible that the dice were fair, we could comfortably say that we had sufficient evidence to believe that they were unfair. However, what if 7 or 11 came up 30 times? Is that different enough from 22.2 to raise our suspicions? We'll learn how to answer this question later after studying probability distributions.

➡ EXAMPLE 3.15

Sixty students volunteered to participate in an experiment comparing the effects of coffee, caffeinated cola, and herbal tea on pulse rates. Twenty students are randomly assigned to each of the three treatments. For each student, the change in pulse rate was measured after consuming eight ounces of the treatment beverage. The results are summarized with the parallel boxplots below.

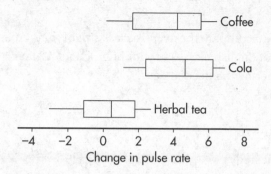

What are reasonable conclusions?

Answer: The median change in pulse rate for the cola drinkers was higher than that for the coffee drinkers; however, looking at the overall spreads, that observed difference does not seem significant. The difference between the coffee and caffeinated cola drinkers with respect to change in pulse rate is likely due to random chance. Now compare the coffee and caffeinated cola drinkers' results to that of the herbal tea drinkers. While there is some overlap, there is not much. It seems reasonable to conclude the difference is statistically significant; that is, drinking coffee or caffeinated cola results in a greater rise in pulse rate than drinking herbal tea.

QUIZ 12

Multiple-Choice Questions

> **Directions:** The questions that follow are each followed by five suggested answers. Choose the response that best answers the question.

1. Suppose you wish to compare the average class size of mathematics classes to the average class size of English classes in your high school. Which is the most appropriate technique for gathering the needed data?

   (A) Census
   (B) Sample survey
   (C) Experiment
   (D) Observational study
   (E) None of these methods is appropriate.

2. Two studies are run to compare the experiences of families living in high-rise public housing to those of families living in townhouse subsidized rentals. The first study interviews 25 families who have been in each government program for at least 1 year, while the second randomly assigns 25 families to each program and interviews them after 1 year. Which of the following is a true statement?

   (A) Both studies are observational studies because of the time period involved.
   (B) Both studies are observational studies because there are no control groups.
   (C) The first study is an observational study, while the second is an experiment.
   (D) The first study is an experiment, while the second is an observational study.
   (E) Both studies are experiments.

3. A study is made to determine whether taking AP Statistics in high school helps students achieve higher GPAs when they go to college. In comparing records of 200 college students, half of whom took AP Statistics in high school, it is noted that the average college GPA is higher for those 100 students who took AP Statistics than for those who did not. Based on this study, guidance counselors begin recommending AP Statistics for college-bound students. Which of the following is *incorrect?*

   (A) While this study indicates a relation, it does not prove causation.
   (B) There could well be a confounding variable responsible for the seeming relationship.
   (C) Self-selection here makes drawing the counselors' conclusion difficult.
   (D) A more meaningful study would be to compare an SRS from each of the two groups of 100 students.
   (E) This is an observational study, not an experiment.

4. When the estrogen-blocking drug tamoxifen was first introduced to treat breast cancer, there was concern that it would cause osteoporosis as a side effect. To test this concern, cancer subjects were randomly selected and given tamoxifen, and their bone density was measured before and after treatment. Which of the following is a true statement?

(A) This study was an observational study.
(B) This study was a sample survey of randomly selected cancer patients.
(C) This study was an experiment in which the subjects were used as their own controls.
(D) With the given procedure, there cannot be a placebo effect.
(E) Causation cannot be concluded without knowing the survival rates.

5. In designing an experiment, blocking is used

(A) to reduce bias.
(B) to reduce variation.
(C) as a substitute for a control group.
(D) as a first step in randomization.
(E) to control the level of the experiment.

Free-Response Questions

> **Directions:** You must show all work and indicate the methods you use. You will be graded on the correctness of your methods and on the accuracy of your final answers.

1. Rock and pop fame is associated with risk taking and premature mortality. A study followed the lives of 100 rock and pop stars, periodically asking them about their frequency of substance abuse, and followed their lives until death, at which time the age at death was noted. The researchers noted that the greater the reported substance abuse, the earlier the age at death.

(a) What are the explanatory and response variables?
(b) Is this a prospective observational study, a retrospective observational study, or an experiment? Explain.
(c) Does this study show that greater frequency of substance abuse leads to early death?

2. Explain how you would design an experiment to evaluate whether subliminal advertising (for example, flashing "BUY POPCORN" on the screen for a fraction of a second) results in more popcorn being sold in a movie theater. Show how you will incorporate comparison, randomization, and blinding.

3. A new vegetable fertilizer is to be tested at two different levels (regular concentration and double concentration). Design an experiment, including a control, for 30 test plots, half of which are in shade. Explain carefully how you will use randomization.

The answers for this quiz can be found in the Appendix on page 594.

QUIZ 13

Multiple-Choice Questions

> **Directions:** The questions that follow are each followed by five suggested answers. Choose the response that best answers the question.

1. Some researchers believe that too much iron in the blood can raise the level of cholesterol. The iron level in the blood can be lowered by making periodic blood donations. A study is performed by randomly selecting half of a group of volunteers to give periodic blood donations while the rest do not. Is this an experiment or an observational study?

   (A) An experiment with a single factor
   (B) An experiment with control group and blinding
   (C) An experiment with blocking
   (D) An observational study with comparison and randomization
   (E) An observational study with little, if any, bias

2. In a 1927–1932 Western Electric Company study on the effect of lighting on worker productivity, productivity increased with each increase in lighting but then also increased with every decrease in lighting. If it is assumed that the workers knew that they were being observed and that a study was in progress, this is an example of

   (A) the effect of a treatment unit.
   (B) the placebo effect.
   (C) the control group effect.
   (D) sampling error.
   (E) voluntary response bias.

3. Which of the following is *incorrect*?

   (A) Blocking is to experiment design as stratification is to sampling design.
   (B) By controlling certain variables, blocking can make conclusions more specific.
   (C) The paired comparison design is a special case of blocking.
   (D) Blocking results in increased accuracy because the blocks have smaller size than the original group.
   (E) In a randomized block design, the randomization occurs within the blocks.

4. A consumer product agency tests miles per gallon for a sample of automobiles using each of four different octanes of gasoline. Which of the following is true?

   (A) There are four explanatory variables and one response variable.
   (B) There is one explanatory variable with four levels of response.
   (C) Miles per gallon is the only explanatory variable, but there are four response variables corresponding to the different octanes.
   (D) There are four levels of a single explanatory variable.
   (E) Each explanatory level has an associated level of response.

5. Do teenagers prefer sports drinks colored blue or green? Two different colorings, which have no effect on taste, are used on an identical drink to result in either a blue or a green beverage. Volunteer teenagers are randomly assigned to drink one or the other colored beverage, and the volunteers then rate the beverage on a one to ten scale. Because of concern that sports interest may affect the outcome, the volunteers are first blocked by whether or not they play on a high school sports team. Is blinding possible in this experiment?

(A) No, because the volunteers know whether they are drinking a blue or a green drink.

(B) No, because the volunteers know whether or not they play on a high school sports team.

(C) Yes, by having the experimenter in a separate room randomly pick one of two containers and remotely having a drink poured from that container.

(D) Yes, by having the statistician analyzing the results not knowing which volunteer sampled which drink.

(E) Yes, by having the volunteers drink out of solid black thermoses so that they don't know the color of the drink they are tasting.

Free-Response Questions

> **Directions:** You must show all work and indicate the methods you use. You will be graded on the correctness of your methods and on the accuracy of your final answers.

1. A random sample of college graduates were asked whether or not they took afternoon power naps, and for how many minutes, and what were their GPAs at graduation. It was noted that, on average, the more minutes spent on power naps, the higher the GPA at graduation.

   (a) What are the explanatory and response variables?
   (b) Is this a prospective observational study, a retrospective observational study, or an experiment? Explain.
   (c) Does this study show that the more minutes spent on power naps, the higher a college student's GPA at graduation will be? Explain.

2. Explain how you would design an experiment to evaluate whether praying for a hospitalized heart attack patient leads to a speedier recovery. Show how you would incorporate comparison, randomization, and blinding.

3. The computer science department plans to offer three introductory-level CS courses: one using Python, one using C++, and one using Java.

(a) The department chairperson plans to give all students the same general programming exam at the end of the year and to compare the relative effectiveness of using each of the programming languages by comparing the mean grades of the students from each course. What is wrong, if anything, with the chairperson's plan?

(b) The chairperson also wishes to determine whether math majors or science majors do better in the courses. Suppose he calculates that the average grade of science majors was higher than the average grade of math majors in each of the courses. Does it follow that the average grade of all the science majors taking the three courses must be higher than the average grade of all the math majors? Explain.

(c) Suppose 300 students wish to take introductory programming. How would you randomly assign 100 students to each of the three courses?

(d) How would you randomly assign students to the three courses if you wanted the assignment to be independent from student to student with each student in turn having a one-third probability of taking each of the three classes?

(e) Name a confounding variable that all the above methods do not account for.

The answers for this quiz can be found in the Appendix on page 596.

SUMMARY

- Bias is the tendency to favor the selection of certain members of a population.
- Voluntary response bias occurs when individuals choose whether to respond (for example, Internet surveys) and typically give too much emphasis to people with strong opinions (particularly strong negative opinions).
- Convenience surveys, like interviews at shopping malls, are based on choosing individuals who are easy to reach. These surveys tend to produce data highly unrepresentative of the entire population.
- Use of randomization is absolutely critical in selecting an unbiased sample.
- Use of randomization in selecting a sample is crucial in being able to generalize from a sample to the population.
- A simple random sample (SRS) is one in which every possible sample of the desired size has an equal chance of being selected.
- Stratified sampling involves dividing the population into homogeneous groups called strata and then picking random samples from each of the strata.
- Cluster sampling involves dividing the population into heterogeneous groups called clusters, randomly selecting one or more of the clusters, and then using everyone in the selected clusters.
- Sampling error (sampling variability) is not an avoidable mistake but, rather, is the natural variability among samples.
- While observational studies involve observing without causing changes, experiments involve actively imposing treatments and observing responses.
- Differences in observed responses are statistically significant when it is unlikely that they can be explained by chance variation.
- Random assignment of subjects to treatment groups is extremely important in handling unknown and uncontrollable differences.
- Random assignment refers to what is done with subjects after they've been picked for a study, whereas random sampling refers to how subjects are selected for a study.
- Variables are said to be confounded when there is uncertainty as to which variable is causing an effect.
- The placebo effect refers to the fact that many people respond to any kind of perceived treatment.
- Blinding refers to subjects not knowing which treatment they are receiving.
- Double-blinding refers to subjects and those evaluating their responses not knowing who received which treatments.
- Completely randomized designs refer to experiments in which everyone has an equal chance of receiving any treatment.
- Blocking is the process of dividing the subjects into representative groups to bring certain differences into the picture (for example, blocking by gender, age, or race).
- Randomized block designs refer to experiments in which the randomization occurs only within blocks.
- Randomized paired comparison designs refer to experiments in which subjects are paired and randomization is used to decide who in each pair receives what treatment.

Probability, Random Variables, and Probability Distributions

4

(10–20% AP EXAM WEIGHTING)

In this unit, you will learn basic probability calculations and how to calculate and interpret means and standard deviations of random variables and of binomial and geometric probability distributions. You will further study patterns and uncertainty and see how an event can be both random and still have predictability.

UNIT LEARNING OBJECTIVES

- To understand how the law of large numbers relates to relative frequencies.
- To be able to make basic probability calculations involving the complement, union, and intersection of events, including the use of the general addition rule.
- To understand when events are mutually exclusive and when they are independent and how that applies to probability calculations.
- To be able to calculate conditional probabilities.
- To be able to calculate reverse conditional probabilities, using a tree diagram if appropriate (multistage probability calculations).

- To be able to calculate and interpret the mean (expected value) and standard deviation of a random variable.
- To be able to calculate and interpret the mean of the sum or difference of random variables.
- To be able to calculate and interpret the standard deviation of the sum or difference of independent random variables.
- To be able to calculate the probability of exactly x successes in n trials in a binomial scenario.
- To be able to calculate the probability of at least or at most x successes in n trials in a binomial scenario.
- To be able to calculate the mean (expected value) and standard deviation of a binomial probability distribution.
- To be able to calculate the probabilities of the possible values of a geometric random variable.
- To be able to calculate the mean (expected value) and standard deviation of a geometric probability distribution.

In the world around us, unlikely events sometimes take place. At other times, events that seem inevitable do not occur. Chance is everywhere! The cards you are dealt in a poker game, the particular genes you inherit from your parents, and the coin toss at the beginning of a tennis match to determine who serves first are examples of chance behavior that mathematics can help us understand. Even though we may not be able to foretell a specific result, we can sometimes assign what is called a *probability* to indicate the likelihood that a particular event will occur.

Probabilities are always between 0 and 1, with a probability close to 0 meaning that an event is unlikely to occur and a probability close to 1 meaning that the event is likely to occur. The sum of the probabilities of all the separate outcomes of an experiment is always 1.

THE LAW OF LARGE NUMBERS

The *relative frequency* of an event is the proportion of times the event happened, that is, the number of times the event happened divided by the total number of trials. Relative frequencies may change every time an experiment is performed. The *law of large numbers* states that when an experiment is performed a large number of times, the relative frequency of an event tends to become closer to the true probability of the event; that is, *probability is long-run relative frequency*.

The law of large numbers says nothing at all about short-run behavior. There is no such thing as a law of small numbers or a law of averages. Gamblers might say that "red" is due on a roulette table, a basketball player is due to make a shot, a player has a hot hand at the craps table, or a certain number is due to come up in a lottery, but if events are independent, the probability of the outcome of the next trial has nothing to do with what happened in previous trials. Even though casinos and life insurance companies might lose money in the short run, they make long-term profits because of their understanding of the law of large numbers.

The law of large numbers has two conditions. First, the chance event under consideration does not change from trial to trial. Second, any conclusion must be based on a large (a very large!) number of observations.

➡ EXAMPLE 4.1

There are two games involving flipping a fair coin. In the first game, you win a prize if you can throw between 45% and 55% heads. In the second game, you win if you can throw more than 60% heads. For each game, would you rather flip 20 times or 200 times?

Answer: The probability of throwing heads is 0.5. By the law of large numbers, the more times you throw the coin, the more the relative frequency tends to become closer to this probability. With fewer tosses, there is greater chance of wide swings in the relative frequency. Thus, in the first game, you would rather have 200 flips, whereas in the second game, you would rather have only 20 flips.

BASIC PROBABILITY RULES

➡ EXAMPLE 4.2

A standard literacy test consists of 100 multiple-choice questions, each with five possible answers. There is no penalty for guessing. A score of 60 is considered passing, and a score of 80 is considered superior. When an answer is completely unknown, test takers employ one of three strategies: guess, choose answer (c), or choose the longest answer. The table below summarizes the results of 1000 test takers.

Strategy	Score 0–59	Score 60–79	Score 80–100	
Guess	40	160	100	300
Choose answer (c)	35	170	115	320
Choose longest answer	45	200	135	380
	120	530	350	1000

Note that in analyzing tables such as the one above, it is usually helpful to sum rows and columns.

a. What is the probability that someone in this group uses the "guess" strategy?

   Answer: $P(guess) = \dfrac{300}{1000} = 0.3$

b. What is the probability that someone in this group scores 60–79?

   Answer: $P(score\ 60\text{–}79) = \dfrac{530}{1000} = 0.53$

c. What is the probability that someone in this group does not score 60–79?

   Answer: $P(does\ not\ score\ 60\text{–}79) = 1 - P(score\ 60\text{–}79) = 1 - 0.53 = 0.47$

[The probability that an event will not occur, that is, the probability of its complement, is equal to 1 minus the probability that the event will occur.]

d. What is the probability that someone in this group chooses strategy "answer (c)" *and* scores 80–100 (sometimes called the *joint probability* of the two events)?

   Answer: $P(answer(c) \cap score\ 80\text{–}100) = \dfrac{115}{1000} = 0.115$

e. What is the probability that someone in this group chooses strategy "longest answer" *or* scores 0–59?

   Answer: $P(longest\ answer \cup 0\text{–}59) = \dfrac{40+35+45+200+135}{1000} = \dfrac{455}{1000} = 0.455$

(continued)

NOTE

Two notations for the complement are A^C and A'. Then $P(A^C) = P(A') = 1 - P(A)$.

TIP

The word "and" here means *both* events.

TIP

The word "or" here means one event or the other event *or* both events.

NOTE

The subtraction of the intersection probability is so that it is not counted twice.

Note that

$$P(longest\ answer \cup score\ 0\text{-}59) =$$
$$P(longest\ answer) + P(score\ 0\text{-}59) - P(longest\ answer \cap score\ 0\text{-}59) =$$
$$0.380 + 0.120 - 0.045 = 0.455$$

[For any pair of events A and B, $P(A \cup B) = P(A) + P(B) - P(A \cap B)$.]

f. What is the probability that someone in this group chooses the strategy "guess" *given that* his or her score was 0-59?

Answer: $P(guess \mid score\ 0\text{-}59) = \dfrac{40}{120} \approx 0.333$

Note that we narrowed our attention to the 120 test takers who scored 0-59 to calculate this *conditional probability*.

TIP

Conditional probability shrinks the population of interest.

g. What is the probability that someone in this group scored 80-100 *given that* the person chose strategy "longest answer"?

Answer: $P(score\ 80\text{-}100 \mid longest\ answer) = \dfrac{135}{380} \approx 0.355$

Note that we narrowed our attention to the 380 test takers who chose "longest answer" to calculate this *conditional probability*.

[A formula for conditional probability is given by $P(A \mid B) = \dfrac{P(A \cap B)}{P(B)}$. This immediately gives the general multiplication rule: $P(A \cap B) = P(B) \cdot P(A \mid B)$ or $P(A \cap B) = P(A) \cdot P(B \mid A)$.]

TIP

If A and B are independent, we also have $P(A \mid not\ B) = P(A)$.

h. Are the strategy "guess" and scoring 0-59 *independent events*? That is, is whether a test taker used the strategy "guess" unaffected by whether the test taker scored 0-59?

Answer: We must check if $P(guess \mid score\ 0\text{-}59) = P(guess)$. From (f) and (a), we see that these probabilities are not equal ($0.333 \neq 0.3$), so the strategy "guess" and scoring 0-59 are *not* independent events.

[Events A and B are *independent* if $P(A \mid B) = P(A)$ or, equivalently, if $P(B \mid A) = P(B)$. It is also true that A and B are independent if and only if $P(A \cap B) = P(A)P(B)$, that is, if and only if the probability of both events happening is the product of their probabilities. Yet another insight into independence is that A and B are independent if and only if $P(A \mid B) = P(A \mid B^C)$.]

TIP

Don't multiply probabilities unless the events are independent.

TIP

Don't confuse independence with mutually exclusive.

i. Are the strategy "longest answer" and scoring 80-100 *mutually exclusive events*? That is, are these two events disjoint and cannot simultaneously occur?

Answer: $longest\ answer \cap score\ 80\text{-}100 \neq \varnothing$ and

$P(longest\ answer \cap score\ 80\text{-}100) = \dfrac{135}{1000} \neq 0$, so the strategy "longest answer" and scoring 80-100 are *not* mutually exclusive events.

Note that if two events are mutually exclusive, the probability that at least one event will occur is equal to the sum of the respective probabilities of the two events:

$$If\ P(A \cap B) = 0,\ then\ P(A \cup B) = P(A) + P(B)$$

TIP

Don't add probabilities unless the events are mutually exclusive.

Whether events are mutually exclusive or are independent are two very different properties! One refers to events being disjoint; the other refers to an event having no effect on whether or not the other occurs. Note that mutually exclusive (disjoint) events are *not* independent (except in the special case that one of the events has probability 0). That is, mutually exclusive gives that $P(A \cap B) = 0$, whereas independence gives that $P(A \cap B) = P(A)P(B)$ (and the only way these are ever simultaneously true is in the very special case when $P(A) = 0$ or $P(B) = 0$).

MULTISTAGE PROBABILITY CALCULATIONS

➡ **EXAMPLE 4.3** _____

On a university campus, 60%, 30%, and 10% of the computers use Windows, Apple, and Linux operating systems, respectively. A new virus affects 3% of the Windows, 2% of the Apple, and 1% of the Linux operating systems. What is the probability a computer on this campus has the virus?

Answer: In such problems, it is helpful to start with a tree diagram.

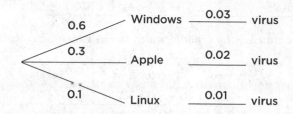

TIP

Tree diagrams can be very useful in working with conditional probabilities.

We then have:

$$P(Windows \cap virus) = (0.6)(0.03) = 0.018$$
$$P(Apple \cap virus) = (0.3)(0.02) = 0.006$$
$$P(Linux \cap virus) = (0.1)(0.01) = 0.001$$

$$P(virus) = P(Windows \cap virus) + P(Apple \cap virus) + P(Linux \cap virus)$$
$$= 0.018 + 0.006 + 0.001 = 0.025$$

We can take the above analysis one stage further and answer such questions as: If a randomly chosen computer on this campus has the virus, what is the probability it is a Windows machine? An Apple machine? A Linux machine?

$$P(Windows \mid virus) = \frac{P(Windows \cap virus)}{P(virus)} = \frac{0.018}{0.025} = 0.72$$

$$P(Apple \mid virus) = \frac{P(Apple \cap virus)}{P(virus)} = \frac{0.006}{0.025} = 0.24$$

$$P(Linux \mid virus) = \frac{P(Linux \cap virus)}{P(virus)} = \frac{0.001}{0.025} = 0.04$$

IMPORTANT

Naked or *bald* answers will receive little or *no* credit. You must show where answers come from.

A second way of analyzing Example 4.3 is to construct a table of counts for a hypothetical population. In this case, let's assume that the original given percentages hold exactly for a population of size 1000.

	Virus	No Virus	Total
Windows			
Apple			
Linux			
Total			1000

Numbers in parentheses indicate the order in which the cells can be calculated.

	Virus	No Virus	Total
Windows	(4) 18	(5) 582	(1) 600
Apple	(6) 6	(7) 294	(2) 300
Linux	(8) 1	(9) 99	(3) 100
Total	(10) 25	(11) 975	1000

Then the conditional probabilities in which we are interested can be calculated straight from the first column of the table:

$$P(Windows \mid virus) = \frac{18}{25} = 0.72$$

$$P(Apple \mid virus) = \frac{6}{25} = 0.24$$

$$P(Linux \mid virus) = \frac{1}{25} = 0.04$$

QUIZ 14

Multiple-Choice Questions

> **Directions:** The questions that follow are each followed by five suggested answers. Choose the response that best answers the question.

Questions 1–5 refer to the following study: One thousand students at a city high school were classified both according to GPA and whether or not they consistently skipped classes.

	GPA			
	<2.0	2.0–3.0	>3.0	
Many skipped classes	80	25	5	110
Few skipped classes	175	450	265	890
	255	475	270	

1. What is the probability that a student has a GPA between 2.0 and 3.0?

(A) $\dfrac{25}{110}$

(B) $\dfrac{450}{890}$

(C) $\dfrac{25}{1,000}$

(D) $\dfrac{450}{1,000}$

(E) $\dfrac{475}{1,000}$

2. What is the probability that a student has a GPA under 2.0 and has skipped many classes?

(A) $\dfrac{80}{110}$

(B) $\dfrac{80}{255}$

(C) $\dfrac{80}{1,000}$

(D) $\dfrac{110 + 255}{1,000}$

(E) $\dfrac{110 + 255 - 80}{1,000}$

3. What is the probability that a student has a GPA under 2.0 or has skipped many classes?

(A) $\dfrac{80}{110}$

(B) $\dfrac{80}{255}$

(C) $\dfrac{80}{1,000}$

(D) $\dfrac{110 + 255}{1,000}$

(E) $\dfrac{110 + 255 - 80}{1,000}$

4. What is the probability that a student has a GPA under 2.0 given that he has skipped many classes?

(A) $\dfrac{80}{110}$

(B) $\dfrac{80}{255}$

(C) $\dfrac{110}{255}$

(D) $\dfrac{80}{1,000}$

(E) $\dfrac{110}{1,000}$

5. Are "GPA between 2.0 and 3.0" and "skipped few classes" independent?

(A) No, because $0.475 \neq 0.506$.
(B) No, because $0.475 \neq 0.890$.
(C) No, because $0.450 \neq 0.475$.
(D) Yes, because of conditional probabilities.
(E) Yes, because of the product rule.

6. The following data are from *The Commissioner's Standard Ordinary Table of Mortality:*

Age	Number Surviving
0	10,000,000
20	9,664,994
40	9,241,359
70	5,592,012

What is the probability that a 20-year-old will survive to be 70?

(A) $\dfrac{5,592,012}{9,664,994}$

(B) $\dfrac{5,592,012}{10,000,000}$

(C) $\dfrac{9,664,994}{10,000,000}$

(D) $1 - \dfrac{5,592,012}{9,664,994}$

(E) $1 - \dfrac{\dfrac{5,592,012}{9,664,994}}{10,000,000}$

7. In a 1974 "Dear Abby" letter, a woman lamented that she had just given birth to her eighth child and all were girls! Her doctor had assured her that the chance of the eighth child being a girl was less than 1 in 100. What was the real probability that the eighth child would be a girl?

(A) 0.0039
(B) 0.5
(C) $(0.5)^7$
(D) $(0.5)^8$
(E) $\dfrac{(0.5)^7 + (0.5)^8}{2}$

8. There are two games involving flipping a fair coin. In the first game, you win a prize if you can throw between 40% and 60% heads. In the second game, you win if you can throw more than 75% heads. For each game, would you rather flip the coin 50 times or 500 times?

(A) 50 times for each game
(B) 500 times for each game
(C) 50 times for the first game, and 500 for the second
(D) 500 times for the first game, and 50 for the second
(E) The outcomes of the games do not depend on the number of flips.

9. Suppose that, for any given year, the probabilities that the stock market declines, that women's hemlines are lower, and that both events occur are, respectively, 0.4, 0.35, and 0.3. Are the two events independent?

(A) Yes, because $(0.4)(0.35) \neq 0.3$.

(B) No, because $(0.4)(0.35) \neq 0.3$.

(C) Yes, because $0.4 > 0.35 > 0.3$.

(D) No, because $0.5(0.3 + 0.4) = 0.35$.

(E) There is insufficient information to answer this question.

10. Suppose that in a certain part of the world, in any 50-year period the probability of a major plague is 0.39, the probability of a major famine is 0.52, and the probability of both a plague and a famine is 0.15. What is the probability of a famine given that there is a plague?

(A) $0.39 - 0.15$

(B) $\dfrac{0.15}{0.52}$

(C) $0.52 - 0.15$

(D) $\dfrac{0.15}{0.39}$

(E) $0.39 + 0.52 - 0.15$

11. Suppose that 2% of a clinic's patients are known to have cancer. A blood test is developed that is positive in 98% of patients with cancer but is also positive in 3% of patients who do not have cancer. If a person who is chosen at random from the clinic's patients is given the test and it comes out positive, what is the probability that the person actually has cancer?

(A) 0.02

(B) $0.02 + 0.03$

(C) $(0.02)(0.98)$

(D) $(0.02)(0.98) + (0.98)(0.03)$

(E) $\dfrac{(0.02)(0.98)}{(0.02)(0.98) + (0.98)(0.03)}$

Free-Response Questions

1. A sample of applicants for a management position yields the following numbers with regard to age and experience:

	Years of experience		
	0–5	6–10	>10
Less than 50 years old	80	125	20
More than 50 years old	10	75	50

   (a) What is the probability that a randomly picked applicant from this sample is less than 50 years old? Has more than 10 years of experience? Is more than 50 years old and has five or fewer years of experience?

   (b) What is the probability that a randomly picked applicant from this sample is less than 50 years old given that she has between 6 and 10 years of experience?

   (c) Are the two events "less than 50 years old" and "more than 10 years of experience" independent events? How about the two events ""more than 50" and "between 6 and 10 years of experience"? Explain.

2. Explain what is wrong with each of the following statements:

   (a) The probability that a student will score high on the AP Statistics exam is 0.43, while the probability that she will not score high is 0.47.

   (b) The probability that a student plays tennis is 0.18, while the probability that he plays basketball is six times as great.

   (c) The probability that a student enjoys her English class is 0.64, while the probability that she enjoys both her English and her social studies classes is 0.71.

   (d) The probability that a student will be accepted by his first choice for college is 0.38, while the probability that he will be accepted by his first or second choice is 0.32.

   (e) The probability that a student fails AP Statistics and will still be accepted by an Ivy League school is −0.17.

The answers for this quiz can be found in the Appendix on page 597.

QUIZ 15

Multiple-Choice Questions

> **Directions:** The questions that follow are each followed by five suggested answers. Choose the response that best answers the question.

Questions 1–4 refer to the following study: Five hundred people used a home test for HIV, and then all underwent more conclusive hospital testing. The accuracy of the home test was evidenced in the following table.

	HIV	Healthy	
Positive test	35	25	60
Negative test	5	435	440
	40	460	

1. What is the *predictive value* of the test? That is, what is the probability that a person has HIV and tests positive?

   (A) $\dfrac{35}{500}$

   (B) $\dfrac{35}{60}$

   (C) $\dfrac{35}{40}$

   (D) $\dfrac{35}{35+25+5}$

   (E) $\dfrac{35+25+5}{500}$

2. What is the *false-positive* rate? That is, what is the probability of testing positive given that the person does not have HIV?

   (A) $\dfrac{25}{460}$

   (B) $\dfrac{25}{60}$

   (C) $\dfrac{35}{40}$

   (D) $\dfrac{25}{500}$

   (E) $\dfrac{35+25+5}{500}$

3. What is the *sensitivity* of the test? That is, what is the probability of testing positive given that the person has HIV?

(A) $\dfrac{35 + 25 + 5}{500}$

(B) $\dfrac{35}{35 + 25 + 5}$

(C) $\dfrac{35}{500}$

(D) $\dfrac{35}{60}$

(E) $\dfrac{35}{40}$

4. What is the *specificity* of the test? That is, what is the probability of testing negative given that the person does not have HIV?

(A) $\dfrac{5}{40}$

(B) $\dfrac{35}{60}$

(C) $\dfrac{35 + 5 + 435}{500}$

(D) $\dfrac{435}{500}$

(E) $\dfrac{435}{460}$

5. Suppose you toss a fair coin ten times and it comes up heads every time. Which of the following is a true statement?

(A) By the law of large numbers, the next toss is more likely to be tails than another heads.

(B) By the properties of conditional probability, the next toss is more likely to be heads given that ten tosses in a row have been heads.

(C) Coins actually do have memories, and thus what comes up on the next toss is influenced by the past tosses.

(D) The law of large numbers tells how many tosses will be necessary before the percentages of heads and tails are again in balance.

(E) The probability that the next toss will again be heads is 0.5.

6. Mathematically speaking, casinos and life insurance companies make a profit because of

(A) their understanding of sampling error and sources of bias.

(B) their use of well-designed, well-conducted surveys and experiments.

(C) their use of simulation of probability distributions.

(D) the central limit theorem.

(E) the law of large numbers.

7. Given that 52% of the U.S. population are female and 15% are older than age 65, can we conclude that $(0.52)(0.15) = 7.8\%$ are women older than age 65?

   (A) Yes, by the multiplication rule.
   (B) Yes, by conditional probabilities.
   (C) Yes, by the law of large numbers.
   (D) No, because the events are not independent.
   (E) No, because the events are not mutually exclusive.

8. A city water supply system involves three pumps, the failure of any one of which crashes the system. The probabilities of failure for each pump in a given year are 0.025, 0.034, and 0.02, respectively. Assuming the pumps operate independently of each other, what is the probability that the system does crash during the year?

   (A) $0.025 + 0.034 + 0.02$
   (B) $1 - (0.025 + 0.034 + 0.02)$
   (C) $1 - (0.025)(0.034)(0.02)$
   (D) $(1 - 0.025)(1 - 0.034)(1 - 0.02)$
   (E) $1 - (1 - 0.025)(1 - 0.034)(1 - 0.02)$

9. Can the function $f(x) = \frac{x+6}{24}$, for $x = 1, 2,$ and 3, be the probability distribution for some random variable taking the values 1, 2, and 3?

   (A) Yes.
   (B) No, because probabilities cannot be negative.
   (C) No, because probabilities cannot be greater than 1.
   (D) No, because the probabilities do not sum to 1.
   (E) Not enough information is given to answer the question.

10. Consider the following table of ages of U.S. senators:

Age (yr):	<40	40–49	50–59	60–69	70–79	>79
Number of senators:	5	30	36	22	5	2

What is the probability that a senator is under 70 years old given that he or she is at least 50 years old?

   (A) 0.580
   (B) 0.624
   (C) 0.643
   (D) 0.892
   (E) 0.969

11. Suppose that 60% of students who take the AP Statistics exam score 4 or 5, 25% score 3, and the rest score 1 or 2. Suppose further that 95% of those scoring 4 or 5 receive college credit, 50% of those scoring 3 receive such credit, and 4% of those scoring 1 or 2 receive credit. If a student who is chosen at random from among those taking the exam receives college credit, what is the probability that she scored 3 on the exam?

(A) 0.25

(B) $(0.25)(0.50)$

(C) $(0.60)(0.95)$

(D) $\dfrac{0.25}{(0.60)(0.95)+(0.25)(0.50)+(0.15)(0.04)}$

(E) $\dfrac{(0.25)(0.50)}{(0.60)(0.95)+(0.25)(0.50)+(0.15)(0.04)}$

Free-Response Questions

Directions: You must show all work and indicate the methods you use. You will be graded on the correctness of your methods and on the accuracy of your final answers.

1. Last year, the acceptance rates at Stanford and MIT were 5% and 7%, respectively.

   (a) A guidance counselor calculates the probability of a student getting into both of these colleges to be $(0.05)(0.07) = 0.0035$. What assumption is she making, and does it seem reasonable?

   (b) She also calculates the probability of a student getting into at least one of these colleges to be $0.05 + 0.07 = 0.12$. What assumption is she making here, and does it seem reasonable?

2. Assume there is no overlap between the 56% of the population who wear glasses and the 4% who wear contacts. And assume 55% of those who wear glasses are women and 63% of those who wear contacts are women. What is the probability that the next person you encounter on the street will be

   (a) a woman with glasses?

   (b) a woman with contacts?

   (c) a man with glasses?

   (d) a man with contacts?

   (e) a person not wearing glasses or contacts?

The answers for this quiz can be found in the Appendix on page 599.

RANDOM VARIABLES, MEANS (EXPECTED VALUES), AND STANDARD DEVIATIONS

Often each outcome of an experiment has not only an associated probability but also an associated *real number*. For example, the probability may be 0.5 that a student is taking 0 AP classes, 0.3 that she is taking 1 AP class, and 0.2 that she is taking 2 AP classes. If X represents the different numbers associated with the potential outcomes of some chance situation, we call X a random variable.

NOTE

The expected value is sometimes referred to as a *weighted average*.

While the mean of the set $\{2, 7, 12, 15\}$ is $\dfrac{2+7+12+15}{4} = 9$, or $2(\frac14) + 7(\frac14) + 12(\frac14) + 15(\frac14) = 9$, the *expected value* or mean of a random variable takes into account that the various outcomes may not be equally likely. The expected value or mean of a random variable X is given by $\mu_X = E(X) = \Sigma x P(x)$, where $P(x)$ is the probability of outcome x. [We also write $\mu_X = \Sigma x_i p_i$.]

➡ **EXAMPLE 4.5** _____

A charity holds a lottery in which 10,000 tickets are sold at $1 each and with a prize of $7500 for one winner. What is the average result for each ticket holder?

Answer: The actual winning payoff is $7499 because the winner paid $1 for a ticket, so we have:

Outcome	Probability	Random Variable
Win	$\dfrac{1}{10{,}000}$	7499
Lose	$\dfrac{9{,}999}{10{,}000}$	−1

IMPORTANT

The expected value of a random variable does not, *in any way*, imply that the value is expected to occur.

$$\text{Expected value} = \sum x P(x) = 7499\left(\frac{1}{10{,}000}\right) + (-1)\left(\frac{9{,}999}{10{,}000}\right) = -0.25$$

Thus, the *average* result for each ticket holder is a $0.25 loss. [Alternatively, we can say that the expected payoff to the charity is $0.25 for each ticket sold.]

Note that the expected value (−$0.25) is not the most likely outcome (−$1). In fact, in this example, the expected value (−$0.25) is not expected at all as it will never occur. The expected value is simply a long-run average.

_____

We have seen that the mean of a random variable is $\Sigma x_i p_i$. In addition to calculating the mean, we would like to measure the variability for the values taken on by a random variable. Since we are dealing with chance events, the proper tool is variance (along with standard deviation). In Unit 1, *variance* was defined to be the mean average of the squared deviations $(x - \mu)^2$. If we regard the $(x - \mu)^2$ terms as the values of some random variable (whose probability is the same as the probability of x), the mean of this new random variable is $\Sigma(x_i - \mu_x)^2 p_i$, which is precisely how we define the variance, σ^2, of a random variable:

NOTE

It is also true that

var(x) =

$\left[\sum x_i^2 p_i\right] - \mu_x^2$.

$$\text{var}(X) = \sigma_x^2 = \sum (x_i - \mu_x)^2 p_i$$

As before, the standard deviation σ is the square root of the variance.

⇒ EXAMPLE 4.6

A highway engineer knows that the workers can lay 5 miles of highway on a clear day, 2 miles on a rainy day, and only 1 mile on a snowy day. Suppose the probabilities are as follows:

Outcome:	Clear	Rain	Snow
Probability:	0.6	0.3	0.1
Random Variable (Miles of Highway):	5	2	1

What are the mean (expected value), standard deviation, and variance?

Answer:

$$\mu_X = \sum x_i p_i = 5(0.6) + 2(0.3) + 1(0.1) = 3.7$$

$$\sigma_X^2 = \sum (x_i - \mu_X)^2 p_i = (5 - 3.7)^2(0.6) + (2 - 3.7)^2(0.3) + (1 - 3.7)^2(0.1) = 2.61$$

$$\sigma_X = \sqrt{2.61} = 1.62$$

TIP

In a problem like this, there is no shortcut. You must show your work. However, the calculator shortcut of using two lists and running `1-Variable` statistics can be useful for checking your answers.

NOTE

Alternatively, you can calculate:
$$\sigma_X^2 = 5^2(0.6)$$
$$+ 2^2(0.3)$$
$$+ 1^2(0.1)$$
$$- 3.7^2 = 2.61$$

MEANS AND VARIANCES FOR SUMS AND DIFFERENCES OF SETS

In many experiments, a pair of numbers is associated with each outcome. This situation leads to a study of pairs of random variables. When the random variables are independent, there is an easy calculation for finding both the mean and standard deviation of the sum (and difference) of the two random variables.

⇒ EXAMPLE 4.7

A casino offers couples at the hotel free chips by allowing each person to reach into a bag and pull out a card stating the number of free chips. One person's bag has a set of four cards with $X = \{1, 9, 20, 74\}$; the other person's bag has a set of three cards with $Y = \{5, 15, 55\}$. Note that the mean of set X is $\mu_X = \dfrac{1 + 9 + 20 + 74}{4} = 26$ and that of set Y is $\mu_Y = \dfrac{5 + 15 + 55}{3} = 25$. What is the average amount a couple should be able to pool together?

Answer: Form the set W of sums: $W = \{1 + 5, 1 + 15, 1 + 55, 9 + 5, 9 + 15, 9 + 55,$ $20 + 5, 20 + 15, 20 + 55, 74 + 5, 74 + 15, 74 + 55\} = \{6, 16, 56, 14, 24, 64, 25, 35, 75,$ $79, 89, 129\}$ and

$$\mu_W = \mu_{X+Y} = \frac{6 + 16 + 56 + 14 + 24 + 64 + 25 + 35 + 75 + 79 + 89 + 129}{12} = 51$$

Finally, $\mu_W = \mu_{X+Y} = \mu_X + \mu_Y = 26 + 25 = 51$. On average, a couple should be able to pool together 51 free chips.

In general, the mean of a set of sums is equal to the sum of the means of the two original sets. Even more generally, if a sum is formed by picking one element from each of several sets, the mean of such sums is simply the sum of the means of the various sets.

(continued)

How is the variance of W related to the variances of the original sets?

$$\sigma^2_X = \frac{(1-26)^2 + (9-26)^2 + (20-26)^2 + (74-26)^2}{4} = 813.50$$

$$\sigma^2_Y = \frac{(5-25)^2 + (15-25)^2 + (55-25)^2}{3} = 466.67$$

$$\sigma^2_W = \sigma^2_{X+Y} = \frac{(6-51)^2 + (16-51)^2 + \cdots + (129-51)^2}{12} = 1280.17$$

Note that $\sigma^2_W = \sigma^2_{X+Y} = \sigma^2_X + \sigma^2_Y$.

This is true for the variance of any set of sums. More generally, if a sum is formed by picking one element from each of several sets, the variance of such sums is simply the sum of the variances of the various sets.

TIP

Remember that you cannot add standard deviations!

We can also calculate the standard deviation $\sigma_W = \sqrt{1280.17} = 35.78$ and conclude that the average pooled number of chips received by a couple is 51 with a standard deviation of 35.78.

➡ EXAMPLE 4.8

In Example 4.7, what is the average difference in number of chips received from bag X and bag Y?

Answer: Form the set of differences: $Z = \{1-5, 1-15, 1-55, 9-5, 9-15, 9-55, 20-5, 20-15, 20-55, 74-5, 74-15, 74-55\} = \{-4, -14, -54, 4, -6, -46, 15, 5, -35, 69, 59, 19\}$.

A quick calculation gives $\mu_Z = 1$. Note that $\mu_Z = \mu_{X-Y} = \mu_X - \mu_Y = 26 - 25 = 1$. In general, the mean of a set of differences is equal to the difference of the means of the two original sets.

How is the variance of Z related to the variance of the original sets?

A calculation (for example, using 1-Var Stats on the TI-84) yields $\sigma^2_Z = 1280.17$. Interestingly, the set of differences has the same variance as the set of sums. This relationship is true for any two sets. We have $\sigma^2_Z = \sigma^2_{X-Y} = \sigma^2_X + \sigma^2_Y$. More generally, if a total is formed by a procedure that adds or subtracts one element from each of several independent sets, the variance of the resulting totals is simply the sum of the variances of the several sets.

Not only can we sum variances as shown above to calculate total variance, but we can also reverse the process and determine how the total variance is split up among its various sources. For example, we can find what portion of the variance in numbers of sales made by a company's sales representatives is due to the individual salesperson, what portion is due to the territory, what portion is due to the particular products sold by each salesperson, and so on.

EXAMPLE 4.9

A company's sports shoes sell for an average of $100 with a standard deviation of $9. The costs going into their sports shoes are as follows:

Retail store costs and profits: average of $45 with a standard deviation of $5.
Brand company costs and profits: average of $24 with a standard deviation of $4.
Advertising and publicity costs: average of $11 with a standard deviation of $3.

Assuming these various costs are independent of one another, what is the mean and standard deviation for the remaining costs, R, together (which includes materials, transportation, factory worker wages, etc.)?

Answer: $45 + 24 + 11 + \mu_R = 100$ gives $\mu_R = \$20$; and $5^2 + 4^2 + 3^2 + (\sigma_R)^2 = 9^2$ gives $(\sigma_R)^2 = 31$ and $\sigma_R = \sqrt{31} = \$5.57$.

MEANS AND VARIANCES FOR SUMS AND DIFFERENCES OF RANDOM VARIABLES

The mean of the sum (or difference) of two random variables is equal to the sum (or difference) of the individual means. *If two random variables are independent*, the variance of the sum (or difference) of the two random variables is equal to the *sum* of the two individual variances.

> **IMPORTANT**
>
> Variances of independent random variables can be added together, but standard deviations cannot!

EXAMPLE 4.10

An insurance salesperson estimates the numbers of new auto and home insurance policies she sells per day as follows:

# of auto policies	0	1	2	3
Probability	0.2	0.4	0.3	0.1

# of home policies	0	1	2
Probability	0.5	0.3	0.2

What is the expected value or mean for the overall number of policies sold per day?

Answer: $\mu_{auto} = (0)(0.2) + (1)(0.4) + (2)(0.3) + (3)(0.1) = 1.3,$

$\mu_{home} = (0)(0.5) + (1)(0.3) + (2)(0.2) = 0.7$, and so

$\mu_{total} = \mu_{auto} + \mu_{home} = 1.3 + 0.7 = 2.0.$

Assuming the selling of new auto policies is independent of the selling of new home policies (which may not be true if some new customers buy both), what would be the standard deviation in the number of new policies sold per day?

Answer: $\sigma^2_{auto} = (0 - 1.3)^2(0.2) + (1 - 1.3)^2(0.4) + (2 - 1.3)^2(0.3) + (3 - 1.3)^2(0.1) = 0.81,$

$\sigma^2_{home} = (0 - 0.7)^2(0.5) + (1 - 0.7)^2(0.3) + (2 - 0.7)^2(0.2) = 0.61$, and so,

assuming independence, $\sigma^2_{total} = \sigma^2_{auto} + \sigma^2_{home} = 0.81 + 0.61 = 1.42$ and

$\sigma_{total} = \sqrt{1.42} = 1.192.$

➡ EXAMPLE 4.11

Vitamin C megadoses are claimed by alternative medicine advocates to have effects on diseases such as cancer and AIDS as well as on the common cold. In one long-term study, participants received either a placebo or a 200 mg vitamin C tablet each day. Suppose that the mean compliance rate was 73% with a standard deviation of 24%. Furthermore, the daily intake of vitamin C through foods among the adult population is 94 g with a standard deviation of 39 g. Among those receiving the supplement, what is the mean and standard deviation for their daily vitamin C intake?

Answer: With a compliance of 73%, the mean intake of vitamin C from the supplement is $(0.73)(200) = 146$ with a standard deviation of $(0.24)(200) = 48$. Thus, the total mean intake of vitamin C is $146 + 94 = 240$ g with a standard deviation of $\sqrt{48^2 + 39^2} = 61.85$ g (with independence, variances add!).

➡ EXAMPLE 4.12

Suppose the mean SAT Math and Evidence-Based Reading + Writing scores for students at a particular high school are 515 and 501, respectively, with standard deviations of 65 and 40, respectively. What can be said about the mean and standard deviation of the two combined scores for these students?

Answer: The mean combined score is $515 + 501 = 1016$. Can we add the two variances, $65^2 + 40^2$, and then take the square root to find the standard deviation? No, because we don't have independence—students with high scores on one of the two sections will tend to have high scores on the other section. The standard deviation of the combined scores cannot be calculated from the given information.

TRANSFORMING RANDOM VARIABLES

NOTE

The results of adding and multiplying by constants is the same as what we observed in Unit 1 about sets of data.

Adding a constant to every value of a random variable will increase the mean by that constant. However, differences between values remain the same, so measures of variability like the standard deviation will remain unchanged. Multiplying every value of a random variable by a constant will increase the mean by the same multiple. In this case, differences are also increased and the standard deviation will increase by the same multiple.

➡ EXAMPLE 4.13

A carnival game of chance has payoffs of $2 with probability 0.5, of $5 with probability 0.4, and of $10 with probability 0.1. What should you be willing to pay to play the game? Calculate the mean and standard deviation for the winnings if you play this game.

$$\mu_X = E(X) = \sum xP(x) = 2(0.5) + 5(0.4) + 10(0.1) = 4$$

$$\sigma_X = \sqrt{\sum (x-m)^2 P(x)} = \sqrt{(2-4)^2(0.5) + (5-4)^2(0.4) + (10-4)^2(0.1)} = \sqrt{6}$$

What if $4 is added to each payoff?

$$\mu_{X+4} = 6(0.5) + 9(0.4) + 14(0.1) = 8$$

$$\sigma_{X+4} = \sqrt{(6-8)^2(0.5) + (9-8)^2(0.4) + (14-8)^2(0.1)} = \sqrt{6}$$

Note that $\mu_{X+4} = \mu_X + 4$ and $\sigma_{X+4} = \sigma_X$.

What if instead each payoff is tripled?

$$\mu_{3X} = 6(0.5) + 15(0.4) + 30(0.1) = 12$$

$$\sigma_{3X} = \sqrt{(6-12)^2(0.5) + (15-12)^2(0.4) + (30-12)^2(0.1)} = \sqrt{54} = 3\sqrt{6}$$

Note that $\mu_{3X} = 3\mu_X$ and $\sigma_{3X} = 3\sigma_X$.

For random variables X and Y, the generalizations for transforming X and for combining X and Y are sometimes written as:

Transforming X:
$$E(X \pm a) = E(X) \pm a \qquad \text{var}(X \pm a) = \text{var}(X) \qquad \text{SD}(X \pm a) = \text{SD}(X)$$
$$E(bX) = bE(X) \qquad\qquad \text{var}(bX) = b^2\,\text{var}(X) \qquad \text{SD}(bX) = |b|\,\text{SD}(X)$$

Combining X and Y:
$$E(X \pm Y) = E(X) \pm E(Y)$$

$$\text{var}(X \pm Y) = \text{var}(X) + \text{var}(Y) \text{ if } X \text{ and } Y \text{ are independent}$$

$$\text{SD}(X \pm Y) = \sqrt{[\text{SD}(X)]^2 + [\text{SD}(Y)]^2} \text{ if } X \text{ and } Y \text{ are independent}$$

Does it make sense that variances *add* for both sums and differences of independent random variables? *Ranges* may illustrate this point more clearly. Suppose prices for primary residences in a community go from $210,000 to $315,000, while prices for summer cottages go from $50,000 to $120,000. Someone wants to purchase a primary residence and a summer cottage. The minimum possible *total* spent is $260,000, and the maximum possible is $435,000. Thus, the range for the total is $435,000 − $260,000 = $175,000. The minimum *difference* spent between the two homes will be $90,000, and the maximum difference will be $265,000. Thus, the range for the difference is $265,000 − $90,000 = $175,000, the same as the range for the total. (Variances and ranges are both measures of spread and behave the same way with regard to the above rule.)

As can be seen here, the rules from algebra do not apply when combining random variables. For example, with random variables, $2X \neq X + X$. From the formulas, we have $E(2X) = 2E(X)$ and $\text{SD}(2X) = 2\text{SD}(X)$; however, while $E(X + X) = E(X) + E(X) = 2E(X)$:

$$\text{SD}(X + X) = \sqrt{[\text{SD}(X)]^2 + [\text{SD}(X)]^2} = \sqrt{2[\text{SD}(X)]^2} = \sqrt{2}\,\text{SD}(X)$$

➡ EXAMPLE 4.14 _____

Let X be what shows when rolling a single die. What are the mean and the standard deviation of X, $2X$, and $X + X$?

X takes the values {1, 2, 3, 4, 5, 6}, each with probability $\frac{1}{6}$, $\mu_X = \sum xP(x) = 3.5$, and $\sigma_X = \sqrt{\sum (x - \mu)^2 P(x)} = 1.7078$.

$2X$ takes the values {2, 4, 6, 8, 10, 12}, each with probability $\frac{1}{6}$, $\mu_{2X} = 7$, and $\sigma_{2X} = 3.4156$. (Note that $\mu_{2X} = 2\mu_X$ and $\sigma_{2X} = 2\sigma_X$.)

But $X + X$ is throwing a die twice and takes the values {2, 3, 4, 5, 6, 7, 8, 9, 10, 11, 12} with probabilities $\left\{ \frac{1}{36}, \frac{2}{36}, \frac{3}{36}, \frac{4}{36}, \frac{5}{36}, \frac{6}{36}, \frac{5}{36}, \frac{4}{36}, \frac{3}{36}, \frac{2}{36}, \frac{1}{36} \right\}$, respectively.

Then $\mu_{X+X} = 7$ and $\sigma_{X+X} = 2.4152$. (Note that $\mu_{X+X} = 2\mu_X$ and $\sigma_{X+X} = \sqrt{2}\sigma_X$.)

$2X$ and $X + X$ are not the same thing, and while they do have the same mean, their standard deviations are different.

_____

More generally, with n independent copies of X, we have:

$$E(X + X + \cdots + X) = E(X) + E(X) + \cdots + E(X) = nE(X)$$

$$\operatorname{var}(X + X + \cdots + X) = \operatorname{var}(X) + \operatorname{var}(X) + \cdots + \operatorname{var}(X) = n \operatorname{var}(X)$$

So, $\operatorname{SD}(X + X + \cdots + X) = \sqrt{n} \operatorname{SD}(X)$, while $\operatorname{SD}(nX) = n\operatorname{SD}(X)$.

QUIZ 16

Multiple-Choice Questions

Directions: The questions that follow are each followed by five suggested answers. Choose the response that best answers the question.

1. At a warehouse sale, 100 customers are invited to choose one of 100 unopened, identical boxes, each containing one item. Five boxes contain $700 flat-screen television sets, 25 boxes contain $540 smartphones, and the remaining boxes contain $260 cameras. What should a customer be willing to pay to participate in the sale?

   (A) $260
   (B) $352
   (C) $500
   (D) $540
   (E) $699

2. An auto dealer offers discounts averaging $2450 with a standard deviation of $575. If 50 autos are sold during one month, what is the expected value and standard deviation for the total discounts given?

   (A) E(Total) = $17,324   SD(Total) = $170
   (B) E(Total) = $17,324   SD(Total) = $4066
   (C) E(Total) = $122,500   SD(Total) = $170
   (D) E(Total) = $122,500   SD(Total) = $4066
   (E) E(Total) = $122,500   SD(Total) = $28,750

3. Science majors in college pay an average of $650 per year for books with a standard deviation of $130, whereas English majors pay an average of $465 per year for books with a standard deviation of $90. What is the mean difference and standard deviation between the amounts paid for books by science and English majors?

   (A) E(Diff) = $92.50   SD(Diff) = $15
   (B) E(Diff) = $185   SD(Diff) = $110
   (C) E(Diff) = $185   SD(Diff) = $158
   (D) E(Diff) = $185   SD(Diff) = $220
   (E) E(Diff) = $557.50   SD(Diff) = $110

4. The auditor working for a veterinary clinic calculates that the mean annual cost of medical care for dogs is $98 with a standard deviation of $25 and that the mean annual cost of medical care for pets is $212 with an average deviation of $39 for owners who have one dog and one cat. Assuming expenses for dogs and cats are independent for those owning both a dog and a cat, what is the mean annual cost of medical care for cats, and what is the standard deviation?

(A) $E(\text{Cats}) = \$114$ $\text{SD}(\text{Cats}) = \$14$

(B) $E(\text{Cats}) = \$114$ $\text{SD}(\text{Cats}) = \$46$

(C) $E(\text{Cats}) = \$114$ $\text{SD}(\text{Cats}) = \$30$

(D) $E(\text{Cats}) = \$155$ $\text{SD}(\text{Cats}) = \$32$

(E) $E(\text{Cats}) = \$188$ $\text{SD}(\text{Cats}) = \$30$

5. Boxes of 50 donut holes weigh an average of 16.0 ounces with a standard deviation of 0.245 ounces. If the empty boxes alone weigh an average of 1.0 ounce with a standard deviation of 0.2 ounces, what are the mean and standard deviation of donut hole weights?

(A) $E(\text{Donut hole}) = 0.3$ oz $\text{SD}(\text{Donut hole}) = 0.0063$ oz

(B) $E(\text{Donut hole}) = 0.3$ oz $\text{SD}(\text{Donut hole}) = 0.02$ oz

(C) $E(\text{Donut hole}) = 0.3$ oz $\text{SD}(\text{Donut hole}) = 0.142$ oz

(D) $E(\text{Donut hole}) = 15$ oz $\text{SD}(\text{Donut hole}) = 0.142$ oz

(E) $E(\text{Donut hole}) = 15$ oz $\text{SD}(\text{Donut hole}) = 0.445$ oz

6. Suppose you are one of 7.5 million people who send in their name for a drawing with 1 top prize of $1 million, 5 second-place prizes of $10,000, and 20 third-place prizes of $100. Is it worth the $0.55 postage it costs you to send in your name?

(A) Yes, because $\dfrac{1{,}000{,}000}{0.55} = 1{,}818{,}182$, which is less than 7,500,000.

(B) No, because your expected winnings are only $0.14.

(C) Yes, because $\dfrac{7{,}500{,}000}{1+5+20} = 288{,}642.$

(D) No, because $1{,}052{,}000 < 7{,}500{,}000.$

(E) Yes, because $\dfrac{1{,}052{,}000}{26} = 40{,}462.$

Free-Response Questions

1. The two most popular women's spring semester sports at a particular school are softball and lacrosse. The softball team plays 0, 1, or 2 days a week with probabilities 0.2, 0.5, and 0.3, respectively.

   (a) What is the mean number of games played by the softball team per week?
   (b) What is the standard deviation for number of games played by the softball team per week?

   The lacrosse team plays a mean of 1.4 days per week with a standard deviation of 0.6.

   (c) What is the mean total number of softball and lacrosse games played per week?
   (d) Assuming independence, what is the standard deviation for the total number of softball and lacrosse games played per week?

2. Weights of coconuts have a mean of 3.2 pounds with a standard deviation of 0.7 pounds.

   (a) If each pound of coconut produces 5 ounces of fruit, calculate the mean and standard deviation for the amount of fruit per coconut.
   (b) If coconuts are packed five per box for shipment, calculate the mean and standard deviation for the total weight of coconuts per box.

The answers for this quiz can be found in the Appendix on page 601.

QUIZ 17

Multiple-Choice Questions

Directions: The questions that follow are each followed by five suggested answers. Choose the response that best answers the question.

1. For an advertising promotion, an auto dealer hands out 1000 lottery tickets with a prize of a new car worth $25,000. For someone with a single ticket, what is the standard deviation for the amount won?

   (A) $7.07
   (B) $25.00
   (C) $49.95
   (D) $790.17
   (E) $624,375

2. An insurance company charges $800 annually for car insurance. The policy specifies that the company will pay $1000 for a minor accident and $5000 for a major accident. If the probability of a motorist having a minor accident during the year is 0.2 and of having a major accident is 0.05, how much can the insurance company expect to make on a policy?

   (A) $200
   (B) $250
   (C) $300
   (D) $350
   (E) $450

3. Each hospital Internet security breach results in the theft of an average of 361 patient records with a standard deviation of 74 records. If there are 45 independent breaches during a one-year period, what is the expected value and standard deviation for the number of patient records stolen?

   (A) $E(\text{thefts}) = \sqrt{45} \times 361$ $SD(\text{thefts}) = \sqrt{45} \times 74$
   (B) $E(\text{thefts}) = \sqrt{45} \times 361$ $SD(\text{thefts}) = 45 \times 74$
   (C) $E(\text{thefts}) = 45 \times 361$ $SD(\text{thefts}) = \sqrt{45} \times 74$
   (D) $E(\text{thefts}) = 45 \times 361$ $SD(\text{thefts}) = 45 \times 74$
   (E) $E(\text{thefts}) = 45 \times 361$ $SD(\text{thefts}) = 45 \times 74^2$

4. An auto dealer pays an average of $8750 with a standard deviation of $1200 for used car trade-ins and sells new cars for an average of $28,500 with a standard deviation of $3100. Assuming independence of trade-in and new car prices for a customer, what is the standard deviation of the revenue the dealer should expect to make if a customer trades in a used car and buys a new one?

   (A) $\sqrt{3100 + 1200}$ dollars

   (B) $\sqrt{3100 - 1200}$ dollars

   (C) $\sqrt{3100^2 + 1200^2}$ dollars

   (D) $\sqrt{3100^2 - 1200^2}$ dollars

   (E) $3100 - 1200$ dollars

5. Suppose a retailer knows that the mean number of broken eggs per carton is 0.3 with a standard deviation of 0.18. In a shipment of 100 cartons, what is the expected number of broken eggs and what is the standard deviation? Assume independence between cartons.

   (A) E(broken) $= 3$ SD(broken) $= 1.8$
   (B) E(broken) $= 30$ SD(broken) $= 1.8$
   (C) E(broken) $= 30$ SD(broken) $= 18$
   (D) E(broken) $= 300$ SD(broken) $= 18$
   (E) E(broken) $= 300$ SD(broken) $= 180$

6. A home-theater projector system technician knows that based on past experience, for unpacking, assembly, and fine tuning, the mean total setup time is 5.6 hours with a standard deviation of 0.886 hours. The mean and standard deviation for the unpacking time are 1.5 hours and 0.2 hours, and for the assembly time are 2.8 hours and 0.85 hours, respectively. If the times for the three steps are independent, what are the mean and standard deviation for the fine tuning time?

   (A) E(Fine tuning) $= 1.3$ hr   SD(Fine tuning) $= 0.15$ hr
   (B) E(Fine tuning) $= 1.3$ hr   SD(Fine tuning) $= 1.24$ hr
   (C) E(Fine tuning) $= 2.15$ hr   SD(Fine tuning) $= 0.164$ hr
   (D) E(Fine tuning) $= 2.15$ hr   SD(Fine tuning) $= 1.24$ hr
   (E) E(Fine tuning) $= 4.61$ hr   SD(Fine tuning) $= 0.15$ hr

Free-Response Questions

> **Directions:** You must show all work and indicate the methods you use. You will be graded on the correctness of your methods and on the accuracy of your final answers.

1. A packaging company sells orchard fruit boxes-containing a variety of in-season fruits. These boxes contain 35, 36, or 37 pieces of fruit with probabilities of 0.1, 0.5, and 0.4, respectively.

   (a) What is the average number of pieces of fruit per box?
   (b) What is the standard deviation for the number of pieces of fruit per box?

   Full boxes of 36 pieces of fruit have an average weight of 10 pounds with a standard deviation of 0.5 pounds, while empty boxes have an average weight of 1 pound with a standard deviation of 0.1 pound.

   (c) What is the mean weight of a piece of fruit in these 36-piece boxes?
   (d) What is the standard deviation for the weight of a piece of fruit in these 36-piece boxes?

2. Weights of pineapples have a mean of 900 grams with a standard deviation of 120 grams.

   (a) If each gram of pineapple contains 0.01 grams of fiber, calculate the mean and standard deviation for the amount of fiber per pineapple.
   (b) If pineapples are packed four per box for shipment, calculate the mean and standard deviation for the total weight of pineapples per box.

The answers for this quiz can be found in the Appendix on page 602.

BINOMIAL DISTRIBUTION

A probability distribution is a list or formula that gives the probability of each outcome.

In many applications, such as coin tossing, there are only *two* possible outcomes. For applications in which a two-outcome situation is repeated a certain number of times and in which the probability of each of the two outcomes remains the same for each repetition, the resulting calculations involve what are known as *binomial probabilities*.

➡ **EXAMPLE 4.15**

Suppose the probability that a lightbulb is defective is 0.1 (so probability of being good is 0.9).

(a) What is the probability that four lightbulbs are all defective?

Answer: Because of independence (i.e., whether one lightbulb is defective is not influenced by whether any other lightbulb is defective), we can multiply individual probabilities of being defective to find the probability that all the bulbs are defective:

$$(0.1)(0.1)(0.1)(0.1) = (0.1)^4 = 0.0001$$

(b) What is the probability that exactly two out of three lightbulbs are defective?

Answer: The probability that the first two bulbs are defective and the third is good is $(0.1)(0.1)(0.9) = 0.009$. The probability that the first bulb is good and the other two are defective is $(0.9)(0.1)(0.1) = 0.009$. Finally, the probability that the second bulb is good and the other two are defective is $(0.1)(0.9)(0.1) = 0.009$. Summing, we find that the probability that exactly two out of three bulbs are defective is $0.009 + 0.009 + 0.009 = 3(0.009) = 0.027$.

(c) What is the probability that exactly three out of eight lightbulbs are defective?

Answer: The probability of any particular arrangement of three defective and five good bulbs is $(0.1)^3(0.9)^5 = 0.00059049$. We need to know the number of such arrangements.

The answer is given by combinations: $\binom{8}{3} = \dfrac{8!}{3!5!} = 56$. Thus, the probability that exactly three out of eight light bulbs are defective is $56 \times 0.00059049 = 0.03306744$. [On the TI-84, `binompdf(8,0.1,3) = 0.03306744`.]

(Note: On the exam, you can write binompdf($n = 8$, $p = 0.1$, $x = 3$) = 0.033 and you will receive full credit.

> That is, "calc-speak" can be used if the parameters and variables are defined.

NOTE

The *pdf* (such as in binompdf) stands for "probability density function." That is, it's a function that tells you the probability of certain events occurring.

NOTE

Or use Bpd on the Casio Prizm or BINOMIAL on the HP Prime.

More generally, if an experiment has two possible outcomes, called *success* and *failure*, with the probability of success equal to p and the probability of failure equal to q (of course, $p + q = 1$) *outcomes are independent from one another* and if the outcome at any particular time has no influence over the outcome at any other time, then when the experiment is repeated n times, the probability of exactly k successes (and thus $n - k$ failures) is

$$\binom{n}{k} p^k q^{n-k} = \frac{n!}{k!(n-k)!} p^k q^{n-k}$$

EXAMPLE 4.16

In cereal boxes that advertised an enclosed miniature toy, one-third of the toys were purple dragons. If six boxes of cereal are purchased, what is the probability of exactly two purple dragons?

Answer: If the probability of a purple dragon is $\frac{1}{3}$, the probability of no purple dragon is $1 - \frac{1}{3} = \frac{2}{3}$. If two out six boxes have a purple dragon, $6 - 2 = 4$ do not.

Thus, the probability of getting two purple dragons when six boxes of cereal are purchased is $\binom{6}{2}\left(\frac{1}{3}\right)^2\left(\frac{2}{3}\right)^4 = 15\left(\frac{1}{3}\right)^2\left(\frac{2}{3}\right)^4 = \frac{80}{243} = 0.329$.

[Or `binompdf(6,1/3,2)` ≈ 0.329.]

TIP

On the exam, it is sufficient to write: Binomial, $n = 6$, $p = \frac{1}{3}$, $P(X = 2) = 0.329$.

In some situations it is easier to calculate the probability of the complementary event and subtract this value from 1.

EXAMPLE 4.17

The baseball player Joe DiMaggio had a career batting average of 0.325. What was the probability that he would get at least one hit in five official times at bat?

Answer: We could sum the probabilities of exactly one hit, two hits, three hits, four hits, and five hits. However, the complement of "at least one hit" is "zero hits." The probability of no hit is

$$\binom{5}{0}(0.325)^0(0.675)^5 = (0.675)^5 = 0.140$$

Thus, the probability of at least one hit in five times at bat is $1 - 0.140 = 0.860$.

[Or `binomcdf(5,0.675,4)` ≈ 0.860.]

TIP

On the exam, it is sufficient to write: Binomial, $n = 5$, $p = 0.325$, $P(X \geq 1) = 0.860$.

To receive full credit for probability calculations using the probability distributions from this course, students need to show:

1. Name of the distribution ("binomial" in the previous example)
2. Parameters ("$n = 5$ and $p = 0.325$" in the previous example)
3. Boundary ("1" in the previous example)
4. Direction ("$\geq$" in the previous example)

NOTE

Or use Bcd on the Casio Prizm or BINOMIAL_CDF on the HP Prime.

Many, perhaps most, applications of probability involve such phrases as *at least*, *at most*, *less than*, and *more than*. In these cases, solutions involve summing two or more cases. For such calculations, the TI-84 `binomcdf` is very useful. The calculator program `binomcdf(n,p,x)` gives the probability of x or fewer successes in a binomial distribution with number of trials n and probability of success p.

EXAMPLE 4.18

A manufacturer has the following quality control check at the end of a production line. If at least eight of ten randomly picked articles meet all specifications, the whole shipment is approved. If, in reality, 85% of a particular shipment meet all specifications, what is the probability that the shipment will make it through the control check?

Answer: The probability of an item meeting specifications is 0.85, and so the probability of it not meeting specifications must be 0.15. We want to determine the probability that at least eight out of ten articles will meet specifications, that is, the probability that exactly eight or exactly nine or exactly ten articles will meet specifications. We can calculate $1 - $ `binomcdf(10,0.85,7)` $= 0.820$ or `binomcdf(10,0.15,2)` $= 0.820$. [Or we can sum the three binomial probabilities:

Exactly 8 of 10 meet specifications	Exactly 9 of 10 meet specifications	Exactly 10 of 10 meet specifications
$\binom{10}{8}(0.85)^8(0.15)^2$	$+\binom{10}{9}(0.85)^9(0.15)^1$	$+\binom{10}{10}(0.85)^{10}0(.15)^0$

$$= \frac{10!}{8!2!}(0.85)^8(0.15)^2 + 10(0.85)^9(0.15) + (0.85)^{10} = 0.820]$$

NOTE

The *cdf* stands for "cumulative distribution function." That is, the probability that *X* is less than or equal to a particular value.

TIP

On the exam, it is sufficient to write: Binomial, $n = 10$, $p = 0.85$, $P(X \geq 8) = 0.820$.

EXAMPLE 4.19

For the problem in Example 4.18, what is the probability that a shipment in which only 70% of the articles meet specifications will make it through the control check?

Answer: Now we have a binomial with $n = 10$ and $p = 0.7$. Then $P(X \geq 8) = 1 - $ `binomcdf(10,0.7,7)` $= 0.383$.

[Or `binomcdf(10,0.3,2)` $= 0.383$.]

[Or we can calculate:

$$\binom{10}{8}(0.7)^8(0.3)^2 + \binom{10}{9}(0.7)^9(0.3)^1 + \binom{10}{10}(0.7)^{10}(0.3)^0$$

$$= 45(0.7)^8(0.3)^2 + 10(0.7)^9(0.3) + (0.7)^{10} = 0.383]$$

TIP

On the exam, it is sufficient to write: binomcdf $(n = 10, p = 0.3, x = 2) = 0.383$.

EXAMPLE 4.20

A grocery store manager notes that 35% of customers who buy a particular product make use of a store coupon to receive a discount. If seven people purchase the product, what is the probability that fewer than four will use a coupon?

Answer: In this situation, "fewer than four" means zero, one, two, or three. So, we have a binomial with $n = 7$ and $p = 0.35$. Then $P(X \leq 3) = $ `binomcdf(7,0.35,3)` $= 0.800$. [Or we could calculate:

$$\binom{7}{0}(0.35)^0(0.65)^7 + \binom{7}{1}(0.35)^1(0.65)^6 + \binom{7}{2}(0.35)^2(0.65)^5 + \binom{7}{3}(0.35)^3(0.65)^4$$

$$= (0.65)^7 + 7(0.35)(0.65)^6 + 21(0.35)^2(0.65)^5 + 35(0.35)^3(0.65)^4$$

$$= 0.800]$$

NOTE

On the TI-Nspire, you set the lower and upper boundaries for binomcdf.

Suppose we have a *binomial random variable,* that is, a random variable whose values are the numbers of "successes" in some binomial probability distribution.

➥ EXAMPLE 4.21

Of the automobiles produced at a particular plant, 40% had a certain defect. Suppose a company purchases five of these cars. What is the expected value for the number of cars with defects?

Answer: We might guess that the mean, or expected value, is 40% of $5 = 0.4 \times 5 = 2$, but let's calculate from the definition. Letting X represent the number of cars with the defect, we have:

$$P(0) = \binom{5}{0}(0.4)^0(0.6)^5 = (0.6)^5 = 0.07776$$

$$P(1) = \binom{5}{1}(0.4)^1(0.6)^4 = 5(0.4)(0.6)^4 = 0.25920$$

$$P(2) = \binom{5}{2}(0.4)^2(0.6)^3 = 10(0.4)^2(0.6)^3 = 0.34560$$

$$P(3) = \binom{5}{3}(0.4)^3(0.6)^2 = 10(0.4)^3(0.6)^2 = 0.23040$$

$$P(4) = \binom{5}{4}(0.4)^4(0.6)^1 = 5(0.4)^4(0.6) = 0.07680$$

$$P(5) = \binom{5}{5}(0.4)^5(0.6)^0 = (0.4)^5 = 0.01024$$

Outcome (no. of cars):	0	1	2	3	4	5
Probability:	0.07776	0.25920	0.34560	0.23040	0.07680	0.01024
Random variable:	0	1	2	3	4	5

$$E(X) = 0(0.07776) + 1(0.25920) + 2(0.34560) + 3(0.23040)$$
$$+ 4(0.07680) + 5(0.01024)$$
$$= 2$$

Thus, the answer turns out to be the same as would be obtained by simply multiplying the probability of a "success" by the number of cases.

The following is true: **For a binomial probability distribution with the probability of a success equal to p and the number of trials equal to n, the expected value or mean number of successes for the n trials is np.**

What is the variance?

Answer: $\sigma^2 = (0 - 2)^2(0.07776) + (1 - 2)^2(0.2592) + (2 - 2)^2(0.3456) + (3 - 2)^2(0.2304) + (4 - 2)^2(0.0768) + (5 - 2)^2(0.01024) = 1.2$

Could we have calculated the above result more easily? Since we have a binomial random variable, **the variance can be simply calculated as $npq = np(1 - p)$.**

How can we use this method to calculate the variance in Example 4.21 more simply?

Answer: $np(1 - p) = 5(0.4)(0.6) = 1.2$

Thus, for a random variable X:

Mean or expected value:
$$\mu_X = \sum x_i p_i$$

Variance:
$$\sigma_X^2 = \sum (x_i - \mu_X)^2 p_i$$

Standard deviation:
$$\sigma_X = \sqrt{\sum (x_i - \mu_X)^2 p_i}$$

In the case of a binomial probability distribution with probability of a success equal to p and number of trials equal to n, if we let X be the number of successes in the n trials, the above equations become:

Mean or expected value:
$$\mu_X = np$$

Variance:
$$\sigma_X^2 = npq = np(1 - p)$$

Standard deviation:
$$\sigma_X = \sqrt{npq} = \sqrt{np(1 - p)}$$

➡ EXAMPLE 4.22 _____

Ninety-five percent of all buyers of new cars choose automatic transmissions. For a group of 50 buyers of new cars, calculate the mean and standard deviation for the number of buyers choosing automatic transmissions.

Answer:

$$\mu_X = np = 50(0.95) = 47.5$$
$$\sigma_X = \sqrt{np(1 - p)} = \sqrt{50(0.95)(0.05)} = 1.54$$

GEOMETRIC DISTRIBUTION

Suppose an experiment has two possible outcomes, called *success* and *failure*, with the probability of success equal to p and the probability of failure equal to $q = 1 - p$ and the trials are independent. Then the probability that the first success is on trial number $X = k$ is

$$q^{k-1}p$$

➡ EXAMPLE 4.23 _____

Suppose only 12% of men in ancient Greece were honest. What is the probability that the first honest man Diogenes encounters will be the third man he meets?

Answer: This is a geometric distribution with $p = 0.12$. Then $P(X=3) = (0.88)^2(0.12) = 0.092928$, where X is the number of men needed to be met in order to encounter an honest man. [Or we can calculate geometpdf $(0.12, 3) = 0.092928$.]

What is the probability that the first honest man he encounters will be no later than the fourth man he meets?

Answer: This is a geometric distribution with $p = 0.12$. Then $P(X \leq 4) =$ geometcdf $(0.12, 4) = 0.40030464$. [Or we could calculate $(0.12) + (0.88)(0.12) + (0.88)^2(0.12) + (0.88)^3(0.12) = 0.40030464$.]

_____

To receive full credit for geometric distribution probability calculations, students need to state:

1. Name of the distribution ("geometric" in the example above)
2. Parameter ("$p = 0.12$" in the example above)
3. The trial on which the first success occurs ("$X = 3$" in the first question above)

In a geometric distribution, the probability of each successive value decreases by a factor of $q = 1 - p$. Thus, the most likely value of a geometric random variable is 1, and the shape of all geometric distributions is skewed to the right.

For a geometric probability distribution with probability of success equal to p, if we let X be the number of trials needed to get one success, the mean (or expected value) of X is $\frac{1}{p}$ and the variance is $\frac{1-p}{p^2}$.

➡ EXAMPLE 4.24 _____

Suppose in a game of cornhole, you have a 25 percent chance of tossing a beanbag into the hole.

(a) What is the probability that the first bag you get into the hole is on your fourth toss?
(b) What is the probability that it takes more than two tosses before you get the bag into the hole?
(c) What is the mean number of tosses you should expect before getting a bag into the hole?
(d) What is the standard deviation for the number of tosses before getting a bag into the hole?

Answer:

(a) This is a geometric distribution with $p = 0.25$. Then $P(X = 4) = (0.75)^3(0.25) = 0.1055$. [Or geometpdf$(0.25, 4) = 0.1055$.]
(b) This is a geometric distribution with $p = 0.25$. Then $P(X > 2) = P$(miss first two tosses) $= (0.75)^2 = 0.5625$. [Or $1 - [0.25 + (0.25)(0.75)] = 1 - 0.4375 = 0.5625$. Or $1 - $ geometcdf$(0.25, 2) = 0.5625$.]
(c) This is a geometric distribution with $p = 0.25$. $\mu = \frac{1}{p} = \frac{1}{0.25} = 4$
(d) This is a geometric distribution with $p = 0.25$. $\sigma = \sqrt{\frac{1-p}{p^2}} = \sqrt{\frac{1-0.25}{(0.25)^2}} = 3.4641$

_____

QUIZ 18

Multiple-Choice Questions

Directions: The questions that follow are each followed by five suggested answers. Choose the response that best answers the question.

1. According to a CBS/*New York Times* poll taken in 1992, 15% of the public have responded to a telephone call-in poll. In a random group of five people, what is the probability that exactly two have responded to a call-in poll?

   (A) $10(0.15)^2(0.85)^3$
   (B) $5(0.15)^2(0.85)^3$
   (C) $(0.15)^2(0.85)^3$
   (D) $(0.15)^2$
   (E) $5(0.15)^2$

2. Suppose we have a random variable X where, for the values $k = 0, \ldots, 10$, the associated probabilities are $\begin{pmatrix} 10 \\ k \end{pmatrix} (0.37)^k (0.63)^{10-k}$. What is the mean of X?

   (A) 0.37
   (B) 0.63
   (C) 3.7
   (D) 6.3
   (E) None of the above

3. One in six adults in the workplace has experienced cyberbullying. In a random sample of five adults in the workplace, what is the distribution for the number who have experienced cyberbullying?

   (A) Binomial with $n = 6$ and $p = \dfrac{1}{6}$
   (B) Binomial with $n = 5$ and $p = \dfrac{1}{6}$
   (C) Binomial with $n = 5$ and $p = \dfrac{1}{5}$
   (D) Geometric with $p = \dfrac{1}{6}$
   (E) Geometric with $p = \dfrac{1}{5}$

Questions 4–5 refer to the following.

It is estimated that two out of five high school students would fall victim to a phishing e-mail (an online scam asking for sensitive information) if it appears to originate from their high school main office.

4. In a random sample of five high school students, what is the probability that exactly two fall victim to a phishing e-mail that appears to originate from their high school main office?

(A) 0.4

(B) 1.0

(C) $\dbinom{5}{2}(0.4)^2(0.6)^3$

(D) $\dbinom{5}{2}(0.4)^3(0.6)^2$

(E) $(0.4)^2(0.6)^3$

5. What is the probability that the first student to fall victim will be the third student who is sent a phishing e-mail that appears to originate from their high school main office?

(A) $(0.4)^3$

(B) $(0.6)^3$

(C) $(0.6)(0.4)^2$

(D) $(0.6)^2(0.4)$

(E) $\dbinom{5}{2}(0.6)^2(0.4)$

6. Suppose that one out of 20 apples from a particular orchard is wormy. What are the mean and standard deviation for the number of apples to be sampled from this orchard before finding a wormy apple?

(A) Mean = 5, standard deviation = 1

(B) Mean = 5, standard deviation = 1 – 0.05

(C) Mean = 10, standard deviation = $\sqrt{\dfrac{(0.05)^2}{1 - 0.05}}$

(D) Mean = 20, standard deviation = $\sqrt{380}$

(E) Mean = 20, standard deviation = $\sqrt{\dfrac{1 - 0.95}{(0.95)^2}}$

Free-Response Questions

> **Directions:** You must show all work and indicate the methods you use. You will be graded on the correctness of your methods and on the accuracy of your final answers.

1. Could hands-free, automatic faucets actually be housing more bacteria than the manual kind? The concern is that decreased water flow may increase the chance that bacteria grow because the automatic faucets are not being thoroughly flushed through. It is known that 15% of water cultures from manual faucets in hospital patient care areas test positive for *Legionella* bacteria. A recent study at Johns Hopkins Hospital found *Legionella* bacteria growing in 10 of cultured water samples from 20 automatic faucets.

   (a) If the probability of *Legionella* bacteria growing in a faucet is 0.15, what is the probability that in a sample of 20 faucets, 10 or more have the bacteria growing?

   (b) Does the Johns Hopkins study provide sufficient evidence that the probability of *Legionella* bacteria growing in automatic faucets is greater than 15%? Explain.

2. Consider the following probability distribution for the number of tattoos college students have:

Tattoos	0	1	2	3	4
Probability	0.85	0.11	0.02	0.014	0.006

   (a) What is the probability a student has two tattoos given that he has at least one tattoo?

   (b) In a random sample of three students, what is the probability that one of the students has at least two tattoos while the other two students have no tattoos?

   (c) What is the mean number of tattoos per student?

3. Suppose 35% of teenagers say that *The Hunger Games* is the best book series they ever read.

   (a) What is the probability that exactly two in a random sample of four teenagers say that *The Hunger Games* is the best series they ever read?

   (b) What is the probability that at least two in a random sample of four teens say that *The Hunger Games* is the best series they ever read?

   (c) What is the probability that the first teen who says that *The Hunger Games* is the best series he or she ever read is the second teen interviewed?

   (d) What are the mean and standard deviation for the number of teens interviewed before finding one who says that *The Hunger Games* is the best series he or she ever read?

The answers for this quiz can be found in the Appendix on page 603.

QUIZ 19

Multiple-Choice Questions

Directions: The questions that follow are each followed by five suggested answers. Choose the response that best answers the question.

1. An inspection procedure at a manufacturing plant involves picking three items at random and then accepting the whole lot if at least two of the three items are in perfect condition. If in reality 90% of the whole lot are perfect, what is the probability that the lot will be accepted?

   (A) $3(0.1)^2(0.9)$
   (B) $3(0.9)^2(0.1)$
   (C) $(0.1)^3 + (0.1)^2(0.9)$
   (D) $(0.1)^3 + 3(0.1)^2(0.9)$
   (E) $(0.9)^3 + 3(0.9)^2(0.1)$

2. Sixty-five percent of all divorce cases cite incompatibility as the underlying reason. If four couples file for a divorce, what is the probability that exactly two will state incompatibility as the reason?

   (A) $2(0.65)(0.35)$
   (B) $2(0.65)^2(0.35)^2$
   (C) $4(0.65)^2(0.35)^2$
   (D) $6(0.65)^2(0.35)^2$
   (E) 0.65

3. It is estimated that 1 out of 10 adults over the age of 50 use recreational marijuana. In a random sample of seven adults over the age of 50, what is the distribution for the number who use recreational marijuana?

   (A) Binomial with $n = 10$ and $p = \dfrac{1}{10}$
   (B) Binomial with $n = 7$ and $p = \dfrac{1}{10}$
   (C) Binomial with $n = 7$ and $p = \dfrac{1}{7}$
   (D) Geometric with $p = \dfrac{1}{10}$
   (E) Geometric with $p = \dfrac{1}{7}$

Questions 4–5 refer to the following.

Sandy Koufax, one of the greatest baseball pitchers ever, struck out approximately one out of every four batters he faced.

4. In a random sample of four batters, what is the probability that Koufax struck out at least two of them?

(A) $\begin{pmatrix} 4 \\ 2 \end{pmatrix}(0.25)^2(0.75)^2$

(B) $(0.25)^2(0.75)^2 + (0.25)^3(0.75) + (0.25)^4$

(C) $(0.25)^2 + (0.25)^3 + (0.25)^4$

(D) $1 - \begin{pmatrix} 4 \\ 2 \end{pmatrix}(0.25)^2(0.75)^2$

(E) $1 - \left[(0.75)^4 + \begin{pmatrix} 4 \\ 1 \end{pmatrix}(0.25)(0.75)^3 \right]$

5. Assuming independence, what is the probability that the first batter Koufax strikes out is before the third batter he faces?

(A) $(0.25)^2$

(B) $(0.75)^2$

(C) $1 - (0.25)^2$

(D) $1 - (0.75)^2$

(E) $0.75 + (0.75)(0.25)$

6. Suppose that two-thirds of all high schools teachers drive foreign-made cars. What are the mean and standard deviation for the number of teacher-driven cars that enter the school parking lot before one of them is foreign made?

(A) Mean $= \frac{2}{3}$, standard deviation $= \frac{\sqrt{3}}{2}$

(B) Mean $= \frac{2}{3}$, standard deviation $= \frac{\sqrt{6}}{6}$

(C) Mean $= 1.5$, standard deviation $= \frac{\sqrt{3}}{2}$

(D) Mean $= 1.5$, standard deviation $= \frac{\sqrt{6}}{6}$

(E) Mean $= 1.5$, standard deviation $= \frac{\sqrt{2}}{3}$

Free-Response Questions

1. A manufacturer is considering two options for a quality control check at the end of a production line:

   Option A: If at least five of six randomly picked articles meet all specifications, the day's production is approved.

   Option B: If at least ten of twelve randomly picked articles meet all specifications, the day's production is approved.

   If you are a buyer wanting the most assurance that a day's production will not be accepted if there were only 75% of the articles meeting all specifications, which option would you request the manufacturer to use? Give statistical justification.

2. Suppose USAir accounted for 20% of all U.S. domestic flights in the years leading up to 1994. As of mid-1994, USAir was involved in four of the previous seven major disasters. "That's enough to begin getting suspicious but not enough to hang them," said Dr. Brad Efton in *The New York Times* (September 11, 1994, Sec. 4, p. 4). Using a binomial distribution, comment on Dr. Efton's remark.

3. If a car is stopped for speeding, the probability that the driver has illegal drugs hidden in the car is 0.14. A police officer operating a speed trap plans to thoroughly search the cars stopped for speeding until one with illegal drugs is found.

   (a) What is the probability that the police officer will find illegal drugs in the third car stopped?

   (b) What is the probability that the police officer will find illegal drugs before the tenth car stopped?

   (c) What is the mean and standard deviation for the number of cars searched before finding one with illegal drugs?

The answers for this quiz can be found in the Appendix on page 604.

CUMULATIVE PROBABILITY DISTRIBUTION

A probability distribution is a function, table, or graph that links outcomes of a statistical experiment with its probability of occurrence. A *cumulative* probability distribution is a function, table, or graph that links outcomes with the probability of less than or equal to that outcome occurring.

EXAMPLE 4.25

The scores received on the 2019 AP Statistics exam are illustrated in the following tables: the probability distribution and the corresponding cumulative probability distribution.

Score	Probability
1	0.211
2	0.197
3	0.267
4	0.180
5	0.145

Score	Cumulative Probability
1	0.211
2	0.408
3	0.675
4	0.855
5	1.000

Note, for example, the probability was $0.211 + 0.197 = 0.408$ that a student received a 1 or 2 and thus did not receive college credit.

EXAMPLE 4.26

The number of minutes that teenagers spend on social media per day is illustrated in the following graphical displays: a probability distribution and the corresponding cumulative probability distribution.

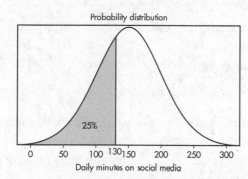
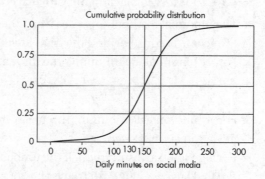

Note how, for example, 25% of the area under the probability distribution graph is to the left of 130 minutes and the point (130, 0.25) is on the cumulative probability distribution graph. Both indicate that the probability is 0.25 that a teenager spends 130 minutes or less per day on social media.

- The law of large numbers states that in the long run, a cumulative relative frequency tends closer and closer to what is called the probability of an event.
- The probability of the complement of an event is equal to 1 minus the probability of the event.
- Two events are mutually exclusive if they cannot occur simultaneously.
- If two events are mutually exclusive, the probability that at least one will occur is the sum of their respective probabilities.
- More generally, $P(A \cup B) = P(A) + P(B) - P(A \cap B)$.
- The conditional probability of A given B is given by: $P(A \mid B) = \dfrac{P(A \cap B)}{P(B)}$; that is, the population of interest is shrunk.
- Two events A and B are independent if $P(A \mid B) = P(A)$; that is, the probability of one event occurring does not influence whether or not the other occurs.
- If two events A and B are independent, $P(A \cap B) = P(A) \times P(B)$.
- A random variable takes various numeric values, each with a given probability.
- The mean or expected value of a random variable is calculated by $\mu = \sum xP(x)$, while the variance is $\sigma^2 = \sum (x - \mu)^2 P(x)$, and the standard deviation is $\sqrt{\sum (x - \mu)^2 P(x)}$.
- When transforming a random variable, $Y = a + bX$, the mean of Y is $\mu_Y = a + b\mu_X$ and the standard deviation of Y is $\sigma_Y = |b|\sigma_X$.
- When combining two random variables, $E(X + Y) = E(X) + E(Y)$ and $E(X - Y) = E(X) - E(Y)$.
- When combining two independent random variables, the variances always add: $\text{var}(X \pm Y) = \text{var}(X) + \text{var}(Y)$.
- It can further be stated that for independent random variables, X and Y, the mean of $aX + bY$ is $a\mu_X + b\mu_Y$ and the variance is $a^2\sigma_X^2 + b^2\sigma_Y^2$.
- The binomial distribution arises when there are two possible outcomes, the probability of success is constant, and the trials are independent.
- For a binomial distribution, the probability of exactly x successes in n trials is $\dbinom{n}{x} p^x (1-p)^{n-x} = \dfrac{n!}{x!(n-x)!} p^x (1-p)^{n-x}$ or binompdf(n, p, x).
- For a binomial distribution, the mean number of successes is np and the standard deviation is $\sqrt{np(1-p)}$.
- For a geometric distribution, if the probability of success is p, the probability of finding the first success in the xth trial is $(1-p)^{x-1} p$ or geometpdf(p, x).
- For a geometric distribution with probability of success p, the mean is $\dfrac{1}{p}$ and the standard deviation is $\sqrt{\dfrac{1-p}{p^2}} = \dfrac{\sqrt{1-p}}{p}$.

Sampling Distributions

5

(7–12% AP EXAM WEIGHTING)

In this unit, you will learn how sample statistics, such as sample means and sample proportions, vary across samples. The population mean μ and population standard deviation σ are examples of *population parameters*. A sample mean $\bar{x}$ and a sample standard deviation s are examples of *statistics*. While a population parameter is a fixed quantity, statistics vary depending on the particular sample chosen. The probability distribution showing how a statistic varies is called a *sampling distribution*. For large enough samples, the normal distribution can be used to approximate these sampling distributions. These concepts will be critical for understanding statistical inference, the key topic that runs through the rest of the curriculum.

> **IMPORTANT**
>
> A *sampling distribution* is not the same thing as the distribution of a sample.

UNIT LEARNING OBJECTIVES

- To be able to find probabilities and percentiles using the normal approximation.
- To be able to find the value that corresponds to a given percentile when the distribution is approximately normal.
- To understand the concept of a sampling distribution.
- To be able to describe the center, spread, and shape of the sampling distribution of a sample proportion.
- To be able to verify that the conditions are met for normal approximation of the sampling distribution of a sample proportion.

- To be able to describe and check the conditions for the sampling distribution for the difference of sample proportions.
- To be able to describe the center, spread, and shape of the sampling distribution of a sample mean.
- To understand the content and importance of the central limit theorem.
- To be able to describe and check the conditions for the sampling distribution for the difference of sample means.
- To understand how simulations can be used to estimate sampling distributions.

NORMAL DISTRIBUTION CALCULATIONS

The normal distribution is valuable in describing various natural phenomena. However, the real importance of the normal distribution in statistics is that it can be used to describe sampling distributions, the topic of this unit. Calculations involving normal distributions are often made through z-scores, which measure standard deviations from the mean. On the TI-84, normalcdf(lowerbound, upperbound) gives the area (probability) between two z-scores, while invNorm(area) gives the z-score with the given area (probability) to the left. The TI-84 also has the capability of working directly with raw scores instead of z-scores. In this case, the mean and standard deviation must be given:

> Normalcdf(lowerbound, upperbound, mean, standard deviation)
> invNorm(area, mean, standard deviation)

➡ **EXAMPLE 5.1** _____

The life expectancy of a particular brand of lightbulb is roughly normally distributed with a mean of 1500 hours and a standard deviation of 75 hours.

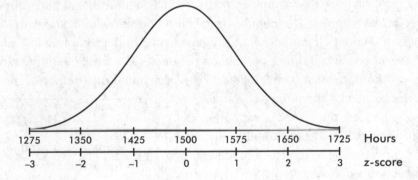

(a) What is the probability that a lightbulb will last less than 1410 hours?

Answer: The z-score of 1410 is $\frac{1410-1500}{75} = -1.2$. On the TI-84, normalcdf (0, 1410, 1500, 75) = 0.1151 and normalcdf(−10, −1.2) = 0.1151.

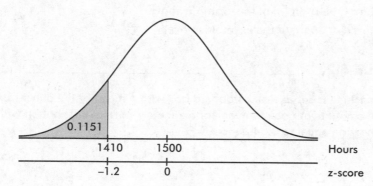

TIP

On the exam, the distribution and parameters must be identified. It is sufficient to write: Normal, $\mu = 1500$, $\sigma = 75$, $P(X < 1410) = 0.1151$.

(b) What is the probability that a lightbulb will last between 1563 and 1648 hours?

Answer: The z-score of 1563 is $\frac{1563-1500}{75} = 0.84$, and the z-score of 1648 is $\frac{1648-1500}{75} = 1.97$. Then we calculate the probability of between 1563 and 1648 hours by normalcdf(1563, 1648, 1500, 75) = 0.1762 or normalcdf(0.84, 1.97) = 0.1760.

TIP

On the exam, it is sufficient to write: Normal, $\mu = 1500$, $\sigma = 75$, $P(1563 < X < 1648) = 0.1762$.

(c) What is the probability that a lightbulb will last between 1416 and 1677 hours?

Answer: The z-score of 1416 is $\frac{1416-1500}{75} = -1.12$, and the z-score of 1677 is $\frac{1677-1500}{75} = 2.36$. Then we calculate the probability of between 1416 and 1677 hours by normalcdf(1416, 1677, 1500, 75) = 0.8595 or normalcdf(−1.12, 2.36) = 0.8595.

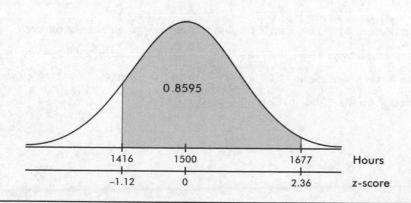

TIP

On the exam, it is also okay to write: normalcdf (lower bd = 1416, upper bd = 1677, $\mu = 1500$, $\sigma = 75$) = 0.8595.

To receive full credit for probability calculations using the probability distributions, you need to show:

1. Name of the distribution ("normal" in the example above)
2. Parameters ("$\mu = 1500$, $\sigma = 75$" in the example above)
3. Boundary ("1410" in (*a*) of the example above)
4. Direction ("$<$" in (*a*) of the example above)

➡ **EXAMPLE 5.2** _____

A packing machine is set to fill a cardboard box with a mean of 16.1 ounces of cereal. Suppose the amounts per box form an approximately normal distribution with a standard deviation equal to 0.04 ounce.

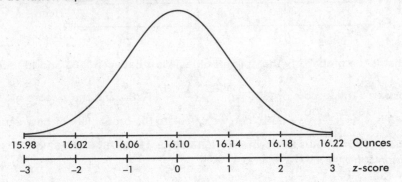

(a) What percentage of the boxes will end up with at least 1 pound of cereal? (Note: 1 pound equals 16 ounces.)

Answer: The *z*-score of 16 is $\dfrac{16 - 16.1}{0.04} = -2.5$. Thus, the probability of having more than 1 pound in a box is calculated by normalcdf(16, 1000, 16.10, 0.04) = 0.9938 or normalcdf(−2.5, 10) = 0.9938. Conclude that 99.38% of the boxes will end up with at least 1 pound of cereal.

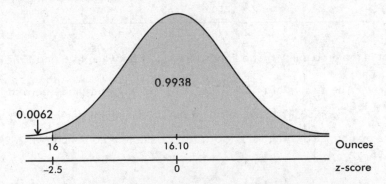

(b) Ten percent of the boxes will contain less than what number of ounces?

Answer: On the TI-84, invNorm(0.1) = −1.2816. Converting the *z*-score of −1.28 into a raw score yields $-1.28 = \dfrac{x - 16.1}{0.04}$ or 16.1 − 1.28(0.04) = 16.049 ounces. Or directly on the TI-84, invNorm(0.1, 16.10, 0.04) = 16.049.

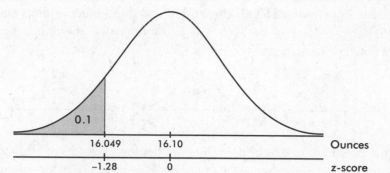

TIP

On the exam, it is also sufficient to write: invNorm (area = 0.1, μ = 16.10, σ = 0.04) = 16.049.

(c) Eighty percent of the boxes will contain more than what number of ounces?

Answer: Eighty percent to the right corresponds to a probability of 0.2 to the left. Then invNorm(0.2) = −0.8416. Converting the z-score of −0.8416 into a raw score yields 16.1 − 0.8416(0.04) = 16.066 ounces. Or directly on the TI-84, invNorm(0.2, 16.10, 0.04) = 16.066.

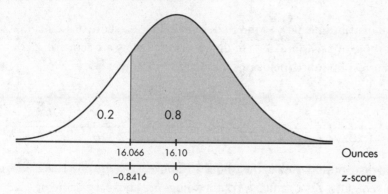

(d) The middle 90% of the boxes will be between what two weights?

Answer: Ninety percent in the middle leaves 5% in each tail. Then invNorm(0.05) = −1.645 and invNorm(0.95) = 1.645. Converting z-scores to raw scores yields 16.1 − 1.645(0.04) = 16.034 ounces and 16.1 + 1.645(0.04) = 16.166 ounces, respectively, for the two weights between which we will find the middle 90% of the boxes. Or we can calculate invNorm(0.05, 16.10, 0.04) = 16.034 and invNorm(0.95, 16.10, 0.04) = 16.166.

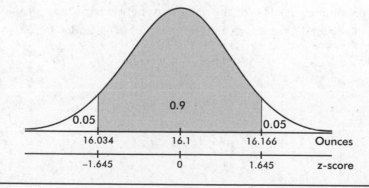

As can be seen from Example 5.2(d), there is often an interest in the limits enclosing some specified middle percentage of the data. For future reference, the limits most frequently asked for are noted below in terms of z-scores.

NOTE

The 2σ corresponding to 95% in the 68–95–99.7 rule is an approximation to the more exact 1.96σ.

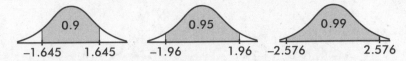

In a normal distribution, 90% of the values are between z-scores of −1.645 and +1.645, 95% of the values are between z-scores of −1.96 and +1.96, and 99% of the values are between z-scores of −2.576 and +2.576.

Sometimes the interest is in values with particular percentile rankings. For example,

NOTE

These critical z-scores can be found at the bottom of Table B in the Appendix.

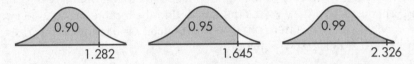

In a normal distribution, 90% of the values are below a z-score of 1.282, 95% of the values are below a z-score of 1.645, and 99% of the values are below a z-score of 2.326.

There are corresponding conclusions for negative z-scores:

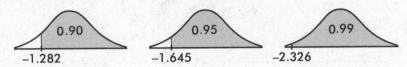

In a normal distribution, 90% of the values are above a z-score of −1.282, 95% of the values are above a z-score of −1.645, and 99% of the values are above a z-score of −2.326.

It is also useful to note the percentages corresponding to values falling between integer z-scores. For example,

NOTE

The Empirical Rule's 68%, 95%, and 99.7% are approximations of 68.26%, 95.44%, and 99.74%.

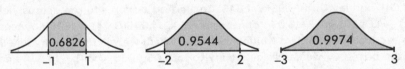

In a normal distribution, approximately 68.26% of the values are between z-scores of −1 and +1, approximately 95.44% of the values are between z-scores of −2 and +2, and approximately 99.74% of the values are between z-scores of −3 and +3.

➡ EXAMPLE 5.3

Suppose that the average height of adult males in a particular locality is 70 inches with a standard deviation of 2.5 inches.

(a) If the distribution is approximately normal, the middle 95% of males are between what two heights?

Answer: As noted above, the critical z-scores in this case are ±1.96, and so the two limiting heights are 1.96 standard deviations from the mean. Therefore, 70 ± 1.96(2.5) = 70 ± 4.9, or from 65.1 to 74.9 inches.

(b) Ninety percent of the heights are below what value?

Answer: The critical z-score is 1.282, and so the value in question is 70 + 1.282(2.5) = 73.205 inches.

(c) Ninety-nine percent of the heights are above what value?

Answer: The critical z-score is −2.326, and so the value in question is 70 − 2.326(2.5) = 64.185 inches.

(d) Approximately what percentage of the heights are between z-scores of ±1? Of ±2? Of ±3?

Answer: 68.26%, 95.44%, and 99.74%, respectively.

If we know that a distribution is normal, we can calculate the mean μ and the standard deviation σ using percentage information from the population.

➡ EXAMPLE 5.4

Given a normal distribution with a mean of 25, what is the standard deviation if 18% of the values are above 29?

Answer: The standard deviation of 18% above corresponds to a probability of 0.82 below. Then invNorm(0.82) gives the critical z-score of 0.9154. Thus, 29 − 25 = 4 is equal to 0.9154 standard deviations; that is, $0.9154\sigma = 4$, and $\sigma = \dfrac{4}{0.9154} = 4.37$.

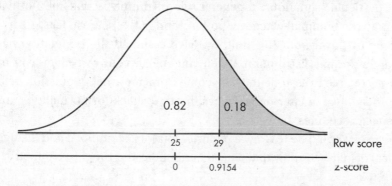

➡ EXAMPLE 5.5

Given a normal distribution with a standard deviation of 10, what is the mean if 21% of the values are below 50?

Answer: Using invNorm(0.21) gives the critical *z*-score of −0.8064. Thus, 50 is −0.8064 standard deviations from the mean, and so $\mu = 50 + 0.8064(10) = 58.06$.

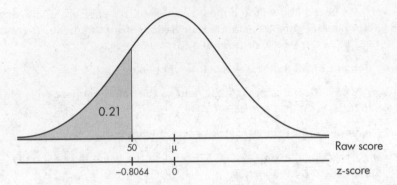

NOTE

To use a normal for a binomial, we would like $\mu - 3\sigma \geq 0$; that is, $np - 3\sqrt{npq} \geq 0$, so $np \geq 3\sqrt{npq}$, $n^2p^2 \geq 9npq$, and $np \geq 9q$. But $9q \leq 9 < 10$, so we ask for $np \geq 10$. Similarly, $nq = n(1 - p)$ should be ≥ 10.

The binomial takes values only at integers, while the normal is continuous with probabilities corresponding to areas over intervals. However, for large enough *n*, binomial distributions can be approximated by a normal distribution.

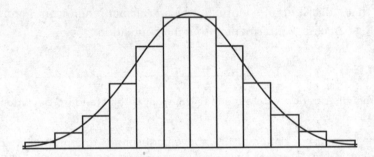

Is the normal a good approximation? The answer, of course, depends on the error tolerances in particular situations. A general rule of thumb is that the normal is a good approximation to the binomial whenever both np and $n(1 - p)$ are at least 10.

In the examples throughout this unit, we have assumed the population has a normal or approximately normal distribution. When you collect your own data, you must decide whether it is reasonable to assume the data come from a normal population before you can apply these procedures. In later units, we will have to check for normality before applying certain inference procedures.

The initial check should be to draw a picture. Dotplots, stemplots, and histograms are all useful graphical displays to show that data are unimodal and roughly symmetric.

➡ **EXAMPLE 5.6** _____

The ages at inauguration of U.S. presidents from Washington to Trump were {57, 61, 57, 57, 58, 57, 61, 54, 68, 51, 49, 64, 50, 48, 65, 52, 56, 46, 54, 49, 51, 47, 55, 55, 54, 42, 51, 56, 55, 51, 54, 51, 60, 61, 43, 55, 56, 61, 52, 69, 64, 46, 54, 47, 70}. Can we conclude that the distribution is roughly normal?

Answer: Entering the 45 data points in a graphing calculator gives the following histogram:

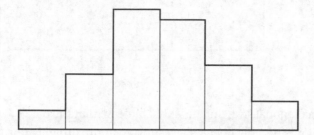

Alternatively, we could have used a dotplot or stemplot:

```
4 | 2 3
4 | 6 6 7 7 8 9 9
5 | 0 1 1 1 1 1 2 2 4 4 4 4 4
5 | 5 5 5 5 6 6 6 7 7 7 7 8
6 | 0 1 1 1 1 4 4
6 | 5 8 9
7 | 0
```

The histogram, dotplot, and stemplot do indicate a distribution that is _roughly_ unimodal and symmetric; note that it is difficult to come to such a conclusion from just a boxplot.

We can check rough normality further by calculating summary statistics and counting values to check against the 68-95-99.7 rule. In this example, 1-VarStat gives $\bar{x} = 54.98$, $s = 6.57$, min $= 42$, and max $= 70$. Counting values (ages) gives the following:

Mean $\pm$ 1 SD	48.4 to 61.1	32 out of 45 = 71.1%
Mean $\pm$ 2 SD	41.8 to 68.1	43 out of 45 = 95.6%
Mean $\pm$ 3 SD	35.3 to 74.7	45 out of 45 = 100%

These results are somewhat close to the 68%, 95%, 99.7% expected from a normal distribution and so do suggest a roughly normal distribution.

A more specialized graphical display to check normality is the _normal probability plot._ When the data distribution is roughly normal, the plot is roughly a diagonal straight line. While this plot more clearly shows deviations from normality, it is not as easy to understand as a histogram. The normal probability plot is difficult to calculate by hand; however, technology such as the TI-84 readily plots the graph. (On the TI-84, in STATPLOT the sixth choice in TYPE gives the normal probability plot.)

NOTE

While the normal probability plot is not part of the AP Statistics curriculum, it can be used on the exam.

The normal probability plot for the data in Example 5.6 is

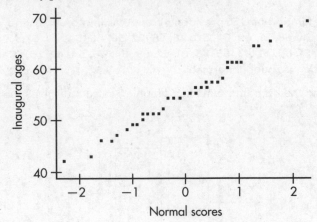

A graph close to a diagonal straight line would indicate that the data have a roughly normal distribution. This plot is roughly linear, suggesting an approximately normal distribution.

QUIZ 20

Multiple-Choice Questions

Directions: The questions that follow are each followed by five suggested answers. Choose the response that best answers the question.

1. Which of the following is a true statement?

   (A) The area under a normal curve is always equal to 1, no matter what the mean and standard deviation are.
   (B) All bell-shaped curves are normal distributions for some choice of μ and σ.
   (C) The smaller the standard deviation of a normal curve, the lower and more spread out the graph.
   (D) Depending upon the value of the standard deviation, normal curves with different means may be centered around the same number.
   (E) Depending upon the value of the standard deviation, the mean and median of a particular normal distribution may be different.

2. Populations P1 and P2 are normally distributed and have identical means. However, the standard deviation of P1 is twice the standard deviation of P2. What can be said about the percentage of observations falling within two standard deviations of the mean for each population?

   (A) The percentage for P1 is twice the percentage for P2.
   (B) The percentage for P1 is greater than but not twice as great as the percentage for P2.
   (C) The percentage for P2 is twice the percentage for P1.
   (D) The percentage for P2 is greater than but not twice as great as the percentage for P1.
   (E) The percentages are identical.

In Questions 3–7, assume the given distributions are approximately normal.

3. A trucking firm determines that its fleet of trucks averages a mean of 12.4 miles per gallon with a standard deviation of 1.2 miles per gallon on cross-country hauls. What is the probability that one of the trucks averages fewer than 10 miles per gallon?

   (A) $P(z < -2.4)$
   (B) $P(z < -2)$
   (C) $P(z < 10)$
   (D) $P\left(z < \dfrac{10}{1.2}\right)$
   (E) $P\left(z < \dfrac{12.4}{1.2}\right)$

4. An electronic product takes an average of 3.4 hours to move through an assembly line. If the standard deviation is 0.5 hours, what is the probability that an item will take between 3 and 4 hours to move through the assembly line?

(A) $P(3 < z < 4)$

(B) $P\left(\dfrac{3}{0.5} < z < \dfrac{4}{0.5}\right)$

(C) $P(3 - 3.4 < z < 4 - 3.4)$

(D) $P((3 - 3.4)(0.5) < z < (4 - 3.4)(0.5))$

(E) $P\left(\dfrac{3 - 3.4}{0.5} < z < \dfrac{4 - 3.4}{0.5}\right)$

5. The mean noise level in a restaurant is 70 decibels with a standard deviation of 4 decibels. Ninety-nine percent of the time the noise level is below what value?

(A) $70 + 1.96(4)$

(B) $70 - 2.576(4)$

(C) $70 + 2.576(4)$

(D) $70 - 2.326(4)$

(E) $70 + 2.326(4)$

6. One company produces movie trailers whose mean length is 150 seconds with a standard deviation of 40 seconds, while a second company produces movie trailers whose mean length is 120 seconds with a standard deviation of 30 seconds. What is the probability that the combined length of two randomly selected trailers, one produced by each company, will be less than three minutes?

(A) 0.000

(B) $P\left(z < \dfrac{180 - 270}{50}\right)$

(C) $P\left(z < \dfrac{180 - 270}{70}\right)$

(D) $P\left(z < \dfrac{180 - 270}{\sqrt{40 + 30}}\right)$

(E) $P\left(z < \dfrac{180 - 270}{\sqrt{\dfrac{1}{40} + \dfrac{1}{30}}}\right)$

7. Cucumbers grown on a certain farm have weights with a standard deviation of 2 ounces. What is the mean weight if 85% of the cucumbers weigh less than 16 ounces?

(A) $16 - 0.518$

(B) $16 - 0.85(2)$

(C) $16 - 1.036(2)$

(D) $16 + 0.85(2)$

(E) $16 + 1.036(2)$

Free-Response Questions

1. Scores on a university entrance exam are approximately normally distributed with a mean of 150 and a standard deviation of 25.

   (a) The university will accept 90% of applicants (based on entrance exam score). What is the cutoff score for acceptance (rounded to the nearest integer)?

   (b) Given that a student scores over 100, what is the probability of acceptance?

   (c) Given that three random students all score over 100, what is the probability that at least one is not accepted?

2. The time it takes Steve to walk to school follows an approximately normal distribution with a mean of 30 minutes and standard deviation of 5 minutes, while the time it takes Jan to walk to school follows an approximately normal distribution with a mean of 25 minutes and standard deviation of 4 minutes. Assume their walking times are independent of each other.

   (a) If they leave at the same time, what is the probability that Steve arrives before Jan?

   (b) How much earlier than Jan should Steve leave so that he has a 90% chance of arriving before Jan?

The answers for this quiz can be found in the Appendix on page 605.

QUIZ 21

Multiple-Choice Questions

> **Directions:** The questions that follow are each followed by five suggested answers. Choose the response that best answers the question.

1. Which of the following is a true statement?

   (A) The area under the standard normal curve between 0 and 2 is twice the area between 0 and 1.
   (B) The area under the standard normal curve between 0 and 2 is half the area between −2 and 2.
   (C) For the standard normal curve, the interquartile range is approximately 3.
   (D) For the standard normal curve, the range is 6.
   (E) For the standard normal curve, the area to the left of 0.1 is the same as the area to the right of 0.9.

2. Consider the following two normal curves:

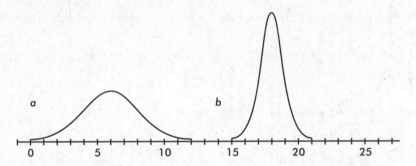

   Which has the larger mean and which has the larger standard deviation?

   (A) Curve "*a*" has the larger mean and the larger standard deviation.
   (B) Curve "*a*" has the larger mean, and curve "*b*" has the larger standard deviation.
   (C) Curve "*b*" has the larger mean, and curve "*a*" has the larger standard deviation.
   (D) Curve "*b*" has the larger mean and the larger standard deviation.
   (E) Curve "*b*" has the larger mean, and curves "*a*" and "*b*" have the same standard deviation.

In Questions 3–7, assume the given distributions are roughly normal.

3. A factory dumps an average of 2.43 tons of pollutants into a river every week. If the standard deviation is 0.88 tons, what is the probability that in a week more than 3 tons are dumped?

   (A) $P\left(z > \dfrac{3 - 2.43}{0.88}\right)$

   (B) $P\left(z > \dfrac{2.43 - 3}{0.88}\right)$

   (C) $P\left(z > \dfrac{\dfrac{3 - 2.43}{0.88}}{\sqrt{7}}\right)$

   (D) $P\left(z > \dfrac{\dfrac{2.43 - 3}{0.88}}{\sqrt{7}}\right)$

   (E) $P(z > (3 - 2.43))$

4. On average, every cigarette you smoke reduces your life expectancy by 11 minutes. Assuming an approximately normal distribution with a standard deviation of 1.5 minutes, 95% of the reductions per cigarette are above what?

   (A) $0.95(11)$
   (B) $11 - 1.645(1.5)$
   (C) $11 + 1.645(1.5)$
   (D) $11 - 1.96(1.5)$
   (E) $11 + 1.96(1.5)$

5. The mean income per household in a certain state is \$9500 with a standard deviation of \$1750. The middle 95% of incomes are between what two values?

   (A) $9500 \pm 1.645(1750)$
   (B) $9500 \pm 1.96(1750)$
   (C) $9500 \pm 1.645\left(\dfrac{1750}{2}\right)$
   (D) $9500 \pm \dfrac{1750}{1.645}$
   (E) $9500 \pm \dfrac{1750}{1.96}$

6. Jay Olshansky, from the University of Chicago, was quoted in *Chance News* as arguing that for the average life expectancy to reach 100, 18% of people would have to live to 120. What standard deviation is he assuming for this statement to make sense?

(A) $\dfrac{120 - 100}{0.915}$

(B) $(0.915)(120 - 100)$

(C) $\dfrac{120 - 100}{-0.915}$

(D) $(0.82)(120 - 100)$

(E) $\dfrac{120 - 100}{0.82}$

7. A coffee machine can be adjusted to deliver any fixed number of ounces of coffee. If the machine has a standard deviation in delivery equal to 0.4 ounce, what should be the mean setting so that an 8-ounce cup will overflow only 0.5% of the time?

(A) $8 - 0.995(0.4)$

(B) $8 + 1.96(0.4)$

(C) $8 - 1.96(0.4)$

(D) $8 + 2.576(0.4)$

(E) $8 - 2.576(0.4)$

Free-Response Questions

Directions: You must show all work and indicate the methods you use. You will be graded on the correctness of your methods and on the accuracy of your final answers.

1. (a) Two components are in series so the failure of either will cause the system to fail. The time to failure for a new component is approximately normally distributed with a mean of 3000 hours and a standard deviation of 400 hours. One of the components has already run for 2500 hours, while the other has run for 2800 hours. Assuming component failures are independent, what is the probability the system survives for 10 more days (240 hours)?

(b) If the components are put in parallel, the system will fail only if both components fail. If the two components in (a) are put in parallel, what is the probability that the system survives for 10 more days?

2. Many people have lactose intolerance leading to digestive problems when they eat dairy products. Substantial relief can be obtained by taking dietary supplements of lactase, an enzyme. One product consists of caplets containing a mean of 9000 FCC lactase units with a standard deviation of 590 units and an approximately normal distribution of lactase units. People with severe lactose intolerance may need to take two such tablets whenever they eat dairy products. Tablets with under 8500 FCC lactase units can produce noticeably less relief.

    (a) Determine the probability that a caplet has less than 8500 FCC lactase units. Round off to the nearest tenth.

    (b) Simulations are run to determine the number of tablets a lab technician analyzes before finding two tablets with a specified number of units. The results of two 100-trial simulations, one looking for two tablets each less than 8500 units and one looking for two tablets each greater than 9000 units, are shown below. Which distribution is which? Explain your answer.

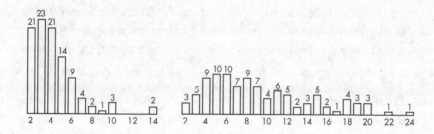

The answers for this quiz can be found in the Appendix on page 606.

CENTRAL LIMIT THEOREM

Few populations are normal, let alone exactly normal. However, it can be shown mathematically that no matter how the original population is distributed, if n is large enough, the set of sample means is approximately normally distributed. For example, there is no reason to suppose that the amounts of money that different people spend in grocery stores are normally distributed. However, if each day we survey 30 people leaving a store and determine the average grocery bill, these daily averages will have an approximately normal distribution.

The following principle forms the basis of much of what we discuss in this unit and in those following. Statement 1 is called the *central limit theorem* of statistics (often simply abbreviated as CLT).

Start with a population with a given mean μ, a standard deviation σ, and any shape distribution whatsoever. Pick n sufficiently large (at least 30), and take all samples of size n. Compute the mean of each of these samples:

1. the set of all sample means is approximately normally distributed (often stated: the distribution of sample means is approximately normal).
2. the mean of the set of sample means equals μ, the mean of the population.
3. the standard deviation $\sigma_{\bar{x}}$ of the set of sample means is equal to $\frac{\sigma}{\sqrt{n}}$, that is, equal to the standard deviation of the whole population divided by the square root of the sample size.

Alternatively, we say that for sufficiently large n, the sampling distribution of $\bar{x}$ is approximately normal with mean μ and standard deviation $\frac{\sigma}{\sqrt{n}}$.

While we mention $n \geq 30$ as a rough rule of thumb (and is sufficient for the AP exam), $n \geq 40$ is often used, and n should be chosen even larger if more accuracy is required or if the original population is far from normal. We have the assumptions of a simple random sample and of sample size n no larger than 10% of the population.

➡ EXAMPLE 5.7

The naked mole rat, a hairless East African rodent that lives underground, has a life expectancy of 21 years with a standard deviation of 3 years. In a random sample of 40 such rats, what is the probability that the mean life expectancy is between 20 and 22 years?

Answer: We have a random sample that is less than 10% of the naked mole rat population. With a sample size of $n = 40 \geq 30$, the central limit theorem applies, and the sampling distribution of $\bar{x}$ is approximately normal with mean $\mu_{\bar{x}} = 21$ and standard deviation $\sigma_{\bar{x}} = \dfrac{3}{\sqrt{40}} = 0.474$. The z-scores of 20 and 22 are $\dfrac{20-21}{0.474} = -2.110$ and $\dfrac{22-21}{0.474} = 2.110$, respectively. The probability of sample mean between 20 and 22 is

`normalcdf(−2.110,2.110)` = 0.965. [Or `normalcdf(20,22,21,0.474)` = 0.965.]

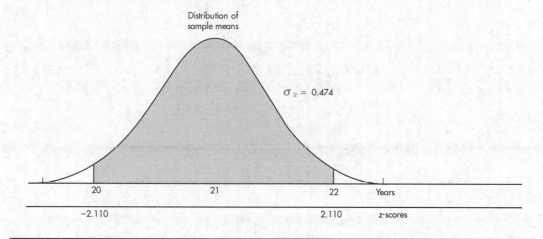

Distribution of sample means

$\sigma_{\bar{x}} = 0.474$

20   21   22   Years

−2.110   2.110   z-scores

➡ EXAMPLE 5.8 _____

Being bilingual appears to slow the onset of dementia by an average of 4.5 years with a standard deviation of 2 years. In a random sample of 75 bilingual people with dementia, the probability is 0.90 that the average delay in the onset of their dementia was at least how many years?

Answer: We have a random sample that is less than 10% of the total bilingual dementia population. With a sample size of $n = 75 \geq 30$, the central limit theorem applies, and the sampling distribution of $\bar{x}$ is approximately normal with mean $\mu_{\bar{x}} = 4.5$ and standard deviation $\sigma_{\bar{x}} = \dfrac{2}{\sqrt{75}} = 0.231$. The critical *z*-score is `invNorm(0.10)` $= -1.282$ with a corresponding raw score of $4.5 - 1.282(0.231) = 4.20$ years. [Or `invNorm(.10,4.5,.231)` $= 4.20$.]

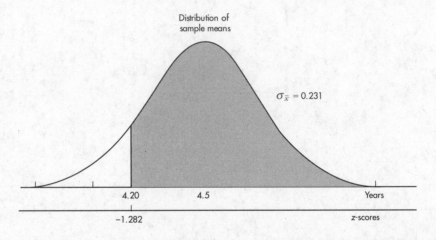

Distribution of sample means

$\sigma_{\bar{x}} = 0.231$

4.20   4.5   Years

−1.282   z-scores

Don't be confused by the several different distributions being discussed! First, there's the distribution of the original population, which may be uniform, bell-shaped, strongly skewed—anything at all. Second, there's the distribution of the data in the sample, and the larger the sample size, the more this will look like the population distribution. Third, there's the distribution of the means of many samples of a given size. The amazing fact is that this *sampling distribution* can be described by a normal model, regardless of the shape of the original population.

Does the central limit theorem say anything about proportions? If the population values are comprised of just 0s and 1s (numerical codes for failures and successes), we have a distribution that is bimodal in the extreme. Means of samples drawn from that population are sample proportions, so the sampling distribution of sample proportions is a special case of the CLT. The CLT doesn't kick in until the sample size gets fairly large, especially if the population is far from symmetric (*p* far from 0.5). Requiring $np \geq 10$ and $nq \geq 10$ is a rule of thumb suggesting how large *n* must be.

BIASED AND UNBIASED ESTIMATORS

In Unit 3, we discussed *bias* in the context of surveys as a tendency to favor the selection of certain members of a population. In the context of sampling distributions, *bias* means that the sampling distribution is not *centered* on the population parameter. The sampling distributions of proportions, means, and slopes are *unbiased*. That is, for a given sample size, the set of all sample proportions, $\hat{p}$, is centered on the population proportion, *p*; the set of all sample means, $\bar{x}$, is centered on the population mean, μ; and the set of all sample slopes, b, is centered on the population slope, β. Here are some illustrative simulations:

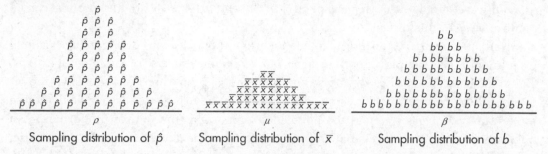

| Sampling distribution of $\hat{p}$ | Sampling distribution of $\bar{x}$ | Sampling distribution of b |

However, the sampling distribution for the maximum is clearly *biased*. That is, for a given sample size, the set of all sample maxima, $\tilde{V}$, is not centered on the population maximum, V. For example, here is one simulation of sample maxima:

Sampling distribution of the sample max $\tilde{v}$

➡ **EXAMPLE 5.9**

Note the following simulations for sampling distributions of $\dfrac{\sum (x - \bar{x})^2}{n}$ and $\dfrac{\sum (x - \bar{x})^2}{n-1}$, both from a population with variance $\sigma^2 = 25$.

NOTE

This shows why we divide by $n - 1$ when calculating a sample variance and standard deviation.

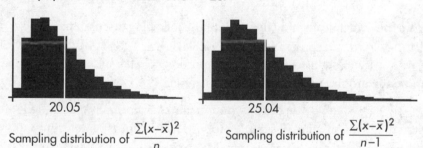

| 20.05 | 25.04 |
| Sampling distribution of $\dfrac{\sum (x-\bar{x})^2}{n}$ | Sampling distribution of $\dfrac{\sum (x-\bar{x})^2}{n-1}$ |

We see that $\dfrac{\sum (x - \bar{x})^2}{n}$ is a *biased* estimator for σ^2 (that is, the distribution, with mean 20.05, *is not* centered at σ^2), while $\dfrac{\sum (x - \bar{x})^2}{n-1}$ is an *unbiased* estimator for σ^2 (that is, the distribution, with mean 25.04, *is* centered at σ^2).

SAMPLING DISTRIBUTION FOR SAMPLE PROPORTIONS

Whereas the mean is a quantitative measurement, the proportion represents essentially a qualitative calculation. The interest is simply in the presence or absence of some attribute. We count the number of yes responses and form a proportion. For example, what proportion of drivers wear seat belts? What proportion of military drones can be intercepted? What proportion of new smartphones have a certain software problem?

This separation of the population into "haves" and "have-nots" suggests that we can make use of our earlier work on binomial distributions. We also keep in mind that, when n (trials, or in this case sample size) is large enough ($np \geq 10$ and $n(1 - p) \geq 10$), the binomial can be approximated by the normal.

We are interested in estimating a population proportion p by considering a single sample proportion $\hat{p}$. This sample proportion is just one of a whole universe of sample proportions, and to judge its significance, we must know how sample proportions vary. Consider the set of proportions from all possible samples of a specified size n. It seems reasonable that these proportions will cluster around the population proportion (the sample proportion is an unbiased statistic of the population proportion) and that the larger the chosen sample size, the tighter the clustering.

How do we calculate the mean and standard deviation of the set of sample proportions? Suppose the sample size is n and the actual population proportion is p. From our work on binomial distributions, we remember that the mean and standard deviation for the number of successes in a given sample are np and $\sqrt{np(1 - p)}$, respectively, and for large values of n the complete distribution begins to look "normal."

Here, however, we are interested in the proportion rather than in the number of successes. From Unit 1, remember that when we multiply or divide every element by a constant, we multiply or divide both the mean and the standard deviation by the same constant. In this case, to change number of successes to proportion of successes, we divide by n:

$$\mu_{\hat{p}} = \frac{np}{n} = p \quad \text{and} \quad \sigma_{\hat{p}} = \frac{\sqrt{np(1-p)}}{n} = \sqrt{\frac{np(1-p)}{n^2}} = \sqrt{\frac{p(1-p)}{n}}$$

NOTE

The standard deviation of the sampling distribution of $\hat{p}$ when sampling without replacement is actually

$$\sigma_{\hat{p}} = \sqrt{\frac{p(1-p)}{n}}\sqrt{1-\frac{n}{N}}.$$

As long as $n < 10\%(N)$, the finite population factor,

$$\sqrt{1-\frac{n}{N}},$$ is close to 1.

Furthermore, if each element in an approximately normal distribution is divided by the same constant, it is reasonable that the result will still be an approximately normal distribution.

Thus, note the principle forming the basis of the following discussion.

Start with a population with a given proportion p. Take all samples of size n. Compute the proportion in each of these samples:

1. the set of all sample proportions is approximately normally distributed (often stated: the distribution of sample proportions is approximately normal).
2. the mean $\mu_{\hat{p}}$ of the set of sample proportions equals p, the population proportion.
3. the standard deviation $\sigma_{\hat{p}}$ of the set of sample proportions is approximately equal to $\sqrt{\frac{p(1-p)}{n}}$.

Alternatively, we say that the sampling distribution of $\hat{p}$ is approximately normal with mean p and standard deviation $\sqrt{\frac{p(1-p)}{n}}$.

NOTE

The sample size should be large enough so that the expected number of "successes" and the expected number of "failures" in the sample are both at least 10.

Since we are using the normal approximation to the binomial, both np and $n(1 - p)$ should be at least 10. Furthermore, in making calculations and drawing conclusions from a specific sample, it is important that the sample be a *simple random sample*.

Finally, because sampling is usually done without replacement, the sample cannot be too large; the sample size n should be no larger than 10% of the population. (We're actually worried about *independence*, but randomly selecting a relatively small sample allows us to assume independence. Of course, it's always better to have larger samples—it's just that if the sample is large relative to the population, the proper inference techniques are different from those taught in introductory statistics classes.)

➡️ **EXAMPLE 5.10**_____

It is estimated that 80% of people with high math anxiety experience brain activity similar to that experienced under physical pain when anticipating doing a math problem. In a simple random sample of 110 people with high math anxiety, what is the probability that less than 75% experience the physical pain brain activity?

Answer: The sample is given to be random, both $np = (110)(0.80) = 88 \geq 10$ and $n(1 - p) = (110)(0.20) = 22 \geq 10$, and our sample is clearly less than 10% of all people with math anxiety. So, the sampling distribution of $\hat{p}$ is approximately normal with

mean 0.80 and standard deviation $\sigma_{\hat{p}} = \sqrt{\dfrac{(0.80)(0.20)}{110}} = 0.0381$. With a z-score of

$\dfrac{0.75 - 0.80}{0.0381} = -1.312$, the probability that the sample proportion is less than 0.75 is

`normalcdf(-1000,-1.312) = 0.0948`.

[Or `normalcdf(−1000, 0.75, .80, .0381) = 0.0947`.]

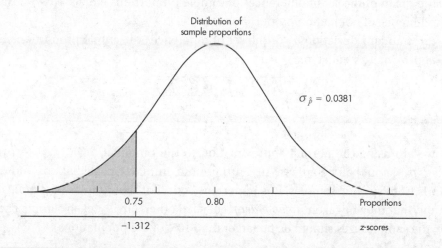

Distribution of sample proportions

$\sigma_{\hat{p}} = 0.0381$

0.75   0.80   Proportions

−1.312   z-scores

> **TIP**
>
> It is sufficient to write: normal with $\mu_{\hat{p}} = 0.80$ and $\sigma_{\hat{p}} = 0.0381$; $P(\hat{p} < 0.75) = 0.0947$.

SAMPLING DISTRIBUTION FOR DIFFERENCES IN SAMPLE PROPORTIONS

Numerous important and interesting applications of statistics involve the comparison of two population proportions. For example, is the proportion of satisfied purchasers of American automobiles greater than that of buyers of Japanese cars? How does the percentage of surgeons recommending a new cancer treatment compare with the corresponding percentage of oncologists? What can be said about the difference between the proportion of single parents on welfare and the proportion of two-parent families on welfare?

Our procedure involves comparing two sample proportions. When is a difference between two such sample proportions significant? Note that we are dealing with one difference from the set of all possible differences obtained by subtracting sample proportions of one population from sample proportions of a second population. To judge the significance of one particular difference, we must first determine how the differences vary among themselves.

From Unit 4, remember that the variance of a set of differences is equal to the sum of the variances of the individual sets:

$$\sigma_d^2 = \sigma_1^2 + \sigma_2^2$$

Now if $\sigma_1 = \sqrt{\dfrac{p_1(1-p_1)}{n_1}}$ and $\sigma_2 = \sqrt{\dfrac{p_2(1-p_2)}{n_2}}$, then $\sigma_d^2 = \dfrac{p_1(1-p_1)}{n_1} + \dfrac{p_2(1-p_2)}{n_2}$ and

$\sigma_d = \sqrt{\dfrac{p_1(1-p_1)}{n_1} + \dfrac{p_2(1-p_2)}{n_2}}$. We have the following about the sampling distribution of $\hat{p}_1 - \hat{p}_2$.

Start with two populations with given proportions p_1 and p_2. Take all samples of sizes n_1 and n_2, respectively. Compute the difference $\hat{p}_1 - \hat{p}_2$ of the two proportions in each pair of samples:

1. the set of all differences of sample proportions is approximately normally distributed (alternatively stated: the distribution of differences of sample proportions is approximately normal).
2. the mean of the set of differences of sample proportions equals $p_1 - p_2$, the difference of population proportions.
3. the standard deviation σ_d of the set of differences of sample proportions is approximately equal to:

$$\sqrt{\frac{p_1(1-p_1)}{n_1} + \frac{p_2(1-p_2)}{n_2}}$$

Since we are using the normal approximation to the binomial, $n_1 p_1$, $n_1(1 - p_1)$, $n_2 p_2$, and $n_2(1 - p_2)$ should all be at least 10. Furthermore, in making calculations and drawing conclusions from specific samples, it is important both that the samples be *simple random samples* and that they be taken *independently* of each other. Finally, each sample cannot be too large; the sample sizes should be no larger than 10% of the populations.

➡ **EXAMPLE 5.11** _____

In a study of how environment affects our eating habits, scientists revamped one of two nearby fast-food restaurants with table cloths, candlelight, and soft music. They then noted that at the revamped restaurant, customers ate more slowly and 25% left at least 100 calories of food on their plates. At the unrevamped restaurant, customers tended to quickly eat their food and only 19% left at least 100 calories of food on their plates. In a random sample of 110 customers at the revamped restaurant and an independent random sample of 120 customers at the unrevamped restaurant, what is the probability that the difference in the percentages of customers in the revamped setting and the unrevamped setting is more than 10% (where the difference is the revamped restaurant percent minus the unrevamped restaurant percent)?

Answer: We have independent random samples, each less than 10% of all fast-food customers, and we note that $n_1p_1 = 110(0.25) = 27.5$, $n_1(1 - p_1) = 110(0.75) = 82.5$, $n_2p_2 = 120(0.19) = 22.8$, and $n_2(1 - p_2) = 120(0.81) = 97.2$ are all ≥ 10. Thus, the sampling distribution of $\hat{p}_1 - \hat{p}_2$ is roughly normal with mean $\mu_{\hat{p}_1 - \hat{p}_2} = 0.25 - 0.19 = 0.06$ and standard deviation $\sigma_{\hat{p}_1 - \hat{p}_2} = \sqrt{\dfrac{(0.25)(0.75)}{110} + \dfrac{(0.19)(0.81)}{120}} = 0.0547$. The z-score of 0.10 is $\dfrac{0.10 - 0.06}{0.0547} = 0.731$, and `normalcdf(0.731,1000) = 0.232`.

[Or `normalcdf(0.10,1.0,0.06,0.0547) = 0.232`.]

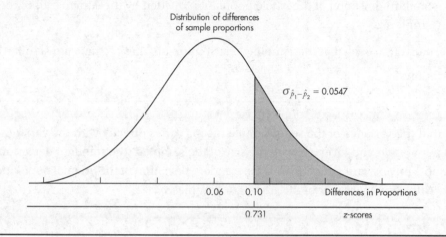

Distribution of differences of sample proportions

$\sigma_{\hat{p}_1 - \hat{p}_2} = 0.0547$

0.06   0.10   Differences in Proportions

0.731   z-scores

SAMPLING DISTRIBUTION FOR SAMPLE MEANS

Suppose we are interested in estimating the mean μ of a population. For our estimate we could simply randomly pick a single member of the population, but then we would have little confidence in our answer. Suppose instead that we pick 100 members and calculate their average. It is intuitively clear that the resulting sample mean has a greater chance of being closer to the mean of the whole population than the value for any individual member of the population does.

When we pick a sample and measure its mean $\bar{x}$, we are finding exactly one sample mean out of a whole universe of sample means. To judge the significance of a single sample mean, we must know how sample means vary. Consider the set of means from all possible samples of a specified size. It is both apparent and reasonable that the sample means are clustered around the mean of the whole population; furthermore, these sample means have a tighter clustering than the members of the original population. In fact, we might guess that the larger the chosen sample size, the tighter the clustering.

How do we calculate the standard deviation $\sigma_{\bar{x}}$ of the set of sample means? Suppose the variance of the population is σ^2 and we are interested in samples of size n. Sample means are obtained by first summing together n elements and then dividing by n. A set of sums has a variance equal to the sum of the variances associated with the original sets. In our case, $\sigma_{sums}^2 = \sigma^2 + \cdots + \sigma^2 = n\sigma^2$. When each element of a set is divided by some constant, the new variance is the old one divided by the square of the constant. Since the sample means are obtained by dividing the sums by n, the variance of the sample means is obtained by dividing the variance of the sums by n^2. Thus, if $\sigma_{\bar{x}}$ symbolizes the standard deviation of the sample means, we find that:

$$\sigma_{\bar{x}}^2 = \frac{\sigma_{sums}^2}{n^2} = \frac{n\sigma^2}{n^2} = \frac{\sigma^2}{n}$$

In terms of standard deviations, we have $\sigma_{\bar{x}} = \dfrac{\sigma}{\sqrt{n}}$.
Thus, we have the following.

> Start with a population with a given mean μ and standard deviation σ. Compute the mean of all samples of size n:
>
> 1. the mean $\mu_{\bar{x}}$ of the set of sample means will equal μ, the mean of the population.
>
> 2. the standard deviation $\sigma_{\bar{x}}$ of the set of sample means will equal $\dfrac{\sigma}{\sqrt{n}}$, that is, the standard deviation of the whole population divided by the square root of the sample size.
>
> Alternatively, we say that the sampling distribution of $\bar{x}$ has mean μ and standard deviation $\dfrac{\sigma}{\sqrt{n}}$.

Note that the variance of the set of sample means varies directly with the variance of the original population and inversely with the size of the samples; the standard deviation of the set of sample means varies directly with the standard deviation of the original population and inversely with the square root of the size of the samples.

➡ **EXAMPLE 5.12**

The number of emergency room visits after drinking energy drinks is skyrocketing. One particular energy drink has an average of 200 mg of caffeine with a standard deviation of 10 mg. A store sells boxes of six bottles each. What is the mean and standard deviation of the average milligrams of caffeine consumers should expect from the six bottles in each box?

Answer: We have samples of size 6. The mean of these sample means will equal the population mean of 200 mg. The standard deviation of these sample means will equal $\sigma_{\bar{x}} = \dfrac{10}{\sqrt{6}} = 4.08$ mg.

Note that while giving the mean and standard deviation of the set of sample means, we did not describe the shape of the distribution. If we are also given that the original population is normal, we can conclude that the set of sample means has a normal distribution. Or if the sample size is large enough ($n \geq 30$) for the central limit theorem to apply, we can conclude that the set of sample means has a roughly normal distribution.

➡ **EXAMPLE 5.13**

It is estimated that domesticated cats who mainly live outdoors kill an average of 10 birds a year. Assuming an approximately normal distribution with a standard deviation of 3 birds, what is the probability an SRS of 25 such cats will kill an average of between 9 and 12 birds in a year?

Answer: The sampling distribution of the sample means is approximately normal (because the original population is approximately normal) with mean $\mu_{\bar{x}} = 10$ and standard deviation $\sigma_X = \dfrac{3}{\sqrt{25}} = 0.6$. The z-scores of 9 and 12 are $\dfrac{9-10}{0.6} = -1.667$ and $\dfrac{12-10}{0.6} = 3.333$, respectively, and the probability of a sample mean between 9 and 12 is `normalcdf(-1.667,3.333) = 0.952`. [Or `normalcdf(9,12,10,0.6) = 0.952`.]

NOTE

On the exam, it is sufficient to write: normalcdf (lower bd = 9, upper bd = 12, $\mu_{\bar{x}}$ = 10, $\sigma_{\bar{x}} = \dfrac{3}{\sqrt{25}}$) = 0.952.

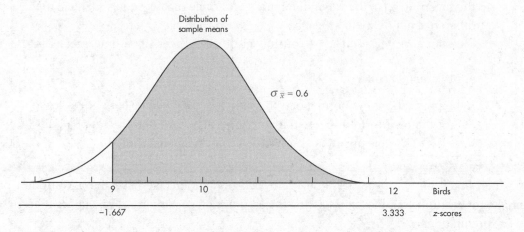

Distribution of sample means

$\sigma_{\bar{x}} = 0.6$

9   10   12   Birds

−1.667   3.333   z-scores

SAMPLING DISTRIBUTION FOR DIFFERENCES IN SAMPLE MEANS

Many real-life applications of statistics involve comparisons of two population means. For example, is the average weight of laboratory rabbits receiving a special diet greater than that of rabbits on a standard diet? Which of two accounting firms pays a higher mean starting salary? Is the life expectancy of a coal miner less than that of a school teacher?

First we consider how to compare the means of samples, one from each population. When is a difference between two such sample means significant? The answer is more apparent when we realize that what we are looking at is one difference from a set of differences. That is, there is the set of all possible differences obtained by subtracting sample means from one set from sample means from a second set. To judge the significance of one particular difference, we must first determine how the differences vary among themselves. The necessary key is the fact that the variance of a set of differences is equal to the sum of the variances of the individual sets. Thus:

TIP

Don't forget to check that the samples are independent!

$$\sigma^2_{\bar{x}_1 - \bar{x}_2} - \sigma^2_{\bar{x}_1} + \sigma^2_{\bar{x}_2}$$

Now if $\sigma_{\bar{x}_1} = \dfrac{\sigma_1}{\sqrt{n_1}}$ and $\sigma_{\bar{x}_2} = \dfrac{\sigma_2}{\sqrt{n_2}}$, then $\sigma^2_{\bar{x}_1 - \bar{x}_2} = \dfrac{\sigma_1^2}{n_1} + \dfrac{\sigma_2^2}{n_2}$ and $\sigma_{\bar{x}_1 - \bar{x}_2} = \sqrt{\dfrac{\sigma_1^2}{n_1} + \dfrac{\sigma_2^2}{n_2}}$.

Then we have the following about the sampling distribution of $\bar{x}_1 - \bar{x}_2$.

Start with two populations with means μ_1 and μ_2 and standard deviations σ_1 and σ_2. Take all samples of sizes n_1 and n_2, respectively. Compute the difference $\bar{x}_1 - \bar{x}_2$ of the two means in each pair of these samples:

1. the mean $\mu_{\bar{x}_1 - \bar{x}_2}$ of the set of differences of sample means equals $\mu_1 - \mu_2$, the difference of population means.

2. the standard deviation $\sigma_{\bar{x}_1 - \bar{x}_2}$ of the set of differences of sample means is approximately equal to $\sqrt{\dfrac{\sigma_1^2}{n_1} + \dfrac{\sigma_2^2}{n_2}}$.

3. if the populations are normal or if the sample sizes are large enough (at least 30) for the central limit theorem to apply, we also have that the set of all differences of sample means is approximately normally distributed.

When analyzing two samples, one from each population, we have the assumption of independent simple random samples and the assumption of sample sizes no larger than 10% of the populations.

➡ **EXAMPLE 5.14**

It is estimated that 40-year-old men contribute an average of 65 genetic mutations to their new children, whereas 20-year-old men contribute an average of only 25. Assuming standard deviations of 15 and 5 mutations, respectively, for the 40- and 20-year-olds, what is the probability that the mean number of mutations in a random sample of thirty-five 40-year-old new fathers is between 35 and 45 more than the mean number in a random sample of forty 20-year-old new fathers?

Answer: We have independent random samples, each less than 10% of their age groups, and both sample sizes are over 30, so the sampling distribution of $\bar{x}_1 - \bar{x}_2$ is roughly normal with mean $\mu_{\bar{x}_1 - \bar{x}_2} = 65 - 25 = 40$ and standard deviation

$\sigma_{\bar{x}_1 - \bar{x}_2} = \sqrt{\dfrac{15^2}{35} + \dfrac{5^2}{40}} = 2.656$. The *z*-scores of 35 and 45 are $\dfrac{35 - 40}{2.656} = -1.883$ and

$\dfrac{45 - 40}{2.656} = 1.883$, respectively, and `normalcdf(-1.883,1.883) = 0.940`.

[Or `normalcdf(35,45,40,2.656) = 0.940`.]

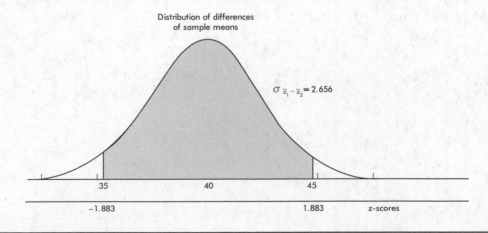

Distribution of differences of sample means

$\sigma_{\bar{x}_1 - \bar{x}_2} = 2.656$

35   40   45

−1.883   1.883   z-scores

➡ **EXAMPLE 5.15**

The average number of missed school days for students going to public schools is 8.5 with a standard deviation of 4.1, while students going to private schools miss an average of 5.3 with a standard deviation of 2.9. In an SRS of 200 public school students and an SRS of 150 private school students, with a probability of 0.95, the difference in average missed days (public average minus private average) is above what number?

Answer: We have large ($n_1 = 200$ and $n_2 = 150$) SRSs that are still smaller than 10% of all public and private school students. The mean of the differences is $8.5 - 5.3 = 3.2$, while the standard deviation is $\sqrt{\dfrac{(4.1)^2}{200} + \dfrac{(2.9)^2}{150}} = 0.374$. An area of 0.95 to the right corresponds to a *z*-score of -1.645. The difference in days is $3.2 - 1.645(0.374) = 2.6$. Thus, there is a 0.95 probability that the difference in average missed days between public and private school samples is over 2.6 days. [invNorm(0.05, 3.2, 0.374) = 2.585.]

SIMULATION OF A SAMPLING DISTRIBUTION

As we've seen, the normal distribution can handle sampling distributions of the statistics we are most interested in, namely the sample proportion and sample mean. If other statistics arise, we can use simulation to obtain a rough idea of the corresponding sampling distributions. For example, a study is made of the number of dreams high school students remember having every night. The median number is 3.41 with a variance of 1.46 and a minimum of 0. Now taking a large number of random samples of 15 students, we calculate the median, variance, and minimum for each sample and graph the resulting simulated sampling distributions.

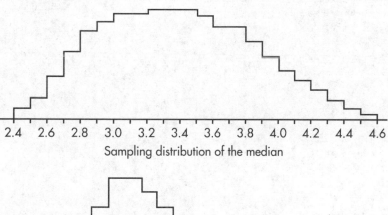

Sampling distribution of the median

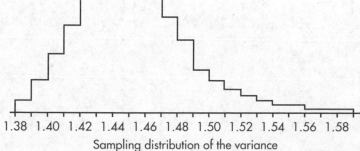

Sampling distribution of the variance

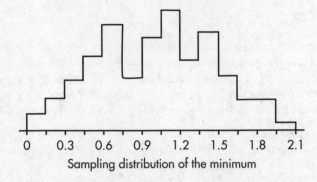

Sampling distribution of the minimum

The simulated sampling distribution of these medians is roughly bell-shaped, the simulated sampling distribution of the variances is skewed right, and the simulated sampling distribution of the minimums is very roughly bell-shaped.

QUIZ 22

Multiple-Choice Questions

Directions: The questions that follow are each followed by five suggested answers. Choose the response that best answers the question.

1. When selecting a random sample from a population, which of the following combinations of n and p satisfies the conditions for the sampling distribution of $\hat{p}$ to be approximately normal?

    (A) $n = 10$ and $p = 0.67$
    (B) $n = 25$ and $p = 0.32$
    (C) $n = 50$ and $p = 0.18$
    (D) $n = 75$ and $p = 0.15$
    (E) $n = 100$ and $p = 0.08$

2. Which of the following is a true statement?

    (A)   The larger the sample, the larger the spread in the sampling distribution is.
    (B)   Bias relates to the spread of a sampling distribution.
    (C)   Provided that the population size is significantly greater than the sample size, the spread of the sampling distribution does not depend on the population size.
    (D)   Sample parameters are used to make inferences about population statistics.
    (E)   Statistics from smaller samples have less variability.

3. Which of the following is an *incorrect* statement?

    (A)   The sampling distribution of $\bar{x}$ has mean equal to the population mean μ even if the population is not normally distributed.
    (B)   The sampling distribution of $\bar{x}$ has standard deviation $\frac{\sigma}{\sqrt{n}}$ even if the population is not normally distributed.
    (C)   The sampling distribution of $\bar{x}$ is normal if the population has a normal distribution.
    (D)   When n is large, the sampling distribution of $\bar{x}$ is approximately normal even if the population is not normally distributed.
    (E)   The larger the value of the sample size n, the closer the standard deviation of the sampling distribution of $\bar{x}$ is to the standard deviation of the population.

4. Which of the following is the best reason that the sample maximum is not used as an estimator for the population maximum?

    (A)   The sample maximum is biased.
    (B)   The sampling distribution of the sample maximum is not binomial.
    (C)   The sampling distribution of the sample maximum is not normal.
    (D)   The sampling distribution of the sample maximum has too large a standard deviation.
    (E)   The sample mean plus three sample standard deviations gives the best estimate for the population maximum.

5. Thirty-four percent of high school administrators say that math is the most important subject in school. In a random sample of 400 high school administrators, what is the probability that between 30% and 35% will say that math is the most important subject?

(A) $P(0.30 < z < 0.35)$

(B) $P(0.30 - 0.34 < z < 0.35 - 0.34)$

(C) $P\left(\dfrac{0.30 - 0.34}{\sqrt{400(0.34)(0.66)}} < z < \dfrac{0.35 - 0.34}{\sqrt{400(0.34)(0.66)}} \right)$

(D) $P\left(\dfrac{0.30 - 0.34}{\sqrt{\dfrac{(0.34)(0.66)}{400}}} < z < \dfrac{0.35 - 0.34}{\sqrt{\dfrac{(0.34)(0.66)}{400}}} \right)$

(E) $P\left(\dfrac{0.30 - 0.34}{\sqrt{\dfrac{(0.30)(0.70)}{400}}} < z < \dfrac{0.35 - 0.34}{\sqrt{\dfrac{(0.35)(0.65)}{400}}} \right)$

6. It is estimated that school CO_2 levels average 1200 ppm with a standard deviation of 300 ppm. In a random sample of 30 schools, what is the probability that the mean CO_2 level is more than 1000 ppm, a level at which some researchers feel will cause a drop in academic performance?

(A) $P\left(z > \dfrac{1000 - 1200}{\dfrac{300}{\sqrt{30}}} \right)$

(B) $P\left(z > \dfrac{1000 - 1200}{300} \right)$

(C) $P\left(z > \dfrac{1200 - 1000}{\dfrac{300}{\sqrt{30}}} \right)$

(D) $P\left(z > \dfrac{1200 - 1000}{300} \right)$

(E) $1 - P\left(z > \dfrac{1000 - 1200}{\dfrac{300}{\sqrt{30}}} \right)$

7. Binge drinking is a serious problem, killing more than 1700 college students per year. Suppose the incidence of binge drinking is 43% at one large university and 37% at a second large university. In a random sample of 75 students at the first school and 80 students at the second, what is the probability that the difference (first minus second) between the percentages of binge drinkers is between 5% and 10%?

(A) $P\left(\dfrac{0.37}{\sqrt{\dfrac{(0.37)(0.63)}{80}}} < z < \dfrac{0.43}{\sqrt{\dfrac{(0.43)(0.57)}{75}}}\right)$

(B) $P\left((0.05-0.06)\sqrt{\dfrac{(0.43)(0.57)}{75} + \dfrac{(0.37)(0.63)}{80}} < z < (0.10-0.06)\sqrt{\dfrac{(0.43)(0.57)}{75} + \dfrac{(0.37)(0.63)}{80}}\right)$

(C) $P\left(\dfrac{0.05-0.06}{\sqrt{\dfrac{(0.43)(0.57)}{75} + \dfrac{(0.37)(0.63)}{80}}} < z < \dfrac{0.10-0.06}{\sqrt{\dfrac{(0.43)(0.57)}{75} + \dfrac{(0.37)(0.63)}{80}}}\right)$

(D) $P\left((0.05-0.06)\sqrt{\dfrac{(0.05)(0.95)}{75} + \dfrac{(0.10)(0.90)}{80}} < z < (0.10-0.06)\sqrt{\dfrac{(0.05)(0.95)}{75} + \dfrac{(0.10)(0.90)}{80}}\right)$

(E) $P\left(\dfrac{0.05-0.06}{\sqrt{\dfrac{(0.05)(0.95)}{75} + \dfrac{(0.10)(0.90)}{80}}} < z < \dfrac{0.10-0.06}{\sqrt{\dfrac{(0.05)(0.95)}{75} + \dfrac{(0.10)(0.90)}{80}}}\right)$

8. A study is to be performed to estimate the proportion of voters who believe the economy is "heading in the right direction." Which of the following pairs of sample size and population proportion p will result in the smallest variance for the sampling distribution of $\hat{p}$?

(A) $n = 100$ and $p = 0.1$
(B) $n = 100$ and $p = 0.5$
(C) $n = 100$ and $p = 0.99$
(D) $n = 1000$ and $p = 0.1$
(E) $n = 1000$ and $p = 0.5$

9. In a recent study of fatal automobile crashes, 32% of drivers ages 21–24 were legally drunk, and 29% of drivers ages 25–34 were legally drunk. A simulation is conducted in which a random sample of 400 drivers ages 21–24 and a random sample of 400 drivers 25–34, all involved in fatal automobile crashes, are selected. The difference in proportions of those who were legally drunk is calculated, and this simulation is repeated 1000 times. Which of the following is most likely to occur in the simulated sampling distribution of the difference between the two sample proportions?

(A)

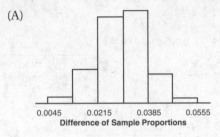

(B)

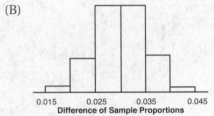

(C)

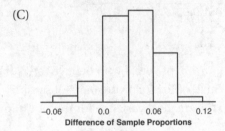

(D)

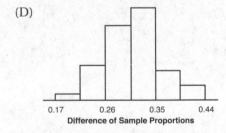

(E)

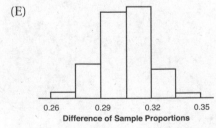

Free-Response Questions

> **Directions:** You must show all work and indicate the methods you use. You will be graded on the correctness of your methods and on the accuracy of your final answers.

1. Suppose that 16% of teenagers who play video games would enjoy playing a new Super Mario game. The manufacturer doesn't know this actual proportion and plans on sampling a random sample of 80 teenagers who play video games.

    (a) Calculate $\sigma_{\hat{p}}$, the standard deviation of the sampling distribution of $\hat{p}$.
    (b) Are the conditions for inference met? Explain.
    (c) If a different sample size would result in $\sigma_{\hat{p}} = 0.037$, is the sample size smaller or larger than 80? Explain.

2. It is estimated that 58% of all Americans sleep on their sides.

    (a) What is the probability that a randomly chosen American sleeps on his or her side?
    (b) In a random sample of five Americans, what is the probability that at least three sleep on their sides?
    (c) In a random sample of 350 Americans, what is the probability that at least 50% sleep on their sides?

3. The amount of fuel used by jumbo jets to take off is approximately normally distributed with a mean of 4000 gallons and a standard deviation of 125 gallons.

    (a) What is the probability that a randomly selected jumbo jet will need more than 4250 gallons of fuel to take off?
    (b) What are the shape, mean, and standard deviation of the sampling distribution of the mean gallons of fuel to take off of 40 randomly selected jumbo jets?
    (c) In a random sample of 40 jumbo jets, what is the probability that the mean number of gallons of fuel needed to take off is less than 3950?
    (d) Would the answer to (a), (b), or (c) be affected if the original population of gallons of fuel used by jumbo jets to take off were skewed instead of normal? Explain.

4. Five AP Statistics students want to extend their knowledge beyond the standard course content and read about the *finite population correction factor*, $\sqrt{\dfrac{N-n}{N-1}}$, where N is the population size and n is the sample size. To study this concept, they each predict winners for each of the 32 games in the first round of the NCAA men's basketball "March Madness" tournament. They note that the number of winners each student picked gives a set of size $N = 5$ consisting of the elements 6, 9, 11, 13, and 21.

   (a) Determine the population mean μ and the population standard deviation σ for this set.

   (b) List all possible samples of size $n = 2$ [there are $C(5, 2) = 10$ of them], and determine the mean $\bar{x}$ of each.

   (c) Calculate the mean $\mu_{\bar{x}}$ and standard deviation $\sigma_{\bar{x}}$ of the distribution in (b).

   (d) Compare $\mu_{\bar{x}}$ to μ and $\sigma_{\bar{x}}$ to $\dfrac{\sigma}{\sqrt{n}}\sqrt{\dfrac{N-n}{N-1}}$.

   (e) More generally, what can be said about $\dfrac{\sigma}{\sqrt{n}}\sqrt{\dfrac{N-n}{N-1}}$ if the population size N is very large compared to the sample size n?

The answers for this quiz can be found in the Appendix on page 607.

QUIZ 23

Multiple-Choice Questions

> **Directions:** The questions that follow are each followed by five suggested answers. Choose the response that best answers the question.

1. Suppose a population is skewed right. For which of the following sample sizes would the sampling distribution of $\bar{x}$ be closest to normal?

   (A) 10
   (B) 30
   (C) 50
   (D) 100
   (E) According to the central limit theorem, all give normal sampling distributions.

2. Which of the following is a true statement?

   (A) The sampling distribution of $\hat{p}$ has a mean equal to the population proportion p.
   (B) The sampling distribution of $\hat{p}$ has a standard deviation equal to $\sqrt{np(1-p)}$.
   (C) The sampling distribution of $\hat{p}$ has a standard deviation that becomes larger as the sample size becomes larger.
   (D) The sampling distribution of $\hat{p}$ is considered close to normal provided that $n \geq 30$.
   (E) The sampling distribution of $\hat{p}$ is always close to normal.

3. In a school of 2500 students, the students in an AP Statistics class are planning a random survey of 100 students to estimate the proportion who would rather drop lacrosse than band during this time of severe budget cuts. Their teacher suggests instead to survey 200 students in order to

   (A) reduce bias.
   (B) reduce variability.
   (C) increase bias.
   (D) increase variability.
   (E) make possible stratification between lacrosse and band.

4. The distribution of ages of people who died last year in the United States is skewed left. What happens to the sampling distribution of sample means as the sample size goes from $n = 50$ to $n = 200$?

   (A) The mean gets closer to the population mean, the standard deviation stays the same, and the shape becomes more skewed left.
   (B) The mean gets closer to the population mean, the standard deviation becomes smaller, and the shape becomes more skewed left.
   (C) The mean gets closer to the population mean, the standard deviation stays the same, and the shape becomes closer to normal.
   (D) The mean gets closer to the population mean, the standard deviation becomes smaller, and the shape becomes closer to normal.
   (E) The mean stays the same, the standard deviation becomes smaller, and the shape becomes closer to normal.

5. Which of the following are unbiased estimators for the corresponding population parameters?

    I. Sample means
    II. Sample proportions
    III. Difference of sample means
    IV. Difference of sample proportions

    (A) None are unbiased.
    (B) I and II only
    (C) I and III only
    (D) III and IV only
    (E) All are unbiased.

6. Researchers found that in the aftermath of the 2011 Fukushima nuclear disaster, 12% of the pale grass blue butterfly larvae developed mutations as adults. What is the probability that in a random sample of 300 of these butterfly larvae, more than 15% developed mutations as adults?

    (A) $P\left(z > \dfrac{0.12 - 0.15}{\sqrt{\dfrac{(0.12)(0.88)}{300}}}\right)$

    (B) $P\left(z > \dfrac{0.15 - 0.12}{\sqrt{\dfrac{(0.12)(0.88)}{300}}}\right)$

    (C) $P\left(z > \dfrac{0.12 - 0.15}{\sqrt{\dfrac{(0.15)(0.85)}{300}}}\right)$

    (D) $P\left(z > \dfrac{0.15 - 0.12}{\left(\dfrac{(0.12)(0.88)}{\sqrt{300}}\right)}\right)$

    (E) $P\left(z > \dfrac{0.12 - 0.15}{\left(\dfrac{(0.15)(0.85)}{\sqrt{300}}\right)}\right)$

7. Suppose that, using accelerometers in helmets, researchers determine that boys playing high school football absorb an average of 355 hits to the head with a standard deviation of 80 hits during a season (including both practices and games). What is the probability on a randomly selected team of 48 players that the average number of head hits per player is between 340 and 360?

(A) $P(340 < z < 360)$

(B) $P(340 - 355 < z < 360 - 355)$

(C) $P\left(\dfrac{340 - 355}{80} < z < \dfrac{360 - 355}{80}\right)$

(D) $P\left(\dfrac{340 - 355}{\sqrt{48}} < z < \dfrac{360 - 355}{\sqrt{48}}\right)$

(E) $P\left(\dfrac{340 - 355}{\frac{80}{\sqrt{48}}} < z < \dfrac{360 - 355}{\frac{80}{\sqrt{48}}}\right)$

8. Suppose "sleep-trained" babies (allowed to cry themselves to sleep) wake up an average of 1.2 times a night with a standard deviation of 0.3 times, while untrained babies wake up an average of 1.8 times a night with a standard deviation of 0.5 times. In a random sample of 80 babies, half of which are sleep-trained, what is the probability that the untrained babies in the sample wake up an average number of times greater than 0.75 more than the average of the sleep-trained babies?

(A) $P\left(z > \dfrac{0.75 - 0.6}{\sqrt{\dfrac{(0.5)^2}{40} - \dfrac{(0.3)^2}{40}}}\right)$

(B) $P\left(z > \dfrac{0.75 - 0.6}{\sqrt{\dfrac{(0.5)^2}{40} + \dfrac{(0.3)^2}{40}}}\right)$

(C) $P\left(z > \dfrac{0.75 - 0.6}{\sqrt{\dfrac{0.5}{40} - \dfrac{0.3}{40}}}\right)$

(D) $P\left(z > \dfrac{0.75 - 0.6}{\sqrt{\dfrac{0.5}{40} + \dfrac{0.3}{40}}}\right)$

(E) $P\left(z > \dfrac{0.75 - 0.6}{\dfrac{0.5}{\sqrt{40}} + \dfrac{0.3}{\sqrt{40}}}\right)$

9. A random sample of 50 high school girls has a mean of 287 friends on a popular social media platform with a standard deviation of 68 friends. An independent random sample of 50 high school boys has a mean of 165 friends on the same platform with a standard deviation of 45 friends. What is the standard error of the difference (girls – boys) between the sample means?

(A) $\sqrt{\dfrac{68 - 45}{100}}$

(B) $\sqrt{\dfrac{68^2 - 45^2}{50}}$

(C) $\sqrt{\dfrac{68^2 - 45^2}{100}}$

(D) $\sqrt{\dfrac{68^2 + 45^2}{50}}$

(E) $\sqrt{\dfrac{68^2 + 45^2}{100}}$

Free-Response Questions

Directions: You must show all work and indicate the methods you use. You will be graded on the correctness of your methods and on the accuracy of your final answers.

1. In a random sample of men working on Wall Street, the following summary statistics are from a question on the number of minutes spent grooming before leaving for the office in the morning:

$$n = 90, \bar{x} = 6.5, \text{median} = 3.8$$

(a) What is probably true about the distribution given that the mean is so much greater than the median?
(b) Are conditions for inference met? Explain.
(c) If a different sample size would result in a larger $\sigma_{\bar{x}}$, is the sample size smaller or larger than 90? Explain.

2. Studies have shown that teenage drivers are three times more likely to be involved in a fatal crash than drivers age 20 and older, and so insurance rates for teenagers are higher. The average yearly cost of teenage auto insurance is $2275 with a standard deviation of $650. A simple random sample is taken of 90 teenage drivers.

(a) Explain why there is insufficient information to determine the probability that a randomly chosen teenage driver pays over $2400.
(b) What are the mean and standard deviation for the sampling distribution for $\bar{x}$, the mean amount paid for insurance?
(c) What is the probability that the average amount paid in the sample is more than $2400?

3. It is estimated that a new baby deprives each of its parents of an average of 374 hours of sleep in the first year with a standard deviation of 38.55 hours. Assume an approximately normal distribution of deprived hours.

    (a) What is the interquartile range (IQR) of the distribution of hours deprived?
    (b) In a random sample of three parents of new babies, what is the probability that a majority are deprived of more than 400 hours?
    (c) In a random sample of three parents of new babies, what is the probability that the mean number of deprived hours is more than 400?

4. Administrators of a national writing exam are concerned about any loss of consistency with regard to the graders they use. A random sample of five of their graders is given the same student's paper to grade. Past such evaluations have shown a variance in scores to be around 3 pts^2. The scores for these graders are 70, 71, 71, 73, and 75.

    (a) Calculate the mean $\bar{x}$ and the variance s^2 for this set of $n = 5$ scores.

    It can be shown that the sampling distribution of $\dfrac{(n-1)s^2}{\sigma^2}$ follows a known distribution.

    (b) With $n = 5$ and $\sigma^2 = 3$, calculate $\dfrac{(n-1)s^2}{\sigma^2}$.

    Using simulation, 200 samples, each of size 5, were randomly generated from a population with mean 72 and variance 3. The dotplot below shows the simulated sampling distribution of $Q = \dfrac{(n-1)s^2}{3}$.

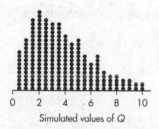

Simulated values of Q

    (c) Given your calculation in (b) and given the above dotplot, does it appear that there is a significant loss of consistency with regard to the graders being used? Justify your answer.

The answers for this quiz can be found in the Appendix on page 610.

SUMMARY

- Drawing pictures of normal curves with horizontal lines showing raw scores and z-scores is usually helpful.
- Areas (probabilities) under a normal curve can be found using calculator functions, such as `normalcdf` on TI calculators, Ncd on the Casio Prizm, and NORMAL_CDF on the HP Prime.
- Critical values corresponding to given probabilities can be found using calculator functions, such as `invNorm` on TI calculators, InvN on the Casio Prizm, and NORMALD_ICDF on the HP Prime.
- While not required, the normal probability plot is very useful in gauging whether a given distribution of data is approximately normal. If the plot is nearly straight, the data are nearly normal.
- Provided that conditions are met, the sampling distribution of sample proportions is approximately normal with mean p and standard deviation $\sqrt{\dfrac{p(1-p)}{n}}$.
- Provided that conditions are met, the sampling distribution of sample means is approximately normal with mean μ and standard deviation $\dfrac{\sigma}{\sqrt{n}}$.
- Provided that conditions are met, the sampling distribution of the difference of sample proportions is approximately normal with mean $p_1 - p_2$ and standard deviation $\sqrt{\dfrac{p_1(1-p_1)}{n_1} + \dfrac{p_2(1-p_2)}{n_2}}$.
- Provided that conditions are met, the sampling distribution of the difference of sample means is approximately normal with mean $\mu_1 - \mu_2$ and standard deviation $\sqrt{\dfrac{\sigma_1^2}{n_1} + \dfrac{\sigma_2^2}{n_2}}$.

Inference for Categorical Data: Proportions

6

(12–15% AP EXAM WEIGHTING)

→ **THE MEANING OF A CONFIDENCE INTERVAL**

→ **CONDITIONS FOR INFERENCE**

→ **CONFIDENCE INTERVAL FOR A PROPORTION**

→ **LOGIC OF SIGNIFICANCE TESTING**

→ **HYPOTHESIS TEST FOR A PROPORTION**

→ **CONFIDENCE INTERVAL FOR THE DIFFERENCE OF TWO PROPORTIONS**

→ **HYPOTHESIS TEST FOR THE DIFFERENCE OF TWO PROPORTIONS**

→ **QUIZ 24**

→ **QUIZ 25**

We want to know some important truth about a population, but in practical terms this truth is unknowable. What's the average adult human body weight? What proportion of people have high cholesterol? What we can do is carefully collect data from as large and representative a group of individuals as practically possible and then use this information to estimate the value of the population parameter. How close are we to the truth? We know that different samples will give different estimates, and so sampling error (sampling variability) is unavoidable. What is wonderful, and what we will learn in this unit, is that we can quantify this sampling error! We can make statements like "the average weight must almost surely be within 5 pounds of 176 pounds" or "the proportion of people with high cholesterol must almost surely be within ±3% of 37%."

When using a measurement from a sample, we are never able to say *exactly* what a population proportion or mean or slope is; rather, we always say we have a certain *confidence* that the population proportion or mean or slope lies in a particular *interval*. The particular interval is centered around a sample statistic and can be expressed as the sample statistic plus or minus an associated *margin of error.*

In this unit, you will use what you learned from the previous unit on sampling distributions in analyzing data to make inferences about a population proportion and about a difference between two population proportions. You will also understand the importance of checking conditions before applying inference procedures and of always making conclusions in context. In addition, you will be able to interpret confidence intervals and confidence levels. You will deepen your understanding of statistical significance. Finally, you will learn to interpret Type I and Type II errors and their possible consequences.

- To be able to check conditions to perform inference involving population proportions.
- To be able to construct and interpret a confidence interval for a population proportion and for the difference of population proportions.
- To understand the relationship among confidence level, sample size, and confidence interval width.
- To be able to interpret confidence levels.
- To be able to perform a hypothesis test to evaluate a claim about a population proportion and about a difference between population proportions.
- To be able to interpret Type I and Type II errors and their possible consequences in context.
- To understand the relationship between significance level, probability of Type II error, and power.

THE MEANING OF A CONFIDENCE INTERVAL

By using what we know about sampling distributions, we are able to establish a certain confidence that a sample proportion or mean lies within a specified interval around the population proportion or mean. However, we then have the same confidence that the population proportion or mean lies within a specified interval around the sample proportion or mean (e.g., the distance from Missoula to Whitefish is the same as the distance from Whitefish to Missoula).

Typically, we consider 90%, 95%, and 99% confidence interval estimates, but any percentage is possible. The percentage is the percentage of samples that would pinpoint the unknown p or μ within plus or minus respective margins of error. We do *not* say there is a 0.90, 0.95, or 0.99 probability that p or μ is within a certain margin of error of a given sample proportion or mean. For a given sample proportion or mean, p or μ either is or isn't within the specified interval, and so the probability is either 1 or 0.

As will be seen, there are two aspects to this concept. First, there is the *confidence interval*, usually expressed in the form:

$$\text{estimate} \pm \text{margin of error}$$

Second, there is the *success rate for the method*, called the *confidence level*, that is, the proportion of times repeated applications of this method would capture the true population parameter.

CONDITIONS FOR INFERENCE

The following are the two standard assumptions for our inference procedures and the "ideal" way they are met. Of course, we seldom, if ever, meet ideals!

1. **INDEPENDENCE ASSUMPTION:** Individuals in a sample or an experiment must be independent of each other, and this is obtained through random sampling or random selection. For samples, the sample should be a simple random sample. For experiments, the subjects should be randomly assigned to treatments. Independence across samples is obtained by selecting two (or more) separate random samples. Always examine how the data were collected to check if the assumption of independence is reasonable. Sample size can also affect independence. Because sampling is usually done without replacement, if the sample is too large, lack of independence becomes a concern. So, we typically require that the sample size n be no larger than 10% of the population (the 10% Rule).

> **NOTE**
>
> Proportions come from the binomial, for which the probability of success shouldn't change. But this does change every time one subject is picked and removed from the population. However, the effect is small if the population is large compared to the sample size.

2. **NORMALITY ASSUMPTION:** Inference for proportions is based on a normal model for the sampling distribution of $\hat{p}$, but actually we have a binomial distribution. Fortunately, the binomial is approximately normal if both np and $nq \geq 10$. Inference for means is based on a normal model for the sampling distribution of $\bar{x}$; this is true if the population is normal and is approximately true (thanks to the CLT) if the sample size is large enough (typically we accept $n \geq 30$). With regard to means, this is referred to as the *Normal/Large Sample* condition.

nearly normal → boxplot

➡ EXAMPLE 6.1

If we pick a simple random sample of size 80 from a large population, which of the following values of the population proportion p would allow use of the normal model for the sampling distribution of $\hat{p}$?

(A) 0.10
(B) 0.15
(C) 0.90
(D) 0.95
(E) 0.99

Answer: **(B)** The relevant condition is that both np and $nq \geq 10$. In (A), $np = (80)(0.10) = 8$; in (C), $nq = (80)(0.10) = 8$; in (D), $nq = (80)(0.05) = 4$; and in (E), $nq = (80)(0.01) = 0.8$. However, in (B), $np = (80)(0.15) = 12$ and $nq = (0.85)(80) = 68$ are both ≥ 10.

➡ EXAMPLE 6.2

It is hypothesized that the ratio $\dfrac{\text{outer edges of lips}}{\text{upper ridges of lips}}$ is most pleasing when equal to the golden ratio, $\varphi \approx 1.618$. A student would like to calculate the mean ratio for students in her high school and plans to randomly sample 50 out of the 780 students at the school. Does this satisfy the 10% Rule, and what is the purpose of the rule?

Answer: 10% of 780 = 78, and 50 < 78, so yes, the 10% Rule is satisfied. In picking the sample of 50 students, the sampling will obviously be done without replacement, because the same student will not be used twice. If the sample were too large compared to the size of the population, lack of independence would have become a concern.

➡ EXAMPLE 6.3

A human resources director is interested in the proportion of employees in an 85-member department who believe they are saving enough for retirement. She randomly selects 12 employees to interview. Explain whether conditions for inference are satisfied.

Answer: Even though it may or may not have been the ideal method of sampling, it is stated that a random sample was taken, satisfying the randomization condition. The 10% condition (related to independence) is violated because 10% of 85 is 8.5, and 12 is greater than 8.5. The normality assumption is also violated because np and nq total 12, so both cannot be ≥ 10.

➡ EXAMPLE 6.4 _____

An English teacher is interested in the mean length of words in a Hemingway novel. She picks a random sample of 25 words. Explain whether conditions for inference are satisfied.

Answer: Even though it may or may not have been the ideal method of sampling, it is stated that a *random* sample was taken, satisfying the randomization condition. The 10% condition (related to independence) is also satisfied because clearly 25 is less than 10% of the number of words in an entire novel. However, the Normal/Large Sample condition is not satisfied because it is not given that the distribution of all words in the novel is approximately normal and the sample size is not large, that is, $n = 25 < 30$.

CONFIDENCE INTERVAL FOR A PROPORTION

We are interested in estimating a population proportion p by considering a single sample proportion $\hat{p}$. This sample proportion is just one of a whole universe of sample proportions, and from Unit 5 we remember the following:

1. The set of all sample proportions is approximately normally distributed.
2. The mean $\mu_{\hat{p}}$ of the set of sample proportions equals p, the population proportion.
3. The standard deviation $\sigma_{\hat{p}}$ of the set of sample proportions is approximately equal to $\sqrt{\dfrac{p(1-p)}{n}}$.

Remember that we are really using a normal approximation to the binomial, so $n\hat{p}$ and $n(1-\hat{p})$ should both be at least 10. Furthermore, in making calculations and drawing conclusions from a specific sample, it is important that the sample be a *simple random sample*. Finally, the population should be large, typically checked by the assumption that the sample is less than 10% of the population. (If the population is small and the sample exceeds 10% of the population, models other than the normal are more appropriate.)

In finding confidence interval estimates of the population proportion p, how do we find $\sigma_{\hat{p}} = \sqrt{\dfrac{p(1-p)}{n}}$ since p is unknown? The reasonable procedure is to use the sample proportion $\hat{p}$:

$$\sigma_{\hat{p}} \approx \sqrt{\frac{\hat{p}(1-\hat{p})}{n}}$$

When the standard deviation is estimated in this way (using the sample), we use the term *standard error*:

$$SE(\hat{p}) = SE_{\hat{p}} = \sqrt{\frac{\hat{p}(1-\hat{p})}{n}}$$

➧ **EXAMPLE 6.5** _____

If 42% of a simple random sample of 550 young adults say that whoever asks for the date should pay for the first date, determine a 99% confidence interval estimate for the true proportion of all young adults who would say that whoever asks for the date should pay for the first date.

Answer: The parameter is p, which represents the proportion of the population of young adults who would say that whoever asks for the date should pay for the first date. We check that $n\hat{p} = 550(0.42) = 231 \geq 10$ and $n(1 - \hat{p}) = 550(0.58) = 319 \geq 10$. We are given that the sample is an SRS, and 550 is clearly less than 10% of all young adults. Since $\hat{p} = 0.42$, the standard error of the set of sample proportions is

$$SE(\hat{p}) = \sqrt{\frac{(0.42)(0.58)}{550}} = 0.021$$

From Unit 5, we know that 99% of the sample proportions should be within 2.576 standard deviations of the population proportion. Equivalently, we are 99% certain that the population proportion is within 2.576 standard deviations of any sample proportion. Thus, the 99% confidence interval estimate for the population proportion is $0.42 \pm 2.576(0.021) = 0.42 \pm 0.054$. We say that the _margin of error_ is ± 0.054. We are 99% confident that the true proportion of young adults who would say that whoever asks for the date should pay for the first date is between 0.366 and 0.474.

NOTE

For 99% confidence, there is 0.005 in each tail and invNorm(0.005) = −2.576 or invNorm(0.995) = 2.576.

IMPORTANT

Verifying assumptions and conditions means more than simply listing them with check marks. You must show work or give some reason to confirm verification.

NOTE

We can also say, using the definition of _confidence level_, that if the interviewing procedure were repeated many times, about 99% of the resulting confidence intervals would contain the true proportion (thus we're 99% confident that the method worked for the interval we got).

➧ **EXAMPLE 6.6** _____

In a simple random sample of machine parts, 18 out of 225 were found to have been damaged in shipment. Establish a 95% confidence interval estimate for the true proportion of machine parts that are damaged in shipment.

Answer:

Parameter: Let p represent the proportion of the population of machine parts that are damaged in shipment.

Procedure: One-sample z-interval for a proportion.

Checks: $n\hat{p} = 18 \geq 10$ and $n(1 - \hat{p}) = 225 - 18 = 207 \geq 10$, we are given that the sample is an SRS, and it is reasonable to assume that 225 is less than 10% of all shipped machine parts.

Mechanics: Calculator software (such as `1-PropZInt` on the TI-84 or `1-Prop ZInterval` on the Casio Prizm) gives (0.04455, 0.11545). [For instructional purposes, we note that $\hat{p} = \frac{18}{225} = 0.08$, $SE(\hat{p}) = \sqrt{\frac{(0.08)(0.92)}{225}} = 0.0181$, and $0.08 \pm 1.96(0.0181) = 0.08 \pm 0.035$.]

Conclusion in context: We are 95% confident that the true proportion of all machine parts damaged in shipment is between 0.045 and 0.115.

TIP

The margin of error accounts for variation inherent in sampling, but it does not account for and is not a redress for bias in sampling.

IMPORTANT

After finding a confidence interval, you must always interpret the interval in _context_, including mentioning the _confidence level_ and _parameter_ and referring to the _population_, not the sample.

NOTE

On the exam, you can write:

normalcdf

$\Big($ lower bd = 0.802,

upper bd = 0.862,

mean = 0.832,

$SD = \sqrt{\dfrac{(0.832)(0.168)}{1000}}\Big)$

= 0.9888, or 98.9%.

➡️ **EXAMPLE 6.7**

In a telephone survey of a random sample of 1000 high school seniors, 832 answered yes to the question: "Do you want to get married someday?" With what confidence can it be asserted that 83.2% ± 3% of the high school senior population want to get married someday?

Answer: The parameter is *p*, which represents the proportion of the population of high school seniors who would answer yes to the question: "Do you want to get married someday?" We check that $n\hat{p} = 832 \geq 10$ and $n(1 - \hat{p}) = 1000 - 832 = 168 \geq 10$. We are given that there was a random sample, and clearly 1000 is <10% of the high school senior population:

$$\hat{p} = \frac{832}{1000} = 0.832 \text{ and so } SE(\hat{p}) = \sqrt{\frac{(0.832)(0.168)}{1000}} = 0.0118$$

The relevant *z*-scores are $\pm\dfrac{0.03}{0.0118} = \pm 2.54$. Normalcdf(-2.54,2.54)=0.9889.

In other words, 83.2% ± 3% is a 98.89% confidence interval estimate for the proportion of all high school seniors who would answer yes to the question: "Do you want to get married someday?"

TIP

On the TI-84, for a confidence interval of a proportion, use 1-PropZInt, *not* Zinterval; and *x* must be an integer.

NOTE

Finding confidence intervals *using formulas* is not necessary on free-response questions, but answer choices using formulas may well appear on multiple-choice questions.

NOTE

Interpreting *confidence level* never involves a specific interval.

As we have seen, there are two types of statements that come out of confidence intervals. First, we can interpret the **confidence interval** and say we are 90% confident that, for example, between 60% and 66% of all voters favor a bond issue. Second, we can interpret the **confidence level** and say that if this survey were conducted many times, about 90% of the resulting confidence intervals would contain the true proportion of voters who favor the bond issue.

Here are three examples of *incorrect* statements: "The percentage of all voters who support the bond issue is between 60% and 66%." "There is a 0.90 probability that the true percentage of all voters who favor the bond issue is between 60% and 66%." "If this survey were conducted many times, about 90% of the sample proportions would be in the interval (0.60, 0.66)."

One important consideration in setting up a survey is the choice of sample size. To obtain a smaller, more precise interval estimate of the population proportion, we must either decrease the degree of confidence or increase the sample size. Similarly, if we want to increase the degree of confidence, we can either accept a wider interval estimate or increase the sample size. Again, while choosing a larger sample size may seem desirable, in the real world this decision involves time and cost considerations.

In setting up a survey to obtain a confidence interval estimate of the population proportion, what should we use for $\sigma_{\hat{p}}$? To answer this question, we first must consider how large $\sqrt{p(1-p)}$ can be. We plot various values of *p*:

p:	0.1	0.2	0.3	0.4	0.5	0.6	0.7	0.8	0.9
$\sqrt{p(1-p)}$:	0.3	0.4	0.458	0.490	0.5	0.490	0.458	0.4	0.3

This results in 0.5, the intuitive answer. Thus, $\sqrt{\dfrac{p(1-p)}{n}}$ is at most $\dfrac{0.5}{\sqrt{n}}$. We make use of this fact to determine sample sizes in questions such as those in Examples 6.8 and 6.9.

➡ EXAMPLE 6.8

An Environmental Protection Agency (EPA) investigator wants to know the proportion of fish that are inedible because of chemical pollution downstream of an offending factory. If the answer must be within ±0.03 at the 96% confidence level, how many fish should be in the sample tested?

Answer: The parameter is p, which represents the proportion of the population of fish that are inedible because of chemical pollution downstream of an offending factory. We want $2.05\sigma_{\hat{p}} \leq 0.03$. From the above remark, $\sigma_{\hat{p}}$ is at most $\frac{0.5}{\sqrt{n}}$, and so it is sufficient to consider $2.05\left(\frac{0.5}{\sqrt{n}}\right) \leq 0.03$. Algebraically, we have $\sqrt{n} \geq \frac{2.05(0.5)}{0.03} = 34.17$ and $n \geq 1167.4$. Therefore, choosing a sample of 1168 fish gives the inedible proportion to within ±0.03 at the 96% level.

NOTE

For 96% confidence, there is 0.02 in each tail and `invNorm(0.02)` `=-2.05` or `invNorm(0.98)` `= 2.05`.

TIP

Noting the inequality, you must round a result "up" if it is a decimal.

Note that the accuracy of the estimate does *not* depend on what fraction of the whole population we have sampled. What is critical is the *absolute size* of the sample. Is some minimal value of n necessary for the procedures we are using to be meaningful? Since we are using the normal approximation to the binomial, both np and $n(1-p)$ should be at least 10 (see Unit 5).

➡ EXAMPLE 6.9

A study is undertaken to determine the proportion of industry executives who believe that workers' pay should be based on individual performance. How many executives should be interviewed if an estimate is desired at the 99% confidence level to within ±0.06? To within +0.03? To within ±0.02?

Answer: The parameter is p, which represents the proportion of the population of industry executives who believe that workers' pay should be based on individual performance. Algebraically, $2.576\left(\frac{0.5}{\sqrt{n}}\right) \leq 0.06$ gives $\sqrt{n} \geq \frac{2.576(0.5)}{0.06} = 21.5$, so $n \geq$ 462.25. Similarly, $2.576\left(\frac{0.5}{\sqrt{n}}\right) \leq 0.03$ gives $\sqrt{n} \geq \frac{2.576(0.5)}{0.03} = 42.9$, so $n \geq 1840.4$. Finally, $2.576\left(\frac{0.5}{\sqrt{n}}\right) \leq 0.02$ gives $\sqrt{n} \geq \frac{2.576(0.5)}{0.02} = 64.4$, so $n \geq 4147.4$. Thus, at the 99% confidence level, at least 463 executives should be interviewed for a margin of error of ±0.06, at least 1841 should be interviewed for a margin of error of ±0.03, and at least 4148 should be interviewed for a margin of error of ±0.02.

Note that to cut the interval estimate in half (from ±0.06 to ±0.03), we would have to increase the sample size fourfold and to cut the interval estimate to a third (from ±0.06 to ±0.02). A ninefold increase in the sample size would be required (answers are not exact because of round-off error).

More generally, to divide the interval estimate by d without affecting the confidence level, we must increase the sample size by a multiple of d^2.

LOGIC OF SIGNIFICANCE TESTING

Closely related to the problem of estimating a population proportion or mean is the problem of testing a hypothesis about a population proportion or mean. For example, a travel agency might determine an interval estimate for the proportion of sunny days in the Virgin Islands or, alternatively, might test a tourist bureau's claim about the proportion of sunny days. A major stockholder in a construction company might ascertain an interval estimate for the proportion of successful contract bids or, alternatively, might test a company spokesperson's claim about the proportion of successful bids. A consumer protection agency might determine an interval estimate for the mean nicotine content of a particular brand of cigarettes or, alternatively, might test a manufacturer's claim about the mean nicotine content of its cigarettes. In each of these cases, the experimenter must decide whether the interest lies in estimating a population proportion or mean or in testing a claimed proportion or mean.

TIP

Understand that the sample statistic almost never equals the claimed population parameter. Either the sample statistic differs by random chance or the claimed parameter is wrong.

The general testing procedure is to choose a specific hypothesis to be tested, called the *null hypothesis*, pick an appropriate random sample, and then use measurements from the sample to determine the likelihood of the null hypothesis. If the sample statistic is far enough away from the claimed population parameter, we say that there is sufficient evidence to reject the null hypothesis. We attempt to show that the null hypothesis is unacceptable by showing that it is improbable.

Consider the context of the population proportion. The null hypothesis H_0 is stated in the form of an equality statement about the *population* proportion (for example, $H_0: p = 0.37$). There is an *alternative hypothesis*, stated in the form of an inequality (for example, $H_a: p < 0.37$ or $H_a: p > 0.37$ or $H_a: p \neq 0.37$). The strength of the sample statistic $\hat{p}$ can be gauged through its associated *P-value*, which is the probability of obtaining a sample statistic as extreme (or more extreme) as the one obtained if the null hypothesis is assumed to be true. The smaller the *P*-value, the more significant the difference between the null hypothesis and the sample results.

Several points worth noting:

- The equality reference in the null hypothesis can be stated using $\geq$ or $\leq$ when the alternative is one-sided in the other direction.
- Never base your hypotheses on what you find in the sample data.
- Note that only population parameter symbols like μ, p, and β appear in H_0 and H_a. Sample statistics like $\bar{x}$, $\hat{p}$, and b do NOT appear in H_0 or H_a.
- For a one-sided test (the alternative is $<$ or $>$), the *P*-value is the proportion of values at or more extreme than the test statistic (in the direction of the alternative), while for a two-sided test (the alternative is $\neq$), the *P*-value is twice the proportion of values at or more extreme than the test statistic.
- The *P*-value is NOT the probability of the null hypothesis being true.

NOTE

The significance level, α, is fixed at the beginning of the hypothesis test. The conclusion to the test will be based upon whether or not the calculated *P*-value is less than or greater than α.

There are two types of possible errors: the error of mistakenly rejecting a true null hypothesis and the error of mistakenly failing to reject a false null hypothesis. The α-risk, also called the *significance level* of the test, is the probability of committing a *Type I error* and mistakenly rejecting a true null hypothesis. A *Type II error*, a mistaken failure to reject a false null hypothesis, has associated probability β. There is a different value of β for each possible correct value for the population parameter p. For each β, $1 - \beta$ is called the "power" of the test against the associated correct value. The *power* of a hypothesis test is the probability that a Type II error is not committed. That is, given a true alternative, the power is the probability of rejecting the false null hypothesis. Increasing the sample size and increasing the significance level are both ways of increasing the power. Also note that a true null that is further away from the hypothesized null is more likely to be detected, thus offering a more powerful test.

		Population truth	
		H_0 true	H_0 false
Decision based on sample	Reject H_0	Type I error	Correct decision
	Fail to reject H_0	Correct decision	Type II error

A simple illustration of the difference between a Type I and a Type II error is as follows. Suppose the null hypothesis is that all systems are operating satisfactorily with regard to a NASA launch. A Type I error would be to delay the launch mistakenly thinking that something was malfunctioning when everything was actually OK. A Type II error would be to fail to delay the launch mistakenly thinking everything was OK when something was actually malfunctioning. The power is the probability of recognizing a particular malfunction. (Note the complementary aspect of power, a "good" thing, with Type II error, a "bad" thing.)

The U.S. justice system provides another often-quoted illustration. If the null hypothesis is that a person is innocent, a Type I error results when an innocent person is found guilty while a Type II error results when a guilty person is not convicted. We try to minimize Type I errors in criminal trials by demanding unanimous jury guilty verdicts. In civil suits, however, many states try to minimize Type II errors by accepting simple majority verdicts.

It should be emphasized that with regard to calculations, questions like "What is the *power* of this test?" and "What is the probability of a *Type II error* in this test?" cannot be answered without reference to a specific alternative hypothesis. Furthermore, AP students are not required to know how to calculate these probabilities. However, they are required to understand the concepts and interactions among the concepts.

TIP

Be able to identify Type I and Type II errors and give possible consequences of each.

HYPOTHESIS TEST FOR A PROPORTION

When conducting a hypothesis test for a proportion, we must check that we have a simple random sample, that both np_0 and $n(1 - p_0)$ are at least 10 (where p_0 is the claimed proportion), and that the sample size is less than 10% of the population.

It is important to understand that because the *P*-value is a conditional probability, calculated based on the assumption that the null hypothesis, H_0: $p = p_0$, is true, we use the claimed proportion p_0 both in checking the $np_0 \geq 10$ and $n(1 - p_0) \geq 10$ conditions and in calculating the standard deviation $\sigma_{\hat{p}} = \sqrt{\dfrac{(p_0)(1 - p_0)}{n}}$.

TIP

It is not enough to simply list the conditions. You must actually check the assumptions and conditions before proceeding with a test.

➡ **EXAMPLE 6.10** _____

A union spokesperson claims that 75% of union members will support a strike if their basic demands are not met. A company negotiator believes the true percentage is lower and runs a hypothesis test. What is the conclusion if 87 out of a simple random sample of 125 union members say they will strike?

Answer:

Parameter: Let *p* represent the proportion of all union members who will support a strike if their basic demands are not met.

Hypotheses: H_0: $p = 0.75$ and H_a: $p < 0.75$.

Procedure: One-sample *z*-test for a population proportion.

(continued)

TIP

Hypotheses are always about the population, never about a sample; note that H_0 is $p = 0.75$, not $\hat{p} = 0.75$.

TIP

For a hypothesis test, we check if np_0 and $n(1 - p_0)$ are ≥ 10. For a confidence interval, we check if $n\hat{p}$ and $n(1 - \hat{p})$ are ≥ 10.

TIP

Although performing hypothesis tests *using formulas* is not necessary on free-response questions, answer choices using formulas may well appear on multiple-choice questions.

TIP

Always give the conclusion in context.

Checks: $np_0 = (125)(0.75) = 93.75$ and $n(1 - p_0) = (125)(0.25) = 31.25$ are both ≥ 10, it is given that we have an SRS, and we must assume that 125 is less than 10% of the total union membership.

Mechanics: Calculator software (such as `1-PropZTest` on the TI-84 or Z-1-PROP on the Casio Prizm) gives $z = -1.394$ and $P = 0.0816$.

[For instructional purposes, we note that the observed sample proportion is

$\hat{p} = \dfrac{87}{125} = 0.696$, and using the claimed proportion, 0.75, we have

$\sigma_{\hat{p}} = \sqrt{\dfrac{(0.75)(0.25)}{125}} = 0.03873$ and $z = \dfrac{0.696 - 0.75}{0.03873} = -1.394$, with a resulting *P*-value

of 0.0817.]

Conclusion in context with linkage to the P-value: There are two possible answers:

1. With this large of a *P*-value, $0.0816 > 0.05$, there is not sufficient evidence to reject H_0; that is, there is not sufficient evidence at the 5% significance level that the true percentage of union members who support a strike is less than 75%.

2. With this small of a *P*-value, $0.0816 < 0.10$, there is sufficient evidence to reject H_0; that is, there is sufficient evidence at the 10% significance level that the true percentage of union members who support a strike is less than 75%.

> **IMPORTANT**
>
> We never "accept" a null hypothesis; we either do or do not have sufficient evidence to reject it.

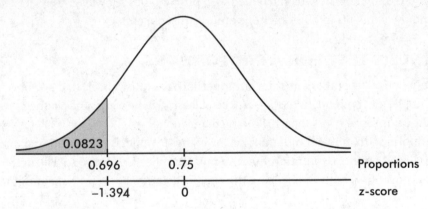

On the TI-Nspire, the result shows as:

zTest_1Prop 0.75,87,125,-1: *stat.results*	"Title"	"1-Prop z Test"
	"Alternate Hyp"	"prop < p0"
	"z"	-1.39427
	"PVal"	0.081617
	"p̂"	0.696
	"n"	125.

For each of the two possible answers above, what error might have been committed, Type I or Type II, and what would be a possible consequence?

Answer: If the *P*-value is considered large, $0.0816 > 0.05$, so that there is not sufficient evidence to reject the null hypothesis, there is the possibility that a false null hypothesis would mistakenly not be rejected and thus a Type II error would be committed. In this case, the union might call a strike thinking they have greater support than they actually do. If the P-value is considered small, 0.0816 < 0.10, so that there is sufficient evidence to reject the null hypothesis, there is the possibility that a true null hypothesis would mistakenly be rejected, and thus a Type I error would be committed. In this case, the union might not call for a strike thinking they don't have sufficient support when they actually do have support.

➡️ **EXAMPLE 6.11**

A cancer research group surveys a random sample of 500 women more than 40 years old to test the hypothesis that 28% of women in this age group have regularly scheduled mammograms. Should the hypothesis be rejected at the 5% significance level if 151 of the women respond affirmatively?

Answer:

Parameter: Let p represent the proportion of the population of women over 40 years old who have regularly scheduled mammograms.

Hypotheses: H_0: $p = 0.28$ and H_a: $p \neq 0.28$. (Since no suspicion is voiced that the 28% claim is low or high, we run a two-sided z-test for proportions.)

Procedure: One-sample z-test for a population proportion.

Checks: $np_0 = (500)(0.28) = 140$ and $n(1 - p_0) = (500)(0.72) = 360$ are both ≥ 10, we are given a random sample, and clearly $500 < 10\%$ of all women over 40 years old.

Mechanics: Calculator software gives $z = 1.0956$ and $P = 0.2732$.

[For instructional purposes, we note that the observed $\hat{p} = \dfrac{151}{500} = 0.302$,

$\sigma_{\hat{p}} = \sqrt{\dfrac{(0.28)(0.72)}{500}} = 0.02008$, and $z = \dfrac{0.302 - 0.28}{0.02008} = 1.0956$, which corresponds to a

tail probability of 0.1366. Doubling this value (because the test is two-sided) gives a P-value of 0.2732.]

Conclusion in context with linkage to the P-value: With this large of a P-value, 0.2732 > 0.05, there is not sufficient evidence to reject H_0; that is, the cancer research group does not have sufficient evidence to dispute the claim that 28% of all women in this age group have regularly scheduled mammograms.

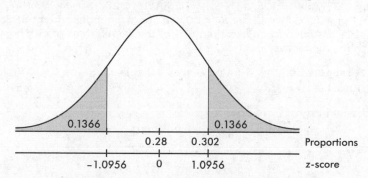

In the answer to Example 6.11, what error might have been committed, Type I or Type II, and what would be a possible consequence?

Answer: There was not sufficient evidence to reject the null hypothesis, so a Type II error might be committed, that is, failing to reject a false null hypothesis. If so, the cancer research group would be doing their research assuming that 28% of women over 40 years old have regularly scheduled mammograms whereas the true figure is different.

CONFIDENCE INTERVAL FOR THE DIFFERENCE OF TWO PROPORTIONS

From Unit 5, we have the following information about the sampling distribution of $\hat{p}_1 - \hat{p}_2$:

1. The set of all differences of sample proportions is approximately normally distributed.
2. The mean of the set of differences of sample proportions equals $p_1 - p_2$, the difference of the population proportions.
3. The standard deviation σ_d of the set of differences of sample proportions is approximately equal to:

$$\sqrt{\frac{p_1(1-p_1)}{n_1} + \frac{p_2(1-p_2)}{n_2}}$$

Remember that we are using the normal approximation to the binomial, so $n_1\hat{p}_1, n_1(1-\hat{p}_1), n_2\hat{p}_2$, and $n_2(1-\hat{p}_2)$ should all be at least 10. In making calculations and drawing conclusions from specific samples, it is important both that the samples be *simple random samples* and that they be taken *independently* of each other. Finally, the original populations should be large compared to the sample sizes, that is, check that $n_1 \leq 10\%N_1$ and $n_2 \leq 10\%N_2$.

➡ EXAMPLE 6.12

Suppose that 84% of a simple random sample of 125 nurses working 7:00 a.m. to 3:00 p.m. shifts in city hospitals express positive job satisfaction, while only 72% of an SRS of 150 nurses on 11:00 p.m. to 7:00 a.m. shifts express similar fulfillment. Establish a 90% confidence interval estimate for the difference.

Answer:

Parameters: Let p_1 represent the proportion of the population of nurses working 7:00 a.m. to 3:00 p.m. shifts in city hospitals who have positive job satisfaction. Let p_2 represent the proportion of the population of nurses working 11:00 p.m. to 7:00 a.m. shifts in city hospitals who have positive job satisfaction.

Procedure: Two-sample z-interval for a difference between population proportions, $p_1 - p_2$.

Checks: $n_1\hat{p}_1 = (125)(0.84) = 105$, $n_1(1-\hat{p}_1) = (125)(0.16) = 20$, $n_2\hat{p}_2 = (150)(0.72) = 108$, and $n_2(1-\hat{p}_2) = (150)(0.28) = 42$ are all ≥ 10; we are given independent SRSs; and the sample sizes are assumed to be less than 10% of the populations of city hospital nurses on the two shifts, respectively.

Mechanics: `2-PropZInt` on the TI-84 or `2-Prop ZInterval` on the Casio Prizm give (0.0391, 0.2009).

[For instructional purposes, we note:

$$n_1 = 125 \quad n_2 = 150$$
$$\hat{p}_1 = 0.84 \quad \hat{p}_2 = 0.72$$
$$SE(\hat{p}_1 - \hat{p}_2) = \sqrt{\frac{(0.84)(0.16)}{125} + \frac{(0.72)(0.28)}{150}} = 0.0492$$

The observed difference is $0.84 - 0.72 = 0.12$, and the critical z-scores are ±1.645. The confidence interval estimate is $0.12 \pm 1.645(0.0492) = 0.12 \pm 0.081$.]

Conclusion in context: We are 90% confident that the true proportion of satisfied nurses on 7:00 a.m. to 3:00 p.m. shifts is between 0.039 and 0.201 higher than the true proportion for nurses on 11:00 p.m. to 7:00 a.m. shifts.

TIP

When interpreting the confidence interval for a difference, you must indicate which direction the difference is taken.

➡ EXAMPLE 6.13 _____

A grocery store manager notes that in a simple random sample of 85 people going through the express checkout line, only 10 paid with checks, whereas, in an SRS of 92 customers passing through the regular line, 37 paid with checks. Find a 95% confidence interval estimate for the difference between the proportion of customers going through the two different lines who use checks.

Answer:

Parameters: Let p_1 represent the proportion of the population of people going through the express checkout line who pay with checks. Let p_2 represent the proportion of the population of people going through the regular checkout line who pay with checks.

Procedure: Two-sample z-interval for a difference between population proportions, $p_1 - p_2$.

Checks: $n_1\hat{p}_1 = 10$, $n_1(1 - \hat{p}_1) = 75$, $n_2\hat{p}_2 = 37$, and $n_2(1 - \hat{p}_2) = 55$ are all at least 10; we are given independent SRSs; and we assume that the sample sizes are less than 10% of the total number of customers.

Mechanics: `2-PropZInt` gives $(-0.4059, -0.1632)$.

[For instructional purposes, we note:

$$n_1 = 85 \qquad n_2 = 92$$

$$\hat{p}_1 = \frac{10}{85} = 0.118 \qquad \hat{p}_2 = \frac{37}{92} = 0.402$$

$$SE(\hat{p}_1 - \hat{p}_2) = \sqrt{\frac{(0.118)(0.882)}{85} + \frac{(0.402)(0.598)}{92}} = 0.0619$$

The observed difference is $0.118 - 0.402 = -0.284$, and the critical z-scores are ± 1.96. Thus, the confidence interval estimate is $-0.284 \pm 1.96(0.0619) = -0.284 \pm 0.121$.]

Conclusion in context: The manager can be 95% confident that the true proportion of all customers passing through the express checkout line who pay with checks is between 0.163 and 0.405 lower than the true proportion going through the regular checkout line who pay with checks.

HYPOTHESIS TEST FOR THE DIFFERENCE OF TWO PROPORTIONS

The null hypothesis for a difference between two proportions is $H_0: p_1 = p_2$, and so the normality condition becomes that $n_1\hat{p}_c$, $n_1(1 - \hat{p}_c)$, $n_2\hat{p}_c$, and $n_2(1 - \hat{p}_c)$ should all be at least 10, where $\hat{p}_c$ is the combined (or pooled) proportion, $\hat{p}_c = \dfrac{x_1 + x_2}{n_1 + n_2}$. The other important conditions to be checked are both that the samples be _simple random samples_ and that they be taken _independently_ of each other. The original populations should also be large compared to the sample sizes, that is, check that $n_1 \leq 10\% N_1$ and $n_2 \leq 10\% N_2$.

The fact that a sample proportion from one population is greater than a sample proportion from a second population does not automatically justify a similar conclusion about the population proportions themselves. Two points need to be stressed. First, sample proportions from the same population can vary from each other. Second, what we are really comparing are confidence interval estimates, not just single points.

For many problems, the null hypothesis states that the population proportions are equal or, equivalently, that their difference is 0:

$$H_0: p_1 - p_2 = 0$$

The alternative hypothesis is then:

$$H_a: p_1 - p_2 < 0, \quad H_a: p_1 - p_2 > 0, \quad \text{or} \quad H_a: p_1 - p_2 \neq 0$$

where the first two possibilities lead to one-sided tests and the third possibility leads to a two-sided test.

Since the null hypothesis is that $p_1 = p_2$, we call this common value p_c and use this pooled value in calculating σ_d:

$$\sigma_d = \sqrt{\frac{p_c(1-p_c)}{n_1} + \frac{p_c(1-p_c)}{n_2}} = \sqrt{p_c(1-p_c)\left(\frac{1}{n_1} + \frac{1}{n_2}\right)}$$

In practice, if $\hat{p}_1 = \frac{x_1}{n_1}$ and $\hat{p}_2 = \frac{x_2}{n_2}$, we use $\hat{p}_c = \frac{x_1 + x_2}{n_1 + n_2}$ as an estimate of p_c in calculating σ_d.

TIP

When using a calculator for this hypothesis test, the pooled p_c is automatically calculated.

➡ EXAMPLE 6.14

In a random sample of 1500 First Nations children in Canada, 162 were in child welfare care, while in an independent random sample of 1600 non-Aboriginal children, 23 were in child welfare care. Many people believe that the large proportion of indigenous children in government care is a humanitarian crisis. Do the above data give significant evidence that a greater proportion of First Nations children in Canada are in child welfare care than the proportion of non-Aboriginal children in child welfare care?

TIP

This is the situation where it is necessary to use a pooled proportion.

Answer:

Parameters: Let p_1 represent the proportion of the population of First Nations children in Canada who are in child welfare care. Let p_2 represent the proportion of the population of non-Aboriginal children in Canada who are in child welfare care.

Hypotheses: $H_0: p_1 - p_2 = 0$ or $H_0: p_1 = p_2$ and $H_a: p_1 - p_2 > 0$ or $H_a: p_1 > p_2$.

Procedure: Two-sample z-test for a difference of two population proportions.

TIP

It is often helpful to use descriptive subscripts like $p_{\text{First Nations}}$ and $p_{\text{non-Aboriginal}}$ instead of p_1 and p_2.

Checks: With $\hat{p}_c = \frac{162 + 23}{1500 + 1600} = 0.0597$, we have $n_1\hat{p}_c = 89.55$, $n_1(1 - \hat{p}_c) = 1410.45$, $n_2\hat{p}_c = 95.52$, and $n_2(1 - \hat{p}_c) = 1504.48$ are all at least 10; the samples are random and independent by design; and it is reasonable to assume the sample sizes are less than 10% of the populations.

Mechanics: Calculator software (such as `2-PropZTest`) gives $z = 11.0$ and $P = 0.000$.

[For instructional purposes, we note that $\hat{p}_1 = \frac{162}{1500} = 0.108$ and $\hat{p}_2 = \frac{23}{1600} = 0.014$.

The observed difference is $0.108 - 0.014 = 0.094$, $\hat{p}_c = \frac{162 + 23}{1500 + 1600} = 0.0597$,

$\sigma_d = \sqrt{(0.0597)(0.9403)\left(\frac{1}{1500} + \frac{1}{1600}\right)} = 0.00852$, $z = \frac{0.094 - 0}{0.00852} = 11.0$, and the tail

probability gives a P-value = `normalcdf(11.0, 1000)` = 0.000.]

TIP

Along with the P-value, you should also give the test statistic, as we have been doing.

Conclusion in context with linkage to the P-value: With this small of a P-value, $0.000 < 0.05$, there is sufficient evidence to reject H_0; that is, there is sufficient evidence that that the true proportion of all First Nations children in Canada in child welfare care is greater than the true proportion of all non-Aboriginal children in Canada in child welfare care.

Does a 95% *confidence interval* for the difference in proportions give a result consistent with the above conclusion?

Answer: Calculator software (such as `2-PropZInt`) gives that we are 95% confident that the true difference in proportions (true proportion of all First Nations children in Canada in child welfare care minus the true proportion of all non-Aboriginal children in Canada in child welfare care) is between 0.077 and 0.110. Since this interval is entirely positive, it is consistent with the conclusion from the hypothesis test.

QUIZ 24

Multiple-Choice Questions

> **Directions:** The questions that follow are each followed by five suggested answers. Choose the response that best answers the question.

1. For a class project, an AP Statistics student wishes to determine a 95% confidence interval for the proportion of students in the school building who have social media profile pages. The total number of students in the building is 3520. Which of the following combinations of n and $\hat{p}$ will satisfy the conditions for inference?

   (A) $n = 15$ and $\hat{p} = 0.67$
   (B) $n = 25$ and $\hat{p} = 0.35$
   (C) $n = 50$ and $\hat{p} = 0.17$
   (D) $n = 75$ and $\hat{p} = 0.15$
   (E) $n = 100$ and $\hat{p} = 0.91$

2. In a simple random sample of 60 teenagers, two-thirds said they would rather text a friend than call. What is the 98% confidence interval for the proportion of all teenagers who would rather text than call a friend?

   (A) $\dfrac{2}{3} \pm 1.96 \sqrt{\dfrac{(2/3)(1/3)}{60}}$

   (B) $\dfrac{2}{3} \pm 2.326 \sqrt{\dfrac{(2/3)(1/3)}{60}}$

   (C) $\dfrac{2}{3} \pm 2.405 \sqrt{\dfrac{(2/3)(1/3)}{60}}$

   (D) $\dfrac{2}{3} \pm 2.96 \dfrac{\sqrt{(2/3)(1/3)}}{60}$

   (E) $\dfrac{2}{3} \pm 3.326 \dfrac{\sqrt{(2/3)(1/3)}}{60}$

3. A 2020 random survey of 1470 adult Americans found that 29% of the American adult population would be willing to pay higher taxes if the government offered "Medicare for All." With what degree of confidence can it be concluded that $29\% \pm 3\%$ of the American adult population would be willing to pay higher taxes if the government offered "Medicare for All"?

   (A)   49.5%
   (B)   90%
   (C)   95%
   (D)   98%
   (E)   99%

4. A concerned-scientists action group wishes to learn the proportion of high school students who believe that humans are the principal cause of observed global warming. What sample size n should be obtained to determine with 98% confidence the proportion to within ±2.5%?

(A) $n \leq \sqrt{\dfrac{0.025}{2.326 \times 0.5}}$

(B) $n \leq \sqrt{\dfrac{2.326 \times 0.5}{0.025}}$

(C) $n \leq \left(\dfrac{2.326 \times 0.5}{0.025}\right)^2$

(D) $n \geq \sqrt{\dfrac{2.326 \times 0.5}{0.025}}$

(E) $n \geq \left(\dfrac{2.326 \times 0.5}{0.025}\right)^2$

5. The National Research Council of the Philippines reported that 210 of 361 members in biology are women, but only 34 of 86 members in mathematics are women. Establish a 95% confidence interval estimate of the difference in proportions of all women in biology and all women in mathematics in the Philippines.

(A) $\left(\dfrac{210}{361} - \dfrac{34}{86}\right) \pm 1.96 \sqrt{\dfrac{\left(\frac{210}{361}\right)\left(1 - \frac{210}{361}\right)}{361} + \dfrac{\left(\frac{34}{86}\right)\left(1 - \frac{34}{86}\right)}{86}}$

(B) $\left(\dfrac{210}{361} - \dfrac{34}{86}\right) \pm 1.96 \sqrt{\left(\dfrac{210 + 34}{361 + 86}\right)\left(1 - \dfrac{210 + 34}{361 + 86}\right)\left(\dfrac{1}{361} + \dfrac{1}{86}\right)}$

(C) $\left(\dfrac{210 + 34}{361 + 86}\right) \pm 1.96 \sqrt{\dfrac{\left(\frac{210}{361}\right)\left(1 - \frac{210}{361}\right)}{361} + \dfrac{\left(\frac{34}{86}\right)\left(1 - \frac{34}{86}\right)}{86}}$

(D) $\left(\dfrac{210 + 34}{361 + 86}\right) \pm 1.96 \sqrt{\left(\dfrac{210 + 34}{361 + 86}\right)\left(1 - \dfrac{210 + 34}{361 + 86}\right)\left(\dfrac{1}{361} + \dfrac{1}{86}\right)}$

(E) $(0.5)\left(\dfrac{210}{361} + \dfrac{34}{86}\right) \pm 1.96 \sqrt{\dfrac{\left(\frac{210}{361}\right)\left(1 - \frac{210}{361}\right)}{361} + \dfrac{\left(\frac{34}{86}\right)\left(1 - \frac{34}{86}\right)}{86}}$

6. Which of the following is a true statement?

(A) A well-planned hypothesis test should result in a statement either that the null hypothesis is true or that it is false.
(B) The alternative hypothesis is stated in terms of a sample statistic.
(C) If a sample is large enough, the necessity for it to be a simple random sample is diminished.
(D) When the null hypothesis is rejected, it is because the null hypothesis is not true.
(E) Hypothesis tests are designed to measure the strength of evidence against the null hypothesis.

7. One extrasensory perception (ESP) test asks the subject to view the backs of cards and identify whether a circle, square, star, or cross is on the front of each card. If p is the proportion of correct answers, this may be viewed as a hypothesis test with H_0: $p = 0.25$ and H_a: $p > 0.25$. The subject is recognized to have ESP when the null hypothesis is rejected. What would a Type II error result in?

(A) Correctly recognizing someone has ESP
(B) Mistakenly thinking someone has ESP
(C) Not recognizing that someone really has ESP
(D) Correctly realizing that someone doesn't have ESP
(E) Failing to understand the nature of ESP

8. A research dermatologist believes that skin cancer presenting on the head or neck will occur most often on the left side, the side next to a window when a person is driving. In a review of 565 cases of skin cancer on the head or neck, 305 occurred on the left side. What is the resulting P-value?

(A) $P\left(z > \dfrac{0.54 - 0.50}{\sqrt{(0.5)(0.5)/565}} \right)$

(B) $2P\left(z > \dfrac{0.54 - 0.50}{\sqrt{(0.5)(0.5)/565}} \right)$

(C) $P\left(z > \dfrac{0.54 - 0.50}{\sqrt{(0.54)(0.46)/565}} \right)$

(D) $P\left(z \geq \dfrac{0.54 - 0.50}{\sqrt{(0.54)(0.46)/565}} \right)$

(E) $2P\left(z > \dfrac{0.54 - 0.50}{\sqrt{(0.54)(0.46)/565}} \right)$

9. Is Internet usage different in the Middle East from Internet usage in Latin America? In a random sample of 500 adults in the Middle East, 151 claimed to be regular Internet users, while in a random sample of 1000 adults in Latin America, 345 claimed to be regular users. What is the P-value for the appropriate hypothesis test?

(A) $P\left(z < \dfrac{0.302 - 0.345}{\sqrt{\dfrac{(0.302)(0.698)}{500} + \dfrac{(0.345)(0.655)}{1000}}} \right)$

(B) $2P\left(z < \dfrac{0.302 - 0.345}{\sqrt{\dfrac{(0.302)(0.698)}{500} + \dfrac{(0.345)(0.655)}{1000}}} \right)$

(C) $P\left(z < \dfrac{0.302 - 0.345}{\sqrt{(0.331)(0.669)\left(\dfrac{1}{500} + \dfrac{1}{1000}\right)}} \right)$

(D) $2P\left(z < \dfrac{0.302 - 0.345}{\sqrt{(0.331)(0.669)\left(\dfrac{1}{500} + \dfrac{1}{1000}\right)}} \right)$

(E) $2(0.095167)$

10. The greater the difference between the null hypothesis claim and the true value of the population parameter,

(A) the smaller the risk of a Type II error and the smaller the power.
(B) the smaller the risk of a Type II error and the greater the power.
(C) the greater the risk of a Type II error and the smaller the power.
(D) the greater the risk of a Type II error and the greater the power.
(E) the greater the probability of no change in Type II error or in power.

Free-Response Questions

Directions: You must show all work and indicate the methods you use. You will be graded on the correctness of your methods and on the accuracy of your final answers.

1. Net neutrality is the idea that all Internet traffic should be treated equally—that is, no Internet service provider (ISP) should have the power to favor one source over another by blocking or paid prioritization. An online survey, responded to by 155,612 Internet users, reports that 93% favor net neutrality with a margin of error of 0.1%.

   (a) Is "93%" a statistic or a parameter?
   (b) Without doing an actual calculation, explain why the margin of error is so small.
   (c) How confident should you be that between 92.9% and 93.1% of all Internet users favor net neutrality?

2. In a random sample of 500 new births in the United States, 41.2% were to unmarried women, while in a random sample of 400 new births in the United Kingdom, 46.5% were to unmarried women.

(a) Calculate a 95% confidence interval for the difference in the proportions of new births to unmarried women in the United States and the United Kingdom.

(b) Does the confidence interval support the belief by a UN health-care statistician that the proportions of new births to unmarried women is different in the United States and the United Kingdom? Explain.

3. In a July 2020 study of 1050 randomly selected smokers, 74% said they would like to give up smoking.

(a) At the 95% confidence level, what is the margin of error?

(b) Explain the meaning of "95% confidence interval" in this example.

(c) Explain the meaning of "95% confidence level" in this example.

(d) Give an example of possible response bias in this example.

(e) If we want instead to be 99% confident, would our confidence interval need to be wider or narrower?

(f) If the sample size were greater, would the margin of error be smaller or greater?

4. It is estimated that 17.4% of all U.S. households own a Roth IRA, a retirement savings account. The American Association of University Professors (AAUP) believes this figure is higher among their members and commissions a study. If 150 out of a random sample of 750 AAUP members own Roth IRAs, is this sufficient evidence to support the AAUP belief?

The answers for this quiz can be found in the Appendix on page 613.

QUIZ 25

Multiple-Choice Questions

> **Directions:** The questions that follow are each followed by five suggested answers. Choose the response that best answers the question.

1. In general, how does doubling the sample size change the confidence interval size?

   (A) Doubles the interval size
   (B) Halves the interval size
   (C) Multiplies the interval size by $\sqrt{2}$
   (D) Divides the interval size by $\sqrt{2}$
   (E) This question cannot be answered without knowing the sample size.

2. One month, the actual unemployment rate in Spain was 13.4%. If during that month you took a simple random sample of 100 Spaniards of working age and constructed a confidence interval estimate of the unemployment rate, which of the following would be true?

   (A) The center of the interval is 13.4.
   (B) The interval contains 13.4.
   (C) A 99% confidence interval estimate contains 13.4.
   (D) The z-score of 13.4 is between ± 2.576.
   (E) None of the above are true statements.

3. The margin of error in a confidence interval estimate using z-scores covers which of the following?

   (A) Sampling error
   (B) Errors due to undercoverage and nonresponse in obtaining sample surveys
   (C) Errors due to using sample standard deviations as estimates for population standard deviations
   (D) Type I errors
   (E) Type II errors

4. In a survey funded by a biomedical research foundation, 750 of 1000 adult Americans said they didn't believe they could come down with a sexually transmitted disease (STD). Construct a 95% confidence interval estimate of the proportion of all adult Americans who don't believe they can contract an STD.

(A) $0.750 \pm 1.645\sqrt{1000(0.750)(0.250)}$

(B) $0.750 \pm 1.645\sqrt{\dfrac{(0.750)(0.250)}{1000}}$

(C) $0.750 \pm 1.96\sqrt{1000(0.750)(0.250)}$

(D) $0.750 \pm 1.96\sqrt{\dfrac{(0.750)(0.250)}{1000}}$

(E) $0.750 \pm 2.576\sqrt{1000(0.750)(0.250)}$

5. A politician wants to know what percentage of all voters support her position on the issue of forced busing for the racial integration of public schools. What size voter sample should be obtained to determine with 90% confidence the support level to within 4%?

(A) 21
(B) 25
(C) 423
(D) 600
(E) 1691

6. In a simple random sample of 300 elderly men, 65% were married, while in an independent simple random sample of 400 elderly women, 48% were married. Determine a 99% confidence interval estimate for the difference between the proportions of all elderly men and women who are married.

(A) $(0.65 - 0.48) \pm 2.326\sqrt{\dfrac{(0.65)(0.35)}{300} + \dfrac{(0.48)(0.52)}{400}}$

(B) $(0.65 - 0.48) \pm 2.576\sqrt{\dfrac{(0.65)(0.35)}{300} + \dfrac{(0.48)(0.52)}{400}}$

(C) $(0.65 - 0.48) \pm 2.576\left(\dfrac{(0.65)(0.35)}{\sqrt{300}} + \dfrac{(0.48)(0.52)}{\sqrt{400}}\right)$

(D) $\left(\dfrac{0.65 + 0.48}{2}\right) \pm 2.576\sqrt{\dfrac{(0.65)(0.35)}{300} + \dfrac{(0.48)(0.52)}{400}}$

(E) $\left(\dfrac{0.65 + 0.48}{2}\right) \pm 2.807\sqrt{(0.565)(0.435)\left(\dfrac{1}{300} + \dfrac{1}{400}\right)}$

7. A hypothesis test comparing two population proportions results in a *P*-value of 0.032. Which of the following is a proper conclusion?

(A) The probability that the null hypothesis is true is 0.032.
(B) The probability that the alternative hypothesis is true is 0.032.
(C) The difference in sample proportions is 0.032.
(D) The difference in population proportions is 0.032.
(E) None of the above are proper conclusions.

8. You plan to perform a hypothesis test with a level of significance of $\alpha = 0.05$. What is the effect on the probability of committing a Type I error if the sample size is increased with everything else unchanged?

(A) The probability of committing a Type I error decreases.
(B) The probability of committing a Type I error is unchanged.
(C) The probability of committing a Type I error increases.
(D) The effect cannot be determined without knowing the relevant standard deviation.
(E) The effect cannot be determined without knowing if a Type II error is committed.

9. A pharmaceutical company claims that 8% or fewer of the patients taking their new statin drug will have a heart attack in a 5-year period. In a government-sponsored study of 2300 patients taking the new drug, 198 have heart attacks in a 5-year period. Is this strong evidence against the company claim?

(A) Yes, because the *P*-value is 0.005657.
(B) Yes, because the *P*-value is 0.086087.
(C) No, because the *P*-value is only 0.005657.
(D) No, because the *P*-value is only 0.086087.
(E) No, because the *P*-value is over 0.10.

10. Choosing a smaller level of significance, that is, a smaller α-risk, results in

(A) a lower risk of Type II error and lower power.
(B) a lower risk of Type II error and higher power.
(C) a higher risk of Type II error and lower power.
(D) a higher risk of Type II error and higher power.
(E) no change in risk of Type II error or in power.

Free-Response Questions

1. During the H1N1 pandemic, one published study concluded that if someone in your family had H1N1, you had a 1 in 8 chance of also contracting the disease. A state health officer tracks a random sample of new H1N1 cases in her state and notes that 129 out of a potential 876 family members later contract the disease.

   (a) Calculate a 90% confidence interval for the proportion of family members who contract H1N1 after an initial family member does in this state.

   (b) Based on this confidence interval, is there evidence that the proportion of family members who contract H1N1 after an initial family member does in this state is different from the 1 in 8 chance concluded in the published study? Explain.

   (c) Would the conclusion in (b) be any different with a 99% confidence interval? Explain.

2. No vaccinations are 100% risk free, and the theoretical risk of rare complications always have to be balanced against the severity of the disease. Suppose the CDC (Center for Disease Control) decides that a risk of one in a million is the maximum acceptable risk of GBS (Guillain-Barré syndrome) complications for a new vaccine for a particularly serious strain of influenza. A large sample study of the new vaccine is conducted with the following hypotheses:

   H_0: The proportion of GBS complications is 0.000001 (one in a million).
   H_a: The proportion of GBS complications is greater than 0.000001 (one in a million).

   The P-value of the test is 0.138.

   (a) Interpret the P-value in the context of this study.

   (b) What conclusion should be drawn at the $\alpha = 0.10$ significance level?

   (c) Given this conclusion, what possible error, Type I or Type II, might be committed? Give a possible consequence of committing this error.

3. A 20-year study of 5000 British adults noted four bad habits: smoking, drinking, inactivity, and poor diet. The study looked to show that there is a higher death rate (proportion who die in a 20-year period) among people with all four bad habits than among people with none of the four bad habits.

   (a) Was this an experiment or observational study? Explain.
   (b) What are the null and alternative hypotheses?
   (c) What would be the result of a Type I error?
   (d) What would be the result of a Type II error?

   Of the 314 people who had all four bad habits, 91 died during the study, while of the 387 people with none of the four bad habits, 32 died during the study.

   (e) Calculate and interpret the P-value in the context of this study.

4. A behavior study of high school students looked at whether a higher proportion of boys than girls meets a recommended level of physical activity (increased heart rate for 60 minutes per day for at least 5 days during the 7 days before the survey). What is the proper conclusion if 370 out of a random sample of 850 boys and 218 out of an independent random sample of 580 girls met the recommended level of activity?

The answers for this quiz can be found in the Appendix on page 615.

SUMMARY

- Important assumptions and conditions always must be checked before calculating a confidence interval or proceeding with a hypothesis test. These include:

  1. The data come from a random sample from the population of interest.
  2. For confidence intervals, both $n\hat{p}$ and $n(1-\hat{p})$ are ≥ 10, and for hypothesis tests, both np_0 and $n(1-p_0)$ are ≥ 10.
  3. When sampling without replacement, $n < 0.10N$.

- We are never able to say exactly what a population parameter is; rather, we say that we have a certain confidence that it lies in a certain interval.

- The width of a confidence interval is proportional to $\dfrac{1}{\sqrt{n}}$.

- The margin of error is exactly one-half the width of the confidence interval.

- If we want a narrower interval, we must either decrease the confidence level or increase the sample size.

- If we want a higher level of confidence, we must either accept a wider interval or increase the sample size.

- The level of confidence refers to the percentage of samples that produce intervals that capture the true population parameter.

- The margin of error formula, $z^*\sqrt{\dfrac{\hat{p}(1-\hat{p})}{n}}$, can be solved for the minimum sample size n needed to achieve a given margin of error (use a supposition or previous value for $\hat{p}$ or use $\hat{p} = 0.5$).

- When scoring a confidence interval free-response question on the exam, readers will look for whether you:

  1. Identify the procedure (such as one-sample z-interval for a population proportion) and check the conditions.
  2. Compute a proper interval (it is sufficient to simply copy the interval from your calculator calculation).
  3. Interpret the interval in context (the conclusion must mention the confidence level and the parameter and must refer to the population, not to the sample).

- The null hypothesis is stated in the form of an equality statement about a population parameter, while the alternative hypothesis is in the form of an inequality.

- We attempt to show that a null hypothesis is unacceptable by showing it is improbable.

- The test statistic for a population proportion is $z = \dfrac{\hat{p} - p_0}{\sqrt{\dfrac{p_0(1-p_0)}{n}}}$.

- The significance level α is the conditional probability of rejecting the null hypothesis given that it is true.

- Lack of sufficient evidence for the alternative hypothesis is not the same as evidence for the null hypothesis.

- The P-value is the probability of obtaining a sample statistic as extreme as (or more extreme than) the one obtained if the null hypothesis is assumed to be true.

- When the P-value is small (typically, less than 0.05 or 0.10), we say we have sufficient evidence to reject the null hypothesis.

- A Type I error is the probability of mistakenly rejecting a true null hypothesis.

- A Type II error is the probability of mistakenly failing to reject a false null hypothesis.
- The power of a hypothesis test is the probability that a Type II error is not committed.
- When scoring a hypothesis test free-response question on the exam, readers will look for whether you:

1. State the hypotheses (both H_0 and H_a) and define the parameter of interest. (The hypotheses must refer to the population, not to the sample.)
2. Identify the procedure (such as one-sample z-test for a population proportion) and check the conditions.
3. Compute the test statistic and the P-value. (It is sufficient to simply copy these from your calculator calculation.)
4. Give a conclusion in context with linkage to the P-value. (The conclusion must refer to the parameter and to the population, not to the sample.)

- A recommended way of formatting conclusions to hypothesis tests is as follows:

1. Because the P-value is small (less than alpha), there is sufficient evidence to reject H_0; that is, there is convincing evidence that (alternative in context).
2. Because the P-value is large (greater than alpha), there is not sufficient evidence to reject H_0; that is, there is not convincing evidence that (alternative in context).
3. Note that this also makes it easier to talk about Type I errors, Type II errors, and power, as the language closely matches the conclusions above.

   - Type I error: finding convincing evidence that (alternative in context) is true, when it really isn't
   - Type II error: not finding convincing evidence that (alternative in context) is true, when it really is
   - Power: the probability that we do find convincing evidence that the (alternative in context) is true, when it really is

- Conditions when constructing a confidence interval or for performing a hypothesis test about a difference in proportions:

1. The data come from two independent random samples or from two groups in a randomized experiment.
2. For confidence intervals, $n_1\hat{p}_1$, $n_1(1-\hat{p}_1)$, $n_2\hat{p}_2$, and $n_2(1-\hat{p}_2)$ are all ≥ 10, and for hypothesis tests, $n_1\hat{p}_c$, $n_1(1-\hat{p}_c)$, $n_2\hat{p}_c$, and $n_2(1-\hat{p}_c)$ are all ≥ 10, where $\hat{p}_c = \dfrac{x_1 + x_2}{n_1 + n_2}$.
3. When sampling without replacement, $n_1 < 0.10N_1$ and $n_2 < 0.10N_2$.

Inference for Quantitative Data: Means

7

(10–18% AP EXAM WEIGHTING)

In this unit, you will learn inference procedures about population means. You will understand when to use t- versus z procedures. Finally, you will see the similarities and differences between the conditions for inference with proportions and means.

UNIT LEARNING OBJECTIVES

- To be able to check conditions to perform inference involving population means.
- To understand that when the population standard deviation σ is unknown and provided that either the original population is normally distributed or the sample size n is large enough, the statistic $\dfrac{\bar{x} - \mu}{s/\sqrt{n}}$ follows a t-distribution with $n - 1$ degrees of freedom.
- To be able to construct and interpret a confidence interval for a population mean and for the difference of population means.
- To understand the relationship between confidence level, sample size, sample standard deviation, and confidence interval width.
- To be able to interpret confidence levels.
- To be able to perform a hypothesis test to evaluate a claim about a population mean and about a difference between population means.
- To be able to interpret Type I and Type II errors and their possible consequences in context.

- To understand the relationships among significance level, probability of Type II error, and power.
- To understand how simulation can be used to estimate a *P*-value.

THE *t*-DISTRIBUTION

NOTE

When we are working with *small* samples from a population that is *not* nearly normal, we must use very different "nonparametric" techniques not discussed in this review book.

NOTE

In the real world, there is no such thing as a truly normal distribution, and instead we assume a *roughly* normally distributed population.

When the population standard deviation σ is unknown, we use the sample standard deviation s as an estimate for σ. But then $\dfrac{\bar{x} - \mu}{s/\sqrt{n}}$ does not follow a normal distribution. There is a distribution that can be used when working with the $s/\sqrt{n}$ ratios. This *Student's t-distribution* was introduced in 1908 by W. S. Gosset, a British mathematician employed by the Guinness Breweries.

For a sample from a *normally distributed population*, we work with the variable

$$t = \frac{\bar{x} - \mu}{s/\sqrt{n}}$$

with a resulting *t*-distribution that is bell-shaped and symmetric but lower at the mean, higher at the tails, and so more spread out than the normal distribution.

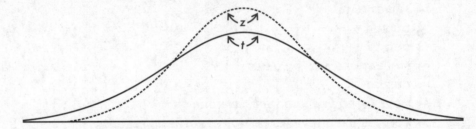

Like the binomial distribution, the *t*-distribution is different for different values of *n*. In the tables, these distinct *t*-distributions are associated with the values for degrees of freedom (*df*). For this discussion, the *df* value is equal to the sample size minus 1. The smaller the *df* value, the larger the dispersion in the distribution. The larger the *df* value, that is, the larger the sample size, the closer the distribution to the normal distribution.

Since there is a separate *t*-distribution for each *df* value, fairly complete tables would involve many pages; therefore, in Table B of the Appendix, we list areas and *t*-values for only the more commonly used percentages or probabilities. The last row of Table B is the normal distribution, which is a special case of the *t*-distribution taken when *n* is infinite. Technology (such as `tcdf` and `invT` on the TI-84) easily handle calculations involving the *t*-distribution.

IMPORTANT

For $\dfrac{\bar{x} - \mu}{s/\sqrt{n}}$ to have a *t*-distribution, we actually need that the sampling distribution of $\bar{x}$ is normal (that is, that $\dfrac{\bar{x} - \mu}{\sigma/\sqrt{n}}$ has a *z*-distribution). This follows either if the original population has a normal distribution or if the sample size is large enough (from the central limit theorem).

Thus, the *t*-distribution is the proper choice whenever the population standard deviation σ is unknown. In the real world, σ is almost always unknown, and so we should almost always use the *t*-distribution. The check that must be satisfied before using the *t*-distribution is that either the original population is roughly normal or the sample size is large enough ($n \geq 30$) for the central limit theorem to apply. What happens if the sample size is less than 30 and we are not told that the parent population is roughly normal? Then we must check that the sample is roughly unimodal and symmetric, showing no outliers and no skewness, so that we can say it is not unreasonable to assume the parent population is roughly normal.

CONFIDENCE INTERVAL FOR A MEAN

We are interested in estimating a population mean μ by considering a single sample mean $\bar{x}$. This sample mean is just one of a whole universe of sample means, and from Unit 5 we remember that if n is sufficiently large:

1. the set of all sample means is approximately normally distributed.
2. the mean of the set of sample means equals μ, the mean of the population.
3. the standard deviation $\sigma_{\bar{x}}$ of the set of sample means is approximately equal to $\frac{\sigma}{\sqrt{n}}$, that is, equal to the standard deviation of the whole population divided by the square root of the sample size.

Frequently we do not know σ, the population standard deviation. In such cases, we must use s, the *standard deviation of the sample*, as an estimate of σ. In this case $\frac{s}{\sqrt{n}}$ is called the *standard error*, $SE(\bar{x})$, and is used as an estimate for $\sigma_{\bar{x}} = \frac{\sigma}{\sqrt{n}}$. (We will use *t*-distributions instead of the standard normal curve whenever σ is unknown, no matter what the sample size.)

Remember that in making calculations and drawing conclusions from a specific sample, it is important that the sample be a *simple random sample* and be no more than 10% of the population.

➡ EXAMPLE 7.1

A new drug results in lowering the heart rate by varying amounts with a standard deviation of 2.49 beats per minute.

(a) Find a 95% confidence interval estimate for the mean lowering of the heart rate in all patients if a 50-person simple random sample averages a drop of 5.32 beats per minute.

> *Answer:* The parameter is μ, which represents the mean lowering of the heart rate in the population of patients taking the new drug. We are given an SRS, and $n = 50$ is less than 10% of all people. The standard deviation of sample means is $\sigma_{\bar{x}} = \frac{\sigma}{\sqrt{n}} = \frac{2.49}{\sqrt{50}} = 0.352$. We are 95% confident that the true mean lowering of the heart rate is $5.32 \pm 1.96(0.352) = 5.32 \pm 0.69$ or between 4.63 and 6.01 heartbeats per minute. [zInterval gives (4.6298, 6.0102).]

(b) How about a 99% confidence interval?

> *Answer:* Here, $5.32 \pm 2.576(0.352) = 5.32 \pm 0.91$ or between 4.41 and 6.23 heartbeats per minute. [zInterval gives (4.4129, 6.2271).] We are 99% confident that the true mean lowering of the heart rate is between 4.41 and 6.23 heartbeats per minute. We can also say, using the definition of confidence *level*, that if the sampling procedure were repeated many times, about 99% of the resulting confidence intervals would contain the true population mean.
>
> Note that when we want a higher confidence (99% instead of 95%), we have to settle for a larger, less specific interval (± 0.9071 instead of ± 0.6902).

NOTE

This is the rare case where the population standard deviation is known, so we can use the normal distribution.

(continued)

(c) What would the 95% confidence interval be if the sample mean of 5.32 had come from a sample of 100 patients?

Answer: The standard deviation of the sample mean would then have

been $\sigma_{\bar{x}} = \frac{\sigma}{\sqrt{n}} = \frac{2.49}{\sqrt{100}} = 0.249$, and the 95% confidence interval would be

$5.32 \pm 1.96(0.249) = 5.32 \pm 0.488$, or between 4.832 and 5.808 heartbeats per minute. [ZInterval also gives (4.832, 5.808).]. Note that when the sample size increased (from 50 to 100), the same sample mean resulted in a narrower, more specific interval (± 0.488 versus ± 0.690).

(d) With what confidence can we assert that the new drug lowers the heart rate by a mean of 5.32 ± 0.75 beats per minute?

Answer: Converting ± 0.75 to *z*-scores yields $\frac{\pm 0.75}{0.352} = \pm 2.13$. Then normalcdf($-2.13$, 2.13) $= 0.9668$.

NOTE

You cannot conclude "nearly normal" or even "roughly symmetric" from a boxplot.

TIP

Understand that you are not trying to prove that the sample data are normal but, rather, trying to infer something about the parent population.

NOTE

The degrees of freedom, *df*, here are the sample size minus the number of parameters being estimated (in this case, just one parameter, the mean), so $df = n - 1$.

Remember, when σ is unknown (which is almost always the case), we use the *t*-distribution instead of the *z*-distribution. The condition to be checked is that either the parent population is approximately normal ("given" or "stated") or the sample size is large enough ($n \geq 30$) for the central limit theorem to apply. If we are not given that the parent population is approximately normal, for small samples ($n < 30$) the sample data should be unimodal and reasonably symmetric with no outliers and no skewness (which we can check using a dotplot, stemplot, histogram, or normal probability plot of the sample data).

 EXAMPLE 7.2

When a random sample of 10 cars of a new model was tested for gas mileage, the results showed a mean of 27.2 miles per gallon with a standard deviation of 1.8 miles per gallon. What is a 95% confidence interval estimate for the mean gas mileage achieved by this model? (Assume that the population of mpg results for all the new model cars is approximately normally distributed.)

Answer:

Parameter: Let μ represent the mean gas mileage (in miles per gallon) in the population of cars of a new model.

Procedure: A one-sample *t*-interval for a population mean.

Checks: The sample is given to be random, 10 cars are assumed to be less than 10% of all cars of the new model, and the population is stated to be approximately normal. So, $\frac{\bar{x} - \mu}{s/\sqrt{n}}$ follows a *t*-distribution.

Mechanics: Calculator software (such as TInterval on the TI-84 or 1-Sample tInterval on the Casio Prizm) gives (25.912, 28.488).

[For instructional purposes, we note that the standard error of the sample means is $SE(\bar{x}) = \frac{1.8}{\sqrt{10}} = 0.569$. With $10 - 1 = 9$ degrees of freedom and 2.5% in each tail, the appropriate *t*-scores are ± 2.262 (from InvT on the TI-84 or Inverse Student-t on the Casio Prizm). Then $27 \pm 2.262(0.569) = 27.2 \pm 1.3$.]

Conclusion in context: We are 95% confident that the true mean gas mileage of all cars of the new model is between 25.9 and 28.5 miles per gallon.

➡ EXAMPLE 7.3 _____

A new process for producing synthetic gems yielded six randomly selected stones weighing 0.43, 0.52, 0.46, 0.49, 0.60, and 0.56 carats, respectively, in its first run. Find a 90% confidence interval estimate for the mean carat weight from this process. (Assume that the population of carats of all gems produced by this new process is approximately normally distributed.)

Answer:

Parameter: Let μ represent the mean carat weight of the population of synthetic gems produced with this new process.

Procedure: A one-sample *t*-interval for a population mean.

Checks: The sample is given to be random, six stones are assumed to be less than 10% of all stones produced by this process, and the population is approximately normal. So, by using *s* instead of the unknown σ, a *t*-interval may be found.

Mechanics: Calculator software (such as `TInterval` with the data put in a `List`) gives (0.45797, 0.56203).

[For instructional purposes, we note the following:

$$\bar{x} = \frac{\sum x}{n} = \frac{0.43 + 0.52 + 0.46 + 0.49 + 0.60 + 0.56}{6} = \frac{3.06}{6} = 0.51$$

$$s = \sqrt{\frac{\sum (x - \bar{x})^2}{n-1}}$$

$$= \sqrt{\frac{(0.08)^2 + (0.01)^2 + (0.05)^2 + (0.02)^2 + (0.09)^2 + (0.05)^2}{5}}$$

$$= 0.0632 \text{ and } SE(\bar{x}) = \frac{s}{\sqrt{n}} = \frac{0.0632}{\sqrt{6}} = 0.0258$$

With $df = 6 - 1 = 5$ and 5% in each tail, the *t*-scores are $\pm \text{invT}(0.95, 5) = \pm 2.015$. Then $0.51 \pm 2.015(0.0258) = 0.51 \pm 0.052$.]

Conclusion in context: We are 90% confident that the true mean carat weight for all stones produced by this procedure is between 0.458 and 0.562 carats.

On the TI-Nspire, the result shows as:

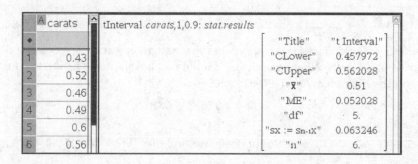

> **NOTE**
>
> **Or,** Inverse Student-t **on the Casio Prizm or** STUDENT_ICDF **on the HP Prime.**

EXAMPLE 7.4 _____

A survey was conducted involving 250 out of 125,000 families living in a city. The average amount of income tax paid per family in the sample was $3540 with a standard deviation of $1150. Establish a 99% confidence interval estimate for the total taxes paid by all the families in the city.

Answer:

Procedure: A one-sample *t*-interval of a population mean.

Checks: $n = 250$ is less than 10% of 125,000, and we must assume that the 250 families was an SRS of the 125,000 families. The sample size, 250, is greater than 30, so the CLT applies.

Mechanics: We are given $\bar{x} = 3540$ and $s = 1150$. Then `TInterval` gives (3351.2, 3728.8).

[For instructional purposes, we can calculate as follows. We use s as an estimate for σ and calculate the standard error of the sample means to be $SE(\bar{x}) = \dfrac{s}{\sqrt{n}} = \dfrac{1150}{\sqrt{250}} = 72.73$.

Since σ is unknown, we use a *t*-distribution. With `invT(0.995, 250 − 1) = 2.596`, we obtain $3540 \pm 2.596(72.73) = 3540 \pm 188.8$.]

Then $125{,}000(3540 \pm 188.8) = 442{,}500{,}000 \pm 23{,}600{,}000$.

Conclusion in context: We are 99% confident that the total tax paid by all families in the city is between $418,900,000 and $466,100,000.

Statistical principles are useful not only in analyzing data but also in setting up experiments. One important consideration is the choice of a sample size. In making interval estimates of population means, we have seen that each inference must go hand in hand with an associated confidence level statement. Generally, if we want a smaller, more precise interval estimate, we either decrease the degree of confidence or increase the sample size. Similarly, if we want to increase the degree of confidence, we either accept a wider interval estimate or increase the sample size. Thus, choosing a larger sample size always seems desirable; in the real world, however, time and cost considerations may limit the sample size.

EXAMPLE 7.5 _____

TIP

Understand that you use a critical *z*-value here because to use a *t*-value, you would need to know *df*, which is unknown because no sample size is given.

Ball bearings are manufactured by a process that results in a standard deviation in diameter of 0.025 inch. What sample size should be chosen if we wish to be 99% sure of knowing the diameter within ±0.01 inch?

Answer: We have $\sigma_{\bar{x}} = \dfrac{\sigma}{\sqrt{n}} = \dfrac{0.025}{\sqrt{n}}$ and $2.576\sigma_{\bar{x}} \le 0.01$.

Thus, $2.576\left(\dfrac{0.025}{\sqrt{n}}\right) \le 0.01$, and algebraically we find that $\sqrt{n} \ge \dfrac{2.576(0.025)}{0.01} = 6.44$, and so $n \ge 41.5$. We choose a sample size of 42.

HYPOTHESIS TEST FOR A MEAN

To conduct a hypothesis test for a mean, we must check that we have a simple random sample, that the sample size is less than 10% of the population, and that either the sample size is large enough ($n \geq 30$) for the CLT to apply or the population has an approximately normal distribution (either stated or if we are given the sample data, a plot should be unimodal and reasonably symmetric, showing no outliers and no skewness).

➡ EXAMPLE 7.6

A manufacturer claims that a new brand of air-conditioning unit uses only 6.5 kilowatts of electricity per day. A consumer agency believes the true figure is higher and runs a test on a random sample of size 50. If the sample mean is 7.0 kilowatts with a standard deviation of 1.4, should the manufacturer's claim be rejected at a significance level of 5%? Of 1%?

Answer:

Parameter: Let μ represent the mean electricity usage (in kilowatts per day) of the population of a new brand of air-conditioning unit.

Hypotheses: H_0: $\mu = 6.5$ and H_a: $\mu > 6.5$.

Procedure: A one-sample *t*-test for a population mean (population SD is unknown).

Checks: We are given a random sample, $n = 50 \geq 30$ is large enough for the CLT to apply, and we can assume that 50 is less than 10% of all of the new AC units.

Mechanics: Calculator software (such as T-Test on the TI-84) gives $t = 2.525$ and $P = 0.0074$.

[For instructional purposes, we note that $SE(\bar{x}) = \dfrac{1.4}{\sqrt{50}} = 0.198$ and $t = \dfrac{7.0 - 6.5}{0.198} = 2.525$ with a resulting *P*-value of tcdf(2.525,1000,49) = 0.0074.]

Conclusion in context with linkage to the P-value: With this small of a *P*-value, $0.0074 < 0.05$, there is sufficient evidence to reject H_0; that is, there is sufficient evidence for the consumer agency to reject the manufacturer's claim that the new unit uses a mean of only 6.5 kilowatts of electricity per day.

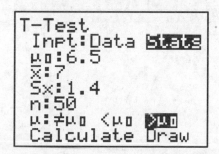

Given the above conclusion, what type of error, Type-I or Type-II, might have been committed, and what would be a possible consequence?

Answer: There was sufficient evidence to reject the null hypothesis. If the null hypothesis were true, we would be committing a Type-I error, that is, mistakenly rejecting a true null hypothesis. A possible consequence here is that the consumer agency would discourage customers from purchasing a new brand of air-conditioning unit that really was saving on electricity consumption as advertised.

EXAMPLE 7.7

A local chamber of commerce claims that the mean sale price for homes in the city is $90,000. A real estate salesperson notes a random sample of eight recent sales of $75,000, $102,000, $82,000, $87,000, $77,000, $93,000, $98,000, and $68,000. How strong is the evidence to reject the chamber of commerce claim?

Answer:

Parameter: Let μ represent the mean sale price for the population of homes in the city.

Hypotheses: H_0: $\mu = 90,000$ and *H_a:* $\mu \neq 90,000$.

Procedure: A one-sample *t*-test for a population mean (population SD is unknown).

Checks: We are given a random sample, we can assume $n = 8$ is less than 10% of all home sales, and we check for normality with a histogram, which yields the roughly unimodal, roughly symmetric:

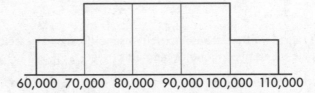

60,000 70,000 80,000 90,000 100,000 110,000

Mechanics: After putting the data into a List, calculator software gives $t = -1.131$ and $P = 0.2953$.

Conclusion in context with linkage to the P-value: With such a large *P*-value, 0.2953 > 0.05, there is not sufficient evidence to reject H_0; that is, there is not sufficient evidence to reject the commerce claim of a mean sale price of $90,000 for homes in the city.

On the TI-Nspire, the result shows as:

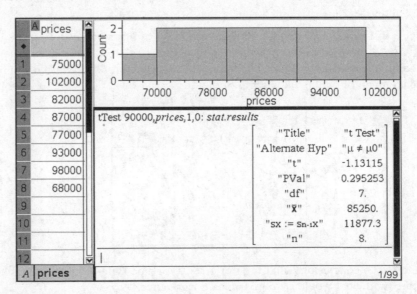

Given the above conclusion, what type of error, Type-I or Type-II, might have been committed, and what would be a possible consequence?

Answer: There was not sufficient evidence to reject the null hypothesis. If the null hypothesis were false, we would be committing a Type-II error, that is, mistakenly failing to reject a false null hypothesis. A possible consequence here is that the chamber of commerce goes on making and advertising a false statement about the mean sales price of homes in the city, misinforming potential home buyers.

CONFIDENCE INTERVAL FOR THE DIFFERENCE OF TWO MEANS

We have the following information about the sampling distribution of $\bar{x}_1 - \bar{x}_2$:

1. The set of all differences of sample means is approximately normally distributed.
2. The mean of the set of differences of sample means equals $\mu_1 - \mu_2$, the difference of population means.
3. The standard deviation $\sigma_{\bar{x}_1 - \bar{x}_2}$ of the set of differences of sample means is approximately equal to $\sqrt{\dfrac{\sigma_1^2}{n_1} + \dfrac{\sigma_2^2}{n_2}}$.

In making calculations and drawing conclusions from specific samples, it is important that the samples be *simple random samples,* that they be taken *independently* of each other, and that each is less than 10% of their respective population.

When the population standard deviations are unknown, use a t-distribution with $SE(\bar{x}_1 - \bar{x}_2) = \sqrt{\dfrac{s_1^2}{n_1} + \dfrac{s_2^2}{n_2}}$. In this case, there is the additional condition that either the original populations are roughly normal or the sample sizes are large enough ($n_1 \geq 30$ and $n_2 \geq 30$) for the CLT to apply.

➡ EXAMPLE 7.8

A 30-month study is conducted to determine the difference in the numbers of accidents per month occurring in two departments in an assembly plant. Suppose the first department averages 12.3 accidents per month with a standard deviation of 3.5, while the second averages 7.6 accidents with a standard deviation of 3.4. Determine a 95% confidence interval estimate for the mean difference in the numbers of accidents per month. (Assume that the two populations are independent and approximately normally distributed.)

Answer:

Parameters: Let μ_1 represent the mean number of accidents in the population of the number of accidents in all months for the first department. Let μ_2 represent the mean number of accidents in the population of the number of accidents in all months for the second department.

Procedure: A two-sample t-interval for a difference between two population means (first department minus second department).

Checks: We are not given that the sample is random, so we must assume that the 30 months are representative. We are given that the two populations are independent and approximately normal, so a t-interval may be found (population SDs are unknown).

Mechanics: Calculator software (such as `2-SampTInt` on the TI-84 or `2-Sample tInterval` on the Casio Prizm) gives (2.9167, 6.4833).

[Note that we can calculate $SE\left(\bar{x}_1 - \bar{x}_2\right) = \sqrt{\dfrac{s_1^2}{n_1} + \dfrac{s_2^2}{n_2}} = \sqrt{\dfrac{3.5^2}{30} + \dfrac{3.4^2}{30}} = 0.891$ and find the interval $(12.3 - 7.6) \pm t^*(0.891)$. The choice for *df* to calculate t^* can be estimated by $df = (n_1 - 1) + (n_2 - 1) = 58$, resulting in $t^* = 2.00$. Using the calculator output, which would lead to a noninteger "*df*," however, is more accurate.]

Conclusion in context: We are 95% confident that the first department has a true mean of between 2.92 and 6.48 more accidents per month than the second department.

HYPOTHESIS TEST FOR THE DIFFERENCE OF TWO MEANS

A hypothesis test for the difference of two means requires us to check that we have two independent simple random samples and that either the original two populations are roughly normal or the sample sizes are each large enough ($n_1 \geq 30$ and $n_2 \geq 30$) for the CLT to apply.

In this situation, the null hypothesis is usually that the means of the populations are the same or, equivalently, that their difference is 0:

$$H_0: \mu_1 - \mu_2 = 0$$

The alternative hypothesis is then:

$$H_a: \mu_1 - \mu_2 < 0, \ \ H_a: \mu_1 - \mu_2 > 0, \ \ \text{or} \ \ H_a: \mu_1 - \mu_2 \neq 0$$

The first two possibilities lead to one-sided tests, and the third possibility leads to two-sided tests.

➡ EXAMPLE 7.9

A sales representative believes that the computer his company sells has more average non-operational time per week than a similar model of computer sold by a competitor. Before taking this concern to his director, the sales representative gathers data and runs a hypothesis test. He determines that in a simple random sample of 40 week-long periods at different firms using his company's product, the average downtime was 125 minutes per week with a standard deviation of 37 minutes. However, 35 week-long periods involving the competitor's computer yield an average downtime of only 115 minutes per week with a standard deviation of 43 minutes. What conclusion should the sales representative draw?

Answer:

Parameters: Let μ_1 represent the mean of the population of non-operational times of the company's model computer. *Let* μ_2 represent the mean of the population of non-operational times of the competitor's model computer.

Hypotheses:
$$H_0: \mu_1 - \mu_2 = 0,$$
$$H_a: \mu_1 - \mu_2 > 0.$$

Procedure: A two-sample t-test for the difference of two population means.

Checks: We are given independent SRSs, and the sample sizes are large enough ($n_1 = 40 \geq 30$ and $n_2 = 35 \geq 30$) for the CLT to apply. The population SDs are unknown, so a t-test is called for.

Mechanics: Calculator software gives $t = 1.0718$ and $P = 0.1438$.

Conclusion in context with linkage to the P-value: With this large of a P-value, $0.1438 > 0.05$, there is not sufficient evidence to reject H_0; that is, the sales representative does not have sufficient evidence that his company's computers have greater mean non-operational time than that of the competitor's computers.

Given the above conclusion, what type of error, Type-I or Type-II, might have been committed, and what would be a possible consequence?

Answer: There was not sufficient evidence to reject the null hypothesis. If the null hypothesis were false, we would be committing a Type-II error, that is, mistakenly failing to reject a false null hypothesis. A possible consequence here is that the company's computers do have a greater mean non-operational time than those of the competitor's, but because the test doesn't show this, the company doesn't make necessary fixes and future sales will suffer.

> **TIP**
>
> It is often advisable to define these variables with descriptive subscripts, like $\mu_{company}$ and $\mu_{competitor}$ instead of μ_1 and μ_2.

➡ EXAMPLE 7.10

A city council member claims that male and female officers wait equal times for promotion in the police department. A women's spokesperson, however, argues that women wait longer than men. If a random sample of five men waited 8, 7, 10, 5, and 7 years, respectively, for promotion while a random sample of four women waited 9, 5, 12, and 8 years, respectively, what conclusion should be drawn?

Answer:

Parameters: Let μ_M represent the mean waiting for promotion in the population of male officers. Let μ_F represent the mean waiting for promotion in the population of female officers.

Hypotheses: H_0: $\mu_M = \mu_F$ and H_a: $\mu_M < \mu_F$.

Procedure: A two-sample *t*-test for the difference of two population means.

Checks: We are given random samples and must assume they are independent and less than 10% of total numbers of male and female officers. The dotplots of the sample data are roughly unimodal and roughly symmetric, indicating that it is not unreasonable to assume the data come from roughly normal distributions:

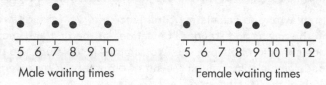

Male waiting times   Female waiting times

Mechanics: The population SDs are unknown, so a *t*-test is called for. By putting the data into `Lists`, calculator software gives $t = -0.6641$ and $P = 0.2685$.

Conclusion In context with linkage to the P-value: With this large of a *P*-value, $0.2685 > 0.05$, there is not sufficient evidence to reject H_0; that is, there is not sufficient evidence to dispute the council member's claim that male and female officers wait equal times for promotion. (There is not sufficient evidence of a difference in the mean waiting times for promotion of all male and female officers.)

Given the above conclusion, what type of error, Type-I or Type-II, might have been committed, and what would be a possible consequence?

Answer: There was not sufficient evidence to reject the null hypothesis. If the null hypothesis were false, we would be committing a Type-II error, that is, mistakenly failing to reject a false null hypothesis. A possible consequence here is that female officers really do have to wait longer times for promotions, but nothing is done to remedy the situation.

The analysis and procedure described above require that the two samples being compared be independent of each other. However, many experiments and tests involve comparing two populations for which the data naturally occur in pairs. In this case, the proper procedure is to run a one-sample test on a single variable consisting of the differences from the paired data.

PAIRED DATA

We've looked at confidence intervals and hypothesis tests about the difference between two means when data come from two independent random samples. However, when we have a quantitative variable measured twice for the same individual or for two very similar individuals, inference on the true mean difference involves one-sample analysis on the single variable consisting of the differences from the paired data.

➥ EXAMPLE 7.11

An SAT preparation class of 30 randomly selected students produces the following total score summary:

	First Score	Second Score	Improvement (2nd Score–1st Score)
Mean	1093.33	1135.58	42.25
Standard Deviation	87.76	85.73	27.92

Find a 90% confidence interval of the mean improvement in test scores.

Answer: It would be wrong to calculate a confidence interval for a difference between two means using the means and standard deviations of the first and second scores. The independence condition between the two samples is violated! The proper procedure is a one-sample *t*-interval on the set of differences (improvement) between the scores for each of the 30 students.

Parameter: Let μ represent the mean improvement (2nd score minus 1st score) in the SAT scores of the population of students who take this SAT preparation class.

Procedure: A one-sample *t*-interval for the mean of a population of differences in paired data.

Checks: We are given a random sample, $n = 30$ is less than 10% of all students, and $n = 30$ is large enough so that the CLT applies.

Mechanics: With an unknown population SD, it's necessary to find a *t*-interval. Using $\bar{x} = 42.25$ and $s = 27.92$, calculator software (such as `TInterval`) gives (33.59, 50.91).

Conclusion in context: We are 90% confident that the true mean improvement in test scores is between 33.59 and 50.91.

The choice between doing a two-sample analysis and doing a matched pair analysis can be confusing. The determining factor is whether or not the two groups are independent. The difference has to do with design. What is the average reaction time of people who are sober compared to that of people who have had two beers? If our experimental design calls for randomly assigning half of a group of volunteers to drink two beers and then comparing reaction times with the remaining sober volunteers, a two-sample analysis is called for. However, if our experimental design calls for testing reaction times of all the volunteers and then giving all the volunteers two beers and retesting, a matched pair analysis is called for. Matching may come about as above because you have made two measurements on the same person, or measurements might be made on sets of identical twins, or between salaries of president and provost at a number of universities, etc. The key is whether the two sets of measurements are independent or related in some way relevant to the question under consideration.

➡ EXAMPLE 7.12 _____

Does a particular drug slow reaction times? If so, the government might require a warning label concerning driving a car while taking the medication. An SRS of 30 people who might benefit from the drug are tested before and after taking the drug, and their reaction times (in seconds) to a standard testing procedure are noted. The resulting histograms and summary statistics are as follows:

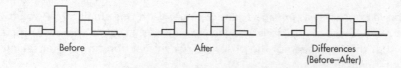

Before   After   Differences
                                (Before–After)

	n	Mean	SD
Before	30	1.46433	0.26267
After	30	1.50200	0.20942
Differences	30	−0.037667	0.11755

If we performed a two-sample test, we would calculate the *P*-value to be 0.2708 and would conclude that with such a large *P*, the observed rise is *not* significant. However, this would not be the proper test or conclusion! The two-sample test works for *independent* sets. Yet, in this case, there is a clear relationship between the data, in pairs, and this relationship is completely lost in the procedure for the two-sample test. The proper procedure is to form the set of 30 differences and to perform a single sample hypothesis test as follows.

Parameter: Let μ represent the mean difference in reaction time in the population of people taking a particular drug.

Hypotheses:

H_0: The reaction times of individuals to a standard testing procedure are the same before and after they take a particular drug; the mean difference is zero: $\mu_d = 0$.

H_a: The reaction time is greater after they take the drug; the mean difference (Before minus After) is less than zero: $\mu_d < 0$.

Procedure: We are using a *paired t-test*, that is, a single sample hypothesis test on the set of differences.

Checks:

1. The *data are paired* because they are measurements on the same individuals before and after taking the drug.
2. The reaction times of any individual are independent of the reaction times of the others, so the *differences are independent*.
3. A *random sample* of people are tested.
4. The histogram of the differences looks *nearly normal* (roughly unimodal and symmetric), or we could use that the sample size, $n = 30$, is large enough for the CLT to apply.
5. The sample size, $n = 30$, is assumed to be less than 10% of the population of all people who may take the drug.

(continued)

> **TIP**
>
> **Use proper terminology! In a two-sample *t*-test, the hypotheses are about the difference of means. In a paired *t*-test, the hypotheses are about the mean difference.**

Mechanics: T-Test gives $t = -1.755$ with $P = 0.0449$.

[Or we can calculate $SE(\bar{x}) = \dfrac{s}{\sqrt{n}} = \dfrac{0.11755}{\sqrt{30}} = 0.021462$ and

$t = \dfrac{\bar{x} - 0}{\sigma_{\bar{x}}} = \dfrac{-0.037667}{0.021462} = -1.755$.

With $df = n - 1 = 29$, the P-value is $P(t < -1.755) = \text{tcdf}(-1000, -1.755, 29) = 0.0449$.]

Conclusion in context with linkage to the P-value: With this small of a P-value, $0.0449 < 0.05$, there is sufficient evidence to reject H_0; that is, there is sufficient evidence that the true mean of the observed rise in reaction times after taking the drug is significant.

Given the above conclusion, what type of error, Type-I or Type-II, might have been committed, and what would be a possible consequence?

Answer: There was sufficient evidence to reject the null hypothesis. If the null hypothesis were true, we would be committing a Type-I error, that is, mistakenly rejecting a true null hypothesis. A possible consequence here is that the drug really doesn't slow reaction times, but the government requires the company to put an unnecessary, and possibly costly, label on the drug. Furthermore, people might choose not to take necessary medication.

QUIZ 26

Multiple-Choice Questions

> **Directions:** The questions that follow are each followed by five suggested answers. Choose the response that best answers the question.

1. Most recent tests and calculations estimate at the 95% confidence level that mitochondrial Eve, the maternal ancestor to all living humans, lived $138,000 \pm 18,000$ years ago. What is meant by "95% confidence" in this context?

   (A) A confidence interval of the true age of mitochondrial Eve has been calculated using z-scores of ± 1.96.
   (B) A confidence interval of the true age of mitochondrial Eve has been calculated using t-scores consistent with $df = n - 1$ and tail probabilities of ± 0.025.
   (C) There is a 0.95 probability that mitochondrial Eve lived between 120,000 and 156,000 years ago.
   (D) If 20 random samples of data are obtained by this method and a 95% confidence interval is calculated from each, the true age of mitochondrial Eve will be in 19 of these intervals.
   (E) Of all random samples of data obtained by this method, 95% will yield intervals that capture the true age of mitochondrial Eve.

2. A confidence interval estimate is determined from the GPAs of a simple random sample of n students. All other things being equal, which of the following will result in a smaller margin of error?

   (A) A smaller confidence level
   (B) A larger sample standard deviation
   (C) A smaller sample size
   (D) A larger population size
   (E) A smaller sample mean

3. One gallon of gasoline is put into each of an SRS of 30 test autos, and the resulting mileage figures are tabulated with $\bar{x} = 28.5$ and $s = 1.2$. Determine a 95% confidence interval estimate of the mean mileage of all comparable autos.

   (A) $28.5 \pm 2.045(1.2)$

   (B) $28.5 \pm 2.045\left(\dfrac{1.2}{\sqrt{29}}\right)$

   (C) $28.5 \pm 2.045\left(\dfrac{1.2}{\sqrt{30}}\right)$

   (D) $28.5 \pm 1.96\left(\dfrac{1.2}{\sqrt{29}}\right)$

   (E) $28.5 \pm 1.96\left(\dfrac{1.2}{\sqrt{30}}\right)$

4. What sample size should be chosen to find the mean number of absences per month for school children to within ± 0.2 at a 95% confidence level if it is known that the standard deviation is 1.1?

(A) $n \leq \sqrt{\dfrac{0.2}{1.96 \times 1.1}}$

(B) $n \leq \sqrt{\dfrac{1.96 \times 1.1}{0.2}}$

(C) $n \leq \left(\dfrac{1.96 \times 1.1}{0.2}\right)^2$

(D) $n \geq \sqrt{\dfrac{1.96 \times 1.1}{0.2}}$

(E) $n \geq \left(\dfrac{1.96 \times 1.1}{0.2}\right)^2$

5. In a study aimed at reducing developmental problems in low-birth-weight babies (under 2500 grams), 347 infants were exposed to a special educational curriculum while 561 did not receive any special help. After 3 years, the children exposed to the special curriculum showed a mean IQ of 93.5 with a standard deviation of 19.1; the other children had a mean IQ of 84.5 with a standard deviation of 19.9. Find a 95% confidence interval estimate for the difference in mean IQs of all low-birth-weight babies who receive special intervention and those who do not.

(A) $(93.5 - 84.5) \pm 1.97\sqrt{\dfrac{(19.1)^2}{347} + \dfrac{(19.9)^2}{561}}$

(B) $(93.5 - 84.5) \pm 1.97\left(\dfrac{19.1}{\sqrt{347}} + \dfrac{19.9}{\sqrt{561}}\right)$

(C) $(93.5 - 84.5) \pm 1.65\sqrt{\dfrac{(19.1)^2}{347} + \dfrac{(19.9)^2}{561}}$

(D) $(93.5 - 84.5) \pm 1.65\left(\dfrac{19.1}{\sqrt{347}} + \dfrac{19.9}{\sqrt{561}}\right)$

(E) $(93.5 - 84.5) \pm 1.65\sqrt{\dfrac{(19.1)^2 + (19.9)^2}{347 + 561}}$

6. Nine subjects, 87 to 96 years old, were given 8 weeks of progressive resistance weight training. Strength before and after training for each individual was measured as maximum weight (in kilograms) lifted by left knee extension:

										Mean	SD
Before:	3	3.5	4	6	7	8	8.5	12.5	15	7.5	4.085
After:	7	17	19	12	19	22	28	20	28	19.1	6.791
Difference:	4	13.5	15	6	12	14	19.5	7.5	13	11.6	4.891

Find a 95% confidence interval estimate for the mean strength gain of all 87- to 96-year-olds who are given the training.

(A) 11.6 ± 3.03

(B) 11.6 ± 3.69

(C) 11.6 ± 3.76

(D) 11.6 ± 5.18

(E) 11.6 ± 5.70

7. A company manufactures a synthetic rubber bungee cord with a braided covering of natural rubber and a minimum breaking strength of 450 kg. If the mean breaking strength of a sample drops below a specified level, the production process is halted and the machinery inspected. Which of the following would result from a Type I error?

(A) Halting the production process when too many cords break

(B) Halting the production process when the breaking strength is below the specified level

(C) Halting the production process when the breaking strength is within specifications

(D) Allowing the production process to continue when the breaking strength is below specifications

(E) Allowing the production process to continue when the breaking strength is within specifications

8. A coffee-dispensing machine is supposed to deliver 12 ounces of liquid into a large paper cup, but a consumer believes that the actual amount is less. As a test, he plans to obtain a random sample of 5 cups of the dispensed liquid and, if the mean content is less than 11.5 ounces, to reject the 12-ounce claim. If the machine operates with a known standard deviation of 0.9 ounces, what is the probability that the consumer will mistakenly reject the 12-ounce claim even though the claim is true? (Assume that all conditions for inference are met.)

(A) $P\left(t > \dfrac{11.5 - 12}{0.9/\sqrt{5}}\right)$

(B) $P\left(t < \dfrac{11.5 - 12}{0.9/\sqrt{5}}\right)$

(C) $P\left(z > \dfrac{11.5 - 12}{0.9/\sqrt{5}}\right)$

(D) $P\left(z < \dfrac{11.5 - 12}{0.9/\sqrt{5}}\right)$

(E) $2P\left(t < \dfrac{11.5 - 12}{0.9/\sqrt{5}}\right)$

9. A fast-food chain advertises that its large bag of french fries has a weight of 150 grams. Some high school students, who enjoy french fries at every lunch, suspect that they are getting less than the advertised amount. With a scale borrowed from their physics teacher, they weigh a random sample of 15 bags. What is the conclusion if the sample mean is 145.8 g and standard deviation is 12.81 g? (Assume that all conditions for inference are met.)

   (A) There is sufficient evidence to prove the fast-food chain advertisement is true.
   (B) There is sufficient evidence to prove the fast-food chain advertisement is false.
   (C) The students have sufficient evidence to reject the fast-food chain's claim.
   (D) The students do not have sufficient evidence to reject the fast-food chain's claim.
   (E) There is not sufficient data to reach any conclusion.

10. Given an experiment with H_0: $\mu = 35$, H_a: $\mu < 35$, and a possible correct value of 32, you obtain a sample statistic of $\bar{x} = 33$. After doing analysis, you realize that the sample size n is actually larger than you first thought. Which of the following results from reworking with the increase in sample size?

   (A) Decrease in probability of a Type I error; decrease in probability of a Type II error; decrease in power
   (B) Increase in probability of a Type I error; increase in probability of a Type II error; decrease in power
   (C) Decrease in probability of a Type I error; decrease in probability of a Type II error; increase in power
   (D) Increase in probability of a Type I error; decrease in probability of a Type II error; decrease in power
   (E) Decrease in probability of a Type I error; increase in probability of a Type II error; increase in power

11. Thirty students volunteer to test which of two strategies for taking multiple-choice exams leads to higher average results. Each student flips a coin, and if heads, uses Strategy A on the first exam and Strategy B on the second, while if tails, uses Strategy B on the first exam and Strategy A on the second. The average of all 30 Strategy A results is then compared to the average of all 30 Strategy B results. What is the conclusion at the 5% significance level if a two-sample hypothesis test, H_0: $\mu_1 = \mu_2$, H_a: $\mu_1 \neq \mu_2$, results in a P-value of 0.18?

   (A) The observed difference in average scores is significant.
   (B) The observed difference in average scores is not significant.
   (C) A conclusion is not possible without knowing the average scores resulting from using each strategy.
   (D) A conclusion is not possible without knowing the average scores and the standard deviations resulting from using each strategy.
   (E) A two-sample hypothesis test should not be used here.

Free-Response Questions

1. A simple random sample of five brands of breakfast cereals is tested for the number of calories per serving. The following data result: 217, 220, 224, 228, 233.

   Establish a 95% confidence interval estimate for the mean number of calories for servings of all breakfast cereals.

2. Next to good brakes, proper tire pressure is the most crucial factor in ensuring the responsive handling of a car. Incorrect tire pressure compromises cornering, braking, and stability. Both underinflation and overinflation can lead to problems. At a roadside vehicle safety checkpoint, officials randomly select 30 cars for which 35 psi is the recommended tire pressure and calculate the average of the actual tire pressure in the front right tires. What is the parameter of interest, and what are the null and alternative hypotheses that the officials are testing?

3. A simple random sample of 40 inner-city gas stations shows a mean price for regular unleaded gasoline to be $3.45 with a standard deviation of $0.05, while a simple random sample of 120 suburban stations shows a mean of $3.38 with a standard deviation of $0.08.

   (a) Construct 95% confidence interval estimates for the mean price of regular gas in inner-city and in suburban stations.
   (b) The confidence interval for the inner-city stations is wider than the interval for the suburban stations even though the standard deviation for inner-city stations is less than that for suburban stations. Explain why this happened.
   (c) Based on your answer in part (a), are you confident that the mean price of inner-city gasoline is less than $3.50? Explain.

4. A particular wastewater treatment system aims at reducing the most probable number per mL (MPN per mL) of *E. coli* to 1000 MPN per 100 mL. A random study of 40 of these systems in current use is conducted with the data showing a mean of 1002.4 MPN per 100 mL and a standard deviation of 7.12 MPN per 100 mL. A test of significance is conducted with:

   H_0: The mean concentration of *E. coli* after treatment under this system is 1000 MPN per 100 mL.

   H_a: The mean concentration of *E. coli* after treatment under this system is greater than 1000 MPN per 100 mL.

   The resulting *P*-value is 0.0197 with $df = 39$ and $t = 2.132$.

   (a) Interpret the *P*-value in the context of this study.
   (b) What conclusion should be drawn at a 5% significance level?
   (c) Given this conclusion, what possible error, Type I or Type II, might be committed? and give a possible consequence of committing this error.

5. A car simulator was used to compare the effect on reaction time between DWI (driving while intoxicated) and DWT (driving while texting). Ten adult volunteers were instructed to drive at 50 mph and then hit the brakes in response to the sudden image of a child darting into the road. A baseline stopping distance was established for each driver. Then one day each driver was tested for stopping distance while driving while texting, and another day the driver was tested after consuming a specified quantity of alcohol. For each driver, which test was done on the first day was decided by coin toss. The following table gives the extra number of feet necessary to stop at 50 mph for each driver for DWI and DWT.

DWI	30	42	36	28	27	37
DWT	30	45	38	30	28	38

The sample means are $\bar{x}_{DWI} = 33.3$, $\bar{x}_{DWT} = 34.8$, and a two-sample t-test $(H_0\colon \mu_{DWI} = \mu_{DWT}, H_a\colon \mu_{DWI} \neq \mu_{DWT})$ gives a P-value of 0.687 and a conclusion that there is no evidence of a difference between the effect on reaction time between DWI and DWT. Explain why this is not the proper hypothesis test, and then perform the proper test.

The answers for this quiz can be found in the Appendix on page 619.

QUIZ 27

Multiple-Choice Questions

> **Directions:** The questions that follow are each followed by five suggested answers. Choose the response that best answers the question.

1. A consumer testing agency plans to calculate a 99% confidence interval for the mean mpg for all cars on the road in 2019. Suppose the mpg measurements for the population of interest is actually sharply skewed right. For which of the sample sizes, $n = 30$, 50, or 70, would the sampling distribution of $\bar{x}$ be closest to normal?

   (A) 30
   (B) 50
   (C) 70
   (D) Because of skewness of the population, none of the sampling distributions can be approximately normal.
   (E) Because of the central limit theorem, all sampling distributions with $n \geq 30$ are equally approximately normal.

2. Suppose (25, 30) is a 90% confidence interval estimate for a population mean μ. Which of the following is a true statement?

   (A) There is a 0.90 probability that $\bar{x}$ is between 25 and 30.
   (B) Of the sample values, 90% are between 25 and 30.
   (C) There is a 0.90 probability that μ is between 25 and 30.
   (D) If 100 random samples of the given size are picked and a 90% confidence interval estimate is calculated from each, μ will be in 90 of the resulting intervals.
   (E) If 90% confidence intervals are calculated from all possible samples of the given size, μ will be in 90% of these intervals.

3. In a simple random sample of 80 teenagers, the average number of texts handled in a day was 50 with a standard deviation of 15. What is the 96% confidence interval for the average number of texts handled by all teens daily?

   (A) $50 \pm 2.054(15)$

   (B) $50 \pm 2.054 \dfrac{15}{\sqrt{79}}$

   (C) $50 \pm 2.054 \dfrac{15}{\sqrt{80}}$

   (D) $50 \pm 2.088 \dfrac{15}{\sqrt{79}}$

   (E) $50 \pm 2.088 \dfrac{15}{\sqrt{80}}$

4. The number of accidents per day at a large factory is noted for each of 64 days with $\bar{x} = 3.58$ and $s = 1.52$. With what degree of confidence can we assert that the mean number of accidents per day at the factory is between 3.20 and 3.96?

   (A) 48%
   (B) 63%
   (C) 90%
   (D) 95%
   (E) 99%

5. Hospital administrators wish to learn the average length of stay of all surgical patients. A statistician determines that, for a 95% confidence level estimate of the average length of stay to within ±0.5 days, 50 surgical patients' records will have to be examined. How many records should be looked at to obtain a 95% confidence level estimate to within ±0.25 days?

   (A) 25
   (B) 50
   (C) 100
   (D) 150
   (E) 200

6. Does socioeconomic status relate to age at the time of HIV infection? For 274 high-income HIV-positive individuals, the average age of infection was 33.0 years with a standard deviation of 6.3, while for 90 low-income individuals, the average age was 28.6 years with a standard deviation of 6.3 (*The Lancet*, October 22, 1994, page 1121). Find a 90% confidence interval estimate for the difference in ages of all high- and low-income people at the time of HIV infection.

   (A) 4.4 ± 0.963
   (B) 4.4 ± 1.27
   (C) 4.4 ± 2.51
   (D) 30.8 ± 2.51
   (E) 30.8 ± 6.3

7. A pharmaceutical company claims that a medicine will produce a desired effect for a mean time of 58.4 minutes. A government researcher runs a hypothesis test of 40 patients and calculates a mean of $\bar{x} = 59.5$ with a standard deviation of $s = 8.3$. What is the P-value?

(A) $P\left(t > \dfrac{59.5 - 58.4}{8.3/\sqrt{40}}\right)$ with $df = 39$

(B) $P\left(t > \dfrac{59.5 - 58.4}{8.3/\sqrt{40}}\right)$ with $df = 40$

(C) $2P\left(t > \dfrac{59.5 - 58.4}{8.3/\sqrt{40}}\right)$ with $df = 39$

(D) $2P\left(t > \dfrac{59.5 - 58.4}{8.3/\sqrt{40}}\right)$ with $df = 40$

(E) $2P\left(z > \dfrac{59.5 - 58.4}{8.3/\sqrt{40}}\right)$

8. A researcher believes a new diet should improve weight gain in laboratory mice. If ten control mice on the old diet gain an average of 4 ounces with a standard deviation of 0.3 ounces, while the average gain for ten mice on the new diet is 4.8 ounces with a standard deviation of 0.2 ounces, what is the P-value? (Assume that all conditions for inference are met.)

(A) $P\left(t < \dfrac{4 - 4.8}{\sqrt{\dfrac{(0.3)^2}{10} + \dfrac{(0.2)^2}{10}}}\right)$

(B) $P\left(z < \dfrac{4 - 4.8}{\sqrt{\dfrac{(0.3)^2}{10} + \dfrac{(0.2)^2}{10}}}\right)$

(C) $P\left(t < \dfrac{4 - 4.8}{\dfrac{0.3}{\sqrt{10}} + \dfrac{0.2}{\sqrt{10}}}\right)$

(D) $P\left(z < \dfrac{4 - 4.8}{\dfrac{0.3}{\sqrt{10}} + \dfrac{0.2}{\sqrt{10}}}\right)$

(E) $P\left(t < \dfrac{4 - 4.8}{\dfrac{0.3}{\sqrt{10} - 1} + \dfrac{0.2}{\sqrt{10} - 1}}\right)$

9. Suppose you do five independent tests of the form H_0: $\mu = 38$ versus H_a: $\mu > 38$, each at the $\alpha = 0.01$ significance level. What is the probability of committing a Type I error and incorrectly rejecting a true null hypothesis with at least one of the five tests?

(A) 0.01

(B) 0.049

(C) 0.05

(D) 0.226

(E) 0.951

10. A factory is located close to a city high school. The manager claims that the plant's smokestacks emit an average of no more than 350 pounds of pollution per day. As an AP Statistics project, the class plans a one-sided hypothesis test with a critical value of 375 pounds. Suppose the standard deviation in daily pollution poundage is known to be 150 pounds and the true mean is 385 pounds. If the sample size is 100 days, what is the probability that the class will mistakenly fail to reject the factory manager's false claim?

(A) 0.0475

(B) 0.2525

(C) 0.7475

(D) 0.7514

(E) 0.9525

11. Do high school girls apply to more colleges than high school boys? A two-sample t-test of the hypotheses H_0: $\mu_{girls} = \mu_{boys}$ versus H_a: $\mu_{girls} > \mu_{boys}$ results in a P-value of 0.02.

Which of the following statements must be true?

I. A 90% confidence interval for the difference in means contains 0.
II. A 95% confidence interval for the difference in means contains 0.
III. A 99% confidence interval for the difference in means contains 0.

(A) I only

(B) III only

(C) I and II only

(D) II and III only

(E) I, II, and III

Free-Response Questions

1. In a simple random sample of 30 subway cars during rush hour, the average number of riders per car was 83.5 with a standard deviation of 5.9. Assume the sample data are unimodal and reasonably symmetric with no extreme values and little, if any, skewness.

   (a) Establish a 90% confidence interval estimate for the average number of riders per car during rush hour. Show your work.

   (b) Assuming the same standard deviation of 5.9, how large of a sample of cars would be necessary to determine the average number of riders to within ± 1 at the 90% confidence level? Show your work.

2. In a sample of ten basketball players, the mean income was $196,000 with a standard deviation of $315,000.

   (a) Assuming all necessary assumptions are met, find a 95% confidence interval estimate of the mean salary of basketball players.

   (b) What assumptions are necessary for the above estimate? Do they seem reasonable here?

3. Bisphenol A (BPA), a synthetic estrogen found in packaging materials, has been shown to leach into infant formula and beverages. More recently, BPA has been detected in high concentrations in cash register receipts. Random sample biomonitoring data to see if retail workers carry higher amounts of BPA concentration in their bodies than nonretail workers is as follows:

		BPA concentration (μg/L)	
	Sample size	Mean	Standard deviation
Nonretail workers	528	2.43	0.45
Retail workers	197	3.28	0.48

   (a) Calculate a 99% confidence interval for the difference in mean BPA body concentrations of nonretail and retail workers.

   (b) Does the confidence interval support the belief that retail workers carry higher amounts of BPA in their bodies than nonretail workers? Explain.

4. In a random sample of 35 NFL games, the average attendance was 68,729 with a standard deviation of 6,110, while in a random sample of 30 Big Ten Conference football games, the average attendance was 70,358 with a standard deviation of 9,139. Is there evidence that the average attendance at Big Ten Conference football games is greater than that at NFL games?

5. An engineer wishes to test which of two drills can more quickly bore holes in various materials. He assembles a random sampling of ten materials of various hardnesses and thicknesses.

   (a) Given that a drill's efficiency is influenced by how long and how hard it has been operating, the engineer decided to randomly choose the order in which the materials will be tested. Design and implement a scheme to place the materials in random order using the following random number table:

   51844 73424 84380 82259 28273 58102 18727 69708

   (b) Suppose the drilling times (in seconds) are summarized as follows:

Material	A	B	C	D	E	F	G	H	I	J
Drill 1	4.2	5.1	8.8	1.5	0.8	7.4	3.4	4.7	6.2	4.8
Drill 2	4.2	5.4	8.7	1.7	0.9	7.8	3.4	4.4	6.3	4.9

   What is the mean drilling time for each drill? Is the difference significant? Justify your answer.

The answers for this quiz can be found in the Appendix on page 622.

SIMULATIONS AND *P*-VALUES

When our learned test procedures do not apply, either because conditions are not met or because we are working with a parameter for which we have not learned a test procedure, we may be able to proceed with a simulation. We can use a simulation to determine what values of a test statistic are likely to occur by random chance alone, assuming the null hypothesis is true. Then looking at where our test statistic falls, we can estimate a *P*-value.

➡ EXAMPLE 7.13

One measure of variability is the median absolute deviation (MAD). It is defined as the median deviation from the median, that is, as the median of the absolute values of the deviations from the median. A particular industrial product has the following quality control check. Random samples of size 6 are gathered periodically, and measurements are taken of the dimension under observation. If the MAD calculation is significantly greater than what is expected during proper operation of the machinery, a recalibration is necessary. In a simulation of 100 such samples of size 6 from when the machinery is working properly, the resulting MAD calculations are summarized in the following dotplot:

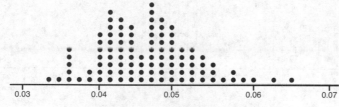

**MAD Calculations from Random Samples of Size 6
When Machinery Is Operating Properly**

Suppose in a random sample of 6 products the measurements are {8.04, 8.06, 8.10, 8.14, 8.18, 8.19}. Is there sufficient evidence to necessitate a recalibration of the machinery?

Answer: The median is $\dfrac{8.10 + 8.14}{2} = 8.12$. The absolute deviations from the median are {0.08, 0.06, 0.02, 0.02, 0.06, 0.07} with median $\dfrac{0.06 + 0.06}{2} = 0.06$ (the MAD calculation). In the simulation, there were 3 values out of 100 that were 0.06 or greater. This gives an estimated *P*-value of 0.03. With this small of a *P*-value, $0.03 < 0.05$, there is sufficient evidence to necessitate a recalibration of the machinery.

➡ EXAMPLE 7.14

Researchers want to study whether Paul the octopus, who lives in a tank at Sea Life Centre in Oberhausen, Germany, could correctly predict the winner of soccer matches. In a random sample of 14 matches, Paul correctly predicted the winner in 12 out of 14 soccer matches (by choosing to eat mussels from the boxes labeled with national flags of the eventual winning teams).

(a) What are the null and alternative hypotheses?

(b) If you conduct a simulation to investigate whether the observed result provides strong evidence that Paul can correctly predict winners, what would you use for the probability of success, for the sample size, and for the number of samples?

(c) Which of the following graphs could reasonably have come from the simulation?

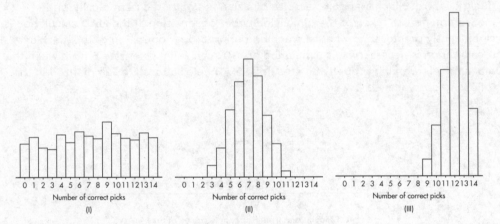

(d) Estimate the *P*-value, and interpret it in context.

(e) Make a conclusion based on the simulation and *P*-value.

Answers:

(a) H_0: Paul can do no better than guesswork in predicting winners, H_0: $p = 0.5$.

H_a: Paul can correctly predict winners better than guesswork, H_a: $p > 0.5$.

(b) $p = 0.5$, $n = 14$, and a large number for number of samples.

(c) Graph (II). The simulation should be centered at the null hypothesis expected value, not the observed value.

(d) Looking at the correct graph, (II), 12 or greater correct picks seems to have a probability of 0. If the null hypothesis were true, that is, if Paul can do no better than guesswork, the probability of Paul correctly predicting 12 or more out of 14 soccer matches is 0.00.

(e) With this small of a *P*-value, $0.00 < 0.05$, there is strong evidence to reject H_0; that is, there is strong evidence that Paul can pick the winner of soccer matches better than guesswork.

QUIZ 28

Multiple-Choice Questions

Directions: The questions that follow are followed by five suggested answers. Choose the response that best answers the question.

1. A company engineer creates a diagnostic measurement, $G = \dfrac{Max + Min}{2}$, which should be at least 34.60 in a sample of size 10 if certain machinery is operating correctly. To explore this diagnostic measurement, the machine is perfectly calibrated, and 100 random samples of size 10 of the product are taken from the assembly line. For each of these 100 samples, the diagnostic measurement G is calculated and shown plotted below.

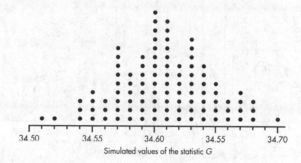

Simulated values of the statistic G

Each day, one sample of size 10 is taken from the assembly line and the diagnostic measurement G is calculated. If G drops too low, a decision to recalibrate the machinery is made.

One day the random sample is {34.89, 35.55, 33.93, 34.4, 35.51, 33.51, 35.0, 34.73, 35.19, 33.71}.

For the hypothesis test H_0: $G = 34.60$, H_a: $G < 34.60$, what is the estimated P-value?

(A) 0.01
(B) 0.02
(C) 0.03
(D) 0.04
(E) 0.05

2. A dietitian compares the effects of two diet pills. Using 25 participants taking each pill, she calculates a statistic W by computing a mean difference in weight loss expressed in pooled standard deviation units. The dietitian calculates $W = -0.7$ in her experiment. One thousand simulations assuming no real difference between the two pills gives the plot below for W.

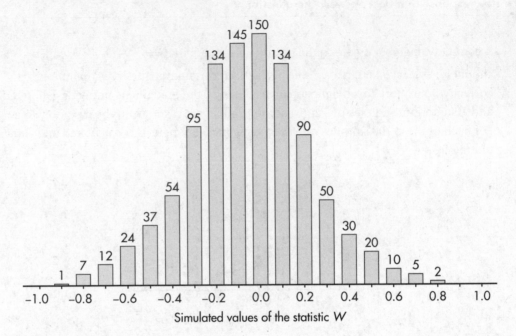

Simulated values of the statistic W

What is the estimated P-value for her observed value of $W = -0.7$ for the null hypothesis of no difference in effect versus the alternative hypothesis that there is a difference in effect?

(A) 0.007

(B) 0.008

(C) 0.01

(D) 0.02

(E) 0.027

Free-Response Questions

> **Directions:** You must show all work and indicate the methods you use. You will be graded on the correctness of your methods and on the accuracy of your final answers.

1. A battery manufacturer advertises that a 12-volt lawn and garden battery it sells has an average cranking power of 350 CCA (cold cranking amps). A consumer group believes that the true figure is lower. The consumer group obtains a random sample of 5 of the company's batteries and determines an average of 320 CCA. Is this significant? The company claims that perfectly good batteries do vary in CCA. To illustrate its point, the company runs a simulation by randomly picking 5 batteries 160 times from a group of new batteries manufactured to have 350 CCA strength and calculating the resulting averages to show that the consumer group's sample average of 320 CCA was possible. The company makes the following dotplot.

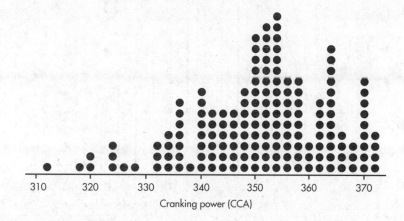

    (a) What are the hypotheses for an appropriate test?
    (b) Explain why or why not the conditions for a *t*-test of a mean are met.
    (c) What is the estimated *P*-value?
    (d) Based on the estimated *P*-value, what is the appropriate conclusion in context?

2. A medical researcher believes that the antidepressant desipramine is more effective than the mood stabilizer lithium in treating cocaine addiction. In a randomized experiment, 50 cocaine addicts were randomly assigned to take either desipramine or lithium. Fourteen out of the 25 addicts taking desipramine had not relapsed after one year, while only 6 of the 25 addicts taking lithium had not relapsed. These numbers were too small to use in our standard proportion hypothesis tests. In a simulation of 500 sets of 50 addicts undergoing this experiment, assuming the null hypothesis of no difference in the treatments, the difference in proportions of addicts not relapsing (desipramine minus lithium) yields:

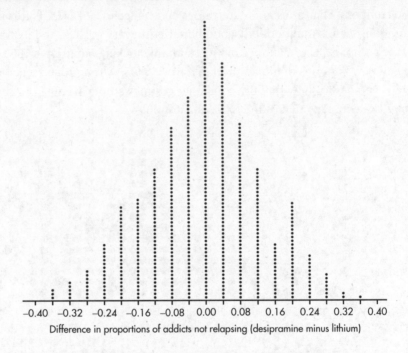

Difference in proportions of addicts not relapsing (desipramine minus lithium)

(a) In terms of appropriate proportions, what are the hypotheses?
(b) What is the observed difference in proportions?
(c) What is the estimated *P*-value?
(d) What is the appropriate conclusion in context?

The answers for this quiz can be found in the Appendix on page 625.

MORE ON POWER AND TYPE II ERRORS

Given a specific alternative hypothesis, a Type II error is a mistaken failure to reject the false null hypothesis, while the *power* is the probability of rejecting that false null hypothesis.

➡ EXAMPLE 7.15 _____

A candidate claims to have the support of 70% of the people, but you believe that the true figure is lower. You plan to gather an SRS and will reject the 70% claim if your sample shows 65% or less support. What if in reality only 63% of the people support the candidate?

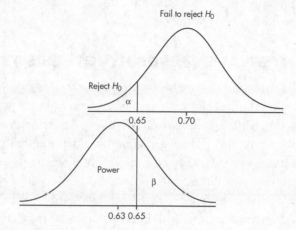

The upper graph shows the null hypothesis model with the claim that $p_0 = 0.70$ and the plan to reject H_0 if $\hat{p} < 0.65$. The lower graph shows the true model with $p = 0.63$. When will we fail to recognize that the null hypothesis is incorrect? Answer: precisely when the sample proportion is greater than 0.65. This is a Type II error with probability β. When will we rightly conclude that the null hypothesis is incorrect? Answer: when the sample proportion is less than 0.65. This is the "power" of the test and has probability $1 - \beta$.

The following points should be emphasized:

- Power gives the probability of avoiding a Type II error.
- Power has a different value for different possible correct values of the population parameter; thus, it is actually a *function* where the independent variable ranges over specific alternative hypotheses.
- Choosing a smaller α (that is, a tougher standard to reject H_0) results in a higher risk of Type II error and a lower power—observe in the above graphs how making α smaller (in this case moving the critical cutoff value to the left) makes the power less and β more!
- The greater the difference between the null hypothesis p_0 and the true value p, the smaller the risk of a Type II error and the greater the power—observe in the above picture how moving the lower graph to the left makes the power greater and β less. (The difference between p_0 and p is sometimes called the *effect*—thus the greater the effect, the greater is the power.)
- A larger sample size n will reduce the standard deviations, making both graphs narrower, and—if the rejection value is unchanged—the result is smaller α, smaller β, and larger power!

TIP

You will not have to calculate actual probabilities of Type II errors or power.

TIP

While calculations are not called for, you must understand the interactions among the errors, confidence level, power, *effect*, and sample size.

TIP

Both blocking in experiments and stratification in sampling reduce variability and so can improve power.

- A general conclusion is that if the other three values below are unchanged, the power will increase (and the probability of a Type II error will decrease) if any of the following occur:

1. Significance level, *alpha*, increases (but this also increases the chance of a Type I error).
2. Standard error decreases (for example, blocking on something strongly related to the response variable can do this).
3. Sample size, *n*, increases (of course, more data also means greater expense in time and money).
4. True parameter is farther from the null (this is called the *effect size* and is something you really have no control over).

CONFIDENCE INTERVALS VERSUS HYPOTHESIS TESTS

A question sometimes arises as to whether a problem calls for calculating a confidence interval or performing a hypothesis test. Generally, a claim about a population parameter indicates a hypothesis test, while an estimate of a population parameter asks for a confidence interval. Sometimes, however, it is possible to conduct a hypothesis test by constructing a confidence interval estimate of the parameter and then checking whether or not the null parameter value is in the interval. While, when possible, this alternative approach is accepted on the AP exam, it does require very special care in still remembering to state hypotheses, in dealing with one-sided versus two-sided, and in how standard errors are calculated in problems involving proportions. So, the recommendation is to avoid the alternative approach and to conduct a hypothesis test like a hypothesis test. Carefully read the question. If it asks whether or not there is evidence, the question requires a hypothesis test; if it involves how much or how effective, the question requires a confidence interval calculation. However, it should be noted that there have been AP free-response questions where part (a) calls for a confidence interval calculation and part (b) asks if this calculation provides evidence relating to a hypothesis test.

➡ EXAMPLE 7.16 _____

Suppose it is reported that random samples of 3-point shots by basketball players Stephen Curry and Michael Jordan show a 43% rate for Curry and a 33% rate for Jordan.

(a) Are these numbers parameters or statistics?

(b) State appropriate hypotheses for testing whether the difference is statistically significant.

(c) Suppose the sample sizes were both 100. Find the z-statistic and the P-value, and give an appropriate conclusion at the 5% significance level.

(d) Calculate and interpret a 95% confidence interval for the difference in population proportions.

(e) Are the test decision and confidence interval consistent with each other?

(f) Repeat (c), (d), and (e) as if the sample size had been 200 for each.

TIP

Sample size plays a substantial role in inferential statistics!

Answers: (a) These two numbers are statistics because they describe samples, not all 3-point shots ever taken by these two players.

(b) $H_0: p_{Curry} = p_{Jordan}$ and $H_a: p_{Curry} \neq p_{Jordan}$.

(c) Calculator software gives $z = 1.46$ and $P = 0.145$. With this large of a P-value, $0.145 > 0.05$, there is not sufficient evidence to reject H_0; that is, there is not sufficient evidence of a difference in the true 3-point percentage rates of Curry and Jordan.

(d) Calculator software gives $(-0.034, 0.234)$. We are 95% confident that the true difference in 3-point percentage rates (Curry minus Jordan) is between -3.4% and 23.4%.

(e) Yes, they are consistent. We did not conclude that the two percentages differ, and the confidence interval (for the difference in population proportions) includes the value zero.

(f) With a sample size of 200, calculator software gives $z = 2.06$ and $P = 0.039$. With this small a P-value, $0.039 < 0.05$, there is sufficient evidence to reject H_0; that is, there is sufficient evidence of a difference in the true 3-point percentage rates of Curry and Jordan. With a sample size of 200, calculator software gives a confidence interval of $(0.005, 0.195)$. We are 95% confident that the true difference in 3-point percentage rates (Curry minus Jordan) is between 0.5% and 19.5%. This interval does not include zero, so it is sufficient evidence of a difference in the true 3-point percentage rates of Curry and Jordan. This is again consistent with the hypothesis test.

SUMMARY

- Important assumptions and conditions always must be checked before calculating a confidence interval or proceeding with a hypothesis test. These include:

  1. The data come from a random sample from the population of interest.
  2. Normal/Large Sample: Either the population has a roughly normal distribution or the sample size is large ($n \geq 30$). If the population distribution is unknown and $n < 30$, use a graph of the sample data to assess whether it is reasonable to assume the data come from a roughly normal distribution.
  3. When sampling without replacement, $n < 0.10N$.

- When the population standard deviation σ is unknown (which is almost always the case) and provided that the original population is normally distributed or the sample size n is large enough, the statistic $\dfrac{x - \mu}{s / \sqrt{n}}$ follows a t-distribution with $n - 1$ degrees of freedom.

- When scoring a confidence interval free-response question on the exam, readers will look for whether you:

  1. Identify the procedure (such as one-sample t-interval for a population mean) and check the conditions.
  2. Compute a proper interval (it is sufficient to simply copy the interval from your calculator calculation).
  3. Interpret the interval in context (the conclusion must mention the confidence level and the parameter and refer to the population, not to the sample).

- When scoring a hypothesis test free-response question on the exam, readers will look for whether you:

  1. State the hypotheses (both H_0 and H_a) and define the parameter of interest. (The hypotheses must refer to the population, not to the sample.)
  2. Identify the procedure (such as one-sample t-test for a population mean) and check the conditions.
  3. Compute the test statistic, P-value, and degrees of freedom if appropriate. (It is sufficient to simply copy all of these from your calculator calculation.)
  4. Give a conclusion in context with linkage to the P-value. (The conclusion must refer to the parameter and to the population, not to the sample.)

- Conditions when constructing a confidence interval or for performing a hypothesis test about a difference in means:

  1. The data come from two independent random samples or from two groups in a randomized experiment.
  2. Normal/Large Sample: For *each* sample, the population distribution is roughly normal or the sample size is large ($n \geq 30$). For *each* sample, if the population distribution is unknown and $n < 30$, use a graph of the sample data to assess whether it is reasonable to assume the data come from a roughly normal distribution.
  3. When sampling without replacement, $n_1 < 0.10N_1$ and $n_2 < 0.10N_2$.

- A general conclusion about power is that the power will increase (and the probability of a Type II error will decrease) if any of the following occur:

  1. Significance level, *alpha*, increases.
  2. Standard error decreases.
  3. Sample size, n, increases.
  4. True parameter is farther from the null.

- When our learned test procedures do not apply, we may be able to proceed with a simulation.

Inference for Categorical Data: Chi-Square

8

(2–5% AP Exam Weighting)

- → **CHI-SQUARE TEST FOR GOODNESS-OF-FIT**
- → **CHI-SQUARE TEST FOR INDEPENDENCE**
- → **CHI-SQUARE TEST FOR HOMOGENEITY**
- → **QUIZ 29**
- → **QUIZ 30**

In this unit, you will be introduced to the chi-square statistic, which gives a statistical measurement of the difference between observed and expected counts. You will learn when to use each of three different chi-square tests: the goodness-of-fit test, which regards the distribution of proportions of a categorical variable; the test for independence, which looks for an association between two categorical variables in a single population; and the test for homogeneity, which compares distributions of a categorical variable across two or more populations.

UNIT LEARNING OBJECTIVES

- To be able to conduct a chi-square test for goodness-of-fit (for a distribution of proportions of one categorical variable).
- To be able to conduct a chi-square test for independence (for associations between two categorical variables within a single population).
- To be able to conduct a chi-square test for homogeneity (for comparing distributions of a categorical variable across two or more populations).

CHI-SQUARE TEST FOR GOODNESS-OF-FIT

A critical question is often whether or not an observed pattern of data fits some given distribution. A perfect fit cannot be expected, and so we must look at discrepancies and make judgments as to the *goodness-of-fit*.

Our approach is similar to that developed earlier. There is the null hypothesis of a good fit, that is, the hypothesis that a given theoretical distribution correctly describes the situation, problem, or activity under consideration. Our observed data consist of one possible sample from a whole universe of possible samples. We ask about the chance of obtaining a sample with the observed discrepancies if the null hypothesis is really true. Finally, if the chance is too small, we reject the null hypothesis and say that the fit is not a good one.

How do we decide about the significance of observed discrepancies? It should come as no surprise that the best information is obtained from squaring the discrepancy values, as this has been our technique for studying variances from the beginning. Since, for example, an observed difference of 23 is more significant if the original values are 105 and 128 than if they are 10,602 and 10,625, we must appropriately *weight* each difference. Such weighting is accomplished by dividing each difference by the expected value. The sum of these weighted differences or discrepancies is called *chi-square* and is denoted as χ^2 (χ is the lowercase Greek letter chi):

$$\chi^2 = \sum \frac{(\text{obs} - \text{exp})^2}{\text{exp}}$$

The smaller the resulting χ^2-value, the better the fit. The *P*-value is the probability of obtaining a χ^2-value as extreme as (or as more extreme than) the one obtained if the null hypothesis is assumed true. If the χ^2-value is large enough, that is, if the *P*-value is small enough, we say there is sufficient evidence to reject the null hypothesis and to claim that the fit is poor.

To decide how large a calculated χ^2-value must be to be significant, that is, to choose a critical value, we must understand how χ^2-values are distributed. A χ^2-distribution is not symmetric and is always skewed to the right. There are distinct χ^2-distributions, each with an associated number of degrees of freedom (*df*). The larger the *df* value, the closer the χ^2-distribution is to a normal distribution. Note, for example, that squaring the often-used *z*-scores 1.645, 1.96, and 2.576 results in 2.71, 3.84, and 6.63, respectively, which are entries found in the first row of the χ^2-distribution table.

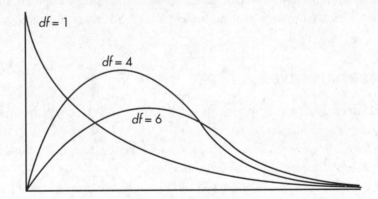

A large χ^2-value may or may not be significant—the answer depends on which χ^2-distribution we are using. Table C in the Appendix gives critical χ^2-values for the more commonly used percentages or probabilities, and probabilities can be calculated using χ^2cdf (on the TI-84). To use the χ^2-distribution for approximations in goodness-of-fit problems, the individual expected values cannot be too small. Our rule of thumb is that no expected value should be less than 5. Finally, as in all hypothesis tests we've looked at, the sample should be randomly chosen from the given population, and the sample size should be less than 10% of the population size.

A large city is divided into four distinct socioeconomic regions, one where the upper class lives, one for the middle class, one for the lower class, and one mixed-class region. Area percentages of the regions are 12%, 38%, 32%, and 18%, respectively. In a random sample of 55 liquor stores in the city, the numbers from each region are 4, 16, 26, and 9, respectively. The following shows how to determine if there is statistical evidence that region makes a difference with regard to numbers of liquor stores.

(a)   If liquor stores in the city were distributed among the four regions in the same proportions as the areas of those regions, what number (out of 55) would be expected in each region?

Answer: $(0.12)(55) = 6.6$, $(0.38)(55) = 20.9$, $(0.32)(55) = 17.6$, and $(0.18)(55) = 9.9$, and we have:

	Upper Class	Middle Class	Lower Class	Mixed Class
Observed	4	16	26	9
Expected	6.6	20.9	17.6	9.9

(b)   Are the sample numbers 4, 16, 26, and 9 significantly different from the expected values 6.6, 20.9, 17.6, and 9.9 to indicate that region makes a difference with regard to numbers of liquor stores?

Answer: First, state the hypotheses.

H_0: Liquor stores are distributed over the four city regions in the same proportions as the areas of those regions, that is, in the percentages 12, 38, 32, and 18, respectively.

H_a: Liquor stores are not distributed over the four city regions in the same proportions as the areas of those regions (at least one proportion is different from claimed).

Second, name the procedure and check the conditions:

Procedure: A chi-square test for goodness-of-fit.

Checks:

1.   *Randomization:* We are given that the sample is random.

2.   We note that the expected values (6.6, 20.9, 17.6, 9.9) are all ≥ 5.

3.   We assume the sample size, 55, is less than 10% of the number of all liquor stores in the large city.

Third, calculate the χ^2 statistic:

$$\chi^2 = \sum \frac{(\text{obs} - \text{exp})^2}{\text{exp}} = \frac{(4 - 6.6)^2}{6.6} + \frac{(16 - 20.9)^2}{20.9} + \frac{(26 - 17.6)^2}{17.6} + \frac{(9 - 9.9)^2}{9.9} = 6.264$$

The *P*-value is $P = P(\chi^2 > 6.264) = 0.099$. [If n is the number of classes, $df = n - 1 = 3$, and $\chi^2 \text{cdf}(6.264, 1000, 3) = 0.099$.] We also note that putting the observed and expected numbers in Lists, calculator software (such as χ^2GOF-Test on the TI-84 or on the Casio Prizm) quickly gives $\chi^2 = 6.262$ and $P = 0.099$.

> **NOTE**
>
> Sometimes it is useful to recognize which terms in the summation are larger and which are smaller, that is, which terms contribute the most to the final sum.

(continued)

Fourth, give a conclusion in context with linkage to the *P*-value:

With this large a *P*-value, 0.099 > 0.05, there is not sufficient evidence to reject H_0. That is, there is not sufficient evidence that liquor stores in this city are distributed over the four city regions (upper, middle, lower, and mixed class) in different proportions to the areas of those regions.

TIP

With the *t*-distribution, the *df* depends on the sample size, but here, *df* depends on the number of classes or categories.

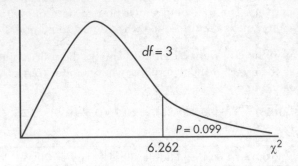

➡ EXAMPLE 8.2

A grocery store manager wishes to determine whether a certain product will sell equally well in any of five locations in the store. Five displays are set up, one in each location, and the resulting numbers of the product sold are noted.

	Location				
	1	2	3	4	5
Actual number sold	43	29	52	34	48

TIP

Remember: the hypotheses are always about the population, never about the sample; it would be *wrong* to say H_0: the sample data are uniformly distributed over the five locations.

Is there enough evidence that location makes a difference? Test at both the 5% and 10% significance levels.

Answer:

H_0: Sales of the product are uniformly distributed over the five locations.

H_a: Sales are not uniformly distributed over the five locations.

A total of $43 + 29 + 52 + 34 + 48 = 206$ units were sold. If location doesn't matter, we would expect $\frac{206}{5} = 41.2$ units sold per location (uniform distribution).

TIP

Even though the observed values are counts and thus integers, the expected values might not be integers and should *not* be rounded to integers.

	Location				
	1	2	3	4	5
Expected number sold	41.2	41.2	41.2	41.2	41.2

Check the conditions:

1. *Randomization:* We must assume that the 206 units sold are a representative sample.

2. We note that the expected values (all 41.2) are all ≥ 5.

By putting the observed and expected numbers in Lists, and with $df = 5 - 1 = 4$, calculator software (such as χ^2GOF-Test on the TI-84 or on the Casio Prizm) gives $\chi^2 = 8.903$ and $P = 0.0636$. [For instructional purposes, we note that

$$\chi^2 = \frac{(43 - 41.2)^2}{41.2} + \frac{(29 - 41.2)^2}{41.2} + \frac{(52 - 41.2)^2}{41.2} + \frac{(34 - 41.2)^2}{41.2} + \frac{(48 - 41.2)^2}{41.2}$$
$$= 8.903$$

and χ^2cdf$(8.903, 1000, 4) = 0.0636.$]

With $P = 0.0636$, there is sufficient evidence to reject H_0 at the 10% level but not at the 5% level. If the grocery store manager is willing to accept a 10% chance of committing a Type I error, there is enough evidence to claim location makes a difference.

CHI-SQUARE TEST FOR INDEPENDENCE

In the previous goodness of fit problems, a set of expectations was based on an assumption about how the distribution should turn out. We then tested whether an observed sample distribution could reasonably have come from a larger set based on the assumed distribution.

There is a class of real-world problems in which we want to determine if there is a significant association between two categorical variables. The procedure is called a *test of independence*. For example, do nonsmokers, light smokers, and heavy smokers all have the same likelihood of eventually being diagnosed with cancer, heart disease, or emphysema? Is there a relationship (association) between smoking status and being diagnosed with one of these diseases?

We classify our **single sample** observations in two ways and then ask whether the two ways are independent of each other. For example, we might consider several age groups and within each group ask how many employees show various levels of job satisfaction. The null hypothesis is that age and job satisfaction are independent, that is, that the proportion of employees expressing a given level of job satisfaction is the same no matter which age group is considered.

Analysis involves calculating a table of *expected values*, assuming the null hypothesis about independence is true. We then compare these expected values with the observed values and ask whether the differences are reasonable if H_0 is true. The significance of the differences is gauged by the same χ^2-value of weighted squared differences. The smaller the resulting χ^2-value, the more reasonable the null hypothesis of independence is. If the χ^2-value is large enough, that is, if the P-value is small enough, we can say that the evidence is sufficient to reject the null hypothesis and to claim that there *is* sufficient evidence of some relationship between the two variables or methods of classification.

As with the goodness-of-fit test, the test for independence requires that we have a simple random sample. The sample should be less than 10% of the population, and expected values should all be > 5.

When testing for independence,

$$df = (r - 1)(c - 1)$$

where df is the number of degrees of freedom, r is the number of rows, and c is the number of columns.

A point worth noting is that even if there is sufficient evidence to reject the null hypothesis of independence, we cannot necessarily claim any direct *causal* relationship. In other words, although we can make a statement about some link or relationship between two variables, we are *not* justified in claiming that one causes the other. For example, we may demonstrate a relationship between salary level and job satisfaction, but our methods would not show that higher salaries cause higher job satisfaction. Perhaps an employee's higher job satisfaction impresses his superiors and thus leads to larger increases in pay. Or perhaps there is a third variable, such as training, education, or personality, that has a direct causal relationship to both salary level and job satisfaction.

➡ **EXAMPLE 8.3**

A growing number of states have legalized marijuana for medical or recreational purposes. In a nationwide telephone poll of 1000 randomly selected adults representing Democrats, Republicans, and Independents, respondents were asked two questions: their party affiliation and if they supported the legalization of marijuana. The answers, cross-classified by party affiliation, are given in the following two-way table (also called a *contingency table*).

Observed	Support Legalization		
	Yes	No	No Opinion
Democrats	280	110	15
Republicans	155	190	10
Independents	180	45	15

Test the null hypothesis that support for legalizing marijuana is independent of party affiliation. Use a 5% significance level.

Answer:

H_0: Party affiliation and support for legalizing marijuana are independent.
H_a: Party affiliation and support for legalizing marijuana are not independent.

Putting the observed data into a "Matrix," calculator software (such as χ^2-Test on the TI-84, Casio Prizm, or HP Prime) gives $\chi^2 = 94.5$ and $P = 0.000$, and stores the expected values in a second matrix:

249.1	139.7	16.2
218.3	122.5	14.2
147.6	82.8	9.6

TIP

It is not enough to just say that all expected cells are ≥5; you must list the expected counts.

We are given a random sample, $n = 1000$ is less than 10% of all adults, and we note that all expected cells are ≥ 5.

With this small of a P-value, $0.000 < 0.05$, there is sufficient evidence to reject H_0; that is, among all adults there is sufficient evidence of a relationship between party affiliation and support for legalizing marijuana.

[For instructional purposes, we note that the expected values, χ^2, and P can be calculated as follows:

Row totals: 280 + 110 + 15 = 405
155 + 190 + 10 = 355
180 + 45 + 15 = 240

Column totals: 280 + 155 + 180 = 615
110 + 190 + 45 = 345
15 + 10 + 15 = 40

Support Legalization				
	Yes	No	No Opinion	Total
Democrats				405
Republicans				355
Independents				240
Total	615	345	40	1000

To calculate, for example, the expected value in the upper left cell, we can proceed in any of several equivalent ways. First, we could note that the proportion of Democrats is $\frac{405}{1000} = 0.405$; and so, if independent, the expected number of Democrat *yes* responses is 0.405(615) = 249.1. Instead, we could note that the proportion of *yes* responses is $\frac{615}{1000} = 0.615$; and so, if independent, the expected number of Democrat *yes* responses is 0.615(405) = 249.1. Finally, we could note that both these calculations simply involve $\frac{(405)(615)}{1000} = 249.1$.

In other words, the expected value of any cell can be calculated by multiplying the corresponding row total by the corresponding column total and then dividing by the table total. Thus, for example, the expected value for the middle cell, which corresponds to Republican *no* responses, is $\frac{(355)(345)}{1000} = 122.5$.

Continuing in this manner, we fill in the table as follows:

Support Legalization				
Expected	Yes	No	No Opinion	Total
Democrats	249.1	139.7	16.2	405
Republicans	218.3	122.5	14.2	355
Independents	147.6	82.8	9.6	240
Total	615	345	40	1000

(An appropriate check at this point is that each expected cell count is at least 5.) Next we calculate the value of chi-square:

$$\chi^2 = \sum \frac{(obs - exp)^2}{exp} = \frac{(280 - 249.1)^2}{249.1} + \frac{(110 - 139.7)^2}{139.7} + \frac{(15 - 16.2)^2}{16.2} + \frac{(155 - 218.3)^2}{218.3}$$

$$+ \frac{(190 - 122.5)^2}{122.5} + \frac{(10 - 14.2)^2}{14.2} + \frac{(180 - 147.6)^2}{147.6} + \frac{(45 - 82.8)^2}{82.8} + \frac{(15 - 9.6)^2}{9.6} = 94.5$$

$$df = (r - 1)(c - 1) = (3 - 1)(3 - 1) = 4$$

The *P*-value is then calculated using calculator software such as $\chi^2\text{cdf}(94.5, 1000, 4) = 0.000$.]

NOTE

While it is important to understand these calculations, on the exam the calculations should be made quickly by putting the observed data into a matrix and using the χ^2-Test on your calculator.

As for conditions to check for chi-square tests for independence, we should check that the sample is randomly chosen, that the expected values for all cells are at least 5, and that the sample size is less than 10% of the population. If a category has one of its expected cell counts less than 5, we can combine categories that are logically similar (for example, "disagree" and "strongly disagree") or combine numerically small categories collectively as "other."

➥ EXAMPLE 8.4

Almost 1 in 10 Americans believes vaccines, such as the MMR vaccine for measles, mumps, and rubella, are not safe for healthy children, indicating a significant number of Americans have misconceptions about the danger of vaccinations. In a nationwide poll of 500 randomly selected adults, respondents were asked two questions: their age and if they believe vaccines are "safe" or "unsafe." The cross-classified answers are given in the following two-way table:

	Age		
	18–29	30–64	Over 65
Believe vaccines safe	138	230	81
Believe vaccines unsafe	22	25	4

NOTE

All *expected* cells should be at least 5. There is no such requirement for *observed* cells.

Test the null hypothesis that belief of whether vaccines are safe is independent of age. Use a 5% significance level.

First, give the hypotheses:

H_0: Age and belief in the safety of vaccines are independent.
H_a: Age and belief in the safety of vaccines are not independent.

Second, name the procedure and check the conditions:

Procedure: A chi-square test for independence.

Checks:

1 We are given a random sample.
2 All expected cells (calculated below) are greater than 5.
3 The sample size, 500, is less than 10% of all adults.

Third, perform the calculations (using Matrix and χ^2-Test on a calculator) to find the expected cells, the χ^2 statistic, and the P-value:

143.7	229.0	76.3
16.3	26.0	8.7

$\chi^2 = 5.05$ and $P = 0.080$.

Fourth, give a conclusion in context with linkage to the P-value:

With this large of a P-value, $0.080 > 0.05$, there is not sufficient evidence to reject the null hypothesis. That is, there is not sufficient evidence of a relationship between age and belief in the safety of vaccines.

CHI-SQUARE TEST FOR HOMOGENEITY

In chi-square goodness-of-fit tests, we work with a single variable in comparing a single sample to a population model. In chi-square independence tests, we work with a single sample classified on two variables. Chi-square procedures can also be used with a single variable to compare samples from two or more populations. Conditions to check include that the samples be *simple random samples*, that they be taken *independently* of each other, that the original populations be large compared to the sample sizes, and that the expected values for all cells be at least 5. The contingency table used has a row for each sample. The resulting procedure is called a *chi-square test for homogeneity*.

EXAMPLE 8.5

In a large city, a group of AP Statistics students work together on a project to determine which group of school employees has the greatest proportion who are satisfied with their jobs. In independent simple random samples of 100 teachers, 60 administrators, 45 custodians, and 55 secretaries, the numbers satisfied with their jobs were found to be 82, 38, 34, and 36, respectively. Is there evidence that the proportion of employees satisfied with their jobs is different in different school system job categories?

Answer:

First, give the hypotheses:

H_0: The proportion of employees satisfied with their jobs is the same across the various school system job categories.

H_a: At least two of the job categories differ in the proportion of employees satisfied with their jobs.

Second, name the procedure and check the conditions:

Procedure: A chi-square test for homogeneity.

Checks:

1 We are given independent simple random samples.
2 The expected cells (calculated below) are all greater than 5.
3 We assume that the four samples are each less than 10% of their respective populations in the large city.

Third, using Matrix and χ^2-Test on a calculator, find the expected cells, the test statistic χ^2, and the *P*-value.

The observed counts are as follows:

	Teachers	Administrators	Custodians	Secretaries
Satisfied	82	38	34	36
Not satisfied	18	22	11	19

Putting the observed data into a Matrix, calculator software gives $\chi^2 = 8.707$ and $P = 0.0335$ and stores the expected values in a second matrix:

73.1	43.8	32.9	40.2
26.9	16.2	12.1	14.8

Fourth, give a conclusion in context with linkage to the *P*-value.

With this small of a *P*-value, $0.0335 < 0.05$, there is sufficient evidence to reject H_0; that is, there is sufficient evidence that the true proportion of employees satisfied with their jobs is not the same across all the school system job categories.

The difference between the test for independence and the test for homogeneity can be confusing. When we're simply given a two-way table, it's not obvious which test should be performed. The crucial difference is in the *design* of the study. Did we pick samples from each of two or more populations to compare the distribution of some variable (one question) among the different populations? If so, we are doing a test for homogeneity. Did we pick one sample from a single population and cross-categorize on two variables (two questions) to see if there is an association between the variables? If so, we are doing a test for independence.

For example, if we separately sample Democrats, Republicans, and Independents to determine whether they are for, against, or have no opinion with regard to stem cell research (several samples, one variable), we do a test for homogeneity with a null hypothesis that the distribution of opinions on stem cell research is the same among Democrats, Republicans, and Independents. However, if we sample the general population, noting the political preference and opinions on stem cell research of the respondents (one sample, two variables), we do a test for independence with a null hypothesis that political preference is independent of opinion on stem cell research.

Finally, it should be remembered that while we used χ^2 for categorical data, the χ^2-distribution is a *continuous* distribution. Applying it to counting data is just an approximation.

QUIZ 29

Multiple-Choice Questions

Directions: The questions that follow are each followed by five suggested answers. Choose the response that best answers the question.

1. To test the claim that dogs bite more or less depending upon the phase of the moon, a university hospital counts admissions for dog bites and classifies it with moon phase.

	New moon	First quarter	Full moon	Last quarter
Dog bite admissions	32	27	47	38

The expected numbers are all $\frac{1}{4}(32 + 27 + 47 + 38) = 36$. Which two categories contribute the largest components to the χ^2-test statistic?

(A) New moon and first quarter
(B) First quarter and full moon
(C) Full moon and last quarter
(D) New moon and full moon
(E) First quarter and last quarter

2. A random sample of 100 former student-athletes are picked from each of the colleges that are members of the Big East conference. Students are surveyed about whether or not they feel they received a quality education while participating in varsity athletics. Which of the following is the most appropriate test to determine whether there is a difference among these schools as to the student-athlete perception of having received a quality education?

(A) A chi-square goodness-of-fit test for a uniform distribution
(B) A chi-square test of independence
(C) A chi-square test of homogeneity
(D) A multiple-sample z-test of proportions
(E) A multiple-population z-test of proportions

3. A disc jockey wants to determine whether middle school students and high school students have similar music tastes. Independent random samples are taken from each group, and each person is asked whether he or she prefers hip-hop, pop, or alternative. A chi-square test of homogeneity of proportions is performed, and the resulting P-value is below 0.05. Which of the following is a proper conclusion?

(A) There is sufficient evidence that for all three music choices, the proportion of middle school students who prefer each choice is equal to the corresponding proportion of high school students.

(B) There is sufficient evidence that the proportion of middle school students who prefer hip-hop is different from the proportion of high school students who prefer hip-hop.

(C) There is sufficient evidence that for all three music choices, the proportion of middle school students who prefer each choice is different from the corresponding proportion of high school students.

(D) There is sufficient evidence that for at least one of the three music choices, the proportion of middle school students who prefer that choice is equal to the corresponding proportion of high school students.

(E) There is sufficient evidence that for at least one of the three music choices, the proportion of middle school students who prefer that choice is different from the corresponding proportion of high school students.

4. Given a two-way table, it's not obvious whether to perform a test for independence or a test for homogeneity. What is the main difference between the tests?

(A) How expected counts are calculated
(B) How df, the degrees of freedom, is calculated
(C) The number of rows in the table
(D) The number of columns in the table
(E) The number of samples

Questions 5–8 refer to the following. In a random sample of 525 teenagers, eye colors are cross-classified with favorite colors from among green, red, blue, and purple. Below is a two-way table of the responses.

	Favorite Color				
	Green	Red	Blue	Purple	Other
Brown eyes	34	21	50	38	20
Hazel eyes	28	20	31	25	13
Blue eyes	15	12	30	21	9
Green eyes	17	13	12	12	7
Other	17	26	18	22	14

5. For performing a chi square test, which of the following is the appropriate null hypothesis?

(A) Favorite color choices among all teenagers are 20% green, 20% red, 20% blue, 20% purple, and 20% other.
(B) There is no difference between the distributions of color choices among teenagers with different eye colors in this sample.
(C) There is no difference between the distributions of color choices among teenagers with different eye colors in the population of all teenagers.
(D) There is no association between eye color and favorite color among teenagers in this sample.
(E) There is no association between eye color and favorite color among teenagers in the population of all teenagers.

6. Which of the following is the expected count of blue-eyed teenagers who choose blue as their favorite color?

(A) 6
(B) 21
(C) 23.4
(D) 28.2
(E) 30

7. What is the degrees of freedom, df, for the chi-square test on these data?

(A) 4
(B) 5
(C) 16
(D) 24
(E) 25

8. For these data, $\chi^2 = 18.591$ with a P-value of 0.290. Assuming a significance level of 0.05, which of the following is true?

(A) A Type-I error may have been committed, but a Type-II error was not committed.
(B) A Type-II error may have been committed, but a Type-I error was not committed.
(C) Both Type-I and Type-II errors may have been committed.
(D) Neither a Type-I nor a Type-II error could have been committed.
(E) There is not sufficient information to make any statements about the possibility of a Type-I or a Type-II error being committed.

Free-Response Questions

> **Directions:** You must show all work and indicate the methods you use. You will be graded on the correctness of your methods and on the accuracy of your final answers.

1. A candy manufacturer advertises that its fruit-flavored sweets have hard sugar shells in five colors with the following distribution: 35% cherry red, 10% vibrant orange, 10% daffodil yellow, 25% emerald green, and 20% royal purple. A random sample of 300 sweets yielded the counts in the following table:

	Cherry red	Vibrant orange	Daffodil yellow	Emerald green	Royal purple
Observed counts	94	34	22	77	73

Is there sufficient evidence that the distribution is different from what is claimed by the manufacturer?

2. You want to study whether smokers and nonsmokers have equal fitness levels (low, medium, high) and are considering two survey designs:

   I. Take a random sample of 200 people, asking each whether or not they smoke and measuring the fitness level of each.
   II. Take a random sample of 100 smokers and measure the fitness level of each, and take a random sample of 100 nonsmokers and measure the fitness level of each.

   (a) Which of the designs is a test of independence appropriate, and for which is a test of homogeneity appropriate? Explain.
   (b) If we are interested in whether the proportion of people who have various fitness levels differs among smokers and nonsmokers, which design should be used? Explain.
   (c) If we are interested in the conditional distribution of people with given fitness levels who are smokers or are not smokers, which design should be used? Explain.

3. Is there a difference in happiness levels between busy and idle people? In one study, after filling out a survey, 175 randomly chosen high school students were told they could either sit 15 minutes while the survey was being tabulated or they could walk 15 minutes to another building where the survey was being tabulated. Then they were given a questionnaire asking how good they felt during the past 15 minutes (on a scale of 1, "not good," to 5, "very good"). The results of the questionnaire were as follows:

	Happiness level				
	1	2	3	4	5
Busy (walking)	8	15	24	26	25
Idle (sitting)	18	20	15	10	14

(a) Does the above data give statistical evidence of a relationship between happiness level and busy/idle choice of the students?

(b) If the answer above is positive for a relationship, is it reasonable to conclude that encouraging high school students to keep busy will lead to higher happiness levels?

The answers for this quiz can be found in the Appendix on page 626.

QUIZ 30

Multiple-Choice Questions

Directions: The questions that follow are each followed by five suggested answers. Choose the response that best answers the question.

1. Is there a relationship between education level and sports interest? A study cross-classified 1500 randomly selected adults in three categories of education level (not a high school graduate, high school graduate, and college graduate) and five categories of major sports interest (baseball, basketball, football, hockey, and tennis). The χ^2-value is 13.95. Is there evidence of a relationship between education level and sports interest?

   (A) The data prove there is a relationship between education level and sports interest.
   (B) The evidence points to a cause-and-effect relationship between education and sports interest.
   (C) There is sufficient evidence at the 5% significance level of a relationship between education level and sports interest.
   (D) There is sufficient evidence at the 10% significance level, but not at the 5% significance level, of a relationship between education level and sports interest.
   (E) The P-value is greater than 0.10, so there is not sufficient evidence of a relationship between education level and sports interest.

2. A geneticist claims that four species of fruit flies should appear in the ratio 1:3:3:9. Suppose that a sample of 2000 flies contained 110, 345, 360, and 1185 flies of each species, respectively. For a chi-square goodness-of-fit test, what is the expected value for the second species?

   (A) 125
   (B) 250
   (C) 345
   (D) 375
   (E) 500

3. Random samples of 25 students are chosen from each high school class level, students are asked whether or not they are satisfied with the school cafeteria food, and the results are summarized in the following table:

	Freshmen	Sophomores	Juniors	Seniors
Satisfied	15	12	9	7
Dissatisfied	10	13	16	18

A chi-square test for homogeneity gives $\chi^2 = 5.998$. Is there sufficient evidence of a difference in cafeteria food satisfaction among the class levels?

(A) The data prove that there is a difference in cafeteria food satisfaction among the class levels.

(B) There is sufficient evidence of a linear relationship between food satisfaction and class level.

(C) There is sufficient evidence at the 1% significance level of a difference in cafeteria food satisfaction among the class levels.

(D) There is sufficient evidence at the 5% significance level, but not at the 1% significance level, of a difference in cafeteria food satisfaction among the class levels.

(E) With $P = 0.1117$, there is not sufficient evidence of a difference in cafeteria food satisfaction among the class levels.

4. With a χ^2-test of independence and a 3 by 4 table, we have $\chi^2 = 12.7$. Assuming a significance level of 0.05, which of the following is true?

(A) While a Type-II error was not committed, a Type-I error may have been committed.

(B) While a Type-I error was not committed, a Type-II error may have been committed.

(C) Either a Type-I or a Type-II error may have been committed.

(D) Neither a Type-I nor a Type-II error may have been committed.

(E) With the information provided, it's not possible to say whether or not a Type I or a Type-II error may have been committed.

Questions 5–8 refer to the following. The proportions of people of various racial/ethnic identities charged with nonviolent crimes in a large city is known to be 58% White, 23% Black, 12% Hispanic, and 7% other. In a random sample of 80 people charged with nonviolent crimes in the city, the numbers receiving especially harsh sentences is tabulated in the following table.

Ethnicity	White	Black	Hispanic	Other
Number of harsh sentences	35	29	12	4

5. For performing a chi-square test, which of the following is the appropriate null hypothesis?

(A) The racial/ethnic distribution of harsh sentences to nonviolent offenders is 25% White, 25% Black, 25% Hispanic, and 25% Other.
(B) The racial/ethnic distribution of harsh sentences to nonviolent offenders is the same as the racial/ethnic distribution of people charged with nonviolent crimes.
(C) The racial/ethnic distribution is independent of receiving a harsh sentence for nonviolent offenders.
(D) The racial/ethnic distribution is not independent of receiving a harsh sentence for nonviolent offenders.
(E) The correlation between race/ethnicity and receiving a harsh sentence for nonviolent offenders is zero.

6. Assuming the null hypothesis is true, what is the expected number of Black nonviolent offenders who would receive a harsh sentence?

(A) 18
(B) 18.4
(C) 19
(D) 20
(E) 23

7. The χ^2-test statistic is 9.96. In which of the following intervals is the P-value?

(A) Less than 0.01
(B) Between 0.01 and 0.05
(C) Between 0.05 and 0.10
(D) Between 0.10 and 0.20
(E) Greater than 0.20

8. The expected numbers are 46.4, 18.4, 9.6, and 5.6. Which two categories contribute the largest components to the χ^2-test statistic?

(A) White and Black
(B) White and Hispanic
(C) White and Other
(D) Black and Hispanic
(E) Black and Other

Free-Response Questions

> **Directions:** You must show all work and indicate the methods you use. You will be graded on the correctness of your methods and on the accuracy of your final answers.

1. A poll, asking a random sample of adults whether or not they eat breakfast and to rate their morning energy level, results in the table:

	Morning energy level		
	Low	Medium	High
Breakfast	22%	24%	24%
No breakfast	12%	10%	8%

   (a) If the sample size was $n = 500$, is there sufficient evidence of a relationship between eating breakfast and morning energy level?

   (b) Does the answer to part (a) change if $n = 1000$ instead? Explain.

2. A survey on acne treatments randomly selected 100 teenagers using topical treatments, 100 using oral medications, and 50 using laser therapy and asked each subject about satisfaction level. The resulting counts were

	Topical	Oral	Laser
Very satisfied	61	54	24
Somewhat satisfied	28	21	10
Unsatisfied	11	25	16

   (a) Do the different treatments lead to different satisfaction levels? Perform an appropriate hypothesis test.

   (b) What is a possible confounding variable that could lead you to jump to a misleading conclusion? Explain.

3. A heavy backpack can cause chronic shoulder, neck, and back pain. Wide, padded shoulder straps, a waist belt, and avoidance of single-strap bags all help, but weight is the main problem. The recommendation for a safe weight for school backpacks is no more than 10% of body weight. In a study of 500 randomly selected high school students, the weights of their backpacks give rise to the following:

Weight (lb)	Below 12.5	12.5–17.5	17.5–22.5	22.5–27.5	Above 27.5
Observed #	28	134	182	112	44

   (a) In a normal distribution with $\mu = 20$ and $\sigma - 5$, what are the probabilities of $x < 12.5$, $12.5 < x < 17.5$, $17.5 < x < 22.5$, $22.5 < x < 27.5$, and $x > 27.5$?

   (b) Given 500 students, if the data follow a normal distribution with $\mu = 20$ and $\sigma = 5$, what are the expected values for the numbers of backpacks in each of the indicated weight ranges?

   (c) Test the null hypothesis that the data follow a normal distribution with $\mu = 20$ and $\sigma = 5$.

The answers for this quiz can be found in the Appendix on page 628.

- Chi-square analysis here is an important tool for inference on distributions of *counts*.
- Chi-square distributions take only positive values and are skewed right. With increasing degrees of freedom, the skew is less pronounced and the distribution looks more like a normal distribution.
- The chi-square statistic is found by summing the weighted squared differences between observed and expected counts, that is, $\chi^2 = \sum \dfrac{(observed - expected)^2}{expected}$.
- A chi-square goodness-of-fit test compares an observed distribution to some expected distribution.
- A chi-square test of independence tests for evidence of an association between two categorical variables from a single sample.
- A chi-square test of homogeneity compares samples from two or more populations with regard to a single categorical variable.
- For these three chi-square tests, the observed counts must be integers; however, this is not true for the calculated expected counts.
- Expected counts for a chi-square goodness-of-fit test are (sample size) × (claimed proportions).
- Expected count for a particular cell of a two-way table is $\dfrac{(row\ total)(column\ total)}{table\ total}$.
- Conditions to check for these chi-square tests are the following:
  1. Random samples (independent random samples for homogeneity tests).
  2. Samples should be less than 10% of the populations.
  3. All expected counts should be greater than 5.
- Just like with proportions and means, the *P*-value for a chi-square test is a conditional probability; it is the probability, given that the null hypothesis is true, of obtaining a test statistic as extreme as, or more extreme than, the observed value.
- Just like with proportions and means, the decision to either reject or fail to reject the null hypothesis for a chi-square test is based on comparison of the *P*-value to the significance level, α.
- When scoring a chi-square free-response question on the exam, readers will look for whether you:
  1. State the hypotheses (both H_0 and H_a) and define the parameter of interest. (The hypotheses must refer to the population, not to the sample.)
  2. Name the test: goodness-of-fit, independence, or homogeneity. (How the hypotheses are stated determines which test you choose.)
  3. Check the conditions. This requires showing the expected values (which can be copied from the calculator).
  4. Report the resulting chi-square value, the *P*-value, and *df* (the number of degrees of freedom).
  5. Give a conclusion in context with linkage to the *P*-value.

Inference for Quantitative Data: Slopes

9

(2–5% AP EXAM WEIGHTING)

→ **CONFIDENCE INTERVAL FOR THE SLOPE OF A LEAST SQUARES**
   REGRESSION LINE
→ **HYPOTHESIS TEST FOR THE SLOPE OF A LEAST SQUARES REGRESSION LINE**
→ **QUIZ 31**
→ **QUIZ 32**

In this unit, you will learn about the sampling distribution for the sample slope. You will then use this concept in finding confidence intervals and performing hypothesis tests on the slope. Just like with inference on means and proportions, you will see conditions to be checked and understand how to give proper conditions in context.

UNIT LEARNING OBJECTIVES

- To be able to read generic computer regression output and identify the quantities needed for inference for the slope of the regression line (sample slope, *SE* of the slope, and the degrees of freedom).
- To be able to state and verify whether or not the conditions are met for inference on the slope of the regression line (based on using the *t*-distribution).
- Conduct a confidence interval procedure for the slope of the regression line.
- Conduct a hypothesis test for the slope of the regression line.

If a scatterplot shows a linear relationship between two quantitative variables, we can calculate the least squares line from the data. As seen in Unit 2, the resulting equation, $\hat{y} = a + bx$, can be used to predict y for a given value of x. Statistical inference will help us answer two questions:

1. What is a confidence interval for the true slope, β? This will give the margin of error for predicting the change in y for a unit increase in x.
2. Is it plausible that the pattern observed in the scatterplot happened by chance, or is there sufficient evidence of a linear relationship between x and y?

Inference on slopes depends on knowing the standard error of the slope, written as s_b or $SE(b)$. While this can be calculated by $s_b = \dfrac{s}{s_x \sqrt{n-1}}$ (where s is the standard deviation of the residuals, and s_x is the standard deviation of the x-values), it is always given in generic computer output.

> **NOTE**
>
> We see that increasing s (that is, increasing spread about the line) *increases* the slope's standard error, while increasing s_x (that is, increasing the spread of x-values) or increasing n (the sample size) *decreases* the slope's standard error.

NOTE

The sampling distribution for the slope of a regression line can be modeled by a *t*-distribution with *df* = *n* − 2.

NOTE

On the TI-84, after Stat → Calc → LinReg, the list of residuals is stored in RESID under LIST NAMES.

TIP

The fourth condition can be checked by a histogram, dotplot, stemplot, or normal probability plot of the residuals.

NOTE

Ideally, the normality check should be made for the residuals at each *x*-value, but we rarely have enough observations for this.

TIP

Practice how to read *generic computer output*, and note that you usually will *not* need to use all the information given.

NOTE

Note the similarities between $b \pm t^* SE(b)$, $\hat{p} \pm z^* SE(\hat{p})$, and $\bar{x} \pm t^* SE(\bar{x})$.

CONFIDENCE INTERVAL FOR THE SLOPE OF A LEAST SQUARES REGRESSION LINE

The slope b of the regression line and the standard error $s_b = SE(b)$ of the slope are listed explicitly in the computer output. A confidence interval for β can be found using t-scores with $df = n - 2$. If given raw data, a confidence interval can readily be found using the statistical software on a calculator.

Conditions for finding a confidence interval for the slope include:

1. The sample must be randomly selected.
2. The scatterplot should be approximately linear.
3. There should be no apparent pattern in the residuals plot.
4. The distribution of the residuals should be approximately normal.
5. The sample size n should be less than 10 percent of the population size N.

➡ **EXAMPLE 9.1** _____

Information concerning SAT verbal scores and SAT math scores was collected from 15 randomly selected students. A linear regression performed on the data using a statistical software package produced the following printout:

```
Dependent variable: Math
Variable          Coef          SE Coef          T          Prob
Constant          92.5724        31.75           2.92        0.012
Verbal            0.763604       0.05597         13.6        0.000
S = 16.69     R-Sq = 93.5%      R-Sq(adj) = 93.0%
```

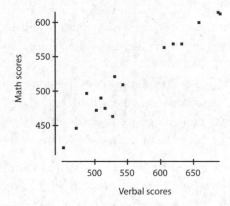

What is the regression equation?

Answer: Assuming that all conditions for regression are met, the *y*-intercept and slope of the equation are found in the Coef column of the above printout.

$$\widehat{Math} = 92.57 + 0.764(\text{Verbal})$$

What is a 95% confidence interval estimate for the slope of the regression line?

Answer: The standard deviation of the residuals is $S = 16.69$ and the standard error of the slope is $s_b = 0.05597$. With 15 data points, $df = 15 - 2 = 13$, and the critical *t*-values are $\pm\text{invT}(0.975, 13) = \pm 2.160$. The 95% confidence interval of the true slope is:

$$b \pm t^* s_b = 0.764 \pm 2.160(0.05597) = 0.764 \pm 0.121 \text{ or } (0.643, 0.885)$$

We are 95% confident that the interval from 0.64 to 0.89 captures the slope of the true regression line relating the SAT math score, y, and SAT verbal score, x. (Or we are 95% confident that for every 1-point increase in verbal SAT score, the average increase in math SAT score is between 0.64 and 0.89.)

➡ **EXAMPLE 9.2**

A random sample of ten high school students produced the following results for number of hours of television watched per week and GPA.

TV hours	12	21	8	20	16	16	24	0	11	18
GPA	3.1	2.3	3.5	2.5	3.0	2.6	2.1	3.8	2.9	2.6

Determine the least squares line and give a 95% confidence interval estimate for the true slope.

Answer: Checking the conditions:

We are told that the data come from a *random* sample of students.

Scatterplot is approximately linear.

No apparent pattern is evident in residuals plot.

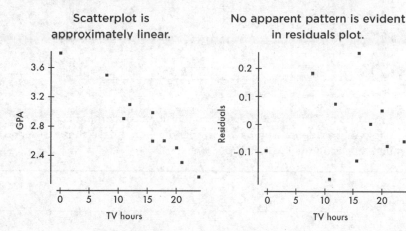

Distribution of residuals is approximately normal.

Checked by histogram.

Checked by normal probability plot.

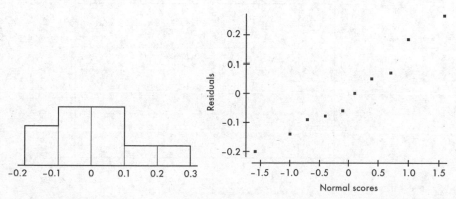

The sample of 10 high school students is less than 10% of all high school students.

(continued)

TIP

If you refer to a graph—whether it is a histogram, boxplot, stemplot, scatterplot, residuals plot, normal probability plot, or some other kind of graph—you should *roughly draw it*. It is not enough to simply say, "I did a normal probability plot of the residuals on my calculator and it looked linear."

NOTE

Remember from the checking normality discussion in the Normal Distribution Calculations section in Unit 5 that if the normal probability plot is nearly straight, the data are nearly normal.

Now using the statistics software on a calculator gives:

$$\hat{y} = 3.892 - 0.07202x$$

Putting the data into two Lists and using calculator software (such as `LinRegTInt` on the TI-84) gives (−0.0887, −0.0554).

We are thus 95% confident that the interval from −0.089 to −0.055 captures the slope of the true regression line relating the GPA, y, and the TV hours, x. (Or we are 95% confident that each additional hour before watching the television each week is associated with a mean drop in GPA of between 0.055 and 0.089.)

On the TI-Nspire, we can show the following:

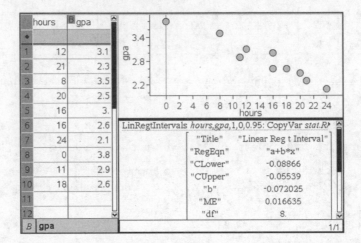

Remember, we must be careful about drawing conclusions concerning cause and effect; it is possible that students who have lower GPAs would have these GPAs no matter how much television they watched per week.

➡ EXAMPLE 9.3

The distances a caterpillar crawls in a random sample of 15 time periods are plotted in the following scatterplot with some regression analysis given:

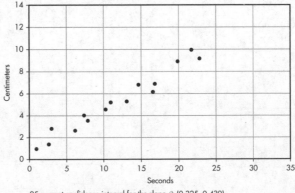

95 percent confidence interval for the slope β: (0.325, 0.430)

(a) Another measurement of 14 centimeters in 35 seconds is obtained. With the addition of this data point, would the confidence interval change, and, if so, would the margin of error increase or decrease?

Answer: (35, 14) appears to follow the linear trend of the rest of the data. The slope of the regression line will change little, if at all. However, with more points following the same trend, the standard error of the slope will be smaller, and thus the margin of error will decrease. In this example, the 95% confidence interval for the slope with the additional data point is actually (0.339, 0.415).

(b) What would happen if instead the data point (12, 1) is added?

Answer: With the x-coordinate, 12, appearing close to the mean of all the x-coordinates, the slope of the regression line will change little, if at all; however, the y-intercept will change. Both the standard error of the slope and the margin of error of the confidence interval will also increase. In this example, the 95% confidence interval for the slope with the additional data point is actually (0.272, 0.479).

HYPOTHESIS TEST FOR SLOPE OF LEAST SQUARES REGRESSION LINE

In addition to finding a confidence interval for the true slope, we can also perform a hypothesis test for the value of the slope. Often we use the null hypothesis $H_0: \beta = 0$, that is, that there is no linear relationship between the two variables.

Assumptions for inference for the slope of the least squares line include the following:

1. The sample must be randomly selected.
2. The scatterplot should be approximately linear.
3. There should be no apparent pattern in the residuals plot.
4. The distribution of the residuals should be approximately normal.
5. The sample size n should be less than 10 percent of the population size N.

Note that a low *P*-value tells us that if the two variables did not have some linear relationship, it would be highly unlikely to find such a random sample. However, **strong evidence that there is some linear association does not mean the association is strong**.

> **TIP**
>
> *Be careful* about abbreviations. For example, your teacher might use LOBF (line of best fit), but the grader may have no idea what this means.

➡️ **EXAMPLE 9.4** _____

A company offers a 10-lesson program of study to improve students' SAT scores. A survey is made of a random sampling of 25 students. A scatterplot of improvement in total (Math and Verbal) SAT score versus number of lessons taken is as follows:

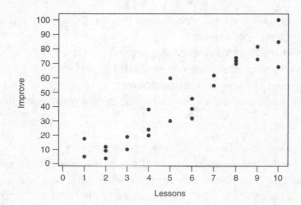

(continued)

Some computer output of the regression analysis follows, along with a plot and a histogram of the residuals.

Regression Analysis: Improvement Versus Lessons

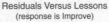

Dependent variable: Improvement

Predictor	Coef	SE Coef	T	P
Constant	−7.718	4.272	−1.81	0.084
Lessons	9.2854	0.6789	13.68	0.000

S = 9.744   R−Sq = 89.1%

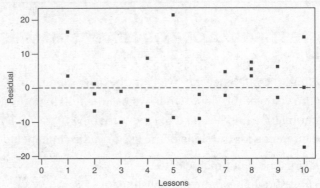

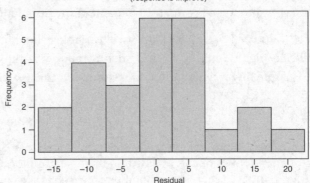

(a) What is the equation of the regression line that predicts score improvement as a function of number of lessons?

(b) Interpret, in context, the slope of the regression line.

(c) What is the meaning of R−Sq in the context of this study?

(d) Give the value of the correlation coefficient.

(e) Perform a test of significance for the slope of the regression line.

Answers:

(a) Predicted Improvement = −7.718 + 9.2854(Lessons)

(b) The slope of the regression line is 9.2854, meaning, that on the average, each additional lesson is predicted to improve one's total SAT score by 9.2854.

(c) R-Sq = 89.1% means that 89.1% of the variation in total SAT score improvement can be explained by variation in the number of lessons taken.

(d) $r = +\sqrt{0.891} = 0.944$, where the sign is positive because the slope is positive.

(e) A test of significance of the slope of the regression line with $H_0: \beta = 0$ and $H_a: \beta \neq 0$ (test of significance of correlation) is as follows:

Hypotheses: H_0: $\beta = 0$, H_a: $\beta \neq 0$.

Procedure: t-test for the slope of a regression model.

Checks: We are told the data came from a random sample, the scatterplot appears to be approximately linear, there is no apparent pattern in the residuals plot, the histogram of residuals is roughly unimodal and symmetric, and we assume the sample of 25 is less than 10% of all students taking the 10-lesson program of study.

Mechanics: $t = \dfrac{9.2854 - 0}{0.6789} = 13.68$, and with $df = 25 - 2 = 23$, $P = 0.000$. (Note that the values for t and P are given directly in the computer output. All you have to do is copy them down!)

Conclusion in context with linkage to the P-value: With this small of a P-value, $0.000 < 0.05$, there is very strong evidence to reject H_0 and conclude that there is strong evidence of a linear relationship between improvement in total SAT score and number of lessons taken.

➡ EXAMPLE 9.5

The following table gives serving speeds in mph (using a flat or "cannonball" serve) of ten randomly selected professional tennis players before and after using a newly developed tennis racket.

TIP

If the data don't look straight, do not try to fit the data with a straight line.

| With old racket | 125 | 133 | 108 | 128 | 115 | 135 | 125 | 117 | 130 | 121 |
| With new racket | 133 | 134 | 112 | 139 | 123 | 142 | 140 | 129 | 139 | 126 |

(a) Is there evidence of a *straight-line* relationship with positive slope between serving speeds of professionals using their old and the new rackets?

Answer:

Hypotheses: H_0: $\beta = 0$, H_a: $\beta > 0$.

Procedure: t-test for the slope of a regression line.

Checks: We are told that the data come from a *random* sample of professional players, the scatterplot appears to be approximately linear, there is no apparent pattern in the residuals plot, the histogram of residuals appears to be approximately normal, and the sample of size 10 is less than 10% of all professional players.

Scatterplot is approximately linear.   No apparent pattern in residuals plot.

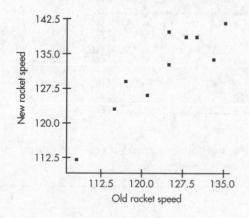
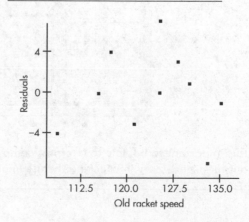

(continued)

Distribution of residuals is approximately normal.

Checked by histogram.   *or*   Checked by normal probability plot.

Mechanics: Using the statistics software on a calculator (for example, LinRegTTest on the TI-84 or LinearReg tTest on the Casio Prizm) gives:

$$\widehat{\text{New speed}} = 8.76 + 0.99(\text{Old speed}) \quad \text{with } t = 5.853 \text{ and } P = 0.00019$$

With such a small P-value, $0.00019 < 0.05$, there is very strong evidence to reject H_0; that is, there is very strong evidence of a straight-line relationship with positive slope between serving speeds of professionals using their old and the new rackets.

(b) Interpret in context the least squares line.

Answer: With a slope of approximately 1 and a y-intercept of 8.76, the regression line indicates that use of the new racket increases serving speed on the average by 8.76 mph regardless of the old racket speed. That is, players with lower and higher old racket speeds experience on the average the same numerical (rather than percentage) increase when using the new racket.

On the TI-Nspire, the result shows as:

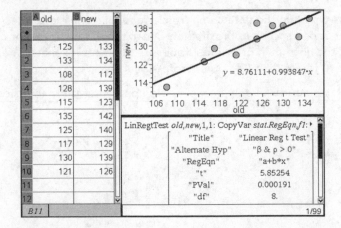

It is important to be able to interpret computer output, and appropriate computer output is necessary in checking the assumptions for inference.

QUIZ 31

Multiple-Choice Questions

Directions: The questions that follow are each followed by five suggested answers. Choose the response that best answers the question.

1. Inference about the slope of a least squares regression line is based on the sampling distribution of b being

   (A) approximately normal.
   (B) a chi-square distribution with $df = n - 1$.
   (C) a chi-square distribution with $df = n - 2$.
   (D) a t-distribution with $df = n - 1$.
   (E) a t-distribution with $df = n - 2$.

2. A statistician investigated the relationship between ages and bowling scores for participants in an adult bowling league. The regression analysis for her data is shown below.

Predictor	Coef	SE Coef	T	P
Constant	122.79	94.44	1.30	0.218
Age	3.342	1.200	2.78	0.017

   S = 37.3598   R—Sq = 84.0%

   By how much do bowling score estimates produced by this model typically differ from the actual scores of the bowlers?

   (A) 1.200
   (B) 3.342
   (C) 37.3598
   (D) 94.44
   (E) 122.79

3. A statistician wonders if dress size can be predicted from a woman's height. In a random sample of 20 female high school students, dress size versus height (cm) gives the following regression results:

```
The regression equation is
Size = -48.8 + 0.374 Height

Predictor      Coef      SE Coef        T        P
Constant      -48.81       30.57     -1.60     0.128
Height        0.3736      0.1898      1.97     0.065

S = 4.46720       R-Sq = 17.7%        R-Sq(adj) = 13.1%
```

Is there statistical evidence of a linear relationship between dress size and height ($H_0: \beta = 0$, $H_a: \beta \neq 0$)?

(A) No, because r^2, the coefficient of determination, is too small.
(B) No, because 0.128 is above any reasonable significance level.
(C) Yes, because by any reasonable observation, taller women tend to have larger dress sizes.
(D) Yes, because the computer printout does give a regression equation.
(E) There is sufficient evidence at the 10% significance level but not at the 5% level.

4. A 95% confidence interval for the slope of a regression line is calculated to be $(-0.783, 0.457)$. Which of the following must be true?

(A) The slope of the regression line is 0.
(B) The slope of the regression line is -0.326.
(C) A scatterplot of the data would show a linear pattern.
(D) A residual plot would show no pattern.
(E) The correlation is negative.

Questions 5–8 refer to the following setting. In a random sample of 25 professional baseball players, their salaries (in millions of dollars) and batting averages result in the following regression analysis:

```
Predictor       Coef       SE Coef           T            P
Constant       0.2336      0.005883        39.71        0.0000
Salary        0.008051     0.002825        2.850        0.0091

S = 0.01695      R-Sq = 71.6%        R-Sq(adj) = 70.3%
```

5. What is the equation of the least squares regression line?

(A) Batting Average = 0.008051 + 0.2336(Salary)
(B) Predicted Batting Average = 0.008051 + 0.2236(Salary)
(C) Predicted Batting Average = 0.2336 + 0.008051(Salary)
(D) Predicted Batting Average = 0.002825 + 0.008051(Salary)
(E) Predicted Salary = 0.2336 + 0.008051(Batting Average)

6. Which of the following gives a 95 percent confidence interval for the slope of the regression line?

   (A) $0.008051 \pm 1.96(0.002825)$
   (B) $0.008051 \pm 1.96(0.01695)$
   (C) $0.008051 \pm 2.0639(0.002825)$
   (D) $0.008051 \pm 2.0687(0.002825)$
   (E) $0.008051 \pm 2.85(0.002825)$

7. In testing the hypothesis $H_0: \beta = 0$ versus $H_a: \beta > 0$, what is the t-statistic?

   (A) 2.850
   (B) $(0.5)(2.850)$
   (C) 39.71
   (D) $(0.5)(39.71)$
   (E) This cannot be answered without first deciding on an α-risk.

8. In testing the hypothesis $H_0: \beta = 0$ versus $H_a: \beta > 0$, what is the P-value?

   (A) 0.0045
   (B) 0.0091
   (C) 0.0250
   (D) 0.7030
   (E) 0.7160

Free-Response Questions

> **Directions:** You must show all work and indicate the methods you use. You will be graded on the correctness of your methods and on the accuracy of your final answers.

1. Information with regard to the assessed values (in $1000) and the selling prices (in $1000) of a random sample of homes sold in a Northeast market yields the following computer output:

    ```
    Dependent variable is   Price

    R squared = 89.8%   R squared (adjusted) = 89.2%
    s = 9.508 with   20 - 2 = 18  degrees of freedom

    Variable      Coeff      s.e. of coeff    t-ratio      prob
    Constant     0.890087       16.16          0.055       0.956
    Assessed     1.0292         0.08192        12.6        0.000
    ```

    Assume all conditions for inference are met.

    (a) Determine the equation of the least squares regression line.
    (b) Construct a 99% confidence interval estimate for the slope of the regression line, and interpret this in context.
    (c) What conclusion can be reached given where the interval lies in relation to the value of 0?

2. A study of 100 randomly selected teenagers, ages 13–17, looked at number of texts per waking hour versus age, yielding the computer regression output below:

    ```
    Predictor      Coef       SE Coef        T          P
    Constant      -1.055       2.815        -0.37      0.709
    Age            0.4577      0.1866        2.45       0.016

    S = 2.66501      R-Sq = 5.8%      R-Sq(adj) = 4.8%
    ```

    (a) Interpret the slope in context.
    (b) What three graphs should be checked with regard to conditions for a test of significance for the slope of the regression line?
    (c) Assuming all conditions for inference are met, perform this test of significance.
    (d) Give a conclusion in context, taking into account both your answer to the hypothesis test as well as the value of R-Sq.

The answers for this quiz can be found in the Appendix on page 631.

QUIZ 32

Multiple-Choice Questions

> **Directions:** The questions that follow are each followed by five suggested answers. Choose the response that best answers the question.

1. Researchers are investigating the association between humidity (percent) and attendance (1000s) at major league baseball games for a random sample of days and stadiums. A 95 percent confidence interval for the slope is constructed for predicting attendance based on humidity, obtaining (−1.32, 0.18). Which statement is correct?

   (A) The association is probably negative because most of the interval is below 0.
   (B) The association must be negative because the interval contains −1.
   (C) The association is not significant because 0 is in the interval.
   (D) The relatively high margin of error, 0.75, makes any conclusion suspect.
   (E) A calculation error was made because the lower endpoint is below −1.

2. Which of the following is a necessary assumption for performing inference analysis on the slope of a least squares regression line?

   (A) There is no strong skew or outliers in the data.
   (B) A straight line can be drawn through the set of paired observations in the scatterplot.
   (C) The distribution of the residuals is approximately uniform.
   (D) The distribution of the residuals is approximately linear.
   (E) The distribution of the residuals is approximately normal.

3. Can points per game (PPG) be predicted based on NBA players' salaries? For the New York Knicks 2019–2020 season, a linear association study yields the following computer output:

```
Regression Analysis: PPG versus Salary (in Millions of $)

Variable     N      Mean    SE Mean    StDev
Salary      14      7.34      1.45      5.43
PPG         14      8.08      1.52      5.70

The regression equation is PPG = 4.39 + 0.656 Salary

Predictor      Coef        SE Coef
Constant      3.6648        1.5246
Salary        0.6013        0.2488

S = 4.8695    R-Sq = 32.7%
```

Which of the following is an appropriate test statistic for testing the null hypothesis that the slope of the regression line is greater than 0? (Assume all conditions for inference are met.)

(A) $\dfrac{8.08}{1.52}$

(B) $\dfrac{8.08}{5.70}$

(C) $\dfrac{3.6648}{1.5246}$

(D) $\dfrac{0.6013}{0.2488}$

(E) $\dfrac{0.6013}{4.8695}$

Questions 4–8 refer to the following setting. Smoking is a known risk factor for cardiovascular disease and cancer. A study analyzing smoking levels (seven men smoking at levels from 0.5 to 3.0 packs per day) and risk of dementia (as measured by the Cox hazard ratio) resulted in the following regression analysis:

```
Predictor       Coef      SE Coef         T          P
Constant       0.7737     0.07833       9.88     0.0000
Packs          0.4341     0.07675       5.656    0.0012

S = 0.09817  R-Sq = 95.3%  R-Sq(adj) = 94.4%
```

4. What is the equation of the least squares regression line?

(A) Predicted Cox Ratio = 0.4341 + 0.07675(Packs)
(B) Predicted Cox Ratio = 0.07675 + 0.4341(Packs)
(C) Predicted Cox Ratio = 0.4341 + 0.7737(Packs)
(D) Predicted Cox Ratio = 0.7737 + 0.4341(Packs)
(E) Predicted Cox Ratio = 0.7737 + 0.07675(Packs)

5. By how much do Cox ratio estimates produced by this model typically differ from the actual ratios of the smokers?

   (A) 0.07675
   (B) 0.07833
   (C) 0.09817
   (D) 0.4341
   (E) 0.7737

6. Which of the following gives a 90 percent confidence interval for the slope of the regression line?

   (A) $0.4341 \pm 1.645(0.07675)$
   (B) $0.4341 \pm 1.645(0.09817)$
   (C) $0.4341 \pm 1.943(0.07675)$
   (D) $0.4341 \pm 1.943(0.09817)$
   (E) $0.4341 \pm 2.015(0.07675)$

7. In testing the hypothesis $H_0: \beta = 0$ versus $H_a: \beta > 0$, what is the P-value?

   (A) 0.0006
   (B) 0.0012
   (C) 0.0024
   (D) 0.07675
   (E) 0.09817

8. Which of the following is a correct interpretation of the P-value?

   (A) The probability of making a Type I error
   (B) The probability that there is a linear relationship between smoking levels and risk of dementia of the seven smokers in the sample
   (C) The probability that there is a linear relationship between smoking levels and risk of dementia
   (D) If there is a linear relationship between smoking levels and risk of dementia, the probability of getting a random sample of seven smokers that yields a least squares regression line with a slope of 0.4341 or greater
   (E) If there is no linear relationship between smoking levels and risk of dementia, the probability of getting a random sample of seven smokers that yields a least squares regression line with a slope of 0.4341 or greater

Free-Response Questions

> **Directions:** You must show all work and indicate the methods you use. You will be graded on the correctness of your methods and on the accuracy of your final answers.

1. A sociologist is researching a possible link between household income and self-reported life satisfaction. A least squares regression on "Self-reported life satisfaction" measured on a 10-point scale versus "Household income" in $1000 among 50 randomly selected adults yields the following computer printout:

	Coef	SE Coef
Intercept	1.415	0.0943
Income	0.054	0.00358
S = 0.977	R-Sq = 0.862	

   Assume all conditions for inference are met.

   (a) Determine the equation of the least squares regression line.
   (b) Interpret R-Sq = 0.826 in context.
   (c) Interpret $S = 0.977$ in context.
   (d) What is the test statistic for testing the null hypothesis that the slope of the regression line is greater than 0?
   (e) Find the 90 percent confidence interval for the slope of the regression line, and interpret it in context.

2. Can the average women's life expectancy in a country be predicted from the average fertility rate (children per woman) in that country? The following are recent data from 11 randomly selected countries:

Country	Fertility rate (children per woman)	Life expectancy (women)
Afghanistan	6.25	45.5
Angola	5.33	51.4
Argentina	2.16	80.0
Australia	1.85	84.4
France	1.85	85.1
Liberia	4.69	61.5
Nepal	2.66	69.0
Netherlands	1.77	82.6
Pakistan	3.58	68.3
Poland	1.29	80.4
Singapore	1.29	83.4

   Perform a test of significance for the slope of the regression line that relates fertility rate (children/woman) to life expectancy (women).

   The answers for this quiz can be found in the Appendix on page 633.

SUMMARY

- The sampling distribution for the slope of a regression line can be modeled by a t-distribution with $df = n - 2$; that is, the sampling distribution of $t = \dfrac{b - \beta}{SE(b)}$ has a t-distribution with $df = n - 2$.

- Conditions for linear regression inference include:

  1. The sample must be randomly selected.
  2. The scatterplot should be approximately linear.
  3. There should be no apparent pattern in the residuals plot.
  4. The distribution of the residuals should be approximately normal (actually should be true for each x-value but can be relaxed if the sample size is greater than 30).
  5. The sample size n should be less than 10 percent of the population size N.

- The standard error of the slope can be read directly from generic computer output of regression.

- With regard to confidence intervals, note the similarities between $b \pm t^* SE(b)$, $\hat{p} \pm z^* SE(\hat{p})$, and $\bar{x} \pm t^* SE(\bar{x})$.

- The t-statistic and the P-value (for a two-sided test) for a hypothesis test of the slope can be read directly from generic computer output of regression.

Part Three: Final Review

Final Review

After completing Units 1–9, their associated quizzes, and the following review sections, you'll be ready to try the practice tests. Wishing you good luck will not be appropriate, because you'll be so well prepared that you won't need luck!

SELECTING AN APPROPRIATE INFERENCE PROCEDURE

Of the six free-response questions on the AP Statistics exam, at least one—and often two—involves inference. For these problems, the first thought should be to identify the appropriate inference method. While knowing how to perform the different procedures is relatively easy when the question appears on a page with the unit heading at the top, the exam provides no such help! Thus, time devoted to recognition and decision-making should be an important part of any review. This unit provides that valuable practice.

To increase success, stop and think before proceeding. Hasty decisions often lead to choosing the wrong statistical procedure needed to solve the problem. Take a few moments to answer several questions that should be at the forefront of your mind. Use the test booklet to jot down notes. The following are important questions and tips:

Are there key terms or phrases? (Underline them!)

- Words such as *proportion, mean,* or *association*?
- Expressions such as *find a confidence interval* or *do a hypothesis test*?

What kind of data are presented?

- Categorical (indicating proportions or chi-square)?
- Quantitative (indicating means or slopes)?

What summary statistics are given?

- A sample proportion?
- A sample mean and standard deviation?
- Sample counts (leading to chi-square inference)?
- A sample slope?

What specifically are you asked to do?

- Estimate a quantity (indicating a confidence interval)?
- Look for evidence to test a claim (indicating a hypothesis test)?

How was the data collected?

- From one sample or two samples?
- Is there pairing of data (is information lost if the data in either set are shuffled around)?

If a hypothesis test is indicated, should you use a one-tailed or two-tailed test?

- Are we looking for a difference from some particular value or to see whether nothing has changed (indicating a two-tailed test)?
- Are we looking for a change in a specific direction (indicating a one-tailed test)?

If the question is about the population mean, which standard deviation is given?

- Is the population standard deviation given (indicating z-procedures should be used)?
- Is the sample standard deviation given (indicating t-procedures should be used)?

What is the context of the question?

- Underline the context!
- A response should always refer to the context in the conclusion.

What conditions need to be addressed?

- Randomization (random sampling or random assignment)?
- Independence (how data are collected and $n \leq 0.10N$)?
- Normality (for proportions, check that np and $nq \geq 10$; for means, check that either the population is roughly normal or the sample size is large)?

➡ **EXAMPLE** _____

What procedure is called for with the data below?

$$29 \quad 32 \quad 50 \quad 41 \quad 32$$
$$34 \quad 50 \quad 48 \quad 44 \quad 40$$

Answer: Of course, this is impossible to answer without knowing context!!!

(a) A random sample of five students signing up for a private tutoring service take a pre-test and a post-test with scores on the 50-point tests given below.

Student	A	B	C	D	E
Pre-test score	29	32	50	41	32
Post-test score	34	50	48	44	40

What procedure should be used to find a 95% confidence interval for the mean difference in score results (post-test score minus pre-test score) for students who use the tutoring service?

Answer: The pre-test and post-test data are not independent. This is a paired data test, so we cannot use two-sample inference. No population standard deviation is given, so we will use a *t*-procedure rather than a *z*-procedure. Use a one-sample *t*-interval on the set of differences.

(b) In a random sample of 400 attendees at a Cubs–White Sox game, favorite team is cross-classified against favorite reading genre. The following table shows the data.

	Science fiction	Romance	Sports	Mystery	History
Cubs	29	32	50	41	32
White Sox	34	50	48	44	40

Which procedure should be used to test whether or not favorite team and favorite reading genre are independent?

Answer: There is a single sample cross-classified on two variables (rather than samples from two or more populations and a single categorical variable), so use a chi-square test of independence (rather than a chi-square test of homogeneity).

(c) A miniature golf course offers two nine-hole courses. A random sample of five players who play both courses shows the following scores.

	A	B	C	D	E
Course 1	29	32	50	41	32
Course 2	34	50	48	44	40

The franchise owner is interested in whether a player's score on one of the courses can be used to predict his or her score on the other course. What procedure should be used?

Answer: Use a linear regression *t*-test to determine if there is a statistically significant linear relationship between how players do on the two courses.

(d) A school's AP coordinator picks a random sample of five students who claim to have stayed up all night studying for a next day's AP exam and an independent random sample of five students who claim to have had a good night's sleep the night before the AP exam. Their scores on the exam are in the table below.

Student	A	B	C	D	E
No sleep	29	32	50	41	32
Sleep	34	50	48	44	40

What procedure should the AP coordinator use to determine if there is a statistically significant difference in the mean scores obtained by students who stay up all night studying and those who have a good night's sleep before the exam?

Answer: We have two independent samples, are asked to determine if there is a significant difference (rather than a confidence interval) between two means, and are not given any population standard deviation (so we must use a *t*-procedure rather than a *z*-procedure). Use a two-sample *t*-test for the difference of two population means.

The following quizzes, Quiz 33 and Quiz 34, evaluate your ability to name procedures, define parameters, list conditions to be checked, and state hypotheses if appropriate.

QUIZ 33

> **Directions:** For each of the following problems, name the procedure you would use, define any parameter you use, list the conditions to be checked, and state hypotheses if appropriate.

1. At the high school level, do men and women play varsity sports for the same reasons? Of all students who tried out for varsity sports, a random sample of men and a random sample of women are picked. The selected students are given questionnaires asking which of the following is their primary reason for playing sports: social, health, or status.

2. Estimate the proportion of AP Statistics students who will graduate in the top ten percent of their senior class. There were 150 AP Statistics students tracked.

3. Estimate the average number of books read by high school students during their four years in secondary school. A random sample of 125 students are followed during their high school years.

4. To test the claim that over 70 percent of racially motivated hate crimes are motivated by anti-Black bias, a criminologist obtains data from a random sample of police commissioners.

5. It is believed that in cases of identity theft, 30% of the victims used their mother's maiden name for their banking password, 25% used their pet's name, 20% used "password," and the rest used something else. A study is to be made to test this claimed distribution.

6. Do baseball pitchers throw harder if they participate in ten-minute yoga sessions before games? A pitcher who normally averages 95 mph with his fastballs participated in ten-minute yoga sessions for a random sample of his games. His fastballs for those games were clocked.

7. Does family income have an impact on school performance? In a random sample of 100 students, family income and student GPA were recorded. Is there significant statistical evidence of a linear association?

8. A study is planned to find the difference in the proportion of patients with warts who are cured with a treatment of cryotherapy and the proportion cured with a treatment of duct tape occlusion.

9. A study is planned to determine if there is significant evidence that the mean cholesterol level (measured in milligrams per deciliter) for people living in "Western" countries is higher than that of people living in "non-Western" countries.

10. What is the 95 percent confidence interval for the average increase in CO_2 emissions (in tons per person) for each unit increase in income level (in GDP per capita) among countries of the world?

11. An economist is interested in whether there is an association between race and whether an adult is unbanked (does not have a checking or savings account). A random sample of 500 adults gives the following table:

	White	Black	Hispanic (non-White)	Native American
Banked	203	98	77	49
Unbanked	18	25	20	10

12. For the prevention of strokes, at-risk patients are often given anti-clotting drugs. A study is being designed to compare the proportion of patients receiving rivaroxaban who still have strokes to the proportion receiving warfarin who still have strokes.

13. Malaria is endemic in Liberia. Many people contract and are cured of malaria repeatedly during their lifetimes. A new prophylactic medication was developed as a pill taken weekly to prevent the contraction of malaria. A random sample of adults took the pill for one year and took a placebo pill for another year. The year each adult took the prophylactic was decided randomly for each participant. Summary statistics were as follows.

	n	Mean	Standard deviation
Episodes on the prophylactic	1500	0.22	0.11
Episodes on the placebo	1500	0.84	1.04
Difference (on – off)	1500	−0.62	0.35

14. In a random sample of 5000 live births in the U.S., the infant mortality rate was 5.8 (infant deaths per 1000 live births). In a random sample of 2000 live births in Japan, the infant mortality rate was 2.0 (infant deaths per 1000 live births). What is the 99 percent confidence interval for the difference in mortality rates in the United States and Japan?

15. How much better do students do on national standardized exams if they take them a second time? The scores of 500 students who took a national standardized exam twice are charted.

16. What is the mean waist size of male high school math teachers? The waist sizes of a random sample of 50 male high school math teachers are measured. From past studies, it is known that the standard deviation of such waist sizes is 2.15 inches.

The answers for this quiz can be found in the Appendix on page 635.

QUIZ 34

Directions: For each of the following problems, name the procedure you would use, define any parameter you use, list the conditions to be checked, and state hypotheses if appropriate.

1. College advisers suspect that first-year students study more hours per week than sophomores.

2. Are students willing to report cheating by other students? Students from a random sample are asked to fill out an anonymous questionnaire asking whether they would report cheating by other students.

3. A cardiologist is interested in whether there is a relationship between blood pressure level (low, average, or high) and personality type (high-strung or easy-going).

4. During a previous season, Division 1 college basketball players successfully made 34% of their 3-point shot attempts. A sports magazine is interested in determining if the percentage is lower this year.

5. The director of a health clinic is interested in finding out if participating in a particular month-long workout program will increase the average number of push-ups that can be completed by clinic members. Enrollees will be asked to do as many push-ups as possible in a 90-second period before the program begins and then again a month later when the program ends.

6. It is hypothesized that the average body temperature is less than 98.6°F.

7. By how much does exercise lower resting heart rates for teenagers? Ninety teenagers took part in a survey where their resting heart rates were tested before and after an extensive exercise program.

8. A study about racism in the business world involved sending resumes with stereotypically "White" names and identical resumes except with stereotypically "Black" names. Twelve out of 110 job applications with stereotypically "White" names received callbacks, while only 10 out of 155 job applications with stereotypically "Black" names received callbacks. Establish a 90 percent confidence interval for the difference in proportions of callbacks between applications with stereotypically "White" and stereotypically "Black" names.

9. Is there a significant difference in the distribution of highest school level attained between Whites, Blacks, Asians, and Hispanics (non-White)? Random samples of 200 people from each population interviewed resulted in the following summary table.

	High school or less	Undergraduate degree	Graduate degree
White	130	58	12
Black	168	27	5
Asian	98	82	20
Hispanic (non-White)	172	24	4

10. A reporter for a national car magazine plans to sample used cars of a certain model in hopes of being able to predict the average decrease in selling price for each additional 1000 miles noted on the odometer.

11. After answering your phone, if there is an initial pause on the other end, it often means that the call is from a telemarketer. A study is conducted to estimate the average pause length for telemarketing calls.

12. In a random sample of 550 White students, 4.1 percent reported that, at times, they felt unsafe to go to school, while in an independent random sample of 425 non-White students, 7.8 percent reported that, at times, they felt unsafe to go to school. A sociologist believes that the difference is significant, meaning non-White students are more likely than White students to feel unsafe to go to school, at times.

13. In 2010, the U.S. population of incarcerated adults by race was as follows: 39 percent White (non-Hispanic), 19 percent Hispanic, 40 percent Black, 2 percent other. A random sample of 1000 incarcerated adults from January 2020 had the following distribution: 451 White (non-Hispanic), 147 Hispanic, 400 Black, and 2 other. Is there sufficient evidence that the proportions of the U.S. incarcerated adult population by race has changed?

14. Conservationists plan to determine the mean difference in weights between elephants living in captivity and those in the wild.

15. Is there a positive linear relationship between the quantity of caffeine consumed in a sitting rate and pulse rate?

16. It is hypothesized that the mean hourly wage of summer work available for high school students is $10.50. A city planner believes the true mean is lower. She randomly surveys 75 high school students with summer jobs. From past studies, it is known that the standard deviation of such jobs is $1.83.

The answers for this quiz can be found in the Appendix on page 638.

STATISTICAL INSIGHTS INTO SOCIAL ISSUES

Quiz 35 Topics

1. **DISTRIBUTION OF WEALTH**
2. **INCARCERATION BY AGE AND RACE**
3. **FAMILY INCOME AND SAT SCORES**
4. **REPORTING CONFLICT VIOLENCE**
5. **JURY SELECTION**
6. **ENVIRONMENTAL RACISM**
7. **FLAWED FORENSIC TESTIMONY**
8. **MODERN-DAY SLAVERY**
9. **HATE CRIMES**

Quiz 36 Topics

1. **BULLYING**
2. **FIREARM SUICIDES**
3. **DANGEROUS LEAD LEVELS**
4. **LAW ENFORCEMENT–ASSISTED DIVERSION**
5. **POLICE PROFILING**
6. **INEQUALITY AND HEALTH**
7. **WHY JAILS ARE FULL**
8. **VACCINES AND AUTISM**
9. **GLOBAL WARMING**

In the spirit of AP Statistics, the following two quizzes encompass comprehensive review questions that aim to give an appreciation of the power of statistics to better understand some of society's most pressing issues. As you review and apply the statistical tools you have learned, you should also reflect on the content of the examples and the empowerment that statistics gives you to be aware of, to discuss, and maybe someday to help overcome some of the ills present in the broad social world in which we live. More specifically, statistics can help you understand the relationships among power, resource inequities, and uneven opportunities among different social groups and to understand explicit discrimination based on race, class, and gender. In many ways, the AP Statistics curriculum is just as much a civics course as it is a math course. Of course, in addition to reflecting on these significant issues, don't forget to read carefully what the questions ask, to check assumptions and conditions when appropriate, to show your work, and to communicate your answers clearly.

There is great interest in how statistics can help better society. Statistics can be an incredibly powerful tool both in helping you understand social issues in terms of our current information age and in helping you effect social change. From disparities in income levels to disparities in the distribution of wealth, from unequal access to education to unequal access to health care, from unfair criminal justice enforcement to unfair housing market discrimination, statistics can be the key for you to become an engaged participant in our democracy. Based on your background, your interests, and your available time, you can decide how far to take the opportunities for deeper discussion raised in the following exercises.

Following are two quizzes, each with nine multi-section, free response questions. Quiz 35 reviews the concepts and skills from Units 1–5 on exploratory analysis, collecting data, and probability, and Quiz 36 reviews the concepts and skills from Units 6–9 on statistical inference. All the quiz questions are set in the context of contemporary social justice issues, and they are specifically designed to evaluate your understanding of the entire curriculum and reinforce your communication skills. The answers and grading rubrics are in the Appendix.

QUIZ 35

1. A 2016 study found that the richest 62 people in the world have as much wealth as the poorest half of the global population.[1] One way in which the distribution of wealth throughout the world can be analyzed is illustrated below (where GDP refers to Gross Domestic Product and Oceania is a region centered on the islands of the tropical Pacific Ocean).

   (a) Fill in the blank cells in the table below.

Region	Population (in millions)	% of World Population	Wealth (GDP) (in billions)	% of World GDP
Africa	1111		2600	
Asia	4299		18,500	
Oceania	38		1800	
Europe	742		24,400	
U.S./Canada	355		20,300	
Latin America	617		4200	
World Total	7162	100.0	71,800	100.0

   (b) Construct a segmented bar chart with two bars from the above table, and comment on its significance.

   (c) Consider the following graph:

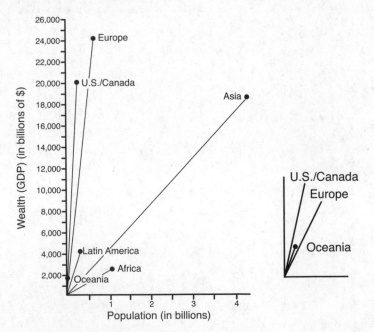

   What do the slopes of the six slanted lines represent? Use these slopes to give an appropriate ranking of the six regions.

[1]Oxfam International, using data from Credit Suisse.

(d) The particular quantitative index used in a statistical study can make a marked difference. For example, an alternative to the GDP for measuring the wealth of a nation is the GPI (Genuine Progress Indicator).[2] The GPI takes into account negative factors, such as industrial pollution, depletion of natural resources, and crime rates, as well as positive factors, such as socially productive time uses like volunteerism. Give an example of how changing the quantitative index from GDP to GPI might affect the relative position of U.S./Canada and Oceania in part (c).

2. The United States has the second highest incarceration rate in the world (second only to the small African country of Seychelles).[3] One study[4] of incarceration rates and ages summarized their findings in the following graph:

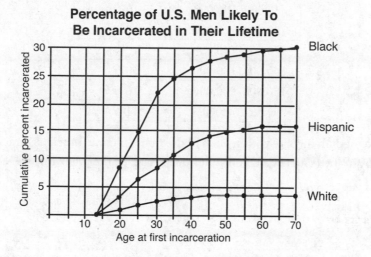

Percentage of U.S. Men Likely To Be Incarcerated in Their Lifetime

(a) What is the meaning of the point (25, 15) on the Black graph?

(b) What is the meaning of the horizontal segment on the White graph from ages 60 to 70?

(c) What is the median age at first incarceration of Hispanics who ever went to prison?

(d) Some states are considering legislation to address the disproportionate suspensions among students of color. How might this affect the above graphs?

[2]Talberth, Cobb, and Slattery, "Gross Domestic Product, A Tool for Sustainable Development," February 2007.
[3]"U.S. Prison Boom: Impact on Age of State Prisoners," *Journalist's Resource*, February 2016.
[4]"The Use of Incarceration in the United States," *American Society of Criminology*, November 2000.

3. SAT used to stand for "Scholastic Aptitude Test," but some researchers suggest it also stands for "Student Affluence Test."[5] There appears to be an association between mean total SAT scores and family income level. One study is summarized in the following table:

2015 College-Bound High School Students

Income Bracket	Total SAT Score
0–20,000	876
20,000–40,000	933
40,000–60,000	972
60,000–80,000	1000
80,000–100,000	1029
100,000–120,000	1054
120,000–140,000	1061
140,000–160,000	1077
160,000–200,000	1091
Over 200,000	1147

Using the median income within each of the first 9 income brackets and a $210,000 income from the last bracket, computer output yields the following:

```
Predictor        Coef      SE Coef        T        P
Constant        903.36      112.57     71.85    0.000
Income          0.00116    0.000104    11.17    0.000

    S = 20.936    R-Sq = 94.0%
```

(a) What is the equation of the regression line?

(b) Interpret the slope of the regression line in context.

(c) What is the coefficient of determination? Interpret it in context.

(d) What is the residual if a family has $63,500 income and a student has a score of 970? What does this mean in context?

(e) What is the meaning of $S = 20.936$?

(f) Why is it not reasonable to conclude that the best way to raise SAT scores is to raise family incomes? Explain, giving a possible confounding variable that could suggest an alternative approach to raising SAT scores.

[5]"SAT Scores and Income Inequality: How Wealthier Kids Rank Higher," *Wall Street Journal*, October 2014.

4. In analyzing conflict violence, the data are often incomplete, as obtaining data in such situations is very difficult if not impossible. However, accurate data are crucial for systematic studies of the origins of conflict, conflict dynamics, and conflict resolution.

(a) Suppose during one conflict month in Syria, the Syrian Observation for Human Rights (SOHR) reports 3060 deaths, the United Nations (UN) reports 3627 deaths, and 2635 of the deaths are found on *both* lists. Letting D be the actual number of deaths, fill in the following table:

SOHR

	Reported Deaths	Unreported Deaths	*Totals*
Reported Deaths			3627
Unreported Deaths			$D - 3627$
Totals	3060	$D - 3060$	D

(UN labels the left-side rows "Reported Deaths", "Unreported Deaths", "*Totals*".)

(b) Assuming the two reports above represent independent random samples of all deaths that month, solve for D.

(c) In reality, what are available are only *convenience samples*—perhaps posted videos of public executions, perhaps text messages sent out by protesters, or perhaps victims' testimonies recorded by local reporters. In general, why do convenience samples often lead to misleading conclusions? In our context, how are assumptions made in part (b) above violated?

(d) Another source of *undercoverage bias* found in conflict situations can be called *size bias*. Events involving large numbers of casualties are much more likely to be documented than events involving only one victim. For example, a cluster bombing of a marketplace with many victims will attract media attention while targeted assassinations may be unreported because the victim's family is afraid to make a report or because the body is never found. Explain how this affects the independence assumption made in part (b) above.

(e) Would you expect the answer found for D above to be an underestimate or an overestimate? Explain.

5. In 1986, the Supreme Court ruled that it is unconstitutional to exclude potential jurors because of race.[6] However, many studies since then have shown that racial discrimination in jury selection remains widespread, particularly in murder trials. A university law department plans a study using mock trials to determine if outcomes are different with all-White versus mixed-race juries. The law department will first randomly select two counties from among its state's 67 counties.

   (a) What is this sampling method called, and what is an advantage of using this method over a simple random sample?

   (b) Give a procedure for randomly selecting the two counties.

   (c) Forty-five people will then be randomly selected from among the pool of all eligible White jurors in each of the two counties to obtain a sample of size 90. What is the purpose of randomization in this two-step process?

   (d) The proposed study will involve 10 mock cases involving a young Black man charged with murdering an elderly White woman. The sample of 90 White jurors will be randomly assigned to two groups—60 to make up five all-White juries (each of size 12) and 30 to be part of five mixed-race juries (each with 6 White jurors and 6 jurors of color). What is the purpose of random assignment here? How might allowing jurors to self-select whether or not to be part of mixed-race juries lead to bias?

6. "Environmental racism" is said to occur when race is associated with the location of hazardous sites. A 2015 California study was designed to assess who was at the greatest risk from oil trains, which can derail and explode. Of the 5.5 million Californians who live within a blast zone (defined as within one mile of the railroad tracks), 75% are Latino and only 10% are White. California's population of 38.8 million is 39.0% Latino and 38.9% White.

   (a) What is the probability a randomly selected Californian is a Latino living within a blast zone?

   (b) What is the probability that a randomly selected Californian Latino lives within a blast zone?

   (c) In California, are being Latino and living within a blast zone independent events? Justify your answer mathematically.

   (d) In a random sample of 5 Californians, what is the probability that at least 2 of them live in a blast zone?

   (e) What is another explanatory variable that could account for the relatively high probability that someone living in a blast zone is a person of color?

[6]*Batson v. Kentucky*, 476 U.S. 79 (1986).

7. Although there is no accepted research on how often hair from different people may appear the same, for over two decades the FBI used visual hair comparison matches to help obtain convictions in major criminal cases. Acknowledging possible flawed forensic testimony based on unsound science, the FBI agreed to review over 2500 cases. Flawed forensic testimony was found in 257 of the first 268 cases reviewed.[7] In the following, assume that the sample proportion of cases involving flawed testimony is an accurate measure of the population proportion.

   (a) The FBI agreed to review 2500 cases involving possible flawed forensic testimony. Assuming this is a random sample of relevant cases, what is the expected number of these 2500 cases with actual flawed forensic testimony? What is the standard deviation for the expected number of cases with actual flawed forensic testimony?

   (b) Among the 257 cases with flawed forensic testimony, 32 resulted in death penalty convictions. Assume this is illustrative of the overall proportion. In examining a random sequence of cases involving flawed testimony, what is the probability that the first case with a resulting death penalty doesn't occur until at least the third case examined?

   (c) According to the Innocence Project, the misapplication of forensic science is the second-leading contributor to wrongful conviction, playing a role in 50% of known wrongful convictions, while eyewitness misidentification has played a role in nearly 75% of known wrongful convictions.[8] In wrongful conviction cases, can "the misapplication of forensic science" and "eyewitness misidentification" be mutually exclusive? Can they be independent? Explain.

8. We often think that slavery ended in the 19th century, but the sad fact is that there are more people in slavery today than at any other time in human history. According to the not-for-profit Force4Compassion, an average of 3287 people are sold or kidnapped and forced into slavery every day.[9] Assume this daily total has a roughly normal distribution with a standard deviation of 250 people.

   (a) What is the probability that on a randomly selected day, over 3500 people are forced into slavery?

   (b) In a random sample of 5 days, what is the probability that on a majority of the days over 3500 people are forced into slavery?

   (c) In a random sample of 5 days, what is the probability that the average number of people forced into slavery is over 3500?

   (d) Are there any statistics that, if more widely disseminated, might mobilize the world to fight this social blight?

---

[7] *National Association of Criminal Defense Lawyers and Innocence Project Analysis of FBI and Justice Department Data as of March 2015.*

[8] *http://www.innocenceproject.org*

[9] *http://www.f-4-c.org/statistics/*

9. When bias-motivated crimes (triggered by prejudice and hate) are committed, whole communities can feel victimized. A study[10] of profiles on America's most popular online hate site revealed that the site members are very young. Consider the following boxplot of self-reported ages of the site members:

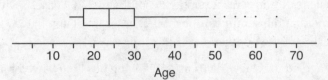

(a) Describe the distribution of ages.

(b) Is the quotient $\dfrac{\text{Mean age}}{\text{Median age}}$ greater than 1, less than 1, or approximately 1? Explain.

(c) Members of online social networks tend to be younger than the general population. What does this possibly indicate about the median age of all hate group members? Explain.

(d) According to recent FBI statistics,[11] 18.6% of bias incidents are religious based. In a random sample of 3 bias incidents, what is the probability that a majority are religious based?

The answers for this quiz can be found in the Appendix on page 642.

[10]The Data of Hate," *The New York Times*, 7/13/2014.

[11]"Bias Breakdown," *www.fbi.gov*, November 2015.

QUIZ 36

1. Schoolyard and online bullying are leading to increased levels of violence and depression among students. In a national survey[12] of 15,686 students in grades six through ten, 19% were involved in moderate to frequent bullying as perpetrators, 17% as victims, and 6% as both victims and perpetrators.

   (a) What proportion of the students were involved in moderate to frequent bullying?
   (b) Find a 95% confidence interval for the proportion of students in grades six through ten who are victims of moderate to frequent bullying.
   (c) Based on this confidence interval, is there evidence that the proportion of students who are victims of moderate to frequent bullying is different from the 22% claimed in one study?[13] Explain.
   (d) School administrators tend to underestimate the rates of bullying because students rarely report it and it often happens when adults are not present. However, knowing the numbers and types of bullying at all grade levels in your school is critical before a plan for bullying prevention and intervention can be developed. How would you plan a survey at your school to assess the frequency and types of bullying present?

2. Using the Second Amendment to the U.S. Constitution, gun rights proponents have argued for nearly or completely unchecked rights of personal gun ownership. However, because the U.S. has one of the highest rates of gun violence in the world, others have sought to regulate gun ownership. In the news, deaths by firearms are most often associated with murders; however, most firearm-related deaths in the United States are suicides, and most suicides are by guns. According to the CDC,[14] in 2017 there were 23,854 suicides in the U.S. using firearms.

   (a) Assume we have a random sample of 7 years that are considered representative with regard to numbers of firearm suicides:

   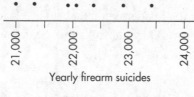

   Yearly firearm suicides

   $\bar{x} = 22{,}128.6, \ s = 779.6$

   Calculate a 95% confidence interval estimate for the mean number of firearm suicides in the U.S. each year.

   (b) According to a major recognized survey,[15] the U.S. has the highest rate of gun ownership in the world with 88.8 guns per 100 people. The same survey reports Israel with 7.3 guns per 100 people. Given that these figures report only privately

---

[12]Nansel, T. R., Overpeck, M., Pilla, R. S., Ruan, W. J., Simons-Morton, B., & Scheidt, P. "Bullying Behaviors Among U.S. Youth: Prevalence and Association with Psychosocial Adjustment," *Journal of the American Medical Association*, 285(16), 2094–2100.

[13]National Center for Educational Statistics, 2015.

[14]*https://www.cdc.gov/nchs/fastats/suicide.htm*

[15]UNODC (United Nations Office on Drugs and Crime) and the Small Arms Survey (Geneva).

owned guns and exclude weapons that are technically owned by the government, how might the apparent wide discrepancy between the U.S. and Israel not be as wide as indicated?

(c) When presented with any data set, we should always try to understand what and whose values are implicitly represented. Israel rejects 40% of gun applications and requires yearly renewals for continuing gun ownership. How would this affect the conversation from the statistics quoted in part (b)?

(d) The average U.S. gun owner owns 3 guns.[16] Comment on measuring rates of suicide by firearms in terms of guns per 100 people (88.8 in the U.S.) versus measuring rates of suicide by firearms in terms of gun owners per 100 people.

3. As a cost-cutting measure in April 2014, Flint, Michigan, began temporarily drawing its drinking water from the Flint River. The result was dangerous lead contamination in the water supply. (Lead poisoning especially impacts a child's brain and nervous system.)

(a) An independent research team[17] tested the Flint water supply for lead. The resulting cumulative frequency plot of the lead levels in ppb (parts per billion) in the water supply of sampled homes is as follows:

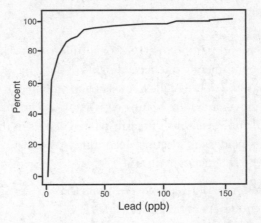

What is the shape of the histogram of the lead levels in the water supply of sampled homes? How would you expect the mean and median lead levels to compare to each other? Explain.

(b) It is recognized that if over 10% of a city's homes have lead levels in their water supply of over 5 ppb, the city has "a very serious problem with lead."[18] The research team determined that 15 ppb corresponded to the 75th percentile of Flint home water supply lead levels. Does this indicate a very serious problem?

(c) Suppose the concern is whether or not the mean lead level in the water supply of the homes is greater than 10 ppb. (The EPA regards 15 ppb in the water supply as a very dangerous level.[19]) Describe Type I and Type II errors in context. Which is of greater concern, a Type I or Type II error, and what would be a possible consequence? Explain.

[16] https://qz.com/1095899/gun-ownership-in-america-in-three-charts/
[17] 2015 Virginia Tech Study
[18] http://flintwaterstudy.org
[19] CDC 24/7: "Saving Lives, Protecting People."

4. Law Enforcement Assisted Diversion (LEAD) is a prebooking diversion program whereby low-level drug and prostitution offenders are redirected to community-based services instead of jail and prosecution.[20] There were 318 offenders who participated in a pilot program in which they were assigned to either the LEAD program ($n = 203$) or control (i.e., booking as usual; $n = 115$). All 318 offenders were deemed eligible for participation in the LEAD program.

   (a) The average yearly criminal and legal system costs of those participating in the LEAD program were $4763. For the control group, the average was $11,695. (Assume standard deviations of $875 and $2125, respectively.) Is this statistically significant evidence that the LEAD program has lower costs than the usual booking program?

   (b) After participation in the program, the LEAD group showed significantly fewer average number of days per year in jail than the control group ($P < 0.05$). Which error, Type I or Type II, might have been committed and what would be a possible consequence?

   (c) Even though it was felt that both the LEAD and the control groups were representative of the targeted offender population, a nonrandomized design was used. This may be of concern. What is the purpose of *random assignment* in such studies?

   (d) Using your answer from part (a), make an argument in favor of the LEAD program to a city council whose members are unfamiliar with statistical arguments.

5. Local activists and community leaders have long maintained that local law enforcement "overpolice" people of color in Berkeley, California. One claim is that people of color are being stopped at a higher rate than their population would indicate and that the stops are often for no legitimate reasons. A study[21] of the Police Department found that during the period from January 26 through August 12, 2015, 4658 civilians were stopped. Of that total, 1710 were White, 1423 were African American, and 543 were Hispanic. Of those stopped, 652 of the Whites, 942 of the African Americans, and 306 of the Hispanics were released without arrest or citation.

   According to 2013 census data, the population of Berkeley was 116,774, of which 64,412 were White, 10,076 were African American, and 11,600 were Hispanic.

   (a) Sketch side-by-side bar charts to illustrate the population percentage distribution and the percentage distribution of who was stopped by race.

   (b) Is there statistically significant evidence that African Americans are being stopped at a higher rate than their population would indicate? (Assume that we have a representative sample of common practices for years.)

   (c) Is there statistically significant evidence of a relationship between race (White, African American, Hispanic) and disposition (arrest/citation or not)? (Assume that we have a representative sample of common practices for years.)

---

[20] "LEAD: Reducing the Role of Criminalization in Local Drug Control," *http://www.drugpolicy.org*

[21] *https://www.nbcbayarea.com/news/local/racial-profiling-rampant-in-berkeley-police-department-report/103331/*

6. Is there a relationship between inequality and health? The Equality Trust, a nonprofit advocacy group, has created an Index of Health and Social Problems (HSP Index)[22] to combine several troublesome indicators (including lower life expectancy, higher infant mortality, and weak social mobility) into a single variable that can be used to describe the overall "health" of a society. A measure of income inequality is the ratio of the income of the top 20% to the bottom 20% in a country. Computer output of a regression analysis of HSP Index versus Income ratio in 21 industrialized countries is as follows:

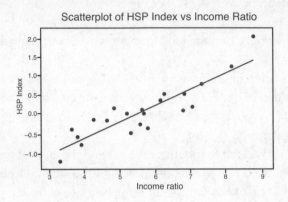

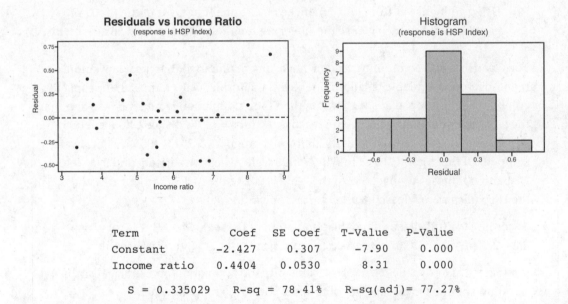

Term	Coef	SE Coef	T-Value	P-Value
Constant	-2.427	0.307	-7.90	0.000
Income ratio	0.4404	0.0530	8.31	0.000

S = 0.335029   R-sq = 78.41%   R-sq(adj)= 77.27%

(a) Find a 95% confidence interval for the slope of the regression line.
(b) Is there sufficient evidence that the higher the income ratio (that is, the higher the income inequality in a country), the greater the HSP Index (that is, the greater the health and social problems of the country)?
(c) Assuming that a lower HSP Index is desirable, what can be concluded about working to lessen the Income ratio, that is, to lessen the income inequality in a country?

[22] http://inequality.org/inequality-health/#sthash.H77M9bp9.dpuf

7. According to the Bureau of Justice Statistics, 1 out of every 35 adults in the United States is under correctional supervision (probation, parole, jail, or prison).[23] According to a 2014 Human Rights Watch report, "tough-on-crime" laws have filled U.S. prisons with mostly nonviolent offenders.[24]

(a) A previous study[25] on the population of local jails showed that 60.6% were people too poor to make bail or pay fines. More specifically, 34.4% were convicted males, 5.0% convicted females, 53.4% unconvicted males, and 7.2% unconvicted females. Suppose a current random sample of local jail populations counts 81 convicted males, 5 convicted females, 110 unconvicted males, and 4 unconvicted females. Is there sufficient evidence that the population distribution has changed from the findings of the earlier study?

(b) Advocacy groups are pushing for counties across the country to reform their local criminal justice systems to make them fairer. The groups argue that these reforms are needed to stop increasing costs, reduce overcrowding, and improve outdated facilities. Local jails hold nearly 700,000 inmates on any given day, with small counties holding 44% of the total.[26] More data are very much needed. How would you proportionately choose 500 inmates to interview and study?

8. Immunizations are a cornerstone of the world's efforts to protect people from infectious diseases. After a fraudulent, discredited 1998 research paper claiming to show a link between vaccines and ASD (autism spectrum disorder), vaccination rates dropped sharply[27]. In 2019, the U.S. experienced a large, multistate measles outbreak, mostly with unvaccinated children.

(a) Is there a link between vaccines and autism? A recent major study[28] concluded that between 30% and 36% of all parents have some such concerns. Assuming this is a 95% confidence interval, what was the minimum sample size necessary for such a conclusion?

(b) After looking at total cumulative immunogens from birth to three months, the researchers produced the graph below. Compare the ASD Case distribution and the Control distribution.

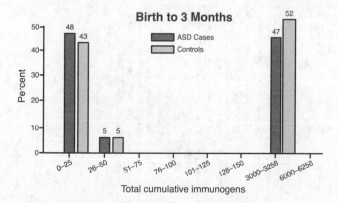

23"Correctional Populations in the United States," Bureau of Justice Statistics, 2013.

24"Nation Behind Bars: A Human Rights Solution," Human Rights Watch, May 2014.

25Bureau of Justice Statistics, 2012.

26Pearson, Jake, "Study: Smaller Counties Driving U.S. Jail Population Growth," December 2015.

27https://www.sciencedaily.com/releases/2012/06/120604142726.htm

28DeStefano, Price, and Weintraub, "Increasing Exposure to Antibody-Stimulating Proteins and Polysaccharides in Vaccines Is Not Associated with Risk of Autism," 2013.

(c)   After looking at the maximum number of immunogens in one day between birth and two years, the researchers produced the graph below. Compare the ASD Case distribution and the Control distribution.

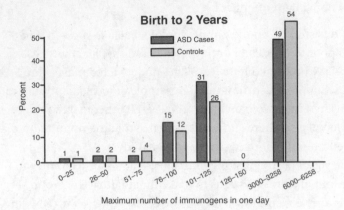

(d)   The study performed a large-scale test with the following:

H_0: There is no association between autism and vaccines.
H_a: There is an association between autism and vaccines.

Given a very large P-value, what was the proper conclusion?

9.   There is a strong consensus among scientists that global surface temperatures have increased markedly in recent years and that the trend is caused by emissions of greenhouse gases. However, in the popular media, especially in the United States, there is much debate over whether there is any problem—and even if there is a problem, the debate continues over whether humans are responsible. (China and the U.S. are the two largest producers of greenhouse gases.)

(a)   Given the following graph, describe the relationship between Temperature Anomaly (deviation) and Time.

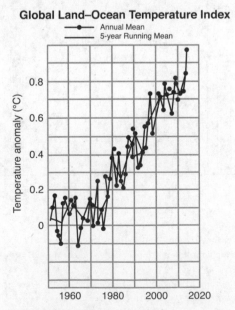

Source: National Aeronautics and Space Administration,
Goddard Institute for Space Studies

(b) In a Pew Global Attitudes Project survey in a random sample of 2029 Indians, 1319 expressed a great deal of worry about global warming. In a random sample of 2180 Chinese, 436 expressed a great deal of worry about global warming. Is the difference statistically significant?

(c) Suppose we wish to conduct a survey among the student body of your high school and calculate a confidence interval for the proportion who worry a great deal about global warming. Comment about bias and direction of bias, from the wording of the following survey:

> Charles, Prince of Wales, has consistently taken a strong stance criticizing both climate change deniers and corporate lobbyists by likening the Earth to a dying patient. Do you have a great deal of worry about global warming?

The answers for this quiz can be found in the Appendix on page 646.

THE INVESTIGATIVE TASK: FREE-RESPONSE QUESTION 6

The last question on the exam is called the Investigative Task. This free-response question counts for one-eighth of the total grade on the AP Statistics exam. The purpose of this question is to test your ability to take what you've learned throughout the course and apply that knowledge in a novel way. The topic of this question often includes material not only unfamiliar to students but also to their teachers.

This question always has several parts. Very often the first parts are straightforward, coming directly out of standard concepts of the curriculum. These are followed by parts that will challenge you to think creatively and require you to use learned concepts in new ways. This "investigative part" is written in such a way as to be accessible to students who have learned sound statistical thinking, not just memorized formulas and procedures.

There is no reason to panic and absolutely no reason to leave a blank response. All students should be able to receive some credit by answering the more standard beginning to this question. Following this with something "reasonable" will earn you even more credit.

It is not possible to list all topics that you might encounter on the Investigative Task. However, working through the following examples will help you understand how to approach the task on the exam. This should give you the confidence to do well on the Investigative Task you will see on the May exam.

Just as the other free-response questions, the Investigative Task is scored on a 0 to 4 scale with 1 point for a *minimal* response, 2 points for a *developing* response, 3 points for a *substantial* response, and 4 points for a *complete* response. Individual parts of this question are scored as E for *essentially* correct, P for *partially* correct, and I for *incorrect*. Note that *essentially* correct does not mean *perfect*. Work is also graded holistically—that is, your complete response is considered as a whole whenever scores do not fall precisely on an integral value on the 0 to 4 scale.

The scoring rubric for the Investigative Task is not standardized, and so the minimal requirements change depending on the question. These rubrics tend to be tough on the things the Investigative Task is intended to address and lenient on peripheral issues. Choosing appropriate methods and demonstrating good communication are the keys to high scores.

Following are three Investigative Tasks with carefully explained solutions illustrating possible thought processes for approaching such problems. Following these is Quiz 37 with seven more Investigative Tasks for which you're on your own. (There are answers and grading rubrics in the Appendix.)

➡ EXAMPLE 1

The R&D division of an automobile company has developed a new fuel additive that they hope will increase gas mileage. For a pilot study, they send 30 cars using the new additive on a road trip from Detroit to Los Angeles. From previous studies, they know that without the additive, the cars will average 32.0 mpg with a standard deviation of $\sigma = 2.5$ mpg. (Assume σ remains the same with or without the additive.)

(a) What is the parameter of interest, and what are the null and alternative hypotheses?

Think: The interest is in gas mileage, and there is mention of the "average" mpg. So the parameter is the mean mpg using the new fuel additive. The parameter is always something about a whole population, so be sure to use μ, not $\bar{x}$. The null hypothesis is some specific claim about this parameter, and the old mean mpg is given to be a specific 32.0 mpg. The alternative hypothesis will have an inequality. Since the R&D division hopes to increase gas mileage, the alternative must be > 32.0.

Answer: The parameter of interest is μ, the mean gas mileage in mpg for cars using the new fuel additive.

$$H_0: \mu = 32.0 \text{ and } H_a: \mu > 32.0$$

(b) What are Type I and Type II errors in this context? Give a possible consequence of each.

Think: A Type I error is when you mistakenly reject a true null hypothesis. In this case, if the null hypothesis was true, that would mean the cars really get a mean of 32.0 mpg. If you reject this, you mistakenly think the mean mpg is > 32.0. A consequence is that you'll waste money on this new additive, thinking it will improve gas mileage when in reality it won't.

TIP

Nothing new yet, but you should probably think for a moment what Type I and Type II errors are in general before putting the problem in context.

Answer: Type I error: The additive doesn't improve gas mileage, but the company mistakenly thinks it does. Consequence: The company invests in something that has no value.

Think: A Type II error is when you mistakenly fail to reject a false null hypothesis. In this case, if the null hypothesis is false, that would mean the cars really get a mean mpg > 32.0. If you fail to reject the null hypothesis claim of 32.0, you don't think the additive helps when it really does.

Answer: Type II error: The additive does improve gas mileage, but the company mistakenly thinks it doesn't. Consequence: The company doesn't invest in a potentially lucrative product.

(c) The company decides to work with a level of significance of $\alpha = 0.05$. What is the probability that a Type I error will be committed?

Think: That's easy. The significance level, or α-risk, gives the cutoff score for rejecting the null hypothesis, but it's based on the assumption that the null hypothesis is true. Rejecting a true null hypothesis is what we call a Type I error.

NOTE

Is there some trick here? These questions are not designed to trick you. Some may be just as straightforward as they seem!

Answer: 0.05 (the probability of a Type I error is the same as the level of significance).

(d) What is the null hypothesis rejection region? In other words, what critical value of the test statistic $\bar{x}$ will lead the company to conclude that there is statistically significant evidence that the new fuel additive increases gas mileage?

Think: You are given the population standard deviation ($\sigma = 2.5$), so you can use a normal distribution for the sampling distribution. The mean is μ and standard deviation is $\frac{\sigma}{\sqrt{n}}$. The 0.05 in a tail of the normal curve corresponds to a z-score of 1.645, and the critical value is 1.645 standard deviations from the mean.

TIP

Don't get nervous if you haven't yet come across anything new in the Investigative Task. Just keep moving along and applying what you know!

Answer: $\sigma_{\bar{x}} = \dfrac{2.5}{\sqrt{30}} = 0.4564$ and $32.0 + 1.645(0.4564) = 32.751$, so any $\bar{x} \geq 32.751$ will lead the company to conclude that there is statistically significant evidence that the new fuel additive increases gas mileage. (Note: with the population standard deviation known, the sampling distribution is roughly normal, thus giving a critical score of 1.645.)

(continued)

(e) The R&D statistician plots the following *power curve* showing the "performance" that can be expected from this hypothesis test. That is, for each possible true value of the population parameter, the probability of rejecting the null hypothesis is shown.

TIP

This is a graph not covered in this course! However, all you're being asked to explain is why the graph goes through one particular point.

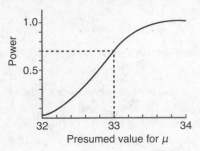

Explain why the graph intersects the pictured vertical axis above $\mu = 32$ at a y-value of 0.05.

Think: If the true value is 32.0, it's actually the null hypothesis. However, you just explained in part (c) that the probability of rejecting a true null hypothesis is 0.05!

Answer: H_0: $\mu = 32.0$, so if the true mean was 32, the probability of rejecting the null hypothesis is simply the level of significance, which is $\alpha = 0.05$.

(f) What is the meaning of the point (33, 0.7) in terms of a Type II error?

TIP

You are almost done! Again, this question is something about that new type of graph, but don't quit! There is some claimed relationship between how this graph is defined and the meaning of a Type II error. Look for that relationship!

Think: A Type II error is when you mistakenly fail to reject a false null hypothesis. What does that have to do with the given graph? According to the graph, (33, 0.7) signifies that if the true mean is 33, the probability of rejecting it is 0.7. Then the probability of not rejecting it must be $1 - 0.7 = 0.3$. Not rejecting it is exactly what the Type II error is all about.

Answer: If the true mean gas mileage using the additive was 33, the probability of correctly rejecting the null hypothesis is 0.7. So, the probability of committing a Type II error, that is, of mistakenly failing to reject the false null hypothesis, is $1 - 0.7 = 0.3$.

SCORING FOR EXAMPLE 1

Part (a) is essentially correct for correct statements of the parameter, the null hypothesis, and the alternative hypothesis. Part (a) is partially correct for two of these three statements.

Part (b) is essentially correct for correctly explaining Type I and Type II errors in context and for giving reasonable consequences of each. Part (b) is partially correct for two of the four parts.

Parts (c) and (d) together are essentially correct for both giving 0.05 in (c) and calculating 32.751 in (d). They are partially correct for one of these two calculations.

Parts (e) and (f) together are essentially correct for correct explanations of the y-value of 0.05 and of why the Type II error is 0.3. They are partially correct for one of these two.

Count partially correct answers as one-half an essentially correct answer.

4   **Complete Answer**   Four essentially correct answers.

3   **Substantial Answer**   Three essentially correct answers.

2   **Developing Answer**   Two essentially correct answers.

1   **Minimal Answer**   One essentially correct answer.

Use a holistic approach to decide a score totaling between two numbers.

➡ EXAMPLE 2

When you have a sore throat, your doctor has two major options to diagnose strep pharyngitis: a *rapid antigen test* to analyze the bacteria in your throat quickly (10–15 minutes) or a much more accurate *throat culture* that tests to see what grows on a Petri dish (takes 1–2 days).

In a random sample of 500 children brought into a pediatrician's office with sore throat symptoms, both rapid antigen tests and throat cultures were performed for this study with the following summary counts:

<table>
<tr><td colspan="2" rowspan="2"></td><td colspan="2">Patients With or Without Strep Pharyngitis as Confirmed by a Throat Culture</td></tr>
<tr><td>Strep Throat</td><td>No Strep Bacteria</td></tr>
<tr><td rowspan="2">Results of Rapid Antigen Test</td><td>Positive Test</td><td>180</td><td>15</td></tr>
<tr><td>Negative Test</td><td>20</td><td>285</td></tr>
</table>

(a) *Sensitivity* is defined as the probability of a positive test given that the subject has the disease. What was the sensitivity of this study?

Think: The phrase "given that" indicates this is a conditional probability. For such probabilities from a two-way table, you should first calculate row and column totals.

	Strep Throat	No Strep Throat	Totals
Positive Test	180	15	195
Negative Test	20	285	305
Totals	200	300	500

You're "given that" the subject has the disease (strep throat in this case). So, the first column of numbers is indicated. Note that 200 subjects had strep throat, and 180 of these had a positive test.

Answer:

$$P(\text{positive test} \mid \text{strep}) = \frac{180}{200} = 0.90$$

(b) *Specificity* is defined as the probability of a negative test given that the subject is healthy. What was the specificity of this study?

Think: Here you're given that the subject is healthy; there is "no strep throat." So, the second column of numbers is indicated. Note that 300 subjects did not have strep throat, and 285 of these had a negative test.

Answer:

$$P(\text{negative test} \mid \text{healthy}) = \frac{285}{300} = 0.95$$

(c) A valuable tool for assessing the value of such diagnostic tests is the *positive diagnostic likelihood ratio* (LR⁺). It gives the ratio of the probability a positive test result will be observed in a diseased person compared to the probability that the same result will be observed in a healthy person.

$$LR^+ = \frac{\text{sensitivity}}{1 - \text{specificity}}$$

Calculate LR⁺ in this study. Explain why the larger the value of LR⁺ is, the more useful the test is.

(continued)

TIP

Nothing new is here. However, you probably should think for a moment about how to calculate conditional probabilities from a two-way table.

TIP

It is not enough simply to give 0.90 for an answer. You must show where the answer comes from. Writing $\frac{180}{200} = 0.90$ is sufficient.

TIP

"Explain" is a key word in this problem— underline it and don't forget to answer it!

Think: To calculate the LR$^+$, it looks as if you simply plug in the answers from parts (a) and (b). However, the expository part of the question involves some thought. LR$^+$ is a fraction. So, it is larger when either its numerator is larger or its denominator is smaller. The numerator, sensitivity, is larger when there is a greater probability of a positive test when a child does have strep, and this clearly makes the test more useful. The denominator, 1 − specificity, is smaller when specificity is larger and closer to 1. (As a probability, it can't be greater than 1.) Thus, there is a greater probability of a negative test when a child is healthy, and this too clearly makes the test more useful.

Answer:

$$LR^+ = \frac{0.90}{1 - 0.95} = 18.0$$

A positive test given a child with strep and a negative test given a healthy child are both desired outcomes, so higher values for both sensitivity and specificity are good. Higher values of sensitivity, the numerator of LR$^+$, lead to greater values of LR$^+$. Higher values of specificity give lower values of 1 − specificity, the denominator of LR$^+$, again leading to greater values of LR$^+$.

(d) Another valuable tool is the *negative diagnostic likelihood ratio* (LR$^-$). It gives the ratio of the probability a negative test result will be observed in a diseased person compared to the probability that the same result will be observed in a healthy person. Give an expression for LR$^-$ in terms of *sensitivity* and *specificity*.

Answer: Since sensitivity is the probability of a positive test given that the subject has the disease, 1 − sensitivity gives the probability of a negative test given that the subject has the disease. The probability that the same result (a negative test) will be observed in a healthy person is the definition of "specificity." The answer is

$$LR^- = \frac{1 - \text{sensitivity}}{\text{specificity}}$$

(e) Suppose in one such sample study, LR$^+$ = 14.7. To determine whether or not this is sufficient evidence that the population LR$^+$ is below the desired value of 15.0, 100 samples from a population with a known LR$^+$ of 15.0 are generated and the resulting simulated values of LR$^+$ are shown in the dotplot:

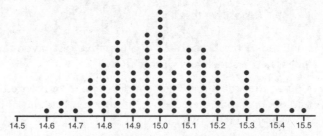

Positive Diagnostic Likelihood Ratio (LR⁺)

TIP

This type of problem has appeared frequently. In other words, you are given a "new" statistic and use simulation and a dotplot to estimate the *P*-value.

TIP

When the question says to use two givens, in this case the dotplot and the sample LR$^+$, that's a very strong hint on what you must base your answer!

Based on this dotplot and the sample LR$^+$ = 14.7, is there evidence that the population LR$^+$ is below the desired value of 15.0? Explain.

Think: There are 4 dots out of 100 that correspond to values of 14.7 or less. However, this is really a *P*-value for the sample LR$^+$. With a *P*-value of 0.04, the LR$^+$ value of 14.7 is statistically significant at the 5% significance level. We have enough evidence to reject the null hypothesis of LR$^+$ = 15.0.

Answer: The estimated *P*-value is the proportion of the simulated statistics that are less than or equal to the sample statistic of 14.7. Counting values in the dotplot, 4 out of 100, gives a *P*-value of 0.04. With this small of a *P*-value, 0.04 < 0.05, there is sufficient evidence that the population LR$^+$ is below the desired value of 15.0.

Parts (a) and (b) together are essentially correct for correctly calculating 0.90 and 0.95. They are partially correct for one of these results.

Part (c) is essentially correct for a correct calculation of LR^+ together with a clear explanation of why larger values are desired. It is partially correct for one of these two.

Part (d) is essentially correct for a correct expression for LR^- in terms of sensitivity and specificity and is incorrect otherwise.

Part (e) is essentially correct for a correct calculation of the P-value together with a correct conclusion in context with linkage to the P-value. It is partially correct if missing one part such as an incorrect P-value, missing context, or missing linkage in the conclusion.

Count partially correct answers as one-half an essentially correct answer.

4	**Complete Answer**	Four essentially correct answers.
3	**Substantial Answer**	Three essentially correct answers.
2	**Developing Answer**	Two essentially correct answers.
1	**Minimal Answer**	One essentially correct answer.

Use a holistic approach to decide a score totaling between two numbers.

➥ EXAMPLE 3

A grocery chain executive in charge of picking locations for future developments is interested in developing a function to predict weekly sales (in $1000) based on store size (in 1000 ft^2), median family income (in $1000) in the neighborhood, median family size in the neighborhood, and number of competitor stores in the neighborhood. Data gathered on 15 existing stores yield the following scatterplot matrix:

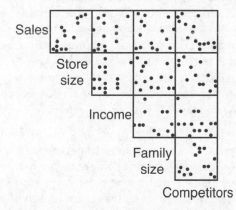

(a) The scatterplot matrix can be used to determine which variables seem to be the best predictors of sales. Explain which variables seem to be the best predictors of sales and the relationship between sales and those variables.

Think: What is going on? There are lots of scatterplots. In fact, you are given 10 of them with 5 different variables. You're asked about predicting "sales," so sales must be the response variable. "Sales" is to the left of the top row. So, the top row of 4 scatterplots must be sales against the other 4 variables. Do any of these 4 scatterplots look as if they could be used for prediction, that is, do any of them look like there is a linear relationship? Yes, it occurs in the first and fourth scatterplots along the top! The first seems to indicate some kind of positive association between sales and store size. The fourth seems to indicate some kind of negative association between sales and competitors.

(continued)

TIP

Wow, this question starts right off with a brand-new, complex picture! If you have no idea how to interpret this, simply move on. You may still be able to make sense out of later problem parts!

Answer: "Store size" and "number of competitor stores" seem to be the best predictors of "weekly sales." The first scatterplot in the top row indicates a strong positive association between weekly sales and store size. The last scatterplot in the top row indicates a strong negative association between weekly sales and number of competitor stores.

(b) Individual regressions of sales on each of the possible predictors give:

Predictor	Coef	SE Coef	T	P
Constant	–115.51	51.74	–2.23	0.044
St Size	5.8801	0.9808	6.00	0.000

Predictor	Coef	SE Coef	T	P
Constant	66.69	86.15	0.77	0.453
Income	3.859	2.691	1.43	0.175

Predictor	Coef	SE Coef	T	P
Constant	214.19	91.31	2.35	0.036
Fam Size	–11.58	36.71	–0.32	0.758

Predictor	Coef	SE Coef	T	P
Constant	374.64	33.41	11.21	0.000
Comp	–24.777	4.109	–6.03	0.000

We will not use any variable whose *t*-test *P*-value is over 0.15 and will first use the variable whose *t*-test *P*-value is least. Under these guidelines, which, if any, variables will be discarded and which variable will be used first? Explain.

Think: The question is about *t*-tests and *P*-values, so this must be about hypothesis tests for regression. You're given a generic computer printout that seems to give separate information about using each of the 4 possible predictors. From working with simpler computer outputs, remember that the line with the predictor first gives the slope, then the *SE* of the slope, followed by the *t*- and *P*-values from a linear regression test. You're told to throw out the predictor variables with *P* over 0.15. You see that Income has $P = 0.175$ and Fam Size has $P = 0.758$. (Looking back, this gives you confidence that your answer to part (a) is correct!) You're asked which predictor variable has the smallest *P*, but both St Size and Comp have $P = 0.000$ rounded to 3 decimals. How can you tell which is smaller? Well, you know the *P*-values come from the *t*-values, and you see that Comp has the larger *t*-value in absolute value ($6.03 > 6.00$), so it must have the smaller *P* if you were given more decimal places!

Answer: The variables "median family income" and "median family size" with *t*-test *P*-values of 0.175 and 0.758, respectively, both over 0.15, should be removed. Although both "store size" and "number of competitor stores" have *t*-test *P*-values of 0.000 to three decimal places, "number of competitor stores" will be used first as it has the smaller *P*-value because it has the larger *t*-statistic in absolute value, 6.03 versus 6.00.

(c) Multiple regression of sales on the two variables, store size and number of competitors, gives:

Predictor	Coef	SE Coef	T	P
Constant	122.79	94.44	1.30	0.218
St Size	3.342	1.200	2.78	0.017
Comp	–14.213	5.051	–2.81	0.016

$S = 37.3598 \quad R\text{-}Sq = 84.0\%$

What sales does this predict for a store of 50,000 ft^2 with 8 competitors?

Think: With a linear regression generic computer printout, you know that the "Coef" column gives the "Constant" and under it the slope or coefficient of the independent variable in the formula for predicting the response variable. Here, though, you have an extra row under the Constant, which means two predictor variables, "St Size" and "Comp." The constant in the formula must be 122.79, and the coefficients of "St Size" and "Comp" must be 3.342 and −14.213, respectively.

Answer: $122.79 + 3.342(50) - 14.213(8) = 176.186$ thousand or $176,186

(d)   What proportion of the variation in weekly sales is explained by its linear relationship with store size and number of competitors?

Think: This sounds like the definition of the "coefficient of determination" or R^2. Sure enough, "R-Sq" is in the computer output!

Answer: 0.84. To be safe, I'll write it out: The proportion of the variation in weekly sales explained by its linear relationship with store size and number of competitors is 0.84.

(e)   Assuming normality of residuals and using the computer output, estimate the probability that the weekly sales will be at least $50,000 over what is predicted by the regression line.

Think: You're being asked about the difference between an actual weekly sale and what is predicted, and this is what you know is a "residual." What is the distribution of the residuals? The question says the distribution is roughly normal. Remember that the sum and thus the mean of the residuals is always 0. By looking at the computer output, you see that the standard deviation of the residuals is $s = 37.3598$. So, you can simply use normalcdf on your calculator to find the probability!

Answer: The sum and thus the mean of the residuals is always 0. The standard deviation of the residuals is estimated with $s = 37.3598$. With a z-score of $\dfrac{50 - 0}{37.3598} = 1.338$ and a roughly normal distribution, we have $P(z > 1.338) \approx 0.090$.

NOTE

The coefficient of determination, the proportion of variation in the response variable that can be explained by the regression model, is a curriculum concept often appearing on the exam.

NOTE

You should recognize this as a straightforward normal probability calculation (where the variable is the residuals).

Part (a) is essentially correct for picking "store size" and "number of competitor stores" as the best predictors and for referring to the two scatterplots together with "strength" and "direction" for each. Part (a) is partially correct for picking these two predictors but giving a weak explanation missing either strength or direction.

Part (b) is essentially correct for picking "median family income" and "median family size" to be removed, for picking "number of competitor stores" to be used first, and for giving correct P-value explanations. Part (b) is partially correct for correctly picking either the two variables to be removed or the variable to be used first together with a correct explanation.

Parts (c) and (d) together are essentially correct for a correction calculation of $176,186 with work and for giving the answer of 0.84. They are partially correct for one of these two parts.

Part (e) is essentially correct for noting that the distribution of residuals is roughly normal with mean 0 and standard deviation 37.3598 and for then using this to calculate the probability correctly. Part (e) is partially correct for correctly noting the distribution of residuals but incorrectly calculating the probability, or for making a calculation based on a normal distribution with mean 0 but using an incorrect standard deviation.

Count partially correct answers as one-half an essentially correct answer.

4	**Complete Answer**	Four essentially correct answers.
3	**Substantial Answer**	Three essentially correct answers.
2	**Developing Answer**	Two essentially correct answers.
1	**Minimal Answer**	One essentially correct answer.

Use a holistic approach to decide a score totaling between two numbers.

QUIZ 37

Directions: Do the following Investigative Tasks on your own, using the methodology illustrated in the previous three examples.

1. A basic problem in ecology is to estimate the number of animals in a wildlife population. Suppose a wildlife management team captures and tags a random sample of 108 springboks in a Southwest African wildlife preserve. The springboks are released and given time to mingle with the population. One month later, a second random sample capture is made of 80 springboks, among which 12 are noted to be tagged from the original capture.

   (a) Assuming that the proportion of marked individuals within the second sample is equal to the proportion of marked individuals in the whole population, estimate the total number of springboks in the wildlife preserve.

   (b) A formula for variance of this estimate is $\text{var}(N) = \dfrac{ac(a-b)(c-b)}{b^3}$, where N = estimate of total population size, a = number of animals originally captured and tagged, b − number of tagged animals that are recaptured, and c − number of animals in the second sample. What is the standard deviation for the population size estimate above?

   The sampling distribution for estimates of N calculated as above is approximately normal under the condition that both samples are random and both $b \geq 10$ and $c - b \geq 10$.

   (c) Determine a 90% confidence interval for the total population size of springboks in this wildlife preserve.

   The above analysis assumes that there is no change to the population during the investigation (the population is closed).

   (d) Suppose that some of the tagged springboks were killed by hunters or escaped from the preserve before they could rejoin the herd. Would this result in an underestimation or an overestimation of the herd? Explain.

   (e) Suppose that the detection device missed some of the wire tags on the springboks in the second random sample. Would this result in an underestimation or an overestimation of the herd? Explain.

2. When operating at full capacity, a stainless steel bottle filler can fill twelve 64-ounce containers per minute. Once a day, a random set of twelve containers is picked, the containers are weighed, and if $W = \dfrac{\text{Max} + \text{Min}}{2}$ varies significantly from 64 ounces, the machine is stopped and adjustments are performed. To explore this quality control procedure further, the machine is perfectly calibrated, and 100 random samples of size 12 bottles are taken from the assembly line. For each of these 100 samples, the W measurement is calculated and shown plotted below.

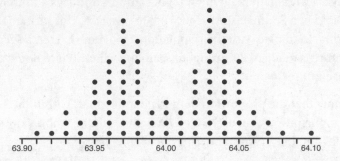

(a) Describe the above sampling distribution.

(b) Assuming a median of 64.005 and quartiles of 63.97 and 64.04, are there any outliers in the above distribution? Explain.

(c) When a sample of size 12 is picked, what are the null and alternative hypotheses being tested?

(d) One day, the random sample is {63.8, 63.84, 63.87, 63.88, 63.91, 63.95, 63.97, 63.97, 63.98, 64.0, 64.02, 64.04}. Based on the dotplot above, is this sufficient evidence at the 5% significance level to conclude that the machine needs adjustment? Justify your answer.

3. In 10 years of measurements, the number of people who died by becoming tangled in their bedsheets (deaths per year) versus per capita consumption of cheese (pounds per year) in the United States is illustrated in the following scatterplot:

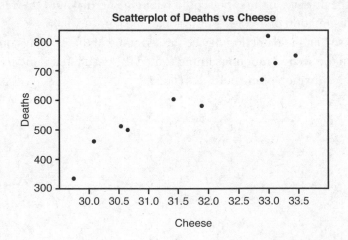

Below is the computer regression output from the 10 data points:

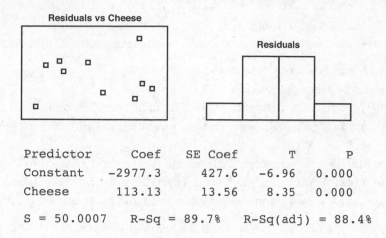

Residuals vs Cheese

Residuals

```
Predictor      Coef    SE Coef        T       P
Constant    -2977.3      427.6    -6.96   0.000
Cheese       113.13      13.56     8.35   0.000

S = 50.0007    R-Sq = 89.7%    R-Sq(adj) = 88.4%
```

(a) Discuss how conditions for inference are met.

(b) During the year when the per capita consumption of cheese was 30.1 pounds, there were 456 deaths from people becoming tangled in their bedsheets. Calculate and interpret the residual for this point.

(c) Determine a 95% confidence interval for the regression slope, and interpret it in context.

(d) What kind of cause-and-effect conclusion is appropriate? Explain.

(e) Using the computer output, give a rough estimate of the probability that the number of deaths in a year is at least 75 over what is predicted by the regression line.

4. Probiotics are microorganisms that are believed to give health benefits, especially with regard to healthy digestion, when consumed. One brand advertises a strength of 30 billion CFU (colony-forming units) per capsule. A company performs a quality check of 100 randomly chosen capsules. The 100 CFU values are arranged in order, the percentile rank of each is noted, and the standard normal value (z-score) corresponding to those percentile ranks is found. (For example, if a value had a percentile rank of 75%, then the corresponding z-score would be invNorm(0.75) = 0.6745.) The resulting plot of z-scores versus CFU values (in billions) is shown below, together with summary statistics:

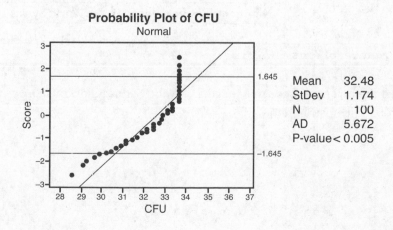

Probability Plot of CFU
Normal

Mean	32.48
StDev	1.174
N	100
AD	5.672
P-value	< 0.005

(a) What is the *range* of CFU values?

(b) What is the 5th percentile of the data? Explain.

The diagonal (sloping) straight line in the plot above is the graph of a theoretical normal distribution with the same mean and standard deviation of the sample data.

(c) In a normal distribution with the mean and SD of these data, what is the 95th percentile? Explain.

(d) What shape will a histogram of the CFU values have? Explain.

(e) Determine a 95% confidence interval for the mean CFU value (in billions).

5. In every MLB World Series since 1904, except for the three years from 1919 to 1921 that featured a best-of-nine contest, the winner has been the first team to win 4 games. Assume that each game is an independent event and the teams are evenly matched.

(a) What is the probability that Team A wins a World Series in 4 straight games? What then is the probability that a series is over in 4 games (Team A or Team B wins the first 4 straight games)?

(b) The probability that a series lasts exactly 5 games is twice the probability that a team wins 3 out of 4 games and then the next game, that is,

$$2\left[\binom{4}{3}(0.5)^3(0.5)\right](0.5) = 0.25$$

Similarly, the probability of a 6-game series is

$$2\left[\binom{5}{3}(0.5)^3(0.5)^2\right](0.5) = 0.3125$$

What is the probability a series lasts 7 games? (Shout out to 2016 Cubs!)

(c) Given a sample of 108 games, with the lengths of the World Series having a binomial distribution with evenly matched teams (as above), what are the expected values for the numbers (4 through 7) of games in a series?

(d) Not counting the 1919–1921 World Series, the counts for lengths of the 108 series from 1909 to 2019 are

	Number of Games Played in the World Series			
	4	5	6	7
Number of Years	20	24	24	40

Test the null hypothesis that lengths of World Series follow the above-calculated probability distribution.

6. At the end of the school year, a state administers a mathematics exam to a sample of fifth-grade students. Census results show that the state population consists of 50% city, 30% suburb, and 20% rural. The state decides to use a proportionate stratified random sample of size $n = 2000$.

(a) How many fifth graders from each of the three regions should be included in the sample?

Partial computer output of the exam scores is as follows:

Variable	Mean	StDev
City	74.3	10.2
Suburb	80.4	9.3
Rural	69.8	12.1

As an estimator for the mean score of all fifth graders, the state is using the following sample statistic:

$$\bar{x}_{\text{overall}} = 0.5\bar{x}_{\text{city}} + 0.3\bar{x}_{\text{suburb}} + 0.2\bar{x}_{\text{rural}}$$

(b) Calculate the sample statistic $\bar{x}_{\text{overall}}$ for these data.

(c) Using part (a) and the above computer output, calculate the three standard errors, one for each region, $SE(\bar{x}_{\text{city}})$, $SE(\bar{x}_{\text{suburb}})$, and $SE(\bar{x}_{\text{rural}})$.

(d) Calculate the standard error, $SE(\bar{x}_{\text{overall}})$, of the sampling distribution of $\bar{x}_{\text{overall}}$.

(e) Assuming all conditions for inference are satisfied and using z as an approximation for t (given the very large sample size), determine a 99% confidence interval for the mean score of all fifth graders if they were to take this exam.

7. The ADA (American Dental Association) recommends that you should brush your teeth for 2 minutes, but over 90 percent of people using a standard toothbrush don't brush that long. A dentist is interested in estimating the mean brushing time of her patients who use electric toothbrushes. Fifteen patients were selected at random, and a histogram of their self-reported brushing times is shown below.

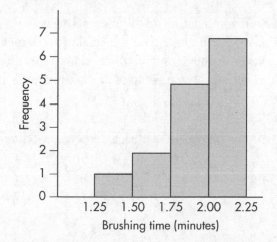

(a) What is an example of possible response bias in this context?

(b) Based on the histogram, why might it not be appropriate to use a one-sample *t*-interval to estimate the mean brushing time for all the dentist's patients who use electric toothbrushes?

If x represents a patient's brushing time, an exponential transformation of the patient's brushing time is given by 10^x. The table below gives some summary statistics for the original brushing times and the respective exponential-transformed brushing times.

	Brushing time, x	10^x
Mean	1.943	99.24
Median	2.000	100.0
Standard deviation	0.247	44.35

A histogram of the 15 exponential-transformed data values is shown below.

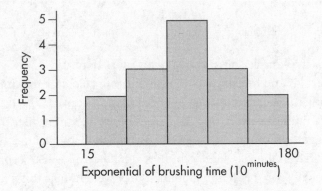

(c) The conditions for inference are met for the exponential-transformed data. Construct and interpret a 95 percent confidence interval for the mean of the exponential of the brushing times.

(d) The mean of the exponential-transformed data is $99.24\left(10^{\text{minutes}}\right)$. This can be converted back to minutes by calculating $\log(99.24) = 2.00$ minutes. Convert the endpoints of your interval in (c) back to minutes to obtain a new interval.

Graph 1 below shows the population distribution of the exponential of brushing times in $\left(10^{\text{minutes}}\right)$, which is approximately normal with the mean, μ. Graph 2 shows the result of converting the population distribution in Graph 1 back to the population of brushing times, in minutes. The median of the distribution is shown in each graph.

Graph 1
Population Distribution of Exponential of Brushing Time

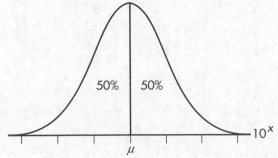

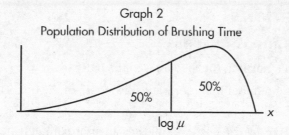

Graph 2
Population Distribution of Brushing Time

50%

50%

$\log \mu$

x

(e) Consider the parameter $\log \mu$ in Graph 2.

   (i) How does the parameter $\log \mu$ compare with the median of the population distribution of brushing times?

   (ii) How does the parameter $\log \mu$ compare with the mean of the population distribution of brushing times?

(f) Using your results from part (e), interpret the interval found in (d).

The answers for this quiz can be found in the Appendix on page 651.

50 MISCONCEPTIONS

MISCONCEPTIONS ABOUT GRAPHS

Misconception #1: In a boxplot, the median is in the IQR box.

> **Fact:** The IQR is a number that represents the length of the box. The box itself is not the IQR.

Misconception #2: A symmetric boxplot indicates an approximately normal distribution.

> **Fact:** Many differently shaped distributions can have identical boxplots; in general, it is very difficult to say anything about the shape of a distribution from its boxplot.

Misconception #3: When comparing two distributions, it is sufficient to list the center, spread, shape, and unusual features of each distribution.

> **Fact:** *Comparative* language must be used to state which distribution has the greater center and which has the greater spread.

MISCONCEPTIONS ABOUT LINEAR REGRESSION

Misconception #4: When $r = 0$, there is no relationship between the variables.

> **Fact:** Correlation measures only linearity. When $r = 0$, there is no LINEAR relationship between the variables, but there might be a strong NONLINEAR relationship between these variables.

Misconception #5: When $r = 0.3$, then 30 percent of the variables are closely related.

> **Fact:** When $r^2 = 0.3$, then 30 percent of the variation in the dependent variable can be explained by the linear relationship with the independent variable.

Misconception #6: When $r = 1$, there is a perfect cause-and-effect relationship between the variables.

> **Fact:** Correlation shows association, not cause and effect.

Misconception #7: A correlation close to 1 means that a linear model will give the best fit to the data.

> **Fact:** Curved data can also have a correlation near 1.

Misconception #8: The slope of a linear regression line gives the change in the dependent variable, y, for each unit change in the independent variable, x.

> **Fact:** The slope gives the "predicted," "estimated," "expected," or "average" change in the y-variable for each one unit increase in the x-variable.

Misconception #9: If $\hat{y} = a + bx$, then we can algebraically solve to get $\hat{x} = -\frac{a}{b} + \frac{y}{b}$.

> **Fact:** The variables x and $\hat{y}$ are not the same as $\hat{x}$ and y. The equation that predicts $\hat{y}$ from x does not correctly predict $\hat{x}$ from y.

Misconception #10: If $\hat{y} = a + bx$ and $\hat{x} = c + dy$ are derived from the same set of data, then b and d are reciprocals.

> **Fact:** $b = r\frac{s_y}{s_x}$ and $d = r\frac{s_x}{s_y}$ are not reciprocals.

MISCONCEPTIONS ABOUT COLLECTING DATA

Misconception #11: An experiment cannot be double-blind if anyone knows who gets which treatment.

> **Fact:** Someone always knows, but as long as this person doesn't interact with the subjects or measure the response, the experiment can still be double-blind.

Misconception #12: There is no difference between a stratified random sample and a randomized block design experiment.

> **Fact:** While the ideas are similar, as both involve forming groups of similar objects, stratified sampling occurs when taking a sample from a population, while blocking occurs before assigning units to treatments in an experiment.

Misconception #13: Blocks and treatment groups are basically the same.

> **Fact:** Blocks are not formed at random but are formed by grouping together similar units. Each block contains units with a certain characteristic. Treatment groups are formed at random with the goal of making the different groups as similar as possible, other than the treatment received.

MISCONCEPTIONS ABOUT PROBABILITY

Misconception #14: The complement rule gives that $0.3^C = 0.7$ or $P(E)^C = 0.7$.

> **Fact:** Probabilities do not have complements; events do. The "C" should appear in the superscript of the event E, that is, $P\left(E^C\right) = 1 - 0.3 = 0.7$.

Misconception #15: If E and F are mutually exclusive, they are independent.

> **Fact:** If E and F are mutually exclusive, $P(E \cap F) = 0$ and $P(E \mid F) = 0$. If E and F are independent, $P(E \cap F) = P(E)P(F)$ and $P(E \mid F) = P(E)$. "Mutually exclusive" and "independent" are two very different concepts.

Misconception #16: The probability a student taking AP Statistics owns a graphing calculator is the same as the probability a student who owns a graphing calculator is taking AP Statistics.

> **Fact:** $P(E \mid F) \neq P(F \mid E)$. This can be seen as $P(E \mid F) = \dfrac{P(E \cap F)}{P(F)}$ while $P(F \mid E) = \dfrac{P(E \cap F)}{P(E)}$.

Misconception #17: If a fair coin comes up "heads" five times in a row, by the "law of averages," the probability that the next toss is "tails" is greater than 0.5.

> **Fact:** There is no "law of averages" or "law of small numbers." By the "law of large numbers," in the long run, the proportion of heads will tend toward 0.5. However, coins have no memories, and the probability the next toss comes up "tails" is still 0.5.

Misconception #18: Writing normalcdf(20, 30, 35, 15) = 0.211 will receive full credit on a Normal calculation question, and writing binompdf(10, 0.2, 3) = 0.201 will receive full credit on a Binomial calculation question.

> **Fact:** If using "calculator-speak," every input must be identified. For example, you may write normalcdf(lower bd = 20, upper bd = 30, mean = 35, SD = 15) = 0.211, and you may write binompdf($n = 10$, $p = 0.2$, $x = 3$) = 0.201.

Misconception #19: If a random process is repeated long enough, eventually the expected value will occur.

> **Fact:** The expected value of a random process may never occur. It is simply a long-run average.

Misconception #20: The expected value of a random variable will always equal one of the possible values of that variable.

> **Fact:** There is no reason why $\sum xP(x)$ should equal one of the x-values. In fact, $\sum xP(x)$ often equals some nonwhole number. Do NOT round the expected value to a whole number after calculation.

MISCONCEPTIONS ABOUT SAMPLING DISTRIBUTIONS

Misconception #21: The larger the sample size, the closer the sample distribution is to a normal distribution.

> **Fact:** The larger the sample size, the closer the sample distribution is to the *population* distribution; the larger the sample size, the closer the *sampling* distribution of the sample mean is to a normal distribution.

Misconception #22: The larger the sample size, the closer the sampling distribution is to the population distribution.

> **Fact:** The larger the sample size, the closer the sampling distribution of the sample mean is to a normal distribution.

Misconception #23: If the sample size is large, the sampling distribution has the same mean as the population and has a standard deviation that is close to the standard deviation of the population.

> **Fact:** No matter what the sample size, the mean of the sampling distribution of $\bar{x}$ equals the population mean, $\mu_{\bar{x}} = \mu$, but the standard deviation of the sampling distribution of $\bar{x}$ equals the population standard deviation divided by the square root of the sample size, $\sigma_{\bar{x}} = \dfrac{\sigma}{\sqrt{n}}$.

Misconception #24: The larger the sample size, the closer the distribution of any statistic will approximate a normal distribution.

> **Fact:** The *central limit theorem* states that for sufficiently large sample sizes, the sampling distribution of the sample mean can be approximated by a normal distribution. This is not true, for example, for the sampling distribution of statistics such as the sample *maximum* or the sample *range*.

Misconception #25: The normal curve is a model that is defined by observational data.

> **Fact:** The normal curve is a theoretical distribution that serves as an approximation to certain sampling distributions.

Misconception #26: The *central limit theorem* has universal validity; that is, it can always be applied to give normal approximations to appropriate sampling distributions.

> **Fact:** There is a critical condition regarding a large enough sample size that must be satisfied (for the sampling distribution of $\bar{x}$, we use $n \geq 30$).

MISCONCEPTIONS ABOUT CONFIDENCE INTERVALS AND LEVELS

Misconception #27: If (35, 40) is a 95% confidence interval of a population mean, there is a 95% probability that the true mean is between 35 and 40.

> **Fact:** The probability that the population mean is between 35 and 40 is either 0 or 1 depending upon whether or not it is in the interval (35, 40).

Misconception #28: If (35, 40) is a 95% confidence interval of a population mean, 95% of the subjects are between 35 and 40.

> **Fact:** The confidence interval is about a population mean, not about individual values.

Misconception #29: If (35, 40) is a 95% confidence interval of a population mean, 95% of sample means from all such samples will be between 35 and 40.

> **Fact:** (35,40) is one specific interval derived from one sample mean and has nothing to do with other samples and other sample means.

Misconception #30: Interpreting a confidence interval is the same as interpreting a confidence level.

> **Fact:** When constructing a confidence interval, the confidence interval must be interpreted in context. Only interpret the confidence level when specifically asked, for example, "What does 95% confidence mean?"

Misconception #31: In general, increasing the sample size makes us more confident.

> **Fact:** In general, the confidence level, for example, 95%, doesn't change. Increasing the sample size while keeping the same confidence level gives a narrower interval, that is, a *more precise* estimate.

Misconception #32: For the same data, a 90% confidence interval is wider than a 95% confidence interval.

> **Fact:** The 95% confidence interval is wider. If we want higher confidence, we must accept a wider, less precise interval.

Misconception #33: The *Normal/Large Sample condition* can be simplified to saying that the sample must have a normal distribution.

> **Fact:** A finite sample cannot have a normal distribution. When samples are too small for the *central limit theorem* to apply, we look to see if the sample is unimodal and roughly symmetric with no outliers in order to conclude that it is not unreasonable to say "the sample came from a roughly normal population."

MISCONCEPTIONS ABOUT HYPOTHESIS TESTING

Misconception #34: H_0 and H_a are alternative conclusions.

> **Fact:** Hypotheses testing is a decision-making process in order to reject or fail to reject a null hypothesis, H_0. You are not deciding that one of H_0 or H_a is true. Your conclusion should be either that there *is* sufficient evidence to reject H_0 (that is, there *is* sufficient evidence in support of H_a) or that there is *not* sufficient evidence to reject H_0 (that is, there is *not* sufficient evidence in support of H_a.)

Misconception #35: The hypotheses can refer to either the population or the sample.

> **Fact:** The hypotheses **always** refer to the population, never to the sample.

Misconception #36: The null hypothesis and the failure-to-reject region are the same thing.

> **Fact:** The null hypothesis is a claim about a population parameter. The failure-to-reject region is an interval of values for the sample statistic that would indicate insufficient evidence to reject the given null hypothesis claim.

Misconception #37: The choice of alternative hypothesis depends on the sample statistic, that is, whether the sample statistic is positive or negative.

> **Fact:** Both the null and alternative hypotheses are decided before the data are collected. One should never decide if a test is one-sided (and in which direction) or two-sided based upon the data.

Misconception #38: The significance level α is the probability that the null hypothesis is true given that it has been rejected.

> **Fact:** The significance level is a fixed value we use to decide how small a P-value is required to be in order to reject the null hypothesis.

Misconception #39: The significance level α is the probability that the alternative hypothesis is true.

> **Fact:** The significance level is a fixed value we use to decide how small a P-value is required to be in order to reject the null hypothesis.

Misconception #40: A hypotheses test, if correctly performed, establishes the truth of one of the two hypotheses, either the null or the alternative.

> **Fact:** Truth is never established. The conclusion is simply whether or not there is sufficient evidence to reject the null hypothesis.

Misconception #41: Statistical significance and practical significance are the same.

> **Fact:** A statistically significant result might have no practical significance, and a practically significant result might turn out to *not* be statistically significant.

Misconception #42: The probability of a Type I error plus the probability of a Type II error should equal 1.

> **Fact:** The probability of a Type I error, α, is a fixed value giving the threshold value of the hypothesis test, while the probability of a Type II error, β, is different for each possible alternative parameter value.

Misconception #43: A well-planned test of significance should result in a statement either that the null hypothesis is true or that it is false.

> **Fact:** Tests of significance are designed only to measure the strength of evidence against the null hypothesis. They do not establish truth.

MISCONCEPTIONS ABOUT THE *P*-VALUE

Misconception #44: If the P-value is 0.16, the probability that the null hypothesis is correct is 0.16.

> **Fact:** The P-value of a test is the conditional probability of obtaining a result as extreme or more extreme as the one obtained assuming the null hypothesis is true.

Misconception #45: When the null hypothesis is rejected, it is because it is not true.

> **Fact:** The null hypothesis is rejected when the P-value is small. The small P-value indicates there is sufficient evidence to doubt the null hypothesis, but it is always possible that we mistakenly reject a true null hypothesis.

Misconception #46: A very large *P*-value provides convincing evidence that the null hypothesis is true.

   Fact: A very large *P*-value simply says that there is not sufficient evidence to reject the null hypothesis; it does not say that the null hypothesis is true and does not say that we can accept the null hypothesis.

Misconception #47: The *P*-value is the probability that the observed event has happened by chance.

   Fact: This misses that the *P*-value is a conditional probability. The *P*-value is the probability of obtaining a statistic as extreme or more extreme than the one actually observed, by chance alone, when assuming that the null hypothesis is true.

MISCONCEPTIONS ABOUT χ^2-TESTS

Misconception #48: The chi-square statistic, $\sum \frac{(\text{obs} - \text{exp})^2}{\text{exp}}$, can be calculated using either counts or proportions.

   Fact: The chi-square statistic is **always** calculated using counts.

Misconception #49: For chi-square goodness-of-fit tests, the null hypothesis is that some observed distribution is the same as some claimed distribution.

   Fact: The hypotheses **never** refer to the sample. The null hypothesis is that the *population* distribution is the same as some claimed distribution.

Misconception #50: Expected counts should always be rounded to integer values.

   Fact: While observed counts are always integers, the expected counts need not be, and students may lose points with this rounding.

50 COMMON ERRORS ON THE AP EXAM

COMMON ERRORS: EXPLORATORY ANALYSIS

1. When asked to draw a graph, many students forget *labeling* and *keys*, which are critical for stemplots, and *labeling* and *scaling*, which are critical for histograms, box-plots, and scatterplots. Proper labeling is more important than getting every point in exactly the right place.

2. When asked to describe a distribution from a graph such as a histogram or dotplot, students address "center, spread, shape, and unusual features" and forget to mention context.

3. When describing a distribution given only a boxplot, students may try to conclude shape like approximate normality or even symmetry; however, shape cannot be concluded from a boxplot.

4. When describing center or spread given only a histogram, definitive language like "the median is 120" will be penalized. One needs to use language that conveys uncertainty, such as "the median is approximately 120" or "the median is in the 115–130 interval."

5. When claiming a value is an outlier or claiming that there are no outliers, some calculation is necessary to communicate how the decision was made, otherwise credit is lost.

6. When asked to *compare* distributions, it is not enough to address center, spread, shape, and unusual features, in context, separately for each distribution. In addition, the characteristics of center and spread must be *compared* using comparative expressions like "greater than" or "less than" or "equal to."

7. When asked to recognize a shape feature, for example of a histogram, which is not apparent in a boxplot, it is not correct to give a feature that is simply *more* apparent.

8. When using words with a statistical meaning, incorrect use will be penalized. For example, in statistics, "range" refers to a single number, the maximum minus the minimum. Writing "the data range from 50 to 100" will be marked wrong. You can say that all the data are in the interval from 50 to 100.

9. When combining independent events, variances add, but it is a mistake to add standard deviations.

10. When writing the equation of a regression line, students forget to include the "hat" on the y-variable (or use the word "expected") and forget to define the variables.

11. When asked if a linear model is appropriate, students need to refer to the residual plot, not to whether or not the correlation is close to ± 1.

12. When a correlation is close to ± 1, students incorrectly think correlation implies causation.

13. When explaining the slope of a regression line, students need to use a word like "predicted," "estimated," "expected," or "average" for the change in the y-variable for each one unit increase in the x-variable. Lack of such a nondeterministic word will be penalized.

14. When asked to describe a scatterplot, students address "form, direction, and strength" but forget context.

15. When interpreting a residual, students address the distance between observed and predicted but forget the *direction* of the error, that is, whether the predicted value is an underestimate or overestimate.

COMMON ERRORS: COLLECTING DATA

16. When describing how to select a sample using a random integer generator or a random integer table, students forget to explicitly state that repeated integers are to be ignored.

17. When asked about a possible source of bias, students name the bias but are penalized for not clearly describing the problem caused by the bias and giving consequences in context.

18. When looking at data collection, students confuse *stratification* and *blocking*. While both involve dividing into homogeneous groups, stratification is a sampling method while blocking is an experimental design.

19. When looking at data collection, students confuse *stratified* and *cluster* sampling. Stratified sampling involves homogeneous groups and random sampling from each strata, while cluster sampling involves heterogeneous groups and a random sample of clusters.

20. When analyzing experimental design, students confuse *treatment groups* and *blocks*. A treatment group is a randomly formed collection of individuals to which a treatment is applied. A block is a group of similar experimental units that is then

randomly allocated into each of the treatment groups. The goal is for treatment groups to be as similar as possible to each other (other than the treatment), while blocks are chosen to be very different from each other.

21. When looking at data collection, students fail to understand that *random selection* and *random assignment* have different purposes. Random selection in sampling is used to generalize to a population, while random assignment in experiments is used to minimize the effect of confounding variables.

22. When looking at data collection, students may correctly state what variable is a confounding variable but are penalized for not explaining *why* the variable is a confounding variable in the context of the problem.

COMMON ERRORS: PROBABILITY

23. When calculating probabilities, minor arithmetic errors might not be penalized; however, reporting that a probability is a negative number or a number greater than 1 will always be penalized.

24. When studying two events, students confuse mutually exclusive and independent events. E and F are mutually exclusive if $P(E \cap F) = 0$. E and F are independent if $P(E \cap F) = P(E)P(F)$.

25. When using the conditional probability formula, students should show the numerator and denominator of the probability, not just a reduced fraction and not just a decimal answer.

26. When interpreting probabilities, there is a "law of large numbers" (in the long run, proportions will tend toward a specific probability); however, students incorrectly think there is a "law of averages" or a "law of small numbers."

27. When calculating probabilities, students confuse "intersection" and "given that."

28. When calculating probabilities, students confuse $P(E \mid F)$ with $P(F \mid E)$.

29. When calculating probabilities, students confuse "independence" with "correlation," "mutually exclusive," and "cause and effect."

30. When using the normal distribution for probability calculations, students forget to state that they are using a normal distribution and forget to give the values of the two parameters, the mean μ and the standard deviation σ.

31. When using the binomial distribution for probability calculations, students forget to state that they are using a binomial distribution and forget to give the values of the two parameters, the number of trials n and the probability of success p.

32. When using the geometric distribution for probability calculations, students forget to state that they are using a geometric distribution and forget to give the value of the parameter p, the probability of success.

33. When calculating probabilities, students forget to clearly identify the *boundary* and *direction* for the outcome of interest.

34. When using "calculator-speak" (often not recommended) like normalcdf(55, 60, 65, 5) or binompdf(25, 0.3, 8), students forget to state what every number refers to.

35. When calculating expected values, rounding to integers will often be penalized. Expected values do not have to be integers!

36. When describing a distribution, students are unclear as to whether they are describing a sample distribution, a population distribution, or a sampling distribution.

COMMON ERRORS: INFERENCE

37. When beginning an inference procedure, students sometimes mistakenly refer to the *sample* instead of the *population* in stating hypotheses. There is nothing to hypothesize about the sample since the sample results are known exactly.

38. When communicating which procedure is being used, it is generally easier to use *words* rather than writing out a *formula*. In free-response inference questions, more mistakes are made when trying to describe using formulas than when using words.

39. When defining a parameter, do so in context and refer to the population (not mistakenly to the sample!).

40. When a formula is used, it is often safer to start with numbers plugged in and not to include the symbolic formula, because including an incorrect symbol will be penalized.

41. When concluding an inference procedure, students confuse interpreting a confidence *interval* and interpreting a confidence *level*.

42. When concluding an inference procedure, students sometimes mistakenly refer to the sample instead of the population in giving the conclusion. For example, students mistakenly refer to a confidence interval of people who gave a particular answer to a survey question rather than to the people who would have answered in a certain way (only those in the sample actually "answered").

43. When analyzing an inference procedure, students confuse *parameters* and *statistics*. A parameter is a fixed, usually unknowable, truth about a population such as p, μ, σ, or β, while statistics are measurable values such as $\hat{p}$, $\bar{x}$, s, or b coming from samples. We use sample statistics to make inferences about population parameters.

44. When conducting an inference procedure, students incorrectly use the expression to "prove" or "accept" the null or alternative hypothesis. Conclusions should always be that there is or is not "convincing evidence" of some claim. (Often, the alternative hypothesis is the claim.)

45. When concluding a confidence interval or hypothesis test question about the population mean, students forget to include the word "mean" in the conclusion.

46. When performing an inference procedure, there is no reason to conduct both a significance test and a confidence interval unless specifically asked to. Students will be graded on whichever is the weaker response.

47. When concluding a hypothesis test, students are often incomplete in their conclusions, forgetting to involve linkage to a P-value (a direct comparison of the P-value to alpha), or forgetting context, or forgetting to refer to the parameter (such as "true proportion" or "true mean").

48. When asked about interpretations of errors on multiple-choice questions, students confuse the definitions of Type I and Type II errors.

49. When asked which type of error might have been committed based on the result of a hypothesis test, students often are unable to pick between Type I or Type II.

50. When asked to explain or analyze possible errors, students often cannot provide or explain possible consequences of Type I and Type II errors.

50 AP EXAM HINTS, ADVICE, AND REMINDERS

1. Let the graders understand *what* you are doing, *why* you are doing it, and *how* you are doing it. **Communication** is just as important as statistical knowledge! Graders want to give you credit—help them! Don't make the readers guess at what you are doing.

2. Read each of the multiple-choice and free-response questions *carefully*! Be sure you understand **exactly what you are being asked to do or find or explain**.

3. *Check assumptions.* **Don't just state them!** Be sure the assumptions to be checked are stated correctly. Verifying assumptions and conditions means more than simply listing them with check marks—you must show work or give some reason to confirm verification.

4. Learn and practice how to read *generic computer output.* When answering questions using computer output, realize that you usually will **not** need to use all the information given.

5. Show where answers come from. *Naked* or *bald answers* will receive little or **no** credit! On the other hand, don't give more than one solution to the same problem—you will receive credit only for the weaker one.

6. Sketch, with labeled axes and numbered scales when possible, any graph to which you refer. This includes histograms, boxplots, stemplots, scatterplots, residuals plots, normal probability plots, or any other kind of graph. It is not enough to simply say, "I did a normal probability plot of the residuals on my calculator and it looked linear."

7. **Use proper terminology!** For example, the language of experiments is different from the language of observational studies—you shouldn't mix up *blocking* and *stratification.* Know what *confounding* means and when it is proper to use this term. Words like *bias, correlation, normal, power, range, skew,* and even *statistic* have specific statistical meanings and should not be used colloquially.

8. **Avoid** "calculator-speak"! For example, do not simply write "2-SampTTest..." or "binomcdf...." There are lots of calculators out there, each with its own abbreviations. Some abbreviated function notation can be referenced IF the parameters used are identified. For example, normalcdf(left bd = 55, right bd = 60, mean = 45, SD = 15) = 0.0938 is okay to write.

9. **Be careful** about using abbreviations in general. For example, your teacher might use LOBF (line of best fit), but the grader may have no idea what this refers to.

10. Write null and alternative hypotheses in terms of the population parameter. If the question refers to sample data, don't automatically parrot the stem of the problem in your hypotheses.

11. Insert new calculator batteries and fully charge your calculator before the test. Bring extra batteries or your charger.

12. Underline key words and phrases while reading questions.

13. Approach each problem systematically. Some problems look scary on first reading but are not overly difficult and are surprisingly straightforward. Other questions might take you beyond the scope of the AP curriculum; however, remember that they will be phrased in ways that you should be able to answer them based on what you have learned in your AP Statistics class.

14. Read through all six free-response questions, underlining key points. Go back and answer the questions you think are easiest (this will usually, but not always, include Problem 1). Then tackle the heavily weighted Problem 6 for a while. Finally, try the remaining problems. Hopefully you will have time at the end to work further on Problem 6!

15. Be sure that your methods, reasoning, and calculations are clear to the reader and that the explanations and conclusions are given *in the context of the problem*. Answers do not have to be in paragraph form. Symbols and algebra are fine.

16. Show calculations carefully. A wrong answer due to a computational error might still result in full credit.

17. For any part of a multi-step problem that you can't answer but a solution is needed in a later step, make up a reasonable answer.

18. State the values that you are substituting into any formula that you use. You do not have to write down the actual formula.

19. State conclusions in clear, proper English. For example, don't use double negatives.

20. Read carefully and recognize that sometimes very different hypothesis tests or very different approaches are required in different parts of the same problem.

21. Realize that there may be several reasonable approaches to a given problem. In such a case, pick the one with which you are most comfortable or the one you feel will require the least amount of time to complete.

22. Realize that there may not be one clear, correct answer. Some questions are designed to give you an opportunity to creatively synthesize a relationship among the problem's statistical components.

23. Punching a long list of data into the calculator does not show statistical knowledge. Be sure it is necessary!

24. Remember that uniform and symmetric are not the same, and that not all symmetric unimodal distributions are normal.

25. Do not determine the shape of a distribution from a boxplot. Differently shaped distributions can have identical boxplots.

26. Use comparative words like "more than," "less than," "equal to," "same as," or "different from" when comparing distributions. A "laundry list" of *each* distribution's shape, center, spread, and unusual features is not enough.

27. Do not refer to the "law of averages" or "law of small numbers" since there are no such things. There *is* a "law of large numbers," which refers to the relative frequency of an event becoming closer to the true probability of the event as the experiment is performed more and more times.

28. Comment on whether or not there are nonlinear patterns, increasing or decreasing spread, and whether the residuals are small or large compared to the different x-values when describing residual plots. "Randomly scattered" is not the same as "half below and half above."

29. Know how to *interpret* the slope, the y-intercept, and the coefficient of determination in context when using a calculator or computer printout to find a regression line.

30. Think about the smallest collection of things to which a single treatment is applied when identifying experimental units.

31. Understand that there are two types of control groups—experimental units that receive a placebo and experimental units that receive no treatment at all.

32. Explain how you will "randomly assign" subjects to treatment groups when asked to design an experiment. For example, you must state that you will use a random number table or a random number generator on a calculator.

33. Follow these three steps when using a random number generator to pick a sample: (1) number the subjects; (2) define the interval of values from which integers are being selected, indicate that repeats are to be ignored, and note the stopping point; and (3) state that the sample is the group of subjects associated with the selected integers.

34. Understand that the expected value of a random variable does *not* mean the value that is expected to occur. In fact, it may have probability 0 of ever occurring. It is simply a long-run average.

35. Know the difference between a *sample distribution* and a *sampling distribution*. The larger the sample, the closer the *sample distribution* is to the population distribution; the larger the sample size, the closer the *sampling distribution* of the sample mean is to a normal distribution.

36. Use the central limit theorem only when there is a large enough sample size. When this critical condition is satisfied, the CLT gives normal approximations to specific sampling distributions.

37. Determine for inference problems whether a variable is *categorical* (leading to proportions or chi-square) or *quantitative* (leading to means or linear regression), whether you are given *raw data* or *summary statistics*, whether there is a *single population* of interest or *two populations* being compared, and, in the case of comparison, whether there are *independent samples* or a *paired comparison*.

38. Define parameters when asked to do so. Otherwise, simply use standard notation for population parameters p, μ, σ, and β.

39. Use clear subscripts if there are two populations. The parameters do not have to be defined if the subscripts make it very clear which is which, for example, $\mu_{\text{new drug}}$ and $\mu_{\text{old drug}}$.

40. Follow four steps when answering a confidence interval problem: (1) Identify the procedure by name or formula (usually name is easier); (2) Check conditions (do not just state them); (3) Calculate the confidence interval (calculator software does this quickly); and (4) State a conclusion with reference to the confidence level, the parameter, and the population, and use context.

41. Follow five steps when answering a hypothesis test problem: (1) State the hypotheses in terms of population parameters; (2) State the test by name or formula (usually by name is easier); (3) Check conditions (do not just state them); (4) Calculate the test statistic and associated P-value (calculator software does this quickly); and (5) State a decision based upon the P-value and write a conclusion *in context* about the alternative hypothesis with reference to the parameter and the population.

42. Follow five steps when answering a chi-square problem: (1) State the hypotheses (both H_0 and H_a in context); (2) Name the test (goodness of fit, independence, or homogeneity), which can often be inferred from the way the hypotheses are stated; (3) Check the conditions, which requires showing the expected values (they can be copied from the calculator); (4) Report the resulting chi-square value, the P-value, and df, the number of degrees of freedom; and (5) Give a conclusion in context with linkage to the P-value.

43. Communicate that there is convincing evidence to support the *alternative hypothesis* if the decision is to reject the null hypothesis.

44. Know which type of error, Type I or Type II, might have been committed based on the conclusion to a hypothesis test, and be able to give a possible reasonable consequence of committing that error.

45. Understand the difference between a simple random sample (SRS) and the random assignment of treatments to subjects.

46. Calculate the expected value of a random variable to be between the lowest and highest values of the random variable. The expected value does not have to be one of the random variable values.

47. Do not assume that the distribution of the sample will be close to a normal distribution just because the sample is large. (If the sample is large, its distribution will tend to look like the population distribution.)

48. Describe both an advantage of one thing and a disadvantage of the other if asked why one thing is better than another.

49. Do not use technical terms in designing a survey or an experiment unless specifically asked to do so. In general, clear, plain English is safer!

50. Follow directions on the Investigative Task (Free-Response Question 6). It often has a flow leading to an overall idea.

ANSWER SHEET
Practice Test 1

1. Ⓐ Ⓑ Ⓒ Ⓓ Ⓔ   11. Ⓐ Ⓑ Ⓒ Ⓓ Ⓔ   21. Ⓐ Ⓑ Ⓒ Ⓓ Ⓔ   31. Ⓐ Ⓑ Ⓒ Ⓓ Ⓔ

2. Ⓐ Ⓑ Ⓒ Ⓓ Ⓔ   12. Ⓐ Ⓑ Ⓒ Ⓓ Ⓔ   22. Ⓐ Ⓑ Ⓒ Ⓓ Ⓔ   32. Ⓐ Ⓑ Ⓒ Ⓓ Ⓔ

3. Ⓐ Ⓑ Ⓒ Ⓓ Ⓔ   13. Ⓐ Ⓑ Ⓒ Ⓓ Ⓔ   23. Ⓐ Ⓑ Ⓒ Ⓓ Ⓔ   33. Ⓐ Ⓑ Ⓒ Ⓓ Ⓔ

4. Ⓐ Ⓑ Ⓒ Ⓓ Ⓔ   14. Ⓐ Ⓑ Ⓒ Ⓓ Ⓔ   24. Ⓐ Ⓑ Ⓒ Ⓓ Ⓔ   34. Ⓐ Ⓑ Ⓒ Ⓓ Ⓔ

5. Ⓐ Ⓑ Ⓒ Ⓓ Ⓔ   15. Ⓐ Ⓑ Ⓒ Ⓓ Ⓔ   25. Ⓐ Ⓑ Ⓒ Ⓓ Ⓔ   35. Ⓐ Ⓑ Ⓒ Ⓓ Ⓔ

6. Ⓐ Ⓑ Ⓒ Ⓓ Ⓔ   16. Ⓐ Ⓑ Ⓒ Ⓓ Ⓔ   26. Ⓐ Ⓑ Ⓒ Ⓓ Ⓔ   36. Ⓐ Ⓑ Ⓒ Ⓓ Ⓔ

7. Ⓐ Ⓑ Ⓒ Ⓓ Ⓔ   17. Ⓐ Ⓑ Ⓒ Ⓓ Ⓔ   27. Ⓐ Ⓑ Ⓒ Ⓓ Ⓔ   37. Ⓐ Ⓑ Ⓒ Ⓓ Ⓔ

8. Ⓐ Ⓑ Ⓒ Ⓓ Ⓔ   18. Ⓐ Ⓑ Ⓒ Ⓓ Ⓔ   28. Ⓐ Ⓑ Ⓒ Ⓓ Ⓔ   38. Ⓐ Ⓑ Ⓒ Ⓓ Ⓔ

9. Ⓐ Ⓑ Ⓒ Ⓓ Ⓔ   19. Ⓐ Ⓑ Ⓒ Ⓓ Ⓔ   29. Ⓐ Ⓑ Ⓒ Ⓓ Ⓔ   39. Ⓐ Ⓑ Ⓒ Ⓓ Ⓔ

10. Ⓐ Ⓑ Ⓒ Ⓓ Ⓔ   20. Ⓐ Ⓑ Ⓒ Ⓓ Ⓔ   30. Ⓐ Ⓑ Ⓒ Ⓓ Ⓔ   40. Ⓐ Ⓑ Ⓒ Ⓓ Ⓔ

Practice Test 1

SECTION I

Questions 1–40

Spend 90 minutes on this part of the exam.

> **Directions:** The questions or incomplete statements that follow are each followed by five suggested answers or completions. Choose the response that best answers the question or completes the statement.

1. The following is a histogram of 87 home sale prices (in thousands of dollars) in one community:

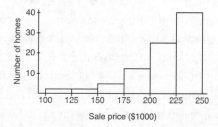

   Which of the following could be the median of the sale prices?

   (A) $162,500
   (B) $175,000
   (C) $187,500
   (D) $212,500
   (E) $237,500

2. Which of the following is most useful in establishing cause-and-effect relationships?

   (A) A complete census
   (B) A least squares regression line showing high correlation
   (C) A simple random sample
   (D) A well-designed, well-conducted survey incorporating chance to ensure a representative sample
   (E) An experiment

3. The average yearly snowfall in a city is 55 inches. What is the standard deviation if 15% of the years have snowfalls above 60 inches? Assume yearly snowfalls are approximately normally distributed.

(A) $\dfrac{60 - 55}{0.15}$

(B) $\dfrac{60 - 55}{0.35}$

(C) $\dfrac{60 - 55}{0.385}$

(D) $\dfrac{60 - 55}{0.85}$

(E) $\dfrac{60 - 55}{1.036}$

4. To determine the average cost of running for a congressional seat, a simple random sample of 50 politicians is chosen and the politicians' records examined. The cost figures show a mean of $1,825,000 with a standard deviation of $132,000. Which of the following is the best interpretation of a 90% confidence interval estimate for the average cost of running for office?

(A) Of all politicians running for a congressional seat, 90% spend between $1,794,000 and $1,856,000.
(B) Of all politicians running for a congressional seat, 90% spend a mean dollar amount that is between $1,794,000 and $1,856,000.
(C) We are 90% confident that politicians running for a congressional seat spend between $1,794,000 and $1,856,000.
(D) We are 90% confident that politicians running for a congressional seat spend a mean dollar amount between $1,794,000 and $1,856,000.
(E) We are 90% confident that in the chosen sample, the mean dollar amount spent running for a congressional seat is between $1,794,000 and $1,856,000.

5. In one study on the effect that eating meat products has on weight level, a simple random sample (SRS) of 500 subjects who admitted to eating meat at least once a day had their weights compared with those of an independent SRS of 500 people who claimed to be vegetarians. In a second study, an SRS of 500 subjects was served at least one meat meal per day for 6 months, while an independent SRS of 500 others were chosen to receive a strictly vegetarian diet for 6 months, with weights compared after 6 months. Which of the following is a true statement?

(A) The first study is a controlled experiment, while the second is an observational study.
(B) The first study is an observational study, while the second is a controlled experiment.
(C) Both studies are controlled experiments.
(D) Both studies are observational studies.
(E) Each study is part controlled experiment and part observational study.

6. A plumbing contractor obtains 60% of her boiler circulators from a company whose defect rate is 0.005 and the rest from a company whose defect rate is 0.010. If a circulator is defective, what is the probability that it came from the first company?

(A) $(0.6)(0.005)$

(B) $(0.6)(0.005) + (0.4)(0.010)$

(C) $\dfrac{0.6}{(0.6)(0.005) + (0.4)(0.010)}$

(D) $\dfrac{0.005}{(0.6)(0.005) + (0.4)(0.010)}$

(E) $\dfrac{(0.6)(0.005)}{(0.6)(0.005) + (0.4)(0.010)}$

7. A kidney dialysis center periodically checks a sample of its equipment and performs a major recalibration if readings are sufficiently off target. Similarly, a fabric factory periodically checks the sizes of towels coming off an assembly line and halts production if measurements are sufficiently off target. In both situations, we have the null hypothesis that the equipment is performing satisfactorily. For each situation, which is the *more* serious concern, a Type I or Type II error?

(A) Dialysis center: Type I error; towel manufacturer: Type I error

(B) Dialysis center: Type I error; towel manufacturer: Type II error

(C) Dialysis center: Type II error; towel manufacturer: Type I error

(D) Dialysis center: Type II error; towel manufacturer: Type II error

(E) This is impossible to answer without making an expected value judgment between human life and accurate towel sizes.

8. A coin is weighted so that the probability of heads is 0.75. The coin is tossed 10 times, and the number of heads is noted. This procedure is repeated a total of 50 times, and the number of heads is recorded each time. What kind of distribution has been simulated?

(A) The sampling distribution of the sample proportion with $n = 10$ and $p = 0.75$

(B) The sampling distribution of the sample proportion with $n = 50$ and $p = 0.75$

(C) The sampling distribution of the sample mean with $\bar{x} = 10(0.75)$ and
    $\sigma = \sqrt{10(0.75)(0.25)}$

(D) The binomial distribution with $n = 10$ and $p = 0.75$

(E) The binomial distribution with $n = 50$ and $p = 0.75$

9. During the years 1886 through 2019, there were an average of 9.0 tropical cyclones per year, of which an average of 5.4 became hurricanes. Assuming that the probability of any cyclone becoming a hurricane is independent of what happens to any other cyclone, if there are five cyclones in one year, what is the probability that at least three become hurricanes?

(A) $\binom{5}{3}(0.6)^3$

(B) $\binom{5}{3}(0.6)^3(0.4)^2$

(C) $(0.6)^3(0.4)^2+(0.6)^4(0.4)+(0.6)^5$

(D) $\binom{5}{3}(0.6)^3(0.4)^2+\binom{5}{4}(0.6)^4(0.4)+(0.6)^5$

(E) $1-\left[(0.4)^5+5(0.6)(0.4)^4+10(0.6)^2(0.4)^3+10(0.6)^3(0.4)^2\right]$

10. Given the probabilities $P(A) = 0.3$ and $P(B) = 0.2$, what is the probability of the union $P(A \cup B)$ if A and B are mutually exclusive? If A and B are independent? If B is a subset of A?

(A) 0.44, 0.5, 0.2
(B) 0.44, 0.5, 0.3
(C) 0.5, 0.44, 0.2
(D) 0.5, 0.44, 0.3
(E) 0, 0.5, 0.3

11. A social science researcher is testing the claim that 69 percent of Americans use social media every day. In a survey involving an SRS of 1100 Americans, 770 say that they use social media every day. With $H_0: p = 0.69$ and $H_a: p \neq 0.69$, what is the value of the test statistic?

(A) $z = \dfrac{0.7 - 0.69}{\sqrt{1100(0.69)(1 - 0.69)}}$

(B) $z = 2\dfrac{0.7 - 0.69}{\sqrt{1100(0.69)(1 - 0.69)}}$

(C) $z = \dfrac{0.7 - 0.69}{\sqrt{\dfrac{(0.69)(1 - 0.69)}{1100}}}$

(D) $z = 2\dfrac{0.7 - 0.69}{\sqrt{\dfrac{(0.69)(1 - 0.69)}{1100}}}$

(E) $z = 2\dfrac{0.7 - 0.69}{\sqrt{1100(0.7)(1 - 0.7)}}$

12. A medical research team tests for tumor reduction in a sample of patients using three different dosages of an experimental cancer drug. Which of the following is true?

   (A) There are three explanatory variables and one response variable.
   (B) There is one explanatory variable with three levels of response.
   (C) Tumor reduction is the only explanatory variable, but there are three response variables corresponding to the different dosages.
   (D) There are three levels of a single explanatory variable.
   (E) Each explanatory level has an associated level of response.

13. The graph below shows cumulative proportion plotted against age for a population.

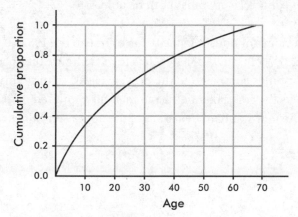

   The median is approximately what age?

   (A) 17
   (B) 30
   (C) 35
   (D) 40
   (E) 45

14. Mr. Bee's statistics class had a standard deviation of 11.2 on a standardized test, while Mr. Em's class had a standard deviation of 5.6 on the same test. Which of the following is the most reasonable conclusion concerning the two classes' performance on the test?

   (A) Mr. Bee's class is less heterogeneous than Mr. Em's.
   (B) Mr. Em's class is more homogeneous than Mr. Bee's.
   (C) Mr. Bee's class performed twice as well as Mr. Em's.
   (D) Mr. Em's class did not do as well as Mr. Bee's.
   (E) Mr. Bee's class had the higher mean, but this may not be statistically significant.

15. A college admissions officer is interested in comparing the SAT math scores of high school applicants who have and have not taken AP Statistics. She randomly pulls the files of five applicants who took AP Statistics and five applicants who did not, and proceeds to run a *t*-test to compare the mean SAT math scores of the two groups. Which of the following is a necessary assumption?

    (A) The population variances from each group are known.
    (B) The population variances from each group are unknown.
    (C) The population variances from the two groups are equal.
    (D) The population of all SAT scores from each group is roughly normally distributed.
    (E) The samples must be independent simple random samples, and for each sample, np and $n(1 - p)$ must both be at least 10.

16. The appraised values of houses in a city have a mean of $195,000 with a standard deviation of $23,000. Because of a new teachers' contract, the school district needs an extra 10% in funds compared to the previous year. To raise this additional money, the city instructs the assessment office to raise all appraised house values by $5,000. What will be the new standard deviation of the appraised values of houses in the city?

    (A) $23,000
    (B) $25,300
    (C) $28,000
    (D) $30,300
    (E) $30,800

17. A psychologist hypothesizes that scores on an aptitude test are normally distributed with a mean of 70 and a standard deviation of 10. In a random sample of size 100, scores are distributed as in the table below. What is the χ^2-statistic for a goodness-of-fit test?

Score:	Below 60	60–70	70–80	Above 80
Number of people:	10	40	35	15

(A) $\dfrac{(10-16)^2}{10}+\dfrac{(40-34)^2}{40}+\dfrac{(35-34)^2}{35}+\dfrac{(15-16)^2}{15}$

(B) $\dfrac{(10-16)^2}{16}+\dfrac{(40-34)^2}{34}+\dfrac{(35-34)^2}{34}+\dfrac{(15-16)^2}{16}$

(C) $\dfrac{(10-25)^2}{25}+\dfrac{(40-25)^2}{25}+\dfrac{(35-25)^2}{25}+\dfrac{(15-25)^2}{25}$

(D) $\dfrac{(10-25)^2}{10}+\dfrac{(40-25)^2}{40}+\dfrac{(35-25)^2}{35}+\dfrac{(15-25)^2}{15}$

(E) $\dfrac{(10-25)^2}{16}+\dfrac{(40-25)^2}{34}+\dfrac{(35-25)^2}{34}+\dfrac{(15-25)^2}{16}$

18. A talk show host recently reported that in response to his on-air question, 82% of the more than 2500 e-mail messages received through his publicized address supported the death penalty for anyone convicted of selling drugs to children. What does this show?

(A) The survey is meaningless because of voluntary response bias.
(B) No meaningful conclusion is possible without knowing something more about the characteristics of his listeners.
(C) The survey would have been more meaningful if he had picked a random sample of the 2500 listeners who responded.
(D) The survey would have been more meaningful if he had used a control group.
(E) This was a legitimate sample, randomly drawn from his listeners, and of sufficient size to be able to conclude that most of his listeners support the death penalty for such a crime.

19. Define a new measurement as the difference between the 60th and 40th percentile scores in a population. This measurement will give information concerning

(A) central tendency.
(B) variability.
(C) symmetry.
(D) skewness.
(E) clustering.

20. Suppose X and Y are random variables with $E(X) = 37$, $\text{var}(X) = 5$, $E(Y) = 62$, and $\text{var}(Y) = 12$. What are the expected value and variance of the random variable $X + Y$?

(A) $E(X + Y) = 99$, $\text{var}(X + Y) = 8.5$
(B) $E(X + Y) = 99$, $\text{var}(X + Y) = 13$
(C) $E(X + Y) = 99$, $\text{var}(X + Y) = 17$
(D) $E(X + Y) = 49.5$, $\text{var}(X + Y) = 17$
(E) There is insufficient information to answer this question.

21. Which of the following can affect the value of the correlation r?

(A) A change in measurement units
(B) A change in which variable is called x and which is called y
(C) Adding the same constant to all values of the x-variable
(D) All of the above can affect the r-value.
(E) None of the above can affect the r-value.

22. The American Medical Association (AMA) wishes to determine the percentage of obstetricians who are considering leaving the profession because of the rapidly increasing number of lawsuits against obstetricians. The AMA would like an answer to within $\pm 3\%$ at the 95% confidence level. Which of the following should be used to find the sample size (n) needed?

(A) $1.645\sqrt{\dfrac{0.5}{n}} \leq 0.03$

(B) $1.645\sqrt{\dfrac{0.5}{n}} \leq 0.06$

(C) $1.645\sqrt{\dfrac{(0.5)(0.5)}{n}} \leq 0.06$

(D) $1.96\sqrt{\dfrac{0.5}{n}} \leq 0.03$

(E) $1.96\sqrt{\dfrac{(0.5)(0.5)}{n}} \leq 0.03$

23. Suppose you did 10 independent tests of the form $H_0: \mu = 25$ versus $H_a: \mu < 25$, each at the $\alpha = 0.05$ significance level. What is the probability of committing a Type I error and incorrectly rejecting a true H_0 with at least one of the 10 tests?

(A) 0.05
(B) 0.40
(C) 0.50
(D) 0.60
(E) 0.95

24. Which of the following sets has the smallest standard deviation? Which has the largest?

   I. 1, 2, 3, 4, 5, 6, 7
   II. 1, 1, 1, 4, 7, 7, 7
   III. 1, 4, 4, 4, 4, 4, 7

(A) Smallest SD: I; largest SD: II
(B) Smallest SD: II; largest SD: III
(C) Smallest SD: III; largest SD: I
(D) Smallest SD: II; largest SD: I
(E) Smallest SD: III; largest SD: II

25. A researcher plans a study to examine long-term confidence in the U.S. economy among the adult population. She obtains a simple random sample of 30 adults as they leave a Wall Street office building one weekday afternoon. All but two of the adults agree to participate in the survey. Which of the following conclusions is correct?

(A) Proper use of chance as evidenced by the simple random sample makes this a well-designed survey.
(B) The high response rate makes this a well-designed survey.
(C) Selection bias makes this a poorly designed survey.
(D) A voluntary response study like this gives too much emphasis to persons with strong opinions.
(E) Lack of anonymity makes this a poorly designed survey.

26. Which among the following would result in the narrowest confidence interval?

(A) Small sample size and 95% confidence
(B) Small sample size and 99% confidence
(C) Large sample size and 95% confidence
(D) Large sample size and 99% confidence
(E) Large sample size with either 95% or 99% confidence

27. Consider the following three scatterplots:

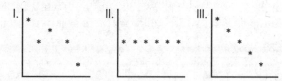

What is the relationship among r_1, r_2, and r_3, the correlations associated with the first, second, and third scatterplots, respectively?

(A) $r_1 < r_2 < r_3$
(B) $r_1 < r_3 < r_2$
(C) $r_2 < r_3 < r_1$
(D) $r_3 < r_1 < r_2$
(E) $r_3 < r_2 < r_1$

28. There are two games involving flipping a fair coin. In the first game, you win a prize if you can throw between 45% and 55% heads. In the second game, you win if you can throw more than 80% heads. For each game, would you rather flip the coin 30 times or 300 times?

(A) 30 times for each game
(B) 300 times for each game
(C) 30 times for the first game and 300 times for the second
(D) 300 times for the first game and 30 times for the second
(E) The outcomes of the games do not depend on the number of flips.

29. City planners are trying to decide among various parking plan options ranging from more on-street spaces to multilevel facilities to spread-out small lots. Before making a decision, they wish to test the downtown merchants' claim that shoppers park for an average of only 47 minutes in the downtown area. The planners have decided to tabulate parking durations for 225 shoppers and to reject the merchants' claim if the sample mean exceeds 50 minutes. If the merchants' claim is wrong and the true mean is 51 minutes, what is the probability that the random sample will lead to a mistaken failure to reject the merchants' claim? Assume that the standard deviation in parking durations is 27 minutes.

(A) $P\left(z < \dfrac{50-47}{27/\sqrt{225}}\right)$

(B) $P\left(z > \dfrac{50-47}{27/\sqrt{225}}\right)$

(C) $P\left(z < \dfrac{50-51}{27/\sqrt{225}}\right)$

(D) $P\left(z > \dfrac{50-51}{27/\sqrt{225}}\right)$

(E) $P\left(z > \dfrac{51-47}{27/\sqrt{225}}\right)$

30. Consider the following parallel boxplots indicating the starting salaries (in thousands of dollars) for blue-collar and white-collar workers at a particular production plant:

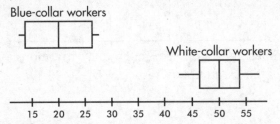

Which of the following is a correct conclusion?

(A) The ranges of the distributions are the same.
(B) In each distribution, the mean is equal to the median.
(C) Each distribution is symmetric.
(D) Each distribution is roughly normal.
(E) The distributions are outliers of each other.

31. The mean Law School Aptitude Test (LSAT) score for applicants to a particular law school is 650 with a standard deviation of 45. Suppose that only applicants with scores above 700 are considered. What percentage of the applicants considered have scores below 740? (Assume the scores are approximately normally distributed.)

(A) 13.3%
(B) 17.1%
(C) 82.9%
(D) 86.7%
(E) 97.7%

32. If all the other variables remain constant, which of the following will increase the power of a hypothesis test?

     I. Increasing the sample size
    II. Increasing the significance level
   III. Increasing the probability of a Type II error

  (A) I only
  (B) II only
  (C) III only
  (D) I and II only
  (E) I, II, and III

33. A researcher planning a survey of school principals in a particular state has lists of the school principals employed in each of the 125 school districts. The procedure is to obtain a random sample of principals from each of the districts rather than grouping all the lists together and obtaining a sample from the entire group. Which of the following is a correct conclusion?

  (A) This is a simple random sample obtained in an easier and less costly manner than procedures involving sampling from the entire population of principals.
  (B) This is a cluster sample in which the population was divided into heterogeneous groups called clusters.
  (C) This is an example of systematic sampling, which gives a reasonable sample as long as the original order of the list is not related to the variables under consideration.
  (D) This is an example of proportional sampling based on sizes of the school districts.
  (E) This is a stratified sample, which may give comparative information that a simple random sample wouldn't give.

34. A simple random sample is defined by

  (A) the method of selection.
  (B) examination of the outcome.
  (C) both of the above.
  (D) how representative the sample is of the population.
  (E) the size of the sample versus the size of the population.

35. Changing from a 90% confidence interval estimate for a population proportion to a 99% confidence interval estimate, with all other things being equal,

  (A) increases the interval size by 9%.
  (B) decreases the interval size by 9%.
  (C) increases the interval size by 57%.
  (D) decreases the interval size by 57%.
  (E) This question cannot be answered without knowing the sample size.

36. Which of the following is a true statement about hypothesis testing?

    (A) If there is sufficient evidence to reject a null hypothesis at the 10% level, there is sufficient evidence to reject it at the 5% level.
    (B) Whether to use a one- or a two-sided test is typically decided after the data are gathered.
    (C) If a hypothesis test is conducted at the 1% level, there is a 1% chance of rejecting the null hypothesis.
    (D) The probability of a Type I error plus the probability of a Type II error always equals 1.
    (E) The power of a test concerns its ability to detect an alternative hypothesis.

37. A population is normally distributed with mean 25. Consider all samples of size 10. The variable $\dfrac{\bar{x} - 25}{\frac{s}{\sqrt{10}}}$

    (A) has a normal distribution.
    (B) has a t-distribution with $df = 10$.
    (C) has a t-distribution with $df = 9$.
    (D) has neither a normal distribution nor a t-distribution.
    (E) has either a normal or a t-distribution depending on the characteristics of the population standard deviation.

38. A correlation of 0.6 indicates that the percentage of variation in y that is explained by the variation in x is how many times the percentage indicated by a correlation of 0.3?

    (A) 2
    (B) 3
    (C) 4
    (D) 6
    (E) There is insufficient information to answer this question.

39. As reported in a national CNN poll, 43% of high school students expressed fear about going to school. Which of the following best describes what is meant by the poll having a margin of error of 5%?

    (A) It is likely that the true proportion of high school students afraid to go to school is between 38% and 48%.
    (B) Five percent of the students refused to participate in the poll.
    (C) Between 38% and 48% of those surveyed expressed fear about going to school.
    (D) There is a 0.05 probability that the 43% result is in error.
    (E) If similar size polls were repeatedly taken, they would be wrong about 5% of the time.

40. In a high school of 1650 students, 132 have personal investments in the stock market. To estimate the total stock investment by students in this school, two plans are proposed. Plan I would sample 30 students at random, find a confidence interval estimate of their average investment, and then multiply both ends of this interval by 1650 to get an interval estimate of the total investment. Plan II would sample 30 students at random from among the 132 who have investments in the market, find a confidence interval estimate of their average investment, and then multiply both ends of this interval by 132 to get an interval estimate of the total investment. Which is the better plan for estimating the total stock market investment by students in this school?

(A) Plan I
(B) Plan II
(C) Both plans use random samples and so will produce equivalent results.
(D) Neither plan will give an accurate estimate.
(E) The resulting data must be seen to evaluate which is the better plan.

STOP

If there is still time remaining, you may check your work on this section.

SECTION II

Part A

QUESTIONS 1–5

Spend about 65 minutes on this part of the exam.
Percentage of Section II grade—75

> **Directions:** You must show all work and indicate the methods you use. You will be graded on the correctness of your methods and on the accuracy of your results and explanations.

1. Cumulative frequency graphs of the heights (in inches) of athletes playing college baseball (A), basketball (B), and football (C) are given below:

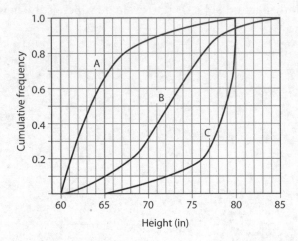

Compare the distributions of heights of athletes playing the three college sports.

2. When students do not understand an author's vocabulary, they often miss subtleties of meaning in the text; thus, learning vocabulary is recognized as one of the most important skills in all subject areas. A school system is planning a study to see which of two programs is more effective in ninth grade. Program A involves teaching students to apply morphemic analysis, the process of deriving a word's meaning by analyzing its meaningful parts, such as roots, prefixes, and suffixes. Program B involves teaching students to apply contextual analysis, the process of inferring the meaning of an unfamiliar word by examining the surrounding text.

   (a) Explain the purpose of incorporating a control group in this experiment.
   (b) To comply with informed consent laws, parents will be asked their consent for their children to be randomly assigned to be taught in one of the new programs or by the local standard method. Students of parents who do not return the consent forms will continue to be taught by the local standard method. Explain why these students should not be considered part of the control group.
   (c) Given 90 students randomly selected from ninth-grade students with parental consent, explain how to assign them to the three groups, Program A, Program B, and control, for a completely randomized design.

3. In the 1960s and 1970s, college students were in the front line of the Civil Rights and anti-war movements. Such activism is one of the most powerful mechanisms for social change. A study looked at the association between social justice advocacy as measured on the ACT activity scale and political interest level as measured on the 1–10 POL scale. The researchers hoped to predict advocacy from political interest. The statistical summary of the data from a random sample of 134 graduate students at a Midwestern university is shown below.

Variable	Mean	SD
ACT	18.92	4.85
POL	5.48	2.24

$r = 0.49$

(a) What is b, the slope of the regression line? Interpret it in context.
(b) What is the equation of the least squares regression line?
(c) Suppose a student has a POL of 4 with an ACT of 15. What is the residual? Interpret it in context.

4. A new anti-spam software program is field tested on 1000 e-mails, and the results are summarized in the following table (positive test = program labels e-mail as spam):

	Spam	Legitimate	
Positive test	205	90	295
Negative test	45	660	705
	250	750	1000

Using the above empirical results, determine the following probabilities.

(a) (i) What is the *predictive value* of the test? That is, what is the probability that an e-mail is spam and the test is positive?
   (ii) What is the *false-positive* rate? That is, what is the probability of testing positive given that the e-mail is legitimate?
(b) (i) What is the *sensitivity* of the test? That is, what is the probability of testing positive given that the e-mail is spam?
   (ii) What is the *specificity* of the test? That is, what is the probability of testing negative given that the e-mail is legitimate?
(c) (i) Given a random sample of five legitimate e-mails, what is the probability that the program labels at least one as spam?
   (ii) Given a random sample of five spam e-mails, what is the probability that the program correctly labels at least three as spam?
(d) (i) If 35% of the incoming e-mails from one source are spam, what is the probability that an e-mail from that source will be labeled as spam?
   (ii) If an e-mail from this source is labeled as spam, what is the probability it really is spam?

5. Cumulative exposure to nitrogen dioxide (NO_2) is a major risk factor for lung disease in tunnel construction workers. In a 2004 study, researchers compared cumulative exposure to NO_2 (in parts per million per year, ppm/yr) for a random sample of drill and blast workers and for an independent random sample of concrete workers. Summary statistics are shown below.

Tunnel worker activity	Sample size	Mean cumulative NO_2 exposure (ppm*/yr)	Standard deviation in cumulative NO_2 exposure (ppm*/yr)
Drill and blast	115	4.1	1.8
Concrete	69	4.8	2.4

(a) Find a 95% confidence interval estimate for the difference in mean cumulative NO_2 exposure for drill and blast workers and for concrete workers.

(b) Using this confidence interval, is there evidence of a difference in mean cumulative NO_2 exposure for drill and blast workers and for concrete workers? Explain.

(c) It turns out that both data sets were somewhat skewed. Does this invalidate your analysis? Explain.

SECTION II

Part B

QUESTION 6

Spend about 25 minutes on this part of the exam.
Percentage of Section II grade—25

6. A physical education teacher compared balance between male and female students at her school. Two independent samples were randomly selected, one of eight men and one of eight women. Each student was timed as to how long (in seconds) he or she could balance on one foot with eyes closed.

Balance times (in seconds)								
Men	19	13	35	20	48	19	24	21	17
Women	42	23	38	19	27	30	34	46	33

(a) Create a back-to-back stemplot to display these data.

(b) Compare the two distributions.

(c) To test whether or not the mean balance times of men and women are different, a two-sample t-test was performed with a resulting t-statistic of -1.816 and P-value of 0.0889. At a 10 percent significance level, what should be concluded?

Also of interest is whether men (M) or women (W) have more variability in their balance times. To test the hypotheses $H_0: \sigma_M = \sigma_W$ versus $H_a: \sigma_M \neq \sigma_W$, the test statistic $\frac{s_M}{s_W}$ (the ratio of the sample standard deviations) will be used. To investigate the sampling distribution of this test statistic when the population standard deviations are equal, a simulation is conducted. Two independent samples of size 9 are selected from the same population, and the ratio of the sample standard deviations is calculated. A histogram of 500 simulated values of $\frac{s_M}{s_W}$ is shown below.

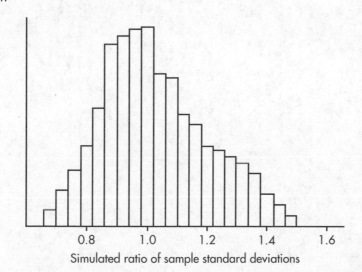

Simulated ratio of sample standard deviations

(d) Explain why the median of the above ratios appears to be about 1.0.

Summary statistics for the original data are below.

	Sample size	Sample mean	Sample standard deviation
Men	9	24.00	10.85
Women	9	32.44	8.76

(e) Using the summary statistics above and the simulation histogram above, is there convincing evidence of a difference in variability of balance times between men and women?

STOP

If there is still time remaining, you may check your work on this section.

Answer Key

1. **D**	9. **D**	17. **B**	25. **C**	33. **E**
2. **E**	10. **D**	18. **A**	26. **C**	34. **A**
3. **E**	11. **C**	19. **B**	27. **D**	35. **C**
4. **D**	12. **D**	20. **E**	28. **D**	36. **E**
5. **B**	13. **A**	21. **E**	29. **C**	37. **C**
6. **E**	14. **B**	22. **E**	30. **A**	38. **C**
7. **C**	15. **D**	23. **B**	31. **C**	39. **A**
8. **D**	16. **A**	24. **E**	32. **D**	40. **B**

Answers and Explanations

SECTION I

1. **(D)** With 87 home sale prices, the median is between the 43rd and 44th values when placed in either ascending or descending order. There appear to be around 40 values between $225,000 and $250,000 and 25 more values between $200,000 and $225,000. So, both the 43rd and 44th values are in the 200 to 225 interval in the histogram.

2. **(E)** Regression lines show association, not causation. Surveys suggest relationships, which experiments can help to show to be cause and effect.

3. **(E)** The critical z-score is `invNorm(0.85)` $= 1.036$. Thus, $60 - 55 = 1.036\sigma$ and $\sigma = \dfrac{60 - 55}{1.036}$.

4. **(D)** Calculator software gives (1,794,000, 1,856,000). We are 90% confident that the population mean is within the interval calculated using the data from the sample.

5. **(B)** The first study was observational because the subjects were not chosen for treatment; their current eating habits were simply noted. In the second study, two treatments were applied (meat meals and vegetarian meals).

6. **(E)**

$$P(\text{def}) = P(\text{1st} \cap \text{def}) + P(\text{2nd} \cap \text{def})$$
$$= (0.6)(0.005) + (0.4)(0.010)$$
$$P(\text{1st} \mid \text{def}) = \frac{P(\text{1st} \cap \text{def})}{P(\text{def})} = \frac{(0.6)(0.005)}{(0.6)(0.005) + (0.4)(0.010)}$$

7. **(C)** At the dialysis center the more serious concern would be a Type II error, which is that the equipment is not performing correctly yet the check does not pick this up; while at the towel manufacturing plant the more serious concern would be a Type I error, which is that the equipment is performing correctly yet the check causes a production halt.

8. **(D)** There are two possible outcomes (heads and tails), with the probability of heads always 0.75 (independent of what happened on the previous toss), and we are interested in the number of heads in 10 tosses. Thus, this is a binomial model with $n = 10$ and $p = 0.75$. Repeating this over and over (in this case 50 times) simulates the resulting binomial distribution.

9. **(D)** In a binomial distribution with $n = 5$ and probability of success $p = \dfrac{5.4}{9.0} = 0.6$, the probability of at least 3 successes is $P(3 \text{ successes}) + P(4 \text{ successes}) + P(5 \text{ successes})$:

$$= \binom{5}{3}(0.6)^3(0.4)^2 + \binom{5}{4}(0.6)^4(0.4) + (0.6)^5$$

10. **(D)** If A and B are mutually exclusive, then $P(A \cap B) = 0$, and so $P(A \cup B) = 0.3 + 0.2 - 0 = 0.5$. If A and B are independent, then $P(A \cap B) = P(A)P(B)$, and so $P(A \cup B) = 0.3 + 0.2 - (0.3)(0.2) = 0.44$. If B is a subset of A, then $A \cup B = A$, and so $P(A \cup B) = P(A) = 0.3$.

11. **(C)** The standard deviation of the test statistic is $\sigma_{\hat{p}} = \sqrt{\dfrac{p_0(1 - p_0)}{n}} = \sqrt{\dfrac{(0.69)(1 - 0.69)}{1100}}$

and $\hat{p} = \dfrac{770}{1100} = 0.7$. Then $z = \dfrac{\hat{p} - p_0}{\sigma_{\hat{p}}} = \dfrac{0.7 - 0.69}{\sqrt{\dfrac{(0.69)(1 - 0.69)}{1100}}}$. Since this is a two-sided test,

the P-value will be twice the tail probability of the test statistic; however, the test statistic itself is not doubled.

12. **(D)** Dosage is the only explanatory variable, and it is being tested at three levels. Tumor reduction is the single response variable.

13. **(A)** The median corresponds to a cumulative proportion of 0.5.

14. **(B)** Standard deviation is a measure of variability. The less variability, the more homogeneity there is.

15. **(D)** Since the sample sizes are small, the samples must come from roughly normally distributed populations. While the samples should be independent simple random samples, np and $n(1 - p)$ refer to conditions for tests involving sample proportions, not means.

16. **(A)** Adding the same constant to all values in a set will increase the mean by that constant but will leave the standard deviation unchanged.

17. **(B)** With about 68% of the values within 1 standard deviation of the mean, the expected numbers for a normal distribution are as follows:

16	34	34	16

$$\chi^2 = \sum \frac{(\text{obs} - \text{exp})^2}{\text{exp}} = \frac{(10 - 16)^2}{16} + \frac{(40 - 34)^2}{34} + \frac{(35 - 34)^2}{34} + \frac{(15 - 16)^2}{16}$$

18. **(A)** This is a good example of voluntary response bias, which often overrepresents strong or negative opinions. The people who chose to respond were very possibly the parents of children facing drug problems or people who had had bad experiences with drugs being sold in their neighborhoods. There is very little chance that the 2500 respondents were representative of the population. Knowing more about his listeners or taking a sample of the sample would not have helped.

19. **(B)** The range (difference between largest and smallest values), the interquartile range $(Q_3 - Q_1)$, and this difference between the 60th and 40th percentile scores all are measures of variability, or how spread out is the population or a subset of the population.

20. **(E)** It is true that $E(X + Y) = E(X) + E(Y) = 37 + 62 = 99$; however, without *independence*, we cannot determine $\text{var}(X + Y)$ from the information given.

21. **(E)** The correlation coefficient r is not affected by changes in units, by which variable is called x or y, or by adding or multiplying all the values of a variable by the same constant.

22. **(E)** The critical z-score for a 95% confidence interval is `invNorm(0.975)` = 1.96. With unknown p, we use $p = 0.5$ in our calculation of n.

23. **(B)** With $\alpha = 0.05$, the probability of committing a Type I error is 0.05, and the probability of not committing a Type I error is 0.95. P(at least one Type I error in 10 tests) = $1 - P$(no Type I errors in 10 tests) $= 1 - (0.95)^{10} = 0.40$.

24. **(E)** Note that all three sets have the same mean and the same range. The third set has most of its values concentrated right at the mean, while the second set has most of its values concentrated far from the mean.

25. **(C)** People coming out of a Wall Street office building are a very unrepresentative sample of the adult population, especially given the question under consideration. Using chance and obtaining a high response rate will not change the selection bias and make this into a well designed survey. This is a convenience sample, not a voluntary response sample.

26. **(C)** Larger samples (so $\frac{\sigma}{\sqrt{n}}$ is smaller) and less confidence (so the critical z or t is smaller) both result in smaller intervals.

27. **(D)** The third scatterplot shows perfect negative association, so $r_3 = -1$. The first scatterplot shows strong, but not perfect, negative correlation, so $-1 < r_1 < 0$. The second scatterplot shows no correlation, so $r_2 = 0$.

28. **(D)** The probability of throwing heads is 0.5. By the law of large numbers, the more times you flip the coin, the more the relative frequency tends to become closer to this probability. With fewer tosses there is more chance for wide swings in the relative frequency.

29. **(C)** $\sigma_{\bar{x}} = \frac{\sigma}{\sqrt{n}} = \frac{27}{\sqrt{225}}$. If the true mean parking duration is 51 minutes, the normal curve should be centered at 51. The critical value of 50 has a z-score of $\frac{50 - 51}{27/\sqrt{225}}$. The city planners will reject the merchants' claim if the sample mean is greater than 50 and so will fail to reject if less than 50.

30. **(A)** The overall lengths (between tips of whiskers) are the same, and so the ranges are the same. Just because the min and max are equidistant from the median, and Q_1 and Q_3 are equidistant from the median, does not imply that a distribution is symmetric or that the mean and median are equal. Even if a distribution is symmetric, this does not imply that it is roughly normal. Particular values, not distributions, may be outliers.

31. **(C)** Critical z-scores are $\frac{700 - 650}{45} = 1.11$ and $\frac{740 - 650}{45} = 2$ with right tail probabilities of 0.1335 and 0.0228, respectively. The percentage below 740 given that the scores are above 700 is $\frac{0.1335 - 0.0228}{0.1335} = 82.9\%$. Or $\frac{\text{normalcdf}(700, 740, 650, 45)}{\text{normalcdf}(700, 1000000, 650, 45)} = 0.82928$.

32. **(D)** There is a different probability of Type II error for each possible correct value of the population parameter, and 1 minus this probability is the power of the test against the associated correct value. Increasing the sample size will make the standard deviations smaller and thus can decrease the probabilities of Type I and Type II errors and so increase the power. Increasing the significance level α (making it easier to reject the null) will lower the probability of a Type II error and thus also increase the power.

33. **(E)** Stratified samples are often easier and less costly to obtain and also make comparative data available. In this case, responses can be compared among various districts. This is not a simple random sample because all possible sets of the required size do not have the same chance of being picked. For example, a set of principals all from just half the school districts has no chance of being picked to be the sample. This is not a cluster sample in that there is no reason to believe that each school district resembles the population as a whole, and furthermore, there was no random sample taken of the school districts. This is not systematic sampling as the districts were not put in some order with every nth district chosen.

34. **(A)** A simple random sample can be any size and may or may not be representative of the population. It is a method of selection in which every possible sample of the desired size has an equal chance of being selected.

35. **(C)** The critical z-scores go from ± 1.645 to ± 2.576, resulting in an increase in the interval size: $\frac{2.576}{1.645} = 1.57$ or an increase of 57%.

36. **(E)** If the P-value is less than 0.10, it does not follow that it is less than 0.05. Decisions such as whether a test should be one- or two-sided are made before the data are gathered. If $\alpha = 0.01$, there is a 1% chance of rejecting the null hypothesis *if* the null hypothesis is true. There is one probability of a Type I error, the significance level, while there is a different probability of a Type II error associated with each possible correct alternative, so the sum does not equal 1. When given a true alternative, power is the probability of rejecting the false null hypothesis.

37. **(C)** If the population is normally distributed, then $\frac{\bar{x} - \mu}{\frac{\sigma}{\sqrt{n}}}$ is normally distributed, but $\frac{\bar{x} - \mu}{\frac{s}{\sqrt{n}}}$ has a t-distribution with $df = n - 1$.

38. **(C)** A correlation of 0.6 explains $(0.6)^2$, or 36%, of the variation in y, while a correlation of 0.3 explains only $(0.3)^2$, or 9%, of the variation in y.

39. **(A)** When using a measurement from a sample, we are never able to say *exactly* what a population proportion is; rather, we always say we have a certain *confidence* that the population proportion lies in a particular *interval*. In this case, that interval is 43% $\pm$ 5% or between 38% and 48%.

40. **(B)** With Plan I, the expected number of students with stock investments is only $30\left(\frac{132}{1650}\right) = 2.4$ out of 30. Plan II allows an estimate to be made using a full 30 investors.

1. A complete answer compares shape, center, and spread and also mentions context.

   Shape: The baseball players (A), for which the cumulative frequency plot rises steeply at first, include more shorter players, and thus the distribution is skewed to the right (toward the greater heights). The football players (C), for which the cumulative frequency plot rises slowly at first and then steeply toward the end, include more taller players, and thus the distribution is skewed to the left (toward the lower heights). The basketball players (B), for which the cumulative frequency plot rises slowly at each end and steeply in the middle, have a more bell-shaped distribution of heights.

   Center: The medians correspond to relative frequencies of 0.5. Reading across from 0.5 and then down to the *x*-axis shows the median heights to be about 63.5 inches for baseball players, about 72.5 inches for basketball players, and about 79 inches for football players. Thus, the center of the baseball height distribution is the least, and the center of the football height distribution is the greatest.

   Spread: The range of the football players is the smallest, $80 - 65 = 15$ inches, then comes the range of the baseball players, $80 - 60 = 20$ inches, and finally the range of the basketball players is the greatest, $85 - 60 = 25$ inches.

   ## SCORING

   The discussion of shape is essentially correct for correctly identifying which distribution is skewed left, skewed right, and more bell-shaped and for giving a correct justification based on the cumulative frequency plots. The discussion of shape is partially correct for correctly identifying which distribution is skewed left, skewed right, and more bell-shaped without giving a good explanation.

   The discussion of center is essentially correct for correctly noting that the baseball players have the lowest median height and the football players have the greatest median height, and giving some numerical justification. The discussion of center is partially correct for correctly noting that the baseball players have the lowest median height and the football players have the greatest median height but without giving a good explanation.

   The discussion of spread is essentially correct for correctly noting that the football players have the smallest range for their heights and the basketball players have the greatest range for their heights, and giving some numerical justification. The discussion of spread is partially correct for correctly noting that the football players have the smallest range for their heights and the basketball players have the greatest range for their heights but without giving a good explanation.

   4 Complete Answer   All three parts essentially correct.

   3 Substantial Answer   Two parts essentially correct and one part partially correct.

   2 Developing Answer   Two parts essentially correct OR one part essentially correct and one or two parts partially correct OR all three parts partially correct.

   1 Minimal Answer   One part essentially correct OR two parts partially correct.

   Lower a 4 to a 3, or a 3 to a 2, if context is never mentioned.

2. (a) A control group would allow the school system to compare the effectiveness of each of the new programs to the local standard method currently being used.

(b) Parents who fail to return the consent form are a special category who may well place less priority on education. The effect of using their children may distort results since their children could be placed only in the control group.

(c) Assign each student a unique number 01–90. Using a random number table or a random number generator, pick numbers between 01 and 90, throwing out repeats. The students corresponding to the first 30 such numbers picked will be assigned to Program A, the next 30 picked to Program B, and the remaining to the control group.

SCORING

Part (a) is essentially correct if the purpose is given for using a control group in this study. Part (a) is partially correct if a correct explanation for the use of a control group is given but not in the context of this study.

Part (b) is essentially correct for a clear explanation in context. Part (b) is partially correct if the explanation is weak.

Part (c) is essentially correct if randomization is used correctly and the method is clear. Part (c) is partially correct if randomization is used but the method is not clearly explained.

4 Complete Answer   All three parts essentially correct.

3 Substantial Answer   Two parts essentially correct and one part partially correct.

2 Developing Answer   Two parts essentially correct OR one part essentially correct and one or two parts partially correct OR all three parts partially correct.

1 Minimal Answer   One part essentially correct OR two parts partially correct.

---

3. (a) $b = r\left(\frac{s_y}{s_x}\right) = 0.49\left(\frac{4.85}{2.24}\right) = 1.0609$. For each unit increase on the POL political interest scale, the expected increase in the ACT activism scale is 1.0609.

(b) The least squares regression line passes through $(\bar{x}, \bar{y})$. Thus, $a + 1.0609(5.48) = 18.92$, which gives $a = 13.11$, and the equation is $\widehat{ACT} = 13.11 + 1.0609(POL)$.

(c) $13.11 + 1.0609(4) = 17.35$. The residual $=$ observed $-$ predicted $= 15 - 17.35 = -2.35$. Thus, the regression model overestimated the ACT activity score for this student.

SCORING

Part (a) is essentially correct for a correct calculation of the slope, an interpretation of the slope in context, and the use of nondeterministic language (such as "expected" or "predicted"). Part (a) is partially correct for two of these three components.

Part (b) is essentially correct for a correct equation, using a "hat" for the dependent variable or the word "Predicted," and defining the variables. Part (b) is partially correct for two of these three components.

Part (c) is essentially correct for a correct calculation of the residual, referring to context in a correct interpretation, and referring to direction (such as an "overestimate"). Part (c) is partially correct for two of these three components.

4 Complete Answer   All three parts essentially correct.

3 Substantial Answer Two parts essentially correct and one part partially correct.

2 Developing Answer   Two parts essentially correct OR one part essentially correct and one or two parts partially correct OR all three parts partially correct.

1 Minimal Answer   One part essentially correct OR two parts partially correct.

4. (a) (i) $P(spam \cap positive) = \dfrac{205}{1000} = 0.205$

   (ii) $P(positive \mid legitimate) = \dfrac{90}{750} = 0.12$

   (b) (i) $P(positive \mid spam) = \dfrac{205}{250} = 0.82$

   (ii) $P(negative \mid legitimate) - \dfrac{660}{750} = 0.88$

   (c) (i) $1 - (0.88)^5 = 0.4723$

   (ii) $10(0.82)^3(0.18)^2 + 5(0.82)^4(0.18) + (0.82)^5 = 0.9563$

   Or `1 - binomcdf(5, 0.82, 3)` $= 1 - 0.0437 = 0.9563$

   (d) (i) $P(positive) = P(positive \cap spam) + P(positive \cap legitimate) =$
   $P(spam)P(positive \mid spam) + P(legitimate)P(positive \mid legitimate) =$
   $(0.35)(0.82) + (0.65)(0.12) = 0.365$

   (ii) $P(spam \mid positive) = \dfrac{P(spam \cap positive)}{P(positive)} = \dfrac{(0.35)(0.82)}{0.365} = 0.7863$

SCORING

There are two probabilities to calculate in each Part (a)–(d). Each Part is essentially correct for both probabilities correctly calculated and is partially correct for one probability correctly calculated. In Parts that use results from previous parts, full credit is given for correctly using the results of the earlier Part, whether that earlier calculation was correct or not. For credit for Part (d), a correct methodology must also be shown.

Count partially correct answers as one-half an essentially correct answer.

4	**Complete Answer**	Four essentially correct answers.
3	**Substantial Answer**	Three essentially correct answers.
2	**Developing Answer**	Two essentially correct answers.
1	**Minimal Answer**	One essentially correct answer.

Use a holistic approach to decide a score totaling between two numbers.

5. (a) *Parameters:* Let μ_{db} represent the mean cumulative NO_2 exposure for the population of drill and blast workers. Let μ_c represent the mean cumulative NO_2 exposure for the population of concrete workers.

Procedure: Two-sample t-interval for $\mu_{db} - \mu_c$.

Checks: We are given that we have independent random samples, 115 and 69 are assumed less than 10% of all "drill and blast" and "concrete" workers, respectively, and we note that the sample sizes are large ($n_{db} = 115 \geq 30$ and $n_c = 69 \geq 30$).

Mechanics: Calculator software (such as 2-SampTInt on the TI-84 or 2-Samplet-Interval on the Casio Prizm) gives $(-1.362, -0.0381)$.

Conclusion in context: We are 95% confident that the true difference (drill and blast minus concrete) in the mean cumulative NO_2 exposure for all drill and blast workers and all concrete workers is between -1.362 and -0.0381 ppm/yr.

(b) Zero is not in the above 95% confidence interval, so at the $\alpha = 0.05$ significance level, there is evidence to reject H_0: $\mu_{db} - \mu_c = 0$ in favor of H_a: $\mu_{db} - \mu_c \neq 0$. That is, there is evidence of a difference in mean cumulative NO_2 exposure for drill and blast workers and for concrete workers.

(c) With sample sizes this large, the central limit theorem applies and our analysis is valid.

Part (a) has two components. The first component, identifying the confidence interval and checking conditions, is essentially correct for naming the confidence interval procedure, noting independent random samples, noting the large sample sizes, and addressing the 10% condition. This component is partially correct for correctly covering two or three of the four points.

The second component of Part (a) is essentially correct for correct mechanics in calculating the confidence interval and for a correct (based on the shown mechanics) conclusion in context. Part (a) is partially correct for one of these two features.

The third component, Parts (b) and (c) together, is essentially correct for 1) noting that zero is not in the interval so the observed difference is significant, 2) stating this in context of the problem, and 3) relating the central limit theorem (CLT) to the samples being large. The third component is partially correct for two out of the above three statements.

4 Complete Answer   All three components essentially correct.

3 Substantial Answer   Two components essentially correct and one component partially correct.

2 Developing Answer   Two components essentially correct OR one component essentially correct and one or two components partially correct OR all three components partially correct.

1 Minimal Answer   One component essentially correct OR two components partially correct.

SECTION II: PART B

6. (a)

```
       Men      Women
              0
    9 9 7 3   1  9
      4 1 0   2  3 7
          5   3  0 3 4 8
          8   4  2 6
```

3 | 1 | 9 represents a man's
time of 13 seconds and a
woman's time of 19 seconds

(b) The distribution of men's balance times is skewed right, while the distribution of women's balance times is more unimodal and roughly symmetric; neither distribution appear to have outliers. The median women's time, 33 seconds, is greater than the median men's time, 20 seconds. The men's times exhibit more variability than the women's times (range of men's times is $48 - 13 = 35$, while range of women's times is $46 - 19 = 27$).

(c) With this small a P-value, $0.0889 < 0.10$, there is sufficient evidence that the mean balance times of men and women are different.

(d) The simulation was conducted under the null hypothesis $H_0: \sigma_M = \sigma_W$, which gives $\frac{s_M}{s_W} = 1$. So, it is reasonable that the simulated sample ratios $\frac{s_M}{s_W}$ will be centered around 1.

(e) In the original samples, $\frac{s_M}{s_W} = \frac{10.85}{8.76} = 1.24$. The value of 1.24 in the distribution of simulated ratios does not fall far into the tail; thus, it could have well occurred by random sampling. So, no, there is not convincing evidence of a difference in variability of balance times between men and women.

SCORING

Section 1 is essentially correct for a correctly drawn back-to-back stemplot, a key for the stemplot, and a comparison addressing shape, noting that the center of the distribution of women's times is greater than that of the men's and noting that the spread of the men's distribution is greater than the women's. Section 1 is partially correct for three out of the five steps above.

Section 2 is essentially correct for 1) in (c) for a correct conclusion in context linked to the *P*-value and is partially correct if just missing context and 2) in (d) for correctly stating that the population standard deviations are equal if the null hypothesis is true, and so the ratios of the simulated standard deviations should be centered around 1. Section 3 is partially correct for one of these two parts correct OR for both correct conclusions but weak explanations.

Section 3 is essentially correct in (e) for a correct conclusion with a clear explanation and is partially correct for a correct conclusion with a weak explanation.

Count partially correct answers as one-half an essentially correct answer.

4 Complete Answer   All three sections essentially correct.

3 Substantial Answer   Two sections essentially correct and one section partially correct.

2 Developing Answer   Two sections essentially correct OR one section essentially correct and one or two sections partially correct OR all three sections partially correct.

1 Minimal Answer   One section essentially correct OR two sections partially correct.

ANSWER SHEET
Practice Test 2

1. Ⓐ Ⓑ Ⓒ Ⓓ Ⓔ
2. Ⓐ Ⓑ Ⓒ Ⓓ Ⓔ
3. Ⓐ Ⓑ Ⓒ Ⓓ Ⓔ
4. Ⓐ Ⓑ Ⓒ Ⓓ Ⓔ
5. Ⓐ Ⓑ Ⓒ Ⓓ Ⓔ
6. Ⓐ Ⓑ Ⓒ Ⓓ Ⓔ
7. Ⓐ Ⓑ Ⓒ Ⓓ Ⓔ
8. Ⓐ Ⓑ Ⓒ Ⓓ Ⓔ
9. Ⓐ Ⓑ Ⓒ Ⓓ Ⓔ
10. Ⓐ Ⓑ Ⓒ Ⓓ Ⓔ

11. Ⓐ Ⓑ Ⓒ Ⓓ Ⓔ
12. Ⓐ Ⓑ Ⓒ Ⓓ Ⓔ
13. Ⓐ Ⓑ Ⓒ Ⓓ Ⓔ
14. Ⓐ Ⓑ Ⓒ Ⓓ Ⓔ
15. Ⓐ Ⓑ Ⓒ Ⓓ Ⓔ
16. Ⓐ Ⓑ Ⓒ Ⓓ Ⓔ
17. Ⓐ Ⓑ Ⓒ Ⓓ Ⓔ
18. Ⓐ Ⓑ Ⓒ Ⓓ Ⓔ
19. Ⓐ Ⓑ Ⓒ Ⓓ Ⓔ
20. Ⓐ Ⓑ Ⓒ Ⓓ Ⓔ

21. Ⓐ Ⓑ Ⓒ Ⓓ Ⓔ
22. Ⓐ Ⓑ Ⓒ Ⓓ Ⓔ
23. Ⓐ Ⓑ Ⓒ Ⓓ Ⓔ
24. Ⓐ Ⓑ Ⓒ Ⓓ Ⓔ
25. Ⓐ Ⓑ Ⓒ Ⓓ Ⓔ
26. Ⓐ Ⓑ Ⓒ Ⓓ Ⓔ
27. Ⓐ Ⓑ Ⓒ Ⓓ Ⓔ
28. Ⓐ Ⓑ Ⓒ Ⓓ Ⓔ
29. Ⓐ Ⓑ Ⓒ Ⓓ Ⓔ
30. Ⓐ Ⓑ Ⓒ Ⓓ Ⓔ

31. Ⓐ Ⓑ Ⓒ Ⓓ Ⓔ
32. Ⓐ Ⓑ Ⓒ Ⓓ Ⓔ
33. Ⓐ Ⓑ Ⓒ Ⓓ Ⓔ
34. Ⓐ Ⓑ Ⓒ Ⓓ Ⓔ
35. Ⓐ Ⓑ Ⓒ Ⓓ Ⓔ
36. Ⓐ Ⓑ Ⓒ Ⓓ Ⓔ
37. Ⓐ Ⓑ Ⓒ Ⓓ Ⓔ
38. Ⓐ Ⓑ Ⓒ Ⓓ Ⓔ
39. Ⓐ Ⓑ Ⓒ Ⓓ Ⓔ
40. Ⓐ Ⓑ Ⓒ Ⓓ Ⓔ

Practice Test 2

SECTION I

Questions 1–40

Spend 90 minutes on this part of the exam.

> **Directions:** The questions or incomplete statements that follow are each followed by five suggested answers or completions. Choose the response that best answers the question or completes the statement.

1. Suppose that the regression line for a set of data, $\hat{y} = mx + 3$, passes through the point (2, 7). If $\bar{x}$ and $\bar{y}$ are the sample means of the x- and y-values, respectively, which of the following is equal to $\bar{y}$?

   (A) $\bar{x}$
   (B) $\bar{x} - 2$
   (C) $\bar{x} + 3$
   (D) $2\bar{x} + 3$
   (E) $3.5\bar{x} + 3$

2. A study is made to determine whether more hours of academic studying leads to a greater number of points scored by basketball players. In surveying 50 basketball players, it is noted that the 25 who claim to study the most hours have a higher point average than the 25 who study less. Based on this study, the coach begins requiring the players to spend more time studying. Which of the following is a correct statement?

   (A) While this study may have its faults, it still does prove causation.
   (B) There could well be a confounding variable responsible for the seeming relationship.
   (C) While this is a controlled experiment, the conclusion of the coach is not justified.
   (D) To get the athletes to study more, it would be more meaningful to have them put in more practice time on the court to boost their point averages, as higher point averages seem to be associated with more study time.
   (E) No proper conclusion is possible without somehow introducing *blinding*.

3. In a social experiment, students were asked to take a selfie and post it to social media. Unknown to the students, what the study was really about was counting how many selfies a student would take before feeling satisfied to post one. The standard deviation for number of selfies taken before posting was 4.3, and 30% of numbers of selfies taken were over 15. Assuming an approximately normal distribution for numbers of selfies taken before posting, what was the mean number?

(A) $15 - 0.30(4.3)$
(B) $15 + 0.30(4.3)$
(C) $15 - 0.4756(4.3)$
(D) $15 + 0.5244(4.3)$
(E) $15 - 0.5244(4.3)$

4. Which of the following is a correct statement about correlation?

(A) If the slope of the regression line is exactly 1, the correlation is exactly 1.
(B) If the correlation is 0, the slope of the regression line is undefined.
(C) Switching which variable is called x and which is called y changes the sign of the correlation.
(D) The correlation r is equal to the slope of the regression line when z-scores for the y-variable are plotted against z-scores for the x-variable.
(E) Changes in the measurement units of the variables may change the correlation.

5. Which of the following are affected by outliers?

   I. Mean
  II. Median
 III. Standard deviation
 IV. Range
  V. Interquartile range

(A) I, III, and V only
(B) II and IV only
(C) I and V only
(D) III and IV only
(E) I, III, and IV only

6. An engineer wishes to determine the quantity of heat being generated by a particular electronic component. She knows that the standard deviation is 2.4 and wants to be 99% sure of knowing the mean quantity to within ±0.6. Which of the following should be used to find the sample size (n) needed?

   (A) $1.645\sqrt{\dfrac{2.4}{n}} \le 0.6$

   (B) $1.96\left(\dfrac{2.4}{\sqrt{n}}\right) \le 0.6$

   (C) $2.576\left(\dfrac{2.4}{\sqrt{n}}\right) \le 0.6$

   (D) $2.326\sqrt{\dfrac{2.4}{n}} \le 0.6$

   (E) $2.576\sqrt{\dfrac{2.4}{n}} \le 0.6$

7. A company that produces facial tissues continually monitors tissue strength. If the mean strength from sample data drops below a specified level, the production process is halted and the machinery inspected. Which of the following would result from a Type I error?

   (A) Halting the production process when sufficient customer complaints are received.
   (B) Halting the production process when the tissue strength is below specifications.
   (C) Halting the production process when the tissue strength is within specifications.
   (D) Allowing the production process to continue when the tissue strength is below specifications.
   (E) Allowing the production process to continue when the tissue strength is within specifications.

8. Two possible wordings for a questionnaire on a proposed school budget increase are as follows:

   I. This school district has one of the highest per student expenditure rates in the state. This has resulted in low failure rates, high standardized test scores, and most students going on to good colleges and universities. Do you support the proposed school budget increase?
   II. This school district has one of the highest per student expenditure rates in the state. This has resulted in high property taxes, with many people on fixed incomes having to give up their homes because they cannot pay the school tax. Do you support the proposed school budget increase?

   One of these questions showed that 58% of the population favor the proposed school budget increase, while the other question showed that only 13% of the population support the proposed increase. Which produced which result and why?

(A) The first showed 58% and the second 13% because of the lack of randomization as evidenced by the wording of the questions.

(B) The first showed 13% and the second 58% because of a placebo effect due to the wording of the questions.

(C) The first showed 58% and the second 13% because of the lack of a control group.

(D) The first showed 13% and the second 58% because of response bias due to the wording of the questions.

(E) The first showed 58% and the second 13% because of response bias due to the wording of the questions.

9. Suppose that for a certain Caribbean island, in any 3-year period the probability of a major hurricane is 0.25, the probability of water damage is 0.44, and the probability of both a hurricane and water damage is 0.22. What is the probability of water damage given that there is a hurricane?

(A) $0.25 + 0.44 - 0.22$

(B) $\dfrac{0.22}{0.44}$

(C) $0.25 + 0.44$

(D) $\dfrac{0.22}{0.25}$

(E) $0.25 + 0.44 + 0.22$

10. A union spokesperson is trying to encourage a college faculty to join the union. She would like to argue that faculty salaries are not truly based on years of service as most faculty believe. She gathers data and notes the following scatterplot of salary versus years of service.

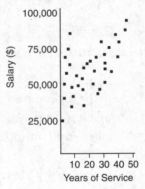

Which of the following most correctly interprets the overall scatterplot?

(A) The faculty member with the fewest years of service makes the lowest salary, and the faculty member with the most years of service makes the highest salary.

(B) A faculty member with more years of service than another has a higher salary than the other.

(C) There is a strong positive correlation with little deviation.

(D) There is no clear relationship between salary and years of service.

(E) While there is a strong positive correlation, there is a distinct deviation from the overall pattern for faculty with fewer than ten years of service.

11. Two random samples of students are chosen, the first set from those taking an AP Statistics class and the second set from those not taking an AP Statistics class. The following back-to-back stemplots compare the GPAs.

	AP Statistics		No AP Statistics	
		1	89	Key: 1 | 8 means 1.8
	97653	2	015688	
	98775332110	3	133344777888	
	1100	4		

Which of the following is true about the ranges and standard deviations?

(A) The first set has both a greater range and a greater standard deviation.
(B) The first set has a greater range, while the second has a greater standard deviation.
(C) The first set has a greater standard deviation, while the second has a greater range.
(D) The second set has both a greater range and a greater standard deviation.
(E) The two sets have equal ranges and equal standard deviations.

12. In a group of 10 scores, the largest score is increased by 40 points. What will happen to the mean?

(A) It will remain the same.
(B) It will increase by 4 points.
(C) It will increase by 10 points.
(D) It will increase by 40 points.
(E) There is not sufficient information to answer this question.

13. Suppose X and Y are random variables with $\mu_X = 32$, $\sigma_X = 5$, $\mu_Y = 44$, and $\sigma_Y = 12$. Given that X and Y are independent, what are the mean and standard deviation of the random variable $X + Y$?

(A) $\mu_{X+Y} = 76$, $\sigma_{X+Y} = 8.5$
(B) $\mu_{X+Y} = 76$, $\sigma_{X+Y} = 13$
(C) $\mu_{X+Y} = 76$, $\sigma_{X+Y} = 17$
(D) $\mu_{X+Y} = 38$, $\sigma_{X+Y} = 17$
(E) There is insufficient information to answer this question.

14. Suppose you toss a fair die three times and it comes up an even number each time. Which of the following is a true statement?

(A) By the law of large numbers, the next toss is more likely to be an odd number than another even number.
(B) Based on the properties of conditional probability, the next toss is more likely to be an even number given that three in a row have been even.
(C) Dice actually do have memories, and thus the number that comes up on the next toss will be influenced by the previous tosses.
(D) The law of large numbers tells how many tosses will be necessary before the percentages of evens and odds are again in balance.
(E) The probability that the next toss will again be even is 0.5.

15. A pharmaceutical company is interested in the association between advertising expenditures and sales for various over-the-counter products. A sales associate collects data on nine products, looking at sales (in $1000) versus advertising expenditures (in $1000). The results of the regression analysis are shown below.

Dependent variable: Sales

Variable	Coefficient	SE Coef	t-ratio	P
Constant	123.800	1.798	68.84	0.000
Advertising	12.633	0.378	33.44	0.000

R—Sq = 99.4%   R—Sq (adj) = 99.3%
s = 2.926 with 9—2 = 7 degrees of freedom

Which of the following gives a 90% confidence interval for the slope of the regression line?

(A) $12.633 \pm 1.415(0.378)$

(B) $12.633 \pm 1.895(0.378)$

(C) $123.800 \pm 1.414(1.798)$

(D) $123.800 \pm 1.895(1.798)$

(E) $123.800 \pm 1.645\left(\dfrac{1.798}{\sqrt{9}}\right)$

16. Suppose you wish to compare the AP Statistics exam results for the male and female students taking AP Statistics at your high school. Which is the most appropriate technique for gathering the needed data?

(A) Census

(B) Sample survey

(C) Experiment

(D) Observational study

(E) None of these is appropriate.

17. Jonathan obtained a score of 80 on a statistics exam, placing him at the 90th percentile. Suppose 5 points are added to everyone's score. Jonathan's new score will be at the

(A) 80th percentile.

(B) 85th percentile.

(C) 90th percentile.

(D) 95th percentile.

(E) There is not sufficient information to answer this question.

18. To study the effect of music on piecework output at a clothing manufacturer, two experimental treatments are planned: day-long classical music for one group versus day-long light rock music for another. Which one of the following groups would serve best as a control for this study?

(A) A third group for which no music is played
(B) A third group that randomly hears either classical or light rock music each day
(C) A third group that hears day-long R & B music
(D) A third group that hears classical music every morning and light rock every afternoon
(E) A third group in which each worker has earphones and chooses his or her own favorite music

19. Suppose H_0: $p = 0.6$ and the power of the test for the alternative $p = 0.7$ is 0.8. Which of the following is a valid conclusion?

(A) The probability of committing a Type I error is 0.1.
(B) If the alternative is true, the probability of failing to reject H_0 is 0.2.
(C) The probability of committing a Type II error is 0.3.
(D) All of the above are valid conclusions.
(E) None of the above are valid conclusions.

20. The following is a histogram of the numbers of ties owned by bank executives.

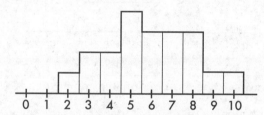

Which of the following is a correct statement?

(A) The median number of ties is five.
(B) More than four executives own over eight ties each.
(C) An executive is equally likely to own fewer than five ties or more than seven ties.
(D) One tie is a reasonable estimate for the standard deviation.
(E) Removing all the executives with three, nine, and ten ties may change the median.

21. Which of the following is a binomial random variable?

(A) The number of tosses before a "5" appears when tossing a fair die
(B) The number of points a hockey team receives in 10 games, where two points are awarded for wins, one point for ties, and no points for losses
(C) The number of hearts out of five cards randomly drawn from a deck of 52 cards, without replacement
(D) The number of motorists not wearing seat belts in a random sample of five drivers
(E) None of the above

22. Company I manufactures demolition fuses that burn an average of 50 minutes with a standard deviation of 10 minutes, while company II advertises fuses that burn an average of 55 minutes with a standard deviation of 5 minutes. Which company's fuse is more likely to last at least 1 hour? Assume roughly normal distributions of fuse times.

    (A) Company I's, because of its greater standard deviation
    (B) Company II's, because of its greater mean
    (C) For both companies, the probability that a fuse will last at least 1 hour is 0.159.
    (D) For both companies, the probability that a fuse will last at least 1 hour is 0.841.
    (E) The problem cannot be solved from the information given.

23. Which of the following is *not* important in the design of experiments?

    (A) Control of confounding variables
    (B) Randomization in assigning subjects to different treatments
    (C) Use of a confounding variable to control the placebo effect
    (D) Replication of the experiment using sufficient numbers of subjects
    (E) All of the above are important in the design of experiments.

24. The travel miles claimed in weekly expense reports of the sales personnel at a corporation are summarized in the following boxplot.

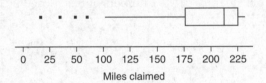

0	25	50	100	125	150	175	200	225

Miles claimed

Which of the following is the most reasonable conclusion?

    (A) The mean and median numbers of travel miles are roughly equal.
    (B) The mean number of travel miles is probably greater than the median number.
    (C) Most of the claimed numbers of travel miles are in the [0, 200] interval.
    (D) Most of the claimed numbers of travel miles are in the [200, 240] interval.
    (E) The left and right whiskers contain the same number of values from the set of personnel travel mile claims.

25. For the least squares regression line, which of the following statements about residuals is true?

    (A) Influential scores have large residuals.
    (B) If the linear model is good, the number of positive residuals will be the same as the number of negative residuals.
    (C) The mean of the residuals is always zero.
    (D) If the correlation is 0, there will be a distinct pattern in the residual plot.
    (E) If the correlation is 1, there will not be a distinct pattern in the residual plot.

26. Four pairs of data are used in determining a regression line $\hat{y} = 3x + 4$. If the four values of the independent variable are 32, 24, 29, and 27, respectively, what is the mean of the four values of the dependent variable?

(A) 68
(B) 84
(C) 88
(D) 100
(E) The mean cannot be determined from the given information.

27. According to one poll, 12% of the public favor legalizing all drugs. In a simple random sample of six people, what is the probability that at least one person favors legalization?

(A) $6(0.12)(0.88)^5$
(B) $(0.88)^6$
(C) $1 - (0.88)^6$
(D) $1 - 6(0.12)(0.88)^5$
(E) $6(0.12)(0.88)^5 + (0.88)^6$

28. Sampling error occurs

(A) when interviewers make mistakes resulting in bias.
(B) because a sample statistic is used to estimate a population parameter.
(C) when interviewers use judgment instead of random choice in picking the sample.
(D) when samples are too small.
(E) in all of the above cases.

29. A telecommunications executive instructs an associate to contact 104 customers using their service to obtain their opinions in regard to an idea for a new pricing package. The associate notes the number of customers whose names begin with *A* and uses a random number table to pick four of these names. She then proceeds to use the same procedure for each letter of the alphabet and combines the $4 \times 26 = 104$ results into a group to be contacted. Which of the following is a correct conclusion?

(A) Her procedure makes use of chance.
(B) Her procedure results in a simple random sample.
(C) Each customer has an equal probability of being included in the survey.
(D) Her procedure introduces bias through *sampling error*.
(E) With this small a sample size, it is better to let the surveyor use the company's data banks to pick representative customers to be surveyed based on features such as gender, political affiliation, income level, race, age, and so on.

30. The graph below shows cumulative proportions plotted against GPAs for high school seniors.

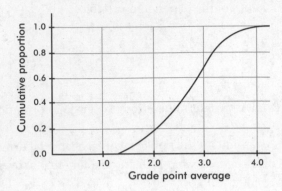

What is the approximate interquartile range?

(A) 0.85
(B) 2.25
(C) 2.7
(D) 2.75
(E) 3.1

31. PCB (polychlorinated biphenyl) contamination of a river by a manufacturer is being measured by amounts of the pollutant found in fish. A company scientist claims that the fish contain only 5 parts per million, but an investigator believes the figure is higher. The investigator catches six fish that show the following amounts of PCB (in parts per million): 6.8, 5.6, 5.2, 4.7, 6.3, and 5.4. In performing a hypothesis test with H_0: $\mu = 5$ and H_a: $\mu > 5$, what is the test statistic?

(A) $t = \dfrac{5.67 - 5}{0.763}$

(B) $t = \dfrac{5.67 - 5}{\sqrt{\dfrac{0.763}{5}}}$

(C) $t = \dfrac{5.67 - 5}{\sqrt{\dfrac{0.763}{6}}}$

(D) $t = \dfrac{5.67 - 5}{\left(\dfrac{0.763}{\sqrt{5}}\right)}$

(E) $t = \dfrac{5.67 - 5}{\left(\dfrac{0.763}{\sqrt{6}}\right)}$

32. The distribution of weights of 16-ounce bags of a particular brand of potato chips is approximately normal with a standard deviation of 0.28 ounce. How does the weight of a bag at the 40th percentile compare with the mean weight?

(A) 0.40 ounce above the mean
(B) 0.25 ounce above the mean
(C) 0.07 ounce above the mean
(D) 0.07 ounce below the mean
(E) 0.25 ounce below the mean

33. In general, how does tripling the sample size change the confidence interval width?

(A) It triples the interval width.
(B) It divides the interval width by 3.
(C) It multiples the interval width by 1.732.
(D) It divides the interval width by 1.732.
(E) This question cannot be answered without knowing the sample size

34. Which of the following statements is *false*?

(A) Like the normal distribution, the t-distributions are symmetric.
(B) The t-distributions are lower at the mean and higher at the tails and so are more spread out than the normal distribution.
(C) The greater the df, the closer the t-distributions are to the normal distribution.
(D) The smaller the df, the better the 68-95-99.7 rule works for t-models.
(E) The areas under all t-distribution curves are 1.

35. A study on school budget approval among people with different party affiliations resulted in the following segmented bar chart:

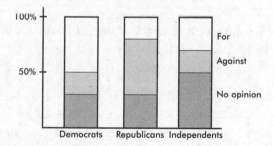

Which of the following is greatest?

(A) Number of Democrats who are for the proposed budget
(B) Number of Republicans who are against the budget
(C) Number of Independents who have no opinion on the budget
(D) The above are all equal.
(E) The answer is impossible to determine without additional information.

36. The sampling distribution of the sample mean is close to the normal distribution

    (A) only if both the original population has a normal distribution and *n* is large.

    (B) if the standard deviation of the original population is known.

    (C) if *n* is large, no matter what the distribution of the original population.

    (D) no matter what the value of *n* or what the distribution of the original population.

    (E) only if the original population is not badly skewed and does not have outliers.

37. What is the probability of a Type II error when a hypothesis test is being conducted at the 10% significance level ($\alpha = 0.10$)?

    (A) 0.05

    (B) 0.10

    (C) 0.90

    (D) 0.95

    (E) There is insufficient information to answer this question.

38.

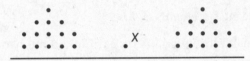

Above is the dotplot for a set of numbers. One element is labeled *X*. Which of the following is a correct statement?

    (A) *X* has the largest *z*-score, in absolute value, of any element in the set.

    (B) A boxplot will plot an outlier like *X* as an isolated point.

    (C) A stemplot will show *X* isolated from two clusters.

    (D) Because of *X*, the mean and median are different.

    (E) The IQR is exactly half the range.

39. A 2019 survey of 500 adults concluded that 82% of adults support increasing the minimum wage by $5. Which of the following best describes what is meant by the poll having a margin of error of ±3%?

    (A) Three percent of those surveyed refused to participate in the poll.

    (B) It would not be unexpected for 3% of adults to begin supporting an increase of the minimum wage by $5 or to stop supporting an increase of the minimum wage by $5.

    (C) Between 395 and 425 of the adults surveyed responded that they support increasing the minimum wage by $5.

    (D) If a similar survey of 500 adults were taken weekly, a 3% change in each week's results would not be unexpected.

    (E) It is likely that between 79% and 85% of all adults support increasing the minimum wage by $5.

40.

College Plans

	Public	Private
Taking AP Statistics:	18	27
Not taking AP Statistics:	26	40

The above two-way table summarizes the results of a survey of a random sample of high school seniors conducted to determine if there is a relationship between whether or not a student is taking AP Statistics and whether the student plans to attend a public or a private college after graduation. Which of the following is the most reasonable conclusion about the relationship between taking AP Statistics and the type of college a student plans to attend?

(A) There appears to be no association since the proportion of AP Statistics students planning to attend public schools is almost identical to the proportion of students not taking AP Statistics who plan to attend public schools.

(B) There appears to be an association since the proportion of AP Statistics students planning to attend public schools is almost identical to the proportion of students not taking AP Statistics who plan to attend public schools.

(C) There appears to be an association since more students plan to attend private than public schools.

(D) There appears to be an association since fewer students are taking AP Statistics than are not taking AP Statistics.

(E) These data do not address the question of association.

STOP

If there is still time remaining, you may check your work on this section.

SECTION II

Part A

QUESTIONS 1–5

Spend about 65 minutes on this part of the exam.
Percentage of Section II grade—75

> **Directions:** You must show all work and indicate the methods you use. You will be graded on the correctness of your methods and on the accuracy of your results and explanations.

1. Ten volunteer male subjects are to be used for an experiment to study four drugs (aloe, camphor, eucalyptus oil, and benzocaine) and a placebo with regard to itching relief. Itching is induced on a forearm with an itch stimulus (cowage), a drug is topically administered, and the duration of itching is recorded.

   (a) If the experiment is to be done in one sitting, how would you assign treatments for a completely randomized design?
   (b) If 5 days are set aside for the experiment, with one sitting a day, how would you assign treatments for a randomized block design where the subjects are the blocks?
   (c) What limits are there on generalization of any results?

2. A comprehensive study of more than 3000 baseball games results in the graph below showing relative frequencies of runs scored by home teams.

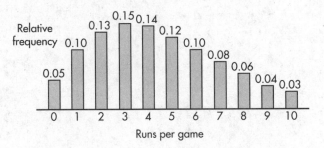

   (a) Calculate the mean and the median.
   (b) Between the mean and the median, is the one that is greater what was to be expected? Explain.
   (c) What is the probability that in 4 randomly selected games, the home team is shut out (scores no runs) in at least one of the games?
   (d) If X is the random variable for the runs scored per game by home teams, its standard deviation is 2.578. Suppose 200 games are selected at random, and $\bar{x}$, the mean number of runs scored by the home teams, is calculated. Describe the sampling distribution of $\bar{x}$.

3. A study is performed to explore the relationship, if any, between 24-hour urinary metabolite 3-methoxy-4-hydroxyphenylglycol (MHPG) levels and depression in bipolar patients. The MHPG level is measured in micrograms per 24 hours while the manic-depression (MD) scale used goes from 0 (manic delirium), through 5 (euthymic), up to 10 (depressive stupor). A partial computer printout of regression analysis with MHPG as the independent variable follows:

```
Average MHPG = 1243.1 with SD = 384.9
Average MD = 5.4 with SD = 2.875
95% confidence interval for slope b = (−0.0096, −0.0016)
```

(a) Calculate the slope of the regression line, and interpret it in the context of this problem.
(b) Find the equation of the regression line.
(c) Calculate and interpret the value of r^2 in the context of this problem.
(d) What does the correlation say about causation in the context of this problem?

4. An experiment is run to test whether daily stimulation of specific reflexes in young infants will lead to earlier walking. Twenty infants were recruited through a pediatrician's service and were randomly split into two groups of ten. One group received the daily stimulation, while the other was considered a control group. The ages (in months) at which the infants first walked alone were recorded.

Ages in months at which first steps alone were taken

With stimulation: 10, 12, 11, 10.5, 11, 11.5, 11.5, 11, 12, 11.5   Mean = 11.2, SD = 0.63246

Control: 10, 13, 12, 11, 11.5, 11.5, 12.5, 12, 11.5, 11   Mean = 11.6, SD = 0.84327

Is there statistical evidence that infants walk earlier with daily stimulation of specific reflexes?

5. A city sponsors two charity runs during the year, one 5 miles and the other 10 kilometers. Both runs attract many thousands of participants. A statistician is interested in comparing the types of runners attracted to each race. She obtains a random sample of 100 runners from each race, calculates five-number summaries of the times, and displays the results in the parallel boxplots below.

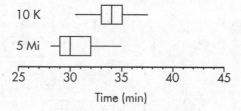

(a) Compare the distributions of times above.

The statistician notes that 5 miles equals 8 kilometers and so decides to multiply the times from the 5-mile event by $\frac{10}{8}$ and then compare box plots. This display is below.

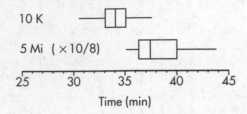

(b) How do the distributions now compare?

(c) Given that the 10-kilometer race was a longer distance than the 5-mile race, was the change from the first set of boxplots to the second set as expected? Explain.

(d) Based on the boxplots, would you expect the difference in mean times in the first set of parallel boxplots to be less than, greater than, or about the same as the difference in mean times in the second set of parallel boxplots? Explain.

SECTION II

Part B

QUESTION 6

Spend about 25 minutes on this part of the exam.
Percentage of Section II grade—25

6. A candy manufacturer sells boxes advertised to contain 30 packs of candy per box. Occasionally a packing machine malfunctions and begins putting fewer than 30 packs in a box. A quality control inspector wants to determine whether or not a machine is functioning properly without opening up boxes. The inspector will weigh a box to test the following hypotheses.

H_0: The machine is putting 30 packs of candy in each box

H_a: The machine is putting fewer than 30 packs of candy in each box

(a) Describe what a Type I error would be, and describe a consequence of making a Type I error.

(b) Describe what a Type II error would be, and describe a consequence of making a Type II error.

The distribution of weights of packs of candy is approximately normal with a mean of 2 ounces and a standard deviation of 0.05 ounce. Assume the packs in a box are a random sample of packs produced.

Let the random variable C represent the total weight of a box of packs of candy, and assume the weight of the packaging is negligible.

(c) Describe the distribution of C if the selected box contains 30 packs.

The inspector decides to use the following rule. If the total weight of a box of packs of candy is 59 ounces or less, he will conclude the box contains fewer than 30 packs, otherwise he will conclude the box contains 30 packs.

(d) Suppose a box actually contains 30 packs. What is the probability the inspector mistakenly concludes that the box contains fewer than 30 packs?

(e) Suppose a box actually contains 29 packs. What is the probability the inspector mistakenly concludes that the box contains 30 packs?

(f) Based on your answers to (d) and (e), comment of the effectiveness of the inspector's rule choice.

STOP

If there is still time remaining, you may check your work on this section.

Answer Key

SECTION I

1. **D**	9. **D**	17. **C**	25. **C**	33. **D**
2. **B**	10. **E**	18. **A**	26. **C**	34. **D**
3. **E**	11. **D**	19. **B**	27. **C**	35. **E**
4. **D**	12. **B**	20. **C**	28. **B**	36. **C**
5. **E**	13. **B**	21. **D**	29. **A**	37. **E**
6. **C**	14. **E**	22. **C**	30. **A**	38. **C**
7. **C**	15. **B**	23. **C**	31. **E**	39. **E**
8. **E**	16. **A**	24. **D**	32. **D**	40. **A**

Answers and Explanations

SECTION I

1. **(D)** Since (2, 7) is on the line $y = mx + 3$, we have $7 = 2m + 3$ and $m = 2$. Thus, the regression line is $y = 2x + 3$. The point $(\bar{x}, \bar{y})$ is always on the regression line, and so we have $\bar{y} = 2\bar{x} + 3$.

2. **(B)** It could well be that conscientious students are the same ones who both study and do well on the basketball court. If students could be randomly assigned to study or not study, the results would be more meaningful. Of course, ethical considerations might make it impossible to isolate the confounding variable in this way.

3. **(E)** `invNorm(0.7)` $= 0.5244$ so $15 - \mu = 0.5244(4.3)$, which gives $\mu = 15 - 0.5244(4.3)$.

4. **(D)** The slope of the regression line and the correlation are related by $b = r\dfrac{S_y}{S_x}$. When using z-scores, the standard deviations s_x and s_y are 1. If $r = 0$, then $b = 0$. Switching which variable is x and which is y, or changing units, will not change the correlation.

5. **(E)** The median and interquartile range are specifically used when outliers are suspected of unduly influencing the mean, range, or standard deviation.

6. **(C)** `invNorm(0.995)` $= 2.576$ is the critical z-score for a 99% confidence interval, and $\sigma_{\bar{x}} = \dfrac{\sigma}{\sqrt{n}} = \dfrac{2.4}{\sqrt{n}}$.

7. **(C)** This is a hypothesis test with H_0: tissue strength is within specifications, and H_a: tissue strength is below specifications. A Type I error is committed when a true null hypothesis is mistakenly rejected.

8. **(E)** The wording of questions can lead to response bias. The neutral way of asking this question would simply have been, "Do you support the proposed school budget increase?"

9. **(D)**
$$P(\text{water} \mid \text{hurricane}) = \frac{P(\text{water} \cap \text{hurricane})}{P(\text{hurricane})} = \frac{0.22}{0.25}$$

10. **(E)** While it is important to look for basic patterns, it is also important to look for deviations from these patterns. In this case, there is an overall positive correlation; however, those faculty with under ten years of service show little relationship between years of service and salary. While (A) is a true statement, it does not give an overall interpretation of the scatterplot.

11. **(D)** The second set has a greater range, $3.8 - 1.8 = 2.0$ as compared to $4.1 - 2.3 = 1.8$, and with its skewness it also has a greater standard deviation.

12. **(B)** With $n = 10$, increasing $\sum x$ by 40 increases $\frac{\sum x}{n}$ by $\frac{40}{10} = 4$.

13. **(B)** Means can always be added, and in this example because of independence, variances can also be added. Thus, the new variance is $5^2 + 12^2 = 169$, and the new standard deviation is $\sqrt{169} = 13$.

14. **(E)** Dice have no memory, so the probability that the next toss will be an even number is 0.5 and the probability that it will be an odd number is 0.5. The law of large numbers says that as the number of tosses becomes larger, the proportion of even numbers tends to become closer to 0.5.

15. **(B)** The sample slope is 12.633, the standard error of the slope is 0.378, and the critical t-scores for 90% confidence with $df = 7$ are $\pm\texttt{invT(0.95, 7)} = \pm 1.895$. The confidence interval is $b \pm t * SE(b) = 12.633 \pm (1.895)(0.378)$.

16. **(A)** Either directly or anonymously, you should be able to obtain the test results for *every* student taking AP Statistics.

17. **(C)** Percentile ranking is a measure of relative position. Adding 5 points to everyone's score will not change the relative positions.

18. **(A)** The control group should have experiences identical to those of the experimental groups except for the treatment under examination. They should not be given a new treatment.

19. **(B)** If the alternative $p = 0.7$ is true, the probability of failing to reject H_0 and thus committing a Type II error is 1 minus the power, that is, $1 - 0.8 = 0.2$.

20. **(C)** In histograms, relative area corresponds to relative frequency. The area from 1.5 to 4.5 (fewer than 5) appears to be the same as the area between 7.5 and 10.5 (more than 7). Five does not split the area in half, so 5 is not the median. Histograms such as these show relative frequencies, not actual frequencies. Given the spread, 1 is too small an estimate of the standard deviation. The area above 3 looks to be the same as the area above 9 and 10, so the median won't change.

21. **(D)** There must be a fixed number of trials, which rules out (A); only two possible outcomes, which rules out (B); and a constant probability of success on any trial, which rules out (C).

22. **(C)** In both cases, 1 hour is one standard deviation from the mean with a right tail probability of 0.159.

23. **(C)** Control, randomization, and replication are all important aspects of well-designed experiments. We try to control confounding variables, not to use them to control something else.

24. **(D)** The median appears to be roughly 215, indicating that the interval [200, 240] probably has more than 50% of the values. While the shape of a distribution is difficult to discern from a boxplot, the data do appear to be skewed to the left, indicating that the mean is less than the median. While in a boxplot without outliers each whisker contains 25% of the values, this boxplot shows four outliers on the left, and so the left whisker has four fewer values than the right whisker.

25. **(C)** For the least squares regression line, the sum and thus the mean of the residuals are always zero. An influential score may have a small residual but still have a great effect on the regression line. If the correlation is 1, all the residuals would be 0, resulting in a very distinct pattern.

26. **(C)** $\bar{x} = \dfrac{32 + 24 + 29 + 27}{4} = 28$. Since $(\bar{x}, \bar{y})$ is a point on the regression line, $\bar{y} = 3(28) + 4 = 88$.

27. **(C)** $P(\text{at least } 1) = 1 - P(\text{none}) = 1 - (0.88)^6$.

28. **(B)** Different samples give different sample statistics, all of which are estimates for the same population parameter, and so error, called *sampling error* (also called *sampling variability*), is naturally present.

29. **(A)** While the associate does use chance, each customer would have the same chance of being selected only if the same number of customers had names starting with each letter of the alphabet. This selection does not result in a simple random sample because each possible set of 104 customers does not have the same chance of being picked as part of the sample. For example, a group of customers whose names all start with A will not be chosen. Sampling error (sampling variability), the natural variation inherent in a survey, is always present and is not a source of bias. Letting the surveyor have free choice in selecting the sample, rather than incorporating chance in the selection process, is a recipe for disaster!

30. **(A)** Corresponding to cumulative proportions of 0.25 and 0.75 are $Q_1 = 2.25$ and $Q_3 = 3.1$, respectively, and so the interquartile range is $3.1 - 2.25 = 0.85$.

31. **(E)** By putting the data in a list, a calculator gives $\bar{x} = 5.67$ and $s = 0.763$. Then $SE(\bar{x}) = \dfrac{s}{\sqrt{n}} = \dfrac{0.763}{\sqrt{6}}$ and the test statistic is $t = \dfrac{\bar{x} - x_0}{SE(\bar{x})} = \dfrac{5.67 - 5}{\left(\dfrac{0.763}{\sqrt{6}}\right)}$.

32. **(D)** The 40th percentile corresponds to a z-score of `invNorm(0.40)` $= -0.2533$, and $-0.2533(0.28) = -0.0709$.

33. **(D)** Increasing the sample size by a multiple of d divides the interval width by $\sqrt{d}$.

34. **(D)** The t-distributions are symmetric; however, they are lower at the mean and higher at the tails and so are more spread out than the normal distribution. The greater the df, the closer the t-distributions are to the normal distribution. The 68-95-99.7 rule applies to the z-distribution and will work for t-models with very large df. All probability density curves have an area of 1 below them.

35. **(E)** The given bar chart shows percentages, not actual numbers.

36. **(C)** This follows from the central limit theorem.

37. **(E)** There is a different Type II error for each possible correct value for the population parameter.

38. **(C)** X and the two clusters will be clearly visible in a stemplot of these data. X is close to the mean and so will have a z-score close to 0. While boxplots do show outliers as isolated points, outliers are from the mean, so X is not an outlier. In symmetric distributions, the mean and median are equal. The IQR here is close to the range.

39. **(E)** When using a measurement from a sample, we are never able to say *exactly* what a population proportion is; rather, we always say we have a certain *confidence* that the population proportion lies in a particular *interval*. In this case, that interval is 82% ± 3% or between 79% and 85%.

40. **(A)** Whether or not students are taking AP Statistics seems to have no relationship to which type of school they are planning to go to. Chi-square is close to 0.

SECTION II: PART A

1. (a) Number the volunteers 1 through 10. Use a random number generator to pick numbers between 1 and 10, throwing out repeats. The volunteers corresponding to the first two numbers chosen will receive aloe, the next two will receive camphor, the next two eucalyptus oil, the next two benzocaine, and the remaining two a placebo.

   (b) Each volunteer (the volunteers are "blocks") should receive all five treatments, one a day, with the time order randomized. For example, label aloe 1, camphor 2, eucalyptus oil 3, benzocaine 4, and the placebo 5. Then, for each volunteer, use a random number generator to pick numbers between 1 and 5, throwing away repeats. The order picked gives the day on which each volunteer receives each treatment.

   (c) Results cannot be generalized to women.

Part (a) is essentially correct for giving a procedure that randomly assigns two volunteers to each of the five treatments. Part (a) is partially correct for giving a procedure that randomly assigns one of the treatments for each volunteer but may not result in *two* volunteers receiving each treatment.

Part (b) is essentially correct for giving a procedure assigns a random order for each of the volunteers to have all five treatments. Part (b) is partially correct for having each volunteer take all five treatments, one a day, but not clearly randomizing the time order.

Part (c) is essentially correct for stating that the results cannot be generalized to women and is incorrect otherwise.

4 Complete Answer   All three parts essentially correct.

3 Substantial Answer   Two parts essentially correct and one part partially correct.

2 Developing Answer   Two parts essentially correct OR one part essentially correct and one or two parts partially correct OR all three parts partially correct.

1 Minimal Answer   One part essentially correct OR two parts partially correct.

2. (a) $\bar{x} = \sum xp(x) = 0(0.05) + 1(0.10) + 2(0.13) + 3(0.15) + 4(0.14) + 5(0.12) + 6(0.10) + 7(0.08) + 8(0.06) + 9(0.04) + 10(0.03) = 4.27.$

With $(0.05 + 0.10 + 0.13 + 0.15) = 0.43$ below 4 runs and with $(0.12 + 0.10 + 0.08 + 0.06 + 0.04 + 0.03) = 0.43$ above 4 runs, the median must be 4.

(b) The mean is greater than the median, as was to be expected because the distribution is skewed to the right.

(c) This is a binomial with $n = 4$ and $p = 0.05$.

$P(\text{at least one shutout in 4 games}) = 1 - P(\text{no shutouts in the 4 games})$

$$= 1 - (1 - 0.05)^4 = 1 - (0.95)^4 = 0.1855$$

(d) The distribution of $\bar{x}$ is approximately normal with mean $\mu_{\bar{x}} = 4.27$ (from above) and standard deviation $\sigma_{\bar{x}} = \dfrac{2.578}{\sqrt{200}} = 0.1823.$

Section 1 is essentially correct for correctly calculating the mean, correctly calculating the median, noting that the mean is greater than the median, and relating this to the skew. Section 1 is partially correct for two or three of these four elements.

Section 2 in Part (c) is essentially correct for recognizing this as a binomial probability calculation and making the correct calculation. Section 2 is partially correct for recognizing this as a binomial probability calculation but with an error such as $1 - (0.05)^4$ or $4(0.05)(0.95)^3$.

Section 3 in Part (d) is essentially correct for "approximately normal," $\mu_{\bar{x}} = 4.27$ and $\sigma_{\bar{x}} = 0.1823$. Section 3 is partially correct for two of these three answers.

4 Complete Answer   All three sections essentially correct.

3 Substantial Answer   Two sections essentially correct and one section partially correct.

2 Developing Answer   Two sections essentially correct OR one section essentially correct and one or two sections partially correct OR all three sections partially correct.

1 Minimal Answer   One section essentially correct OR two sections partially correct.

3. (a) The slope b is in the center of the confidence interval, so $b = \dfrac{-0.0096 - 0.0016}{2} =$ -0.0056. In context, 0.0056 *estimates* the *average* decrease in the manic-depression scale score for each 1-microgram increase in the level of urinary MHPG. (Thus, high levels of MHPG are associated with increased mania, and conversely, low levels of MHPG are associated with increased depression.)

   (b) Recalling that the regression line goes through the point $(\bar{x}, \bar{y})$ or using the AP exam formula page, $a = \bar{y} - b\bar{x} = 5.4 - (-0.0056)(1243.1) = 12.36$, and thus the equation of the regression line is $\widehat{MD} = 12.36 - 0.0056(MHPG)$ (or $\hat{y} = 12.36 - 0.0056x$, where x is the level of urinary MHPG in micrograms per 24 hours and y is the score on a 0–10 manic-depressive scale).

   (c) Recalling that on the regression line, each one SD increase in the independent variable corresponds to an increase of r SD in the dependent variable (or from the AP exam formula page, $b = r\dfrac{S_y}{S_x}$),

$$r = b\frac{S_x}{S_y} = -0.0056\left(\frac{384.9}{2.875}\right) = -0.750$$

   and $r^2 = 56.2\%$. Thus, 56.2% of the variation in the manic-depression scale level is accounted for by the linear regression model with urinary MHPG levels as the explanatory variable.

   (d) Correlation never proves causation. It could be that depression causes biochemical changes leading to low levels of urinary MHPG, or it could be that low levels of urinary MHPG cause depression, or it could be that some other variable (a confounding variable) simultaneously affects both urinary MHPG levels and depression.

Part (a) is essentially correct if the slope is correctly calculated and correctly interpreted in context. Part (a) is partially correct if the slope is not correctly calculated but a correct interpretation is given using the incorrect value for the slope.

Part (b) is essentially correct if the regression equation is correctly calculated (using the slope found in Part (a)) and it is clear what the variables stand for. Part (b) is partially correct if the correct equation is found (using the slope found in Part (a)) but it is unclear what the variables stand for.

Part (c) is essentially correct if the coefficient of determination, r^2, is correctly calculated and correctly interpreted in context. Part (c) is partially correct if r^2 is not correctly calculated but a correct interpretation is given using the incorrect value for r^2.

Part (d) is essentially correct for noting that correlation never proves causation and referring to context. Part (d) is partially correct for a correct statement about correlation and causation but with no reference to context.

Count partially correct answers as one-half an essentially correct answer.

4   Complete Answer   Four essentially correct answers.

3   Substantial Answer   Three essentially correct answers.

2   Developing Answer   Two essentially correct answers.

1   Minimal Answer   One essentially correct answer.

Use a holistic approach to decide a score totaling between two numbers.

4. There are three parts to this solution.

   (a) *Hypotheses:* $H_0: \mu_1 - \mu_2 = 0$ and $H_a: \mu_1 - \mu_2 < 0$

   where μ_1 = mean number of months at which first steps alone are taken for infants receiving daily stimulation, and μ_2 = mean number of months at which first steps are taken for infants not receiving daily simulation. (Other notations are also possible, for example, $H_0: \mu_1 = \mu_2$ and $H_a: \mu_1 < \mu_2$ or $H_0: \mu_{\text{stimulation}} = \mu_{\text{control}}$ and $H_a: \mu_{\text{stimulation}} < \mu_{\text{control}}$.)

   (b) *Procedure:* Two-sample t-test.

   Checks: We are given that the infants were randomly assigned to the two groups, the two groups are independent, and dotplots of the two groups show no outliers and are roughly bell-shaped.

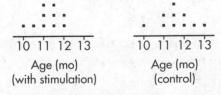

   (c) *Mechanics* and *conclusion in context with linkage to the P-value:* Calculator software gives $t = -1.20$ and $P = 0.1234$. With this large a P-value, $0.1234 > 0.05$, there is not sufficient evidence to reject H_0; that is, there is not sufficient evidence that infants walk earlier with daily stimulation of specific reflexes.

Part (a) is essentially correct for stating the hypotheses and identifying the variables. Part (a) is partially correct for correct hypotheses but missing identification of the variables.

Part (b) is essentially correct for identifying the test and checking the assumptions. Part (b) is partially correct for only one of these two elements.

Part (c) is essentially correct for correctly stating t and P, and a correct conclusion, in context, linked to the P-value. Part (c) is partially correct for everything correct except for an incorrect t-statistic or P-value OR for missing either context or linkage.

4   **Complete Answer**   All three parts essentially correct.

3   **Substantial Answer**   Two parts essentially correct and one part partially correct.

2   **Developing Answer**   Two parts essentially correct OR one part essentially correct and one or two parts partially correct OR all three parts partially correct.

1   **Minimal Answer**   One part essentially correct OR two parts partially correct.

5. (a) A complete answer compares shape, center, and spread and mentions context.

   Shape: Although shape usually cannot be deduced from a boxplot, the 10-kilometer distribution appears very roughly symmetric, while the 5-mile distribution appears skewed right.

   Center: The median of the 10-kilometer distribution is 4 minutes greater than the median of the 5-mile distribution.

   Spread: The ranges of the two distributions are equal, both about 7 minutes.

   (b) *Shape:* Again, the 10-kilometer distribution appears very roughly symmetric, while the 5-mile distribution still appears skewed right.

   Center: Now the median of the 10-kilometer distribution is about 3.5 minutes less than the median of the 5-mile distribution.

   Spread: Now the range of the 10-kilometer distribution is less than the range of the 5-mile distribution.

   (c) One possible answer is that the change from the first set of boxplots to the second was not as expected with the following explanation. One would expect that for 5 miles, the shorter run, the speeds would be faster, so adjusting the 5-mile run times would result in faster times (less minutes) than the 10-kilometer times, but this was not the case. (Perhaps slower, less serious runners participate in the shorter race. So, when their times are adjusted, the minutes are greater than the times for the 10-kilometer run.)

   (d) In each set of parallel boxplots, the 10-kilometer distribution appears roughly symmetric so that the mean will be about the same as the median, while the 5-mile distribution appears to be skewed right so that the mean will in all likelihood be greater than the median. In the first set of boxplots (where the 5-mile median < 10-kilometer median) this will result in the means being closer, while in the second set of boxplots (where the 5-mile median > 10-kilometer median) this will result in the means being further apart. Thus, we would expect the difference in mean times in the first set of parallel boxplots to be less than the difference in mean times in the second set of parallel boxplots.

Section 1 is essentially correct for correctly comparing shape, center, and spread and mentioning context in both Parts (a) and (b). Section 1 is partially correct for correctly comparing two of the three features in both (a) and (b) OR for failing to mention context.

Section 2 is essentially correct for a reasonable statement about the change between the two sets of parallel boxplots together with a correct explanation to go along with the statement. Section 2 is partially correct if the explanation is weak.

Section 3 is essentially correct for a correct prediction about the means together with a reasonable justification based on symmetry of the 10-kilometer distribution and skewness of the 5-mile distribution. Section 3 is partially correct if the justification is weak.

4	**Complete Answer**	All three sections essentially correct.
3	**Substantial Answer**	Two sections essentially correct and one section partially correct.
2	**Developing Answer**	Two sections essentially correct OR one section essentially correct and one or two sections partially correct OR all three sections partially correct.
1	**Minimal Answer**	One section essentially correct OR two sections partially correct.

SECTION II: PART B

6. (a) A Type I error will be committed if the machine is correctly putting 30 packs of candy into each box; however, based on the sample, the inspector concludes that the machine is putting fewer than 30 packs of candy into each box. A consequence could be that the production process is stopped to fix the machine when, in reality, nothing is wrong with the machine.

   (b) A Type II error will be committed if the machine is actually putting fewer than 30 packs of candy into each box; however, based on the sample, the inspector concludes that the machine is functioning properly and is putting 30 packs of candy into each box. A consequence could be that the machine is not fixed even though it is malfunctioning, so it will continue to put fewer than 30 packs into each box.

   (c) The sampling distribution of C is approximately normal with mean $30 \times 2 = 60$ and standard deviation $\sqrt{30} \times 0.05 = 0.27386$.

   (d) $P(C < 59) = P\left(z < \dfrac{59 - 60}{0.27386}\right) = 0.000130$. The probability that the inspector will mistakenly conclude that the box contains fewer than 30 packs when the box really does contain 30 packs is 0.000130.

   (e) Now we have a roughly normal distribution with $\mu = 29 \times 2 = 58$ and $\sigma = \sqrt{29} \times 0.05 = 0.26926$. $P(C > 59) = P\left(z > \dfrac{59 - 58}{0.26926}\right) = 0.000102$. The probability that the inspector will mistakenly conclude that the box contains 30 packs when the box really does contain 29 packs is 0.000102.

(f) The inspector's rule choice is very effective. The probability he commits a Type I error and mistakenly concludes that the machine is putting fewer than 30 packs of candy into each box when it is actually putting the correct 30 packs into each box is only 0.000130. The probability he commits a Type II error and mistakenly concludes that the machine is putting 30 packs of candy into each box when it is actually putting 29 packs into each box is only 0.000102.

SCORING

Section 1 is essentially correct for a correct description of a Type I error, a correct consequence of a Type I error, a correct description of a Type II error, and a correct consequence of a Type II error. Section 1 is partially correct for two or three correct out of the four steps above.

Section 2 is essentially correct for a correct description of the shape of the sampling distribution as approximately normal, a correct calculation of the mean of the sampling distribution, and a correct calculation of the standard deviation of the sampling distribution all in (c), and a correct calculation of the probability in (d). Section 2 is partially correct for three out of these four steps correct.

Section 3 is essentially correct for a correct description of the shape of the sampling distribution as approximately normal, a correct calculation of the mean of the sampling distribution, a correct calculation of the standard deviation of the sampling distribution, and a correct calculation of the probability all in (e). Section 3 is partially correct for two or three out of these four steps correct.

Section 4 is essentially correct for the correction conclusion about the effectiveness of the inspector's rule choice with clear justification linked to both (d) and (e). Section 4 is partially correct if the justification is weak.

Count essentially correct answers as one point and partially correct answers as one-half point.

4   **Complete Answer**   Four points

3   **Substantial Answer**   Three points

2   **Developing Answer**   Two points

1   **Minimal Answer**   One point

Use a holistic approach to decide a score totaling between two numbers, deciding whether to score up or down depending on the strength of the response and communication.

ANSWER SHEET
Practice Test 3

1. Ⓐ Ⓑ Ⓒ Ⓓ Ⓔ
2. Ⓐ Ⓑ Ⓒ Ⓓ Ⓔ
3. Ⓐ Ⓑ Ⓒ Ⓓ Ⓔ
4. Ⓐ Ⓑ Ⓒ Ⓓ Ⓔ
5. Ⓐ Ⓑ Ⓒ Ⓓ Ⓔ
6. Ⓐ Ⓑ Ⓒ Ⓓ Ⓔ
7. Ⓐ Ⓑ Ⓒ Ⓓ Ⓔ
8. Ⓐ Ⓑ Ⓒ Ⓓ Ⓔ
9. Ⓐ Ⓑ Ⓒ Ⓓ Ⓔ
10. Ⓐ Ⓑ Ⓒ Ⓓ Ⓔ

11. Ⓐ Ⓑ Ⓒ Ⓓ Ⓔ
12. Ⓐ Ⓑ Ⓒ Ⓓ Ⓔ
13. Ⓐ Ⓑ Ⓒ Ⓓ Ⓔ
14. Ⓐ Ⓑ Ⓒ Ⓓ Ⓔ
15. Ⓐ Ⓑ Ⓒ Ⓓ Ⓔ
16. Ⓐ Ⓑ Ⓒ Ⓓ Ⓔ
17. Ⓐ Ⓑ Ⓒ Ⓓ Ⓔ
18. Ⓐ Ⓑ Ⓒ Ⓓ Ⓔ
19. Ⓐ Ⓑ Ⓒ Ⓓ Ⓔ
20. Ⓐ Ⓑ Ⓒ Ⓓ Ⓔ

21. Ⓐ Ⓑ Ⓒ Ⓓ Ⓔ
22. Ⓐ Ⓑ Ⓒ Ⓓ Ⓔ
23. Ⓐ Ⓑ Ⓒ Ⓓ Ⓔ
24. Ⓐ Ⓑ Ⓒ Ⓓ Ⓔ
25. Ⓐ Ⓑ Ⓒ Ⓓ Ⓔ
26. Ⓐ Ⓑ Ⓒ Ⓓ Ⓔ
27. Ⓐ Ⓑ Ⓒ Ⓓ Ⓔ
28. Ⓐ Ⓑ Ⓒ Ⓓ Ⓔ
29. Ⓐ Ⓑ Ⓒ Ⓓ Ⓔ
30. Ⓐ Ⓑ Ⓒ Ⓓ Ⓔ

31. Ⓐ Ⓑ Ⓒ Ⓓ Ⓔ
32. Ⓐ Ⓑ Ⓒ Ⓓ Ⓔ
33. Ⓐ Ⓑ Ⓒ Ⓓ Ⓔ
34. Ⓐ Ⓑ Ⓒ Ⓓ Ⓔ
35. Ⓐ Ⓑ Ⓒ Ⓓ Ⓔ
36. Ⓐ Ⓑ Ⓒ Ⓓ Ⓔ
37. Ⓐ Ⓑ Ⓒ Ⓓ Ⓔ
38. Ⓐ Ⓑ Ⓒ Ⓓ Ⓔ
39. Ⓐ Ⓑ Ⓒ Ⓓ Ⓔ
40. Ⓐ Ⓑ Ⓒ Ⓓ Ⓔ

Practice Test 3

SECTION I

Questions 1–40

Spend 90 minutes on this part of the exam.

> **Directions:** The questions or incomplete statements that follow are each followed by five suggested answers or completions. Choose the response that best answers the question or completes the statement.

1. Which of the following is a true statement?

   (A) While properly designed experiments can strongly suggest cause-and-effect relationships, a complete census is the only way of establishing such a relationship.

   (B) If properly designed, observational studies can establish cause-and-effect relationships just as strongly as can properly designed experiments.

   (C) Controlled experiments are often undertaken later to establish cause-and-effect relationships first suggested by observational studies.

   (D) A useful approach to overcome bias in observational studies is to increase the sample size.

   (E) In an experiment, the control group is a self-selected group who choose not to receive a designated treatment.

2. Two classes take the same exam. Suppose a certain score is at the 40th percentile for the first class and at the 80th percentile for the second class. Which of the following is the most reasonable conclusion?

   (A) Students in the first class generally scored higher than students in the second class.

   (B) Students in the second class generally scored higher than students in the first class.

   (C) A score at the 20th percentile for the first class is at the 40th percentile for the second class.

   (D) A score at the 50th percentile for the first class is at the 90th percentile for the second class.

   (E) One of the classes has twice the number of students as the other.

3. In an experiment, the control group should receive

   (A) treatment opposite that given to the experimental group.
   (B) the same treatment given to the experimental group without knowing they are receiving the treatment.
   (C) a procedure identical to that given to the experimental group except for receiving the treatment under examination.
   (D) a procedure identical to that given to the experimental group except for a random decision on receiving the treatment under examination.
   (E) none of the procedures given to the experimental group.

4. In a simple random sample (SRS) of 625 families who do not live near any chemical plant, 10 had children with leukemia. In an SRS of 412 families living near chemical plants, 15 had children with leukemia. A 90% confidence interval of the difference in proportions is reported to be -0.0204 ± 0.0173. What is a proper conclusion?

   (A) The interval is invalid because probabilities cannot be negative.
   (B) The interval is invalid because it does not contain zero.
   (C) Families living near chemical plants are approximately 2.04% more likely to have children with leukemia than families who do not live near any chemical plant.
   (D) Ninety percent of families living near chemical plants are approximately 2.04% more likely to have children with leukemia than families who do not live near any chemical plant.
   (E) We are 90% confident that the difference in proportions between families who do not live near any chemical plant having children with leukemia and families who live near chemical plants having children with leukemia is between -0.0377 and -0.0031.

5. In a study on the effect of music on worker productivity, employees were told that a different genre of background music would be played each day and the corresponding production outputs noted. Every change in music resulted in an increase in production. This is an example of

   (A) the effect of a treatment unit.
   (B) the placebo effect.
   (C) the control group effect.
   (D) sampling error.
   (E) voluntary response bias.

6. A computer manufacturer sets up three locations to provide technical support for its customers. Logs are kept noting whether or not calls about problems are solved successfully. Data from a sample of 1000 calls are summarized in the following table:

| | Location | | | |
	1	2	3	Total
Problem solved	325	225	150	700
Problem not solved	125	100	75	300
Total	450	325	225	1000

Assuming there is no association between location and whether or not a problem is resolved successfully, what is the expected number of successful calls (problem solved) from location 1?

(A) $\dfrac{(325)(450)}{700}$

(B) $\dfrac{(325)(700)}{450}$

(C) $\dfrac{(325)(450)}{1000}$

(D) $\dfrac{(325)(700)}{1000}$

(E) $\dfrac{(450)(700)}{1000}$

7. Which of the following statements about the correlation coefficient is true?

(A) The correlation coefficient and the slope of the regression line may have opposite signs.
(B) A correlation of 1 indicates a perfect cause-and-effect relationship between the variables.
(C) Correlations of $+0.87$ and -0.87 indicate the same degree of clustering around the regression line.
(D) Correlation applies equally well to quantitative and categorical data.
(E) A correlation of 0 shows little or no association between two variables.

8. Suppose X and Y are random variables with $E(X) = 780$, $\text{var}(X) = 75$, $E(Y) = 430$, and $\text{var}(Y) = 25$. Given that X and Y are independent, what is the variance of the random variable $X - Y$?

(A) $75 - 25$
(B) $75 + 25$
(C) $\sqrt{75 - 25}$
(D) $\sqrt{75 + 25}$
(E) $\sqrt{75} - \sqrt{25}$

9. What is a sample?

(A) A measurable characteristic of a population

(B) A set of individuals having a characteristic in common

(C) A value calculated from raw data

(D) A subset of a population

(E) None of the above

10. A histogram of the cholesterol levels of all employees at a large law firm is as follows:

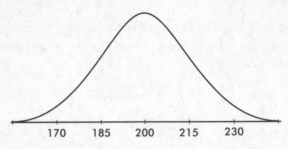

Which of the following is the best estimate of the standard deviation of this distribution?

(A) $\dfrac{230 - 170}{6} = 10$

(B) 15

(C) $\dfrac{200}{6} = 33.3$

(D) $230 - 170 = 60$

(E) $245 - 155 = 90$

11. A soft drink dispenser can be adjusted to deliver any fixed number of ounces. If the machine is operating with a standard deviation in delivery equal to 0.3 ounce, what should be the mean setting so that a 12-ounce cup will overflow less than 1% of the time? Assume an approximately normal distribution for ounces delivered.

(A) $12 - 0.99(0.3)$ ounces

(B) $12 - 2.326(0.3)$ ounces

(C) $12 - 2.576(0.3)$ ounces

(D) $12 + 2.326(0.3)$ ounces

(E) $12 + 2.576(0.3)$ ounces

12. An insurance company wishes to study the number of years drivers in a large city go between automobile accidents. They plan to obtain and analyze the data from a sample of drivers. Which of the following is a true statement?

(A) A reasonable time-and-cost-saving procedure would be to use systematic sampling on an available list of all AAA (Automobile Association of America) members in the city.

(B) A reasonable time-and-cost-saving procedure would be to randomly choose families and include all drivers in each of these families in the sample.

(C) To determine the mean number of years between accidents, randomness in choosing a sample of drivers is not important as long as the sample size is very large.

(D) The larger a simple random sample, the more likely its standard deviation will be close to the population standard deviation divided by the square root of the sample size.

(E) None of the above are true statements.

13. The probability that a person will show a certain gene-transmitted trait is 0.8 if the father shows the trait and 0.06 if the father doesn't show the trait. Suppose that the children in a certain community come from families in 25% of which the father shows the trait. Given that a child shows the trait, what is the probability that her father shows the trait?

(A) $(0.25)(0.8)$

(B) $(0.25)(0.8) + (0.75)(0.06)$

(C) $\dfrac{0.25}{(0.25)(0.8) + (0.75)(0.06)}$

(D) $\dfrac{0.8}{(0.25)(0.8) + (0.75)(0.06)}$

(E) $\dfrac{(0.25)(0.8)}{(0.25)(0.8) + (0.75)(0.06)}$

14. Given an experiment with $H_0: \mu = 10$, $H_a: \mu > 10$, and a possible correct value of 11, which of the following increases as n increases?

   I. The probability of a Type I error
   II. The probability of a Type II error
   III. The power of the test

(A) I only
(B) II only
(C) III only
(D) II and III
(E) None will increase.

15. If all the values of a data set are the same, all of the following must equal zero except for which one?

   (A) Mean
   (B) Standard deviation
   (C) Variance
   (D) Range
   (E) Interquartile range

16. Peer relationships are an important part of the socialization process of children and require the use of many complex social skills. Substantial evidence suggests that children with poor peer relationships are more likely to be truant, repeat grade levels, and drop out of school. In a random sample of 15 fifth-grade students, peer acceptance as measured on a sociometric scale showed a mean of 2.72 with a standard deviation of 0.15. Which of the following gives a 90% confidence interval for the mean peer acceptance of all fifth-grade students?

   (A) $2.72 \pm 1.645 \dfrac{0.15}{\sqrt{14}}$

   (B) $2.72 \pm 1.753 \dfrac{0.15}{\sqrt{14}}$

   (C) $2.72 \pm 1.761 \dfrac{0.15}{\sqrt{14}}$

   (D) $2.72 \pm 1.753 \dfrac{0.15}{\sqrt{15}}$

   (E) $2.72 \pm 1.761 \dfrac{0.15}{\sqrt{15}}$

17. A company has 1000 employees evenly distributed throughout five assembly plants. A sample of 30 employees is to be chosen as follows. Each of the five managers will be asked to place the 200 time cards of their respective employees into a bag, shake them up, and randomly draw out six names. The six names from each plant will be put together to make up the sample. Will this method result in a simple random sample of the 1000 employees?

   (A) Yes, because every employee has the same chance of being selected.
   (B) Yes, because every plant is equally represented.
   (C) Yes, because this is an example of stratified sampling, which is a special case of simple random sampling.
   (D) No, because the plants are not chosen randomly.
   (E) No, because not every group of 30 employees has the same chance of being selected.

18. Given that $P(E) = 0.32$, $P(F) = 0.15$, and $P(E \cap F) = 0.048$, which of the following is a correct conclusion?

(A) The events E and F are both independent and mutually exclusive.
(B) The events E and F are neither independent nor mutually exclusive.
(C) The events E and F are mutually exclusive but not independent.
(D) The events E and F are independent but not mutually exclusive.
(E) The events E and F are independent, but there is insufficient information to determine whether or not they are mutually exclusive.

19. The number of leasable square feet of office space available in a city on any given day has a roughly normal distribution with mean 640,000 square feet and standard deviation 18,000 square feet. What is the interquartile range for this distribution?

(A) 652,000 − 628,000
(B) 658,000 − 622,000
(C) 667,000 − 613,000
(D) 676,000 − 604,000
(E) 694,000 − 586,000

20. Consider the following back-to-back stemplot:

Left	Stem	Right
	0	348
	1	01256
843	2	29
65210	3	2557
92	4	
7552	5	6
	6	1458
6	7	09
8541	8	
90	9	

Key: 2|5|6 represents a value of 52 on the left and 56 on the right

Which of the following is a correct statement?

(A) The distributions have the same mean.
(B) The distributions have the same median.
(C) The interquartile range of the distribution to the left is 20 greater than the interquartile range of the distribution to the right.
(D) The distributions have the same variance.
(E) None of the above is correct.

21. Which of the following is a correct statement?

    (A) A study results in a 99% confidence interval estimate of (34.2, 67.3). This means that in about 99% of all samples selected by this method, the sample means will fall between 34.2 and 67.3.
    (B) A high confidence level may be obtained no matter what the sample size.
    (C) The central limit theorem is most useful when drawing samples from normally distributed populations.
    (D) The sampling distribution for a mean has standard deviation $\frac{\sigma}{\sqrt{n}}$ only when n is sufficiently large (typically one uses $n \geq 30$).
    (E) The center of any confidence interval is the population parameter.

22. The binomial distribution is an appropriate model for which of the following?

    (A) The number of minutes in an hour for which the Dow Jones average is above its beginning average for the day
    (B) The number of cities among the 10 largest in New York State for which the weather is cloudy for most of a given day
    (C) The number of drivers wearing seat belts if 10 consecutive drivers are stopped at a police roadblock
    (D) The number of A's a student receives in five college classes
    (E) None of the above

23. Suppose two events, E and F, have nonzero probabilities p and q, respectively. Which of the following is impossible?

    (A) $p + q > 1$
    (B) $p - q < 0$
    (C) $\frac{p}{q} > 1$
    (D) E and F are neither independent nor mutually exclusive.
    (E) E and F are both independent and mutually exclusive.

24. An inspection procedure at a manufacturing plant involves picking four items at random and accepting the whole lot if at least three of the four items are in perfect condition. If in reality 90% of the whole lot are perfect, what is the probability that the lot will be accepted?

    (A) $(0.9)^4$
    (B) $1 - (0.9)^4$
    (C) $4(0.9)^3(0.1)$
    (D) $0.1 - 4(0.9)^3(0.1)$
    (E) $4(0.9)^3(0.1) + (0.9)^4$

25. A town has one high school, which buses students from urban, suburban, and rural communities. Which of the following samples is recommended in studying attitudes toward tracking of students in honors, regular, and below-grade classes?

(A) Convenience sample
(B) Simple random sample
(C) Stratified sample
(D) Systematic sample
(E) Voluntary response sample

26. Suppose there is a correlation of $r = 0.9$ between number of hours per day students study and GPAs. Which of the following is a reasonable conclusion?

(A) 90% of students who study receive high grades.
(B) 90% of students who receive high grades study a lot.
(C) 90% of the variation in GPAs can be explained by variation in number of study hours per day.
(D) 10% of the variation in GPAs cannot be explained by variation in number of study hours per day.
(E) 81% of the variation in GPAs can be explained by variation in number of study hours per day.

27. To determine the average number of children living in single-family homes, a researcher picks a simple random sample of 50 such homes. However, even after one follow-up visit, the interviewer is unable to make contact with anyone in 8 of these homes. Concerned about nonresponse bias, the researcher picks another simple random sample and instructs the interviewer to keep trying until contact is made with someone in 8 more homes for a total of 50 homes. The average number of children is determined to be 1.73. Is this estimate probably too low or too high?

(A) Too low, because of undercoverage bias
(B) Too low, because convenience samples overestimate average results
(C) Too high, because of undercoverage bias
(D) Too high, because convenience samples overestimate average results
(E) Too high, because voluntary response samples overestimate average results

28. The graph below shows cumulative proportions plotted against land values (in dollars per acre) for farms on sale in a rural community.

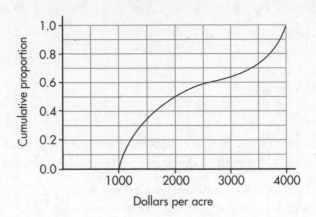

What is the median land value?

(A) $2000
(B) $2250
(C) $2500
(D) $2750
(E) $3000

29. An experiment is to be conducted to determine whether taking fish oil capsules or garlic capsules has more of an effect on cholesterol levels. In past studies, it was noted that daily exercise intensity (low, moderate, high) is associated with cholesterol level, but average sleep length (< 5, $5-8$, > 8 hours) is not associated with cholesterol level. This experiment should be done

(A) by blocking on exercise intensity.
(B) by blocking on sleep length.
(C) by blocking on cholesterol level.
(D) by blocking on capsule type.
(E) without blocking.

30. A confidence interval estimate is determined from the monthly grocery expenditures in a random sample of n families. Which of the following will result in a smaller margin of error?

   I. A smaller confidence level
  II. A smaller sample standard deviation
 III. A smaller sample size

(A) II only
(B) I and II only
(C) I and III only
(D) II and III only
(E) I, II, and III

31. A medical research team claims that high vitamin C intake increases endurance. In particular, 1000 milligrams of vitamin C per day for a month should add an average of 4.3 minutes to the length of maximum physical effort that can be tolerated. Army training officers believe the claim is exaggerated and plan a test on a simple random sample of 400 soldiers in which they will reject the medical team's claim if the sample mean is less than 4.0 minutes. Suppose the standard deviation of added minutes is 3.2. If the true mean increase is only 4.2 minutes, what is the probability that the officers will fail to reject the false claim of 4.3 minutes?

(A) $P\left(z < \dfrac{4.0 - 4.3}{\left(\dfrac{3.2}{\sqrt{400}}\right)}\right)$

(B) $P\left(z > \dfrac{4.0 - 4.3}{\left(\dfrac{3.2}{\sqrt{400}}\right)}\right)$

(C) $P\left(z < \dfrac{4.3 - 4.2}{\left(\dfrac{3.2}{\sqrt{400}}\right)}\right)$

(D) $P\left(z > \dfrac{4.3 - 4.2}{\left(\dfrac{3.2}{\sqrt{400}}\right)}\right)$

(E) $P\left(z > \dfrac{4.0 - 4.2}{\left(\dfrac{3.2}{\sqrt{400}}\right)}\right)$

32. Consider the two sets $X = \{10, 30, 45, 50, 55, 70, 90\}$ and $Y = \{10, 30, 35, 50, 65, 70, 90\}$. Which of the following is false?

(A) The sets have identical medians.
(B) The sets have identical means.
(C) The sets have identical ranges.
(D) The sets have identical boxplots.
(E) None of the above are false.

33. The weight of an aspirin tablet is 300 milligrams according to the bottle label. An FDA investigator weighs a simple random sample of seven tablets, obtains weights of 299, 300, 305, 302, 299, 301, and 303, and runs a hypothesis test of the manufacturer's claim. Which of the following gives the *P*-value of this test?

    (A) $P(t > 1.54)$ with $df = 6$
    (B) $2P(t > 1.54)$ with $df = 6$
    (C) $P(t > 1.54)$ with $df = 7$
    (D) $2P(t > 1.54)$ with $df = 7$
    (E) $0.5P(t > 1.54)$ with $df = 7$

34. A teacher believes that giving her students a practice quiz every week will motivate them to study harder, leading to a greater overall understanding of the course material. She tries this technique for a year, and everyone in the class achieves a grade of at least C. Is this an experiment or an observational study?

    (A) An experiment, but with no reasonable conclusion possible about cause and effect
    (B) An experiment, thus making cause and effect a reasonable conclusion
    (C) An observational study, because there was no use of a control group
    (D) An observational study, but a poorly designed one because randomization was not used
    (E) An observational study, and thus a reasonable conclusion of association but not of cause and effect

35. Which of the following is *not* true with regard to contingency tables for chi-square tests for independence?

    (A) Both variables are categorical.
    (B) Observed frequencies should be whole numbers.
    (C) Expected frequencies should be whole numbers.
    (D) Expected frequencies in each cell should be at least 5, and to achieve this, one sometimes combines categories for one or the other or both of the variables.
    (E) The expected frequency for any cell can be found by multiplying the row total by the column total and dividing by the table total.

36. Which of the following is a correct statement?

    (A) The probability of a Type II error does not depend on the probability of a Type I error.
    (B) In conducting a hypothesis test, it is possible to simultaneously make both a Type I and a Type II error.
    (C) A Type II error will result if one incorrectly assumes the data are normally distributed.
    (D) In medical disease testing with the null hypothesis that the patient is healthy, a Type I error is associated with a *false negative*; that is, the test incorrectly indicates that the patient is disease free.
    (E) When you choose a significance level α, you're setting the probability of a Type I error to exactly α.

37. A random sample of 20 school districts looked at average EOC (end-of-course) test scores versus percent of students living in poverty. A scatterplot of the data is below.

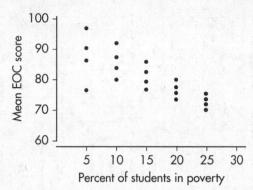

Which of the following would be revealed by a graph of the residuals versus percent of students in poverty?

(A) There is a negative linear relationship between the residuals and the percent in poverty.
(B) The sum of the residuals is less than 0.
(C) The sum of the residuals is greater than 0.
(D) The mean of the residuals is less than 0.
(E) The variation in the mean EOC scores is not the same across the percents in poverty.

38. Studies have shown that only 36% of young adults feel that the death penalty is applied fairly by courts. In a random sample of 50 young adults, what is the expected number of young adults that do not feel that the death penalty is applied fairly by courts?

(A) $50(0.36)$

(B) $50(0.64)$

(C) $50(0.36)(0.64)$

(D) $\sqrt{50(0.36)(0.64)}$

(E) $\sqrt{\dfrac{(0.36)(0.64)}{50}}$

39. The parallel boxplots below show monthly rainfall summaries for Liberia, West Africa.

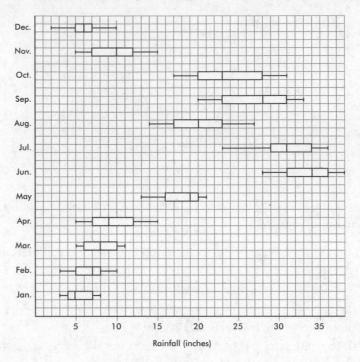

Rainfall (inches)

Which of the following months has the least variability as measured by *interquartile range*?

(A) January
(B) February
(C) March
(D) May
(E) December

40. In comparing the life expectancies of two models of refrigerators, the average years before complete breakdown of 10 model A refrigerators is compared with that of 15 model B refrigerators. The 90% confidence interval estimate of the difference is (6, 12). Which of the following is the most reasonable conclusion?

(A) The mean life expectancy of one model is twice that of the other.
(B) The mean life expectancy of one model is 6 years, while the mean life expectancy of the other is 12 years.
(C) The probability that the life expectancies are different is 0.90.
(D) The probability that the difference in life expectancies is greater than 6 years is 0.90.
(E) We should be 90% confident that the difference in life expectancies is between 6 and 12 years.

If there is still time remaining, you may check your work on this section.

SECTION II

Part A

QUESTIONS 1–5

Spend about 65 minutes on this part of the exam.
Percentage of Section II grade—75

> **Directions:** You must show all work and indicate the methods you use. You will be graded on the correctness of your methods and on the accuracy of your results and explanations.

1. The Information Technology Services division at a university is considering installing a new spam filter software product on all campus computers to combat unwanted advertising and spyware. A sample of 60 campus computers was randomly divided into two groups of 30 computers each. One group of 30 was considered to be a control group, while each computer in the other group had the spam filter software installed. During a two-week period, each computer user was instructed to keep track of the number of unwanted spam e-mails received. The back-to-back stemplot below shows the distribution of such e-mails received for the control and treatment groups.

```
           Control        Treatment
         9 8 7 5      0   2 2 3 5 5 6 6 6 7 8 8 9 9 9
 9 8 8 7 5 5 5 4 2 2 0  1   0 0 1 1 2 2 5 7 9
   8 7 5 4 4 3 2 1 1 0  2   0 2 4 6
         4 3 2 0 0  3   2 5
                    4   1
```

Key: 0|3|2 represents 30 and 32 spam e-mails for the control and treatment groups, respectively.

(a) Compare the distribution of spam e-mails from the control and treatment groups.

(b) The standard deviation of the numbers of e-mails in the control group is 8.1. How does this value summarize variability for the control group data?

(c) A researcher in Information Technology Services calculates a 95% confidence interval for the difference in mean number of spam e-mails received between the control group and the treatment group with the new software and obtains (1.5, 10.9). Assuming all conditions for a two-sample t-interval are met, comment on whether or not there is evidence of a difference in the means for the number of spam e-mails received during a two-week period by computers with and without the software.

(d) The computer users on campus fall into four groups: administrators, staff, faculty, and students. Explain why a researcher might decide to use blocking in setting up this experiment.

2. A game contestant flips three fair coins and receives a score equal to the absolute value of the difference between the number of heads and number of tails showing.

   (a) Construct the probability distribution table for the possible scores in this game.
   (b) Calculate the expected value of the score for a player.
   (c) What is the probability that if a player plays this game three times, the total score will be exactly 3?
   (d) Suppose a player wins a major prize if he or she can average a score of at least 2. Given the choice, should he or she try for this average by playing 10 times or by playing 15 times? Explain.

3. A sociologist is researching a possible link between poverty and lack of education in the United States. A least squares regression on "percent of population 25 years and over who did not graduate from high school" versus "percent of population below poverty level" among a random sample of 45 regions yields the following computer printout.

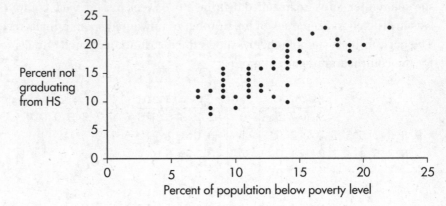

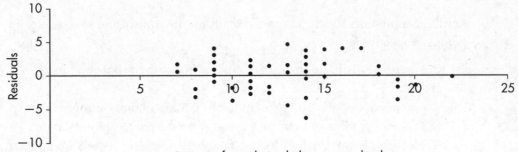

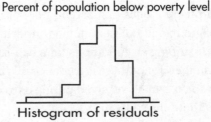

Histogram of residuals

	Coef	SE Coef	t	P
Intercept	4.1929	1.3394	3.1304	0.00148
Poverty	0.8691	0.1010	8.6050	0.00000

(a) Determine the equation of the least squares regression line.

(b) What is the *y*-intercept of the regression line? Interpret in context.

(c) Predict the percentage of students not graduating from high school who are from regions in which the percent of the population below the poverty line are 22% and 42%, and explain your confidence in both of your answers.

(d) Find the 90 percent confidence interval for the slope of the regression line, and interpret in context.

4. State investigators believe that a particular auto repair facility is fraudulently charging customers for repairs they don't need. As part of their investigation, they pick a random sample of ten damaged cars, do their own cost estimate for repair work, and then send the cars to the facility under suspicion for an estimate. The data obtained are shown in the table below.

Car

	1	2	3	4	5	6	7	8	9	10	Mean	SD
Investigator Estimate ($)	2585	3040	560	8250	3800	1575	3590	2830	6550	1820	3460	2324
Facility Estimate ($)	2250	3600	800	9100	4675	1920	3710	4050	6300	2200	3860	2419
Difference	335	−560	−240	−850	−875	−345	−120	−1220	250	−380	−400	494

Is the mean estimate of the facility under suspicion significantly greater than the mean estimate by the investigators? Justify your answer.

5. Five new estimators are being evaluated with regard to quality control in manufacturing professional baseballs of a given weight. Each estimator is tested every day for a month on samples of sizes $n = 10$, $n = 20$, and $n = 40$. The baseballs actually produced that month had a consistent mean weight of 146 grams. The distributions given by each estimator are as follows:

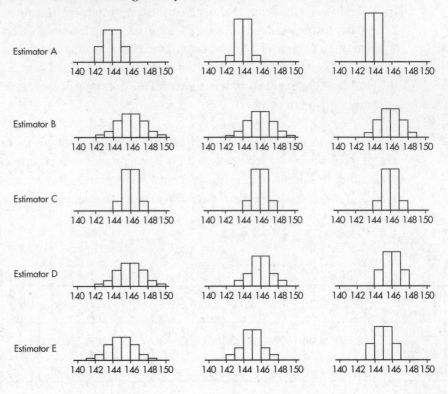

(a) Which of the above appear to be unbiased estimators of the population parameter? Explain.
(b) Which of the above exhibits the lowest variability for $n = 40$? Explain.
(c) Which of the above is the best estimator if the selected estimator will eventually be used with a sample of size $n = 100$? Explain.

SECTION II

Part B

QUESTION 6

Spend about 25 minutes on this part of the exam.
Percentage of Section II grade—25

6. It was recently reported that in the United States, 40.3 percent of all births are to unmarried mothers. A county health administrator is investigating whether births to unmarried women is higher in her county than in the national average. If so, she will propose additional funding to counsel unmarried mothers. A random sample of 100 births in the county will be looked at.

   Let p represent the proportion of the population of women giving birth in the county who are unmarried. Consider the following hypotheses.

   H_0: $p = 0.403$, H_a: $p > 0.403$

   (a) Describe a Type II error in context and a possible consequence.
   (b) What values of the sample proportion $\hat{p}$ would represent sufficient evidence to reject the null hypothesis at a significance level of $\alpha = 0.05$?

   Suppose the actual proportion of all women giving birth in the county who are unmarried is 0.45.

   (c) Using the actual proportion of 0.45 and the answer from (b), find the probability that the null hypothesis will be rejected. Show your work.
   (d) What statistical term describes the probability calculated in (c)?
   (e) Suppose the size of the sample was greater than 100. How would that affect the probability of rejecting the null hypothesis calculated in (c)? Explain.

STOP

Answer Key

SECTION I

1. **C**	9. **D**	17. **E**	25. **C**	33. **B**
2. **A**	10. **B**	18. **D**	26. **E**	34. **A**
3. **C**	11. **B**	19. **A**	27. **C**	35. **C**
4. **E**	12. **E**	20. **D**	28. **A**	36. **E**
5. **B**	13. **E**	21. **B**	29. **A**	37. **E**
6. **E**	14. **C**	22. **E**	30. **B**	38. **B**
7. **C**	15. **A**	23. **E**	31. **E**	39. **E**
8. **B**	16. **E**	24. **E**	32. **E**	40. **E**

Answers and Explanations

SECTION I

1. **(C)** A complete census can give much information about a population, but it doesn't necessarily establish a cause-and-effect relationship among seemingly related population parameters. While the results of well-designed observational studies might suggest relationships, it is difficult to conclude that there is cause and effect without running a well-designed experiment. If bias is present, increasing the sample size simply magnifies the bias. The control group is selected by the researchers making use of chance procedures.

2. **(A)** In the first class, only 40% of the students scored below the given score, while in the second class, 80% scored below the same score.

3. **(C)** The control group should have experiences identical to those of the experimental groups except for the treatment under examination. The control group should not be given a new treatment.

4. **(E)** The negative sign comes about because we are dealing with the difference of proportions. The confidence interval estimate means that we have a certain *confidence* that the difference in population proportions lies in a particular *interval*. Note that $-0.0204 - 0.0173 = -0.0377$ and $-0.0204 + 0.0173 = -0.0031$.

5. **(B)** The workers knew that they were receiving a "treatment" and probably felt that there should be some "outcome." Furthermore, the desire of the workers for the study to be successful could also have had an effect. In this example, the result was increased production no matter what the change in music.

6. **(E)** The proportion of successful calls (problem solved) is $\frac{700}{1000}$, so $\frac{700}{1000}(450)$ is the expected number of calls from location 1 that are successful. Alternatively, the proportion of calls from location 1 is $\frac{450}{1000}$, so $\frac{450}{1000}(700)$ gives the expected number of successful calls from location 1. $\left[\text{Or, Expected} = \frac{(\text{column total})(\text{row total})}{(\text{grand total})} = \frac{(450)(700)}{1000}. \right]$

7. **(C)** The slope and the correlation always have the same sign. Correlation shows association, not causation. Correlation does not apply to categorical data. Correlation measures linear association, so even with a correlation of 0, there may be very strong nonlinear association.

8. **(B)** If two random variables are independent, the mean of the difference of the two random variables is equal to the difference of the two individual means; however, the variance of the difference of the two random variables is equal to the *sum* of the two individual variances.

9. **(D)** A sample is simply a subset of a population.

10. **(B)** The markings, spaced 15 apart, clearly look like the standard deviation spacings associated with a roughly normal curve. The curve seems to be the steepest above the points 185 and 215, and this too indicates that the standard deviation is 15.

11. **(B)** With a right tail having probability 0.01, the critical z-score is `invNorm(0.99)` $= 2.326$. Thus, $\mu + 2.326(0.3) = 12$, giving $\mu = 12 - 2.326(0.3)$.

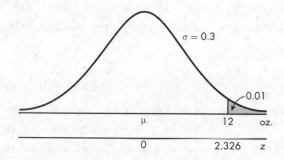

12. **(E)** There is no reason to think that AAA members are representative of the city's drivers. Family members may have similar driving habits, and the independence condition would be violated. Random selection is important regardless of the sample size. The larger a random sample, the closer its standard deviation will be to the population standard deviation.

13. **(E)**

$$P\left(\begin{array}{c}\text{child}\\\text{shows}\end{array}\right) = P\left(\begin{array}{c}\text{father}\\\text{shows}\end{array} \cap \begin{array}{c}\text{child}\\\text{shows}\end{array}\right) + P\left(\begin{array}{c}\text{father}\\\text{doesn't}\end{array} \cap \begin{array}{c}\text{child}\\\text{shows}\end{array}\right)$$
$$= (0.25)(0.8) + (0.75)(0.06)$$

$$P\left(\begin{array}{c|c}\text{father} & \text{child}\\\text{shows} & \text{shows}\end{array}\right) - \frac{P\left(\begin{array}{c}\text{father}\\\text{shows}\end{array} \cap \begin{array}{c}\text{child}\\\text{shows}\end{array}\right)}{P\left(\begin{array}{c}\text{child}\\\text{shows}\end{array}\right)} = \frac{(0.25)(0.8)}{(0.25)(0.8) + (0.75)(0.06)}$$

14. **(C)** As n increases, the power of the test increases and the probabilities of Type I and Type II errors both decrease.

15. **(A)** The mean equals the common value of all the data elements. The other terms all measure variability, which is zero when all the data elements are equal.

16. **(E)** $SE(\bar{x}) = \dfrac{s}{\sqrt{n}} = \dfrac{0.15}{\sqrt{15}}$ and with $df = n - 1 = 15 - 1 = 14$, the critical t-scores are $\pm$`invT` `(0.95, 14)` $= \pm 1.761$. The 90% confidence interval is

$$\bar{x} \pm t^*\left(SE(\bar{x})\right) = 2.72 \pm 1.761\left(\frac{0.15}{\sqrt{15}}\right)$$

17. **(E)** In a simple random sample, every possible group of the given size has to be equally likely to be selected, and this is not true here. For example, with this procedure it is impossible for the employees in the final sample to all be from a single plant. This method is an example of stratified sampling, but stratified sampling does not result in simple random samples.

18. **(D)** $(0.32)(0.15) = 0.048$ so $P(E \cap F) = P(E)P(F)$ and, thus, E and F are independent. $P(E \cap F) \neq 0$, so E and F are not mutually exclusive.

19. **(A)** The quartiles Q_1 and Q_3 have z-scores of $\pm \text{invNorm}(0.75) = \pm 0.67$, so $Q_1 = 640,000 - (0.67)18,000 \approx 628,000$, while $Q_3 = 640,000 + (0.67)18,000 \approx 652,000$. The interquartile range is the difference $Q_3 - Q_1$.

20. **(D)** One set is a shift of 20 units from the other, so they have different means and medians, but they have identical shapes and thus the same variability, including IQR, standard deviation, and variance.

21. **(B)** The wider the confidence interval, the higher the confidence level, so one may obtain as high a confidence level as one wishes if a very wide confidence interval is accepted. A 99% confidence interval estimate means that in about 99% of all samples selected by this method, the population mean will be included in the confidence interval. The central limit theorem applies to any population, no matter if it is normally distributed or not. The sampling distribution for a mean always has standard deviation $\frac{\sigma}{\sqrt{n}}$; large enough sample size n refers to the closer the distribution will be to a normal distribution. The center of a confidence interval is the sample statistic, not the population parameter.

22. **(E)** In none of these are the trials independent. For example, as each consecutive person is stopped at a roadblock, the probability the next person has a seat belt on will quickly increase; if a student has one A, the probability is increased that he or she has another A.

23. **(E)** Independent implies $P(E \cap F) = P(E)P(F)$, while mutually exclusive implies $P(E \cap F) = 0$. So, to be both independent and mutually exclusive in this example would mean $pq = 0$, which is impossible because p and q are given to be nonzero.

24. **(E)** In a binomial with $n = 4$ and $p = 0.9$, $P(\text{at least 3 successes}) = P(\text{exactly 3 successes}) + P(\text{exactly 4 successes}) = 4(0.9)^3(0.1) + (0.9)^4$.

25. **(C)** In stratified sampling, the population is divided into representative groups, and random samples of persons from each group are chosen. In this case, it might well be important to have sufficient numbers from each group and to be able to consider separately the responses from each of the three groups—urban, suburban, and rural.

26. **(E)** The coefficient of determination, r^2, indicates the percentage of variation in the response variable y that is explained by variation in the explanatory variable x.

27. **(C)** It is most likely that the homes at which the interviewer had difficulty finding someone home were homes with fewer children living in them. Replacing these homes with other randomly picked homes will most likely replace homes with fewer children with homes with more children.

28. **(A)** The median corresponds to the 0.5 cumulative proportion.

29. **(A)** Blocking divides the subjects into groups of similar individuals, in this case individuals with similar exercise habits, and runs the experiment on each separate group. This controls the known effect of variation in exercise level on cholesterol level.

30. **(B)** The margin of error varies directly with the critical z-value and directly with the standard deviation of the sample but inversely with the square root of the sample size.

31. **(E)** $\sigma_{\bar{x}} = \dfrac{3.2}{\sqrt{400}}$. With a true mean increase of 4.2, the z-score for 4.0 is $\dfrac{4.0 - 4.2}{\left(\dfrac{3.2}{\sqrt{400}}\right)}$ and the officers fail to reject the claim if the sample mean has a z-score greater than this.

32. **(E)** Both have 50 for their means and medians, both have a range of $90 - 10 = 80$, and both have identical boxplots, with first quartile 30 and third quartile 70.

33. **(B)** Since we are not told that the investigator suspects that the average weight is over 300 mg or is under 300 mg and since a tablet containing too little or too much of a drug clearly should be brought to the manufacturer's attention, this is a two-sided test. Thus, the P-value is twice the tail probability obtained (using the t-distribution with $df = n - 1 = 6$). Note that while t can be calculated by putting the data in a list and using T-Test, this is unnecessary here because every answer choice involves "$t > 1.54$."

34. **(A)** This study was an experiment because a treatment (weekly quizzes) was imposed on the subjects. However, it was a poorly designed experiment with no use of randomization and no control over confounding variables. An example of an experiment in which cause and effect might have been possible would have been if half the class was randomly assigned to take weekly quizzes while the other half did not take the quizzes.

35. **(C)** The expected frequencies, as calculated by $\dfrac{(\text{row total}) \times (\text{column total})}{\text{table total}}$, may not be whole numbers.

36. **(E)** We reject H_0 when the P-value falls below α, and when H_0 is true this rejection will happen precisely with probability α. The probabilities of Type I and Type II errors are related; for example, lowering the Type I error increases the probability of a Type II error. A Type I error can be made only if the null hypothesis is true, while a Type II error can be made only if the null hypothesis is false. In medical testing, with the usual null hypothesis that the patient is healthy, a Type I error is that a healthy patient is diagnosed with a disease, that is, a *false positive.*

37. **(E)** The variation in the mean EOC scores is larger for lower percent poverty and smaller for higher percent poverty, and this will be shown by residuals with larger absolute values and then smaller absolute values. The sum and the mean of the residuals is always exactly 0, and the residual plot will never show either a positive or a negative linear relationship.

38. **(B)** The probability of an adult feeling that the death penalty is not applied fairly by the courts is $1 - 0.36 = 0.64$. The expected value of a binomial with $n = 50$ and $p = 0.64$ is $np = 50(0.64)$.

39. **(E)** A boxplot gives a five-number summary: smallest value, 25th percentile (Q_1), median, 75th percentile (Q_3), and largest value. The interquartile range is given by $Q_3 - Q_1$, or the total length of the two "boxes," ignoring the "whiskers."

40. **(E)** When using a measurement from a sample, we are never able to say *exactly* what a population mean is; rather, we always say we have a certain *confidence* that the population mean lies in a particular *interval.*

SECTION II: PART A

1. (a) A complete answer compares shape, center, and spread and mentions content.
   Shape: The control group distribution is somewhat bell-shaped and symmetric, while the treatment group distribution is somewhat skewed right (skewed toward the higher values).
   Center: The center of the control group distribution is around 20, which is greater than the center of the treatment group distribution, which is somewhere around 10 to 12.
   Spread: The spread of the control group distribution, 5 to 34, is less than the spread of the treatment group distribution, which is 2 to 41.

   (b) For computers in the control group (no spam software), the number of spam e-mails received varies an "average" amount of 8.1 from the mean number of spam e-mails received in the control group.

   (c) Since the 95% confidence interval for the difference does not contain zero, the researcher can conclude the observed difference in mean numbers of spam e-mails received between the control group and the treatment group that received spam software is significant.

   (d) It may well be that the four groups—administrators, staff, faculty, and students—are each exposed to different kinds of spam e-mail risks, and possibly the software will be more or less of a help to each group. In that case, the researcher should in effect run four separate experiments on the homogeneous groups, called blocks. Even though results from the different blocks will be combined at the end, some conclusions may be more specific.

SCORING

Section 1 is essentially correct for correctly comparing shape, center, and spread. Section 1 is partially correct for correctly comparing two of the three features. Lower an E to a P or a P to an I if context is never mentioned.

Section 2 is essentially correct if 1) standard deviation is explained correctly in the context of this problem and 2) for noting that zero is not in the interval so the observed difference, in context, is significant. Section 2 is partially correct for one of these two parts correct.

Section 3 is essentially correct if the purpose of blocking is correctly explained in the context of this problem. Section 3 is partially correct if the general purpose of blocking is correctly explained but not in the context of this problem.

4	**Complete Answer**	All three sections essentially correct.
3	**Substantial Answer**	Two sections essentially correct and one section partially correct.
2	**Developing Answer**	Two sections essentially correct OR one section essentially correct and one or two sections partially correct OR all three sections partially correct.
1	**Minimal Answer**	One section essentially correct OR two sections partially correct.

2. (a) Listing the eight possibilities: {HHH, HHT, HTH, HTT, THH, THT, TTH, TTT} clearly shows that the only possibilities for the absolute value of the difference are 1 with probability $\frac{6}{8} = 0.75$ and 3 with probability $\frac{2}{8} = 0.25$. The table is

Absolute difference	Probability
1	0.75
3	0.25

(b) $E = \sum xP(x) = 1(0.75) + 3(0.25) = 1.5$.

(c) Since the only possible scores for each game are 1 and 3, the only way to have a total score of 3 in three games is to score 1 in each game. The probability of this is $(0.75)^3 = 0.421875$ [or $\left(\frac{3}{4}\right)^3 = \frac{27}{64}$].

(d) The more times the game is played, the closer the average score will be to the expected value of 1.5. The player does not want to average close to 1.5, so he or she should prefer playing 10 times rather than 15 times.

SCORING

Section 1 is essentially correct for the correct probability distribution table. Section 1 is partially correct for one minor error.

Section 2 is essentially correct for 1) the correct calculation of expected value based on the answer given in Part (a) and 2) for the correct probability calculation with some indication of where the answer is coming from. Section 2 is partially correct for one of these two parts correct.

Section 3 is essentially correct for choosing 10 and giving a clear explanation. Section 3 is partially correct for choosing 10 and giving a weak explanation. Section 3 is incorrect for choosing 10 with no explanation or with an incorrect explanation.

4   Complete Answer   All three sections essentially correct.

3   Substantial Answer   Two sections essentially correct and one section partially correct.

2   Developing Answer   Two sections essentially correct OR one section essentially correct and one or two sections partially correct OR all three sections partially correct.

1   Minimal Answer   One section essentially correct OR two sections partially correct.

3. (a) Predicted % not graduating HS = 4.1929 + 0.8691(% below poverty line)

(b) The y-intercept is 4.1929. In regions with 0% of the population below the poverty line, the average (or predicted) percent not graduating from high school is 4.2%.

(c) $4.1929 + 0.8691(22) = 23.3131$ and $4.1929 + 0.8691(42) = 40.6951$.

For regions with 22% of the population below the poverty line, the predicted percent not graduating from high school is 23.3%. For regions with 42% of the population below the poverty line, the predicted percent not graduating from high school is 40.7%. The 22% input is within the domain of the given sample data, while the 42% input is not. So, while we have confidence in the 23.3% calculation, we have very little confidence in the 40.7% calculation because it is an extreme extrapolation.

(d) *Procedure:* Confidence interval for the slope of the regression line.

Check conditions: We are given a random sample, we assume that the sample of 45 regions is less than 10% of all regions, the scatterplot looks roughly linear, there is no major pattern in the residual plot, and the histogram of the residuals is unimodal and roughly symmetric.

Mechanics: With $df = n - 2 = 43$, the critical t-scores are $\pm \texttt{invT(0.95, 43)} = \pm1.681$. Then $0.8691 \pm 1.681(0.1010) = 0.8691 \pm 0.1698$.

Conclusion in context: We are 90% confident that the true slope of the regression line linking percent not graduating from high school to percent below the poverty line is between 0.6993 and 1.0389. That is, we are 90% confident that for each additional 1 percent of the population below the poverty line, the predicted percent of the population not graduating from high school goes up between 0.6993 and 1.0389.

SCORING

Section 1, Parts (a) and (b), is essentially correct for a correct equation, defining the variables, correct interpretation for the intercept, and using nondeterministic language in both parts ("predicted" or "hat," "average"). Section 1 is partially correct for two or three of these four components correct.

Section 2, Part (c), is essentially correct for two correct calculations, answers in context, and a clear explanation of why there is more confidence in the first calculation and is partially correct for two of these three components correct.

Section 3, Part (d), is essentially correct for 1) naming the procedure, 2) referencing all the listed conditions, 3) a correct interval, and 4) the conclusion referencing the confidence level, context, the population (such as "true"), and the parameter "slope" and is partially correct for two or three of the four components correct.

4	**Complete Answer**	All three sections essentially correct.
3	**Substantial Answer**	Two sections essentially correct and one section partially correct.
2	**Developing Answer**	Two sections essentially correct OR one section essentially correct and one or two sections partially correct OR all three sections partially correct.
1	**Minimal Answer**	One section essentially correct OR two sections partially correct.

4. This is a paired data test, not a two-sample test, with three parts to a complete solution.

Part 1:

Parameter: Let μ_d represent the mean difference between what an investigator would estimate and what the facility would estimate.

Hypotheses: H_0: $\mu_d = 0$ and H_a: $\mu_d < 0$

Part 2:

Procedure: This is a paired t-test, a one-sample hypothesis test on the set of differences.

Checks: Random sample (given) and it is reasonable to assume that the 10 data pairs are independent of each other. Rough normality of the population distribution of differences should be checked graphically on the sample data using a histogram, or a dotplot, or a normal probability plot:

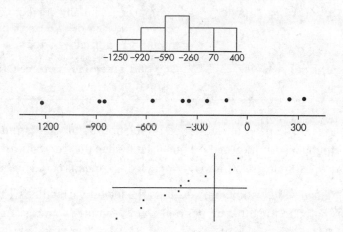

The histogram and dotplot show rough symmetry and no extreme skewness or outliers, while the normal probability plot is roughly linear. So, it is not unreasonable to assume the sample comes from a roughly normal population.

Part 3:

Mechanics: $t = -2.56$ and $P = 0.015$ (from calculator software such as T-Test on the TI-84 or 1-Sample tTest on the Casio Prizm).

Conclusion in context with linkage to the P-value: With this small a P-value, $0.015 < 0.05$, there is sufficient evidence to reject H_0. That is, there is sufficient evidence that the mean estimate of the facility under suspicion is greater than the mean estimate by investigators.

SCORING

Part 1 is essentially correct for a correct statement of the hypotheses (in terms of a population mean). Part 1 is partially correct for a correct statement of the hypotheses but missing identification of the variable.

Part 2 is essentially correct if the test is correctly identified by name or formula and a graphical check of the normality condition is given. Part 2 is partially correct for only one of these two elements.

Part 3 is essentially correct for a correct calculation of both the test statistic t and the P-value, and a correct conclusion in context, linked to the P-value. Part 3 is partially correct for everything correct but missing either context or linkage.

4	**Complete Answer**	All three parts essentially correct.
3	**Substantial Answer**	Two parts essentially correct and one part partially correct.
2	**Developing Answer**	Two parts essentially correct OR one part essentially correct and one or two parts partially correct OR all three parts partially correct.
1	**Minimal Answer**	One part essentially correct OR two parts partially correct.

5. (a) A statistic used to estimate a population parameter is unbiased if the mean of the sampling distribution of the statistic is equal to the true value of the parameter being estimated. Estimators B, C, and D appear to have means equal to the population mean of 146.

(b) For $n = 40$, estimator A exhibits the lowest variability, with a range of only 2 grams compared to the other ranges of 6 grams, 4 grams, 4 grams, and 4 grams.

(c) The estimator should have a distribution centered at 146, thus eliminating A and E. As n increases, D shows tighter clustering around 146 than does B. Finally, while C looks better than D for $n = 40$, the estimator will be used with $n = 100$, and the D distribution is clearly converging as the sample size increases while the C distribution remains the same. Choose D.

SCORING

Part (a) is essentially correct for a correct answer with a good explanation of what unbiased means and is partially correct for a correct answer with a weak explanation. Part (a) is incorrect for a correct answer with no explanation or with an incorrect explanation.

Part (b) is essentially correct for a correct answer together with some numerical justification and is partially correct for a correct answer with a weak explanation. Part (b) is incorrect for a correct answer with no explanation or with an incorrect explanation.

Part (c) is essentially correct for a correct answer with a good explanation and is partially correct for a correct answer with a weak explanation. Part (c) is incorrect for a correct answer with no explanation or with an incorrect explanation.

4	**Complete Answer**	All three parts essentially correct.
3	**Substantial Answer**	Two parts essentially correct and one part partially correct.
2	**Developing Answer**	Two parts essentially correct OR one part essentially correct and one or two parts partially correct OR all three parts partially correct.
1	**Minimal Answer**	One part essentially correct OR two parts partially correct.

SECTION II: PART B

6. (a) A Type II error is mistakenly failing to reject a false null hypothesis. In this situation, it would happen if the proportion of all women giving birth in the county who are unmarried is greater than 0.403, but the sample proportion does not provide sufficient evidence that it is. A possible consequence would be that needed additional funding to counsel unmarried mothers is not provided.

(b) We first note that the standard deviation is $\sqrt{\dfrac{p(1-p)}{n}} = \sqrt{\dfrac{(0.403)(0.597)}{100}} = 0.04905$. This is a one-sided test, and so the critical z-score is `invNorm(0.95)` = 1.645. Then $\dfrac{\hat{p}-0.403}{0.04905} > 1.645$ gives $\hat{p} > 0.4837$.

(c) If the true population proportion is 0.45, the sampling distribution of $\hat{p}$ is approximately normal with mean 0.45 and standard deviation $\sqrt{\dfrac{(0.45)(0.55)}{100}} = 0.04975$. Then $P(\hat{p} > 0.4837) = P\left(z > \dfrac{0.4837-0.45}{0.04975}\right) = 0.249$.

(d) This is called the power of the test.

(e) If the sample size is increased, then, while the rejection region is still $z > 1.645$, the sampling distribution of $\hat{p}$ will have a smaller standard deviation. Thus, the minimum value of $\hat{p}$ for which we would reject H_0 would be lower, and so the probability of rejecting H_0 would be greater.

Section 1 is essentially correct in (a) for a correct explanation of a Type II error in context and including a possible consequence. Section 1 is partially correct for one out of the two steps above.

Section 2 is essentially correct in (b) for calculating $\sigma_{\hat{p}}$, stating the equation $\frac{\hat{p} - 0.403}{0.04905} > 1.645$ to be solved, and correctly solving for $\hat{p}$ and is partially correct for two out of these three steps correct.

Section 3 is essentially correct in (c) and (d) for the correct calculation of the new standard deviation $\sigma_{\hat{p}}$, correctly calculating the probability $P(\hat{p} > 0.4837)$, and correctly recognizing this to be the power of the test and is partially correct for two out of these three steps correct.

Section 4 is essentially correct in (e) for concluding that the probability of rejecting H_0 would be greater and giving a good explanation and is partially correct for concluding that the probability of rejecting H_0 would be greater and giving a weak explanation.

Count essentially correct answers as one point and partially correct answers as one-half point.

4	**Complete Answer**	Four points
3	**Substantial Answer**	Three points
2	**Developing Answer**	Two points
1	**Minimal Answer**	One point

Use a holistic approach to decide a score totaling between two numbers, deciding whether to score up or down depending on the strength of the response and communication.

ANSWER SHEET
Practice Test 4

1. Ⓐ Ⓑ Ⓒ Ⓓ Ⓔ   11. Ⓐ Ⓑ Ⓒ Ⓓ Ⓔ   21. Ⓐ Ⓑ Ⓒ Ⓓ Ⓔ   31. Ⓐ Ⓑ Ⓒ Ⓓ Ⓔ

2. Ⓐ Ⓑ Ⓒ Ⓓ Ⓔ   12. Ⓐ Ⓑ Ⓒ Ⓓ Ⓔ   22. Ⓐ Ⓑ Ⓒ Ⓓ Ⓔ   32. Ⓐ Ⓑ Ⓒ Ⓓ Ⓔ

3. Ⓐ Ⓑ Ⓒ Ⓓ Ⓔ   13. Ⓐ Ⓑ Ⓒ Ⓓ Ⓔ   23. Ⓐ Ⓑ Ⓒ Ⓓ Ⓔ   33. Ⓐ Ⓑ Ⓒ Ⓓ Ⓔ

4. Ⓐ Ⓑ Ⓒ Ⓓ Ⓔ   14. Ⓐ Ⓑ Ⓒ Ⓓ Ⓔ   24. Ⓐ Ⓑ Ⓒ Ⓓ Ⓔ   34. Ⓐ Ⓑ Ⓒ Ⓓ Ⓔ

5. Ⓐ Ⓑ Ⓒ Ⓓ Ⓔ   15. Ⓐ Ⓑ Ⓒ Ⓓ Ⓔ   25. Ⓐ Ⓑ Ⓒ Ⓓ Ⓔ   35. Ⓐ Ⓑ Ⓒ Ⓓ Ⓔ

6. Ⓐ Ⓑ Ⓒ Ⓓ Ⓔ   16. Ⓐ Ⓑ Ⓒ Ⓓ Ⓔ   26. Ⓐ Ⓑ Ⓒ Ⓓ Ⓔ   36. Ⓐ Ⓑ Ⓒ Ⓓ Ⓔ

7. Ⓐ Ⓑ Ⓒ Ⓓ Ⓔ   17. Ⓐ Ⓑ Ⓒ Ⓓ Ⓔ   27. Ⓐ Ⓑ Ⓒ Ⓓ Ⓔ   37. Ⓐ Ⓑ Ⓒ Ⓓ Ⓔ

8. Ⓐ Ⓑ Ⓒ Ⓓ Ⓔ   18. Ⓐ Ⓑ Ⓒ Ⓓ Ⓔ   28. Ⓐ Ⓑ Ⓒ Ⓓ Ⓔ   38. Ⓐ Ⓑ Ⓒ Ⓓ Ⓔ

9. Ⓐ Ⓑ Ⓒ Ⓓ Ⓔ   19. Ⓐ Ⓑ Ⓒ Ⓓ Ⓔ   29. Ⓐ Ⓑ Ⓒ Ⓓ Ⓔ   39. Ⓐ Ⓑ Ⓒ Ⓓ Ⓔ

10. Ⓐ Ⓑ Ⓒ Ⓓ Ⓔ   20. Ⓐ Ⓑ Ⓒ Ⓓ Ⓔ   30. Ⓐ Ⓑ Ⓒ Ⓓ Ⓔ   40. Ⓐ Ⓑ Ⓒ Ⓓ Ⓔ

Practice Test 4

SECTION I

Questions 1–40

Spend 90 minutes on this part of the exam.

> **Directions:** The questions or incomplete statements that follow are each followed by five suggested answers or completions. Choose the response that best answers the question or completes the statement.

1. A company wishes to determine the relationship between the number of days spent training employees and their performances on a job aptitude test. Collected data result in a least squares regression line, $\hat{y} = 12.1 + 6.2x$, where x is the number of training days and $\hat{y}$ is the predicted score on the aptitude test. Which of the following statements best interprets the slope and y-intercept of the regression line?

   (A) The base score on the test is 12.1, and for every day of training one would expect, on average, an increase of 6.2 on the aptitude test.
   (B) The base score on the test is 6.2, and for every day of training one would expect, on average, an increase of 12.1 on the aptitude test.
   (C) The mean number of training days is 12.1, and for every additional 6.2 days of training one would expect, on average, an increase of one unit on the aptitude test.
   (D) The mean number of training days is 6.2, and for every additional 12.1 days of training one would expect, on average, an increase of one unit on the aptitude test.
   (E) The mean number of training days is 12.1, and for every day of training one would expect, on average, an increase of 6.2 on the aptitude test.

2. To survey the opinions of the students at your high school, a researcher plans to select every twenty-fifth student entering the school in the morning. Assuming there are no absences, will this result in a simple random sample of students attending your school?

   (A) Yes, because every student has the same chance of being selected.
   (B) Yes, but only if there is a single entrance to the school.
   (C) Yes, because the 24 out of every 25 students who are not selected will form a control group.
   (D) Yes, because this is an example of systematic sampling, which is a special case of simple random sampling.
   (E) No, because not every sample of the intended size has an equal chance of being selected.

3. Consider a hypothesis test with H_0: $\mu = 70$ and H_a: $\mu < 70$. Which of the following choices of significance level and sample size results in the greatest power of the test when $\mu = 65$?

   (A) $\alpha = 0.05$, $n = 15$
   (B) $\alpha = 0.01$, $n = 15$
   (C) $\alpha = 0.05$, $n = 30$
   (D) $\alpha = 0.01$, $n = 30$
   (E) There is no way of answering without knowing the strength of the given power.

4. In 2018, it was estimated that there were roughly 554,000 homeless people in the United States. The average shelter stay for homeless families with kids is 435 days. Assume a skewed left distribution with a standard deviation of 85 days. Consider random samples of size 100 taken from the distribution with the mean length of stay, $\bar{x}$, recorded for each sample. Which of the following is the best description of the sampling distribution of $\bar{x}$?

   (A) Skewed left with mean 435 days and standard deviation 0.85 days
   (B) Skewed left with mean 435 days and standard deviation 8.5 days
   (C) Skewed left with mean 435 days and standard deviation 85 days
   (D) Approximately normal with mean 435 days and standard deviation 0.85 days
   (E) Approximately normal with mean 435 days and standard deviation 8.5 days

5. The label on a package of cords claims that the breaking strength of a cord is 3.5 pounds, but a hardware store owner believes the real value is less. She plans to test 36 such cords; if their mean breaking strength is less than 3.25 pounds, she will reject the claim on the label. If the standard deviation for the breaking strengths of all such cords is 0.9 pounds, what is the probability of mistakenly rejecting a true claim?

(A) $P\left(z < \dfrac{3.25 - 3.5}{0.9}\right)$

(B) $P\left(z > \dfrac{3.25 - 3.5}{0.9}\right)$

(C) $P\left(z < \dfrac{3.25 - 3.5}{\left(\dfrac{0.9}{\sqrt{36}}\right)}\right)$

(D) $2P\left(z < \dfrac{3.25 - 3.5}{\left(\dfrac{0.9}{\sqrt{36}}\right)}\right)$

(E) $P\left(z > \dfrac{3.25 - 3.5}{\left(\dfrac{0.9}{\sqrt{36}}\right)}\right)$

6. The graph below shows cumulative proportions plotted against numbers of employees working in midsized retail establishments.

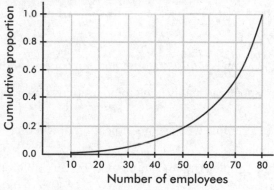

What is the approximate interquartile range?

(A) 18
(B) 35
(C) 57
(D) 68
(E) 75

7. What is a placebo?

(A) A method of selection
(B) An experimental treatment
(C) A control treatment
(D) A parameter
(E) A statistic

8. To study the effect of alcohol on reaction time, subjects were randomly selected and given three beers to consume. Their reaction time to a simple stimulus was measured before and after drinking the alcohol. Which of the following is a correct statement?

(A) This study was an observational study.
(B) Lack of blinding makes this a poorly designed study.
(C) The placebo effect is irrelevant in this type of study.
(D) This study was an experiment with no controls.
(E) This study was an experiment in which the subjects were used as their own controls.

9. Suppose that the regression line for a set of data, $\hat{y} = 7x + b$, passes through the point $(-2, 4)$. If $\bar{x}$ and $\bar{y}$ are the sample means of the x- and y-values, respectively, then which of the following is equal to $\bar{y}$?

(A) $\bar{x}$
(B) $\bar{x} + 2$
(C) $\bar{x} - 4$
(D) $7\bar{x}$
(E) $7\bar{x} + 18$

10. Which of the following is a *false* statement about simple random samples?

(A) A sample must be reasonably large to be properly considered a simple random sample.
(B) Inspection of a sample will give no indication of whether or not it is a simple random sample.
(C) Attributes of a simple random sample may be very different from attributes of the population.
(D) Every element of the population has an equal chance of being picked.
(E) Every sample of the desired size has an equal chance of being picked.

11. A local school has seven math teachers and seven English teachers. When comparing their mean salaries, which of the following is most appropriate?

(A) A two-sample z-test of population means
(B) A two-sample t-test of population means
(C) A one-sample z-test on a set of differences
(D) A one-sample t-test on a set of differences
(E) None of the above are appropriate.

12. The following is a histogram of ages of people applying for a particular high school teaching position.

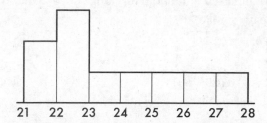

Which of the following is a correct statement?

(A) The median age is between 24 and 25.
(B) The mean age is between 22 and 23.
(C) The mean age is greater than the median age.
(D) More applicants are under 23 years of age than are over 23.
(E) There are a total of 10 applicants.

13. To conduct a survey of which long-distance carriers are used in a particular locality, a researcher opens a telephone book to a random page, closes his eyes, puts his finger down on the page, and then calls the next 75 names. Which of the following is a correct statement?

(A) The procedure results in a simple random sample.
(B) While the survey design does incorporate chance, the procedure could easily result in selection bias.
(C) This is an example of cluster sampling with 75 clusters.
(D) This is an example of stratified sampling with 26 strata.
(E) Given that the researcher truly keeps his eyes closed, this is a good example of blinding.

14. Which of the following is *not* true about *t*-distributions?

(A) There are different *t*-distributions for different values of *df* (degrees of freedom).
(B) *t*-distributions are bell-shaped and symmetric.
(C) *t*-distributions always have mean 0 and standard deviation 1.
(D) *t*-distributions are more spread out than the normal distribution.
(E) The larger the *df* value, the closer the distribution is to the normal distribution.

15. Suppose the probability that a person picked at random has lung cancer is 0.035 and the probability that the person both has lung cancer and is a heavy smoker is 0.014. Given that someone picked at random has lung cancer, what is the probability that the person is a heavy smoker?

(A) $0.035 - 0.014$

(B) $0.035 + 0.014$

(C) $0.035 + 0.014 - (0.035)(0.014)$

(D) $\dfrac{0.035}{1 - 0.014}$

(E) $\dfrac{0.014}{0.035}$

16. Suppose computer science graduates earn an average starting salary of $75,000 with a standard deviation of $12,000. What is the probability that a randomly selected computer science graduate has a starting salary less than $100,000 if it is known that his or her starting salary is over $80,000? Assume a roughly normal distribution of starting salaries of computer science graduates.

(A) 0.06

(B) 0.34

(C) 0.66

(D) 0.68

(E) 0.94

17. A plant manager wishes to determine the difference in number of accidents per day between two departments. She wants to be 90% certain of the difference in daily averages to within 0.25 accidents per day. Assume standard deviations of 0.8 and 0.5 accidents per day in the two departments, respectively. Which of the following should be used to determine how many days' (n) records should be examined?

(A) $1.645\dfrac{\sqrt{0.8^2 + 0.5^2}}{\sqrt{n}} \le 0.25$

(B) $1.645\dfrac{0.8^2 + 0.5^2}{\sqrt{n}} \le 0.25$

(C) $1.645\sqrt{\dfrac{0.8 + 0.5}{n}} \le 0.25$

(D) $1.645\dfrac{\sqrt{0.8^2 + 0.5^2}}{n} \le 0.25$

(E) $1.645\dfrac{\sqrt{0.8 + 0.5}}{n} \le 0.25$

18. Suppose the correlation between two variables is $r = 0.19$. What is the new correlation if 0.23 is added to all values of the x-variable, every value of the y-variable is doubled, and the two variables are interchanged?

(A) 0.84
(B) 0.42
(C) 0.19
(D) −0.19
(E) −0.84

19. Two commercial flights per day are made from a small county airport. The airport manager tabulates the number of on-time departures for a sample of 200 days.

Number of on-time departures	0	1	2
Observed number of days	10	80	110

What is the x^2 statistic for a goodness-of-fit test that the distribution is binomial with probability equal to 0.8 that a flight leaves on time?

(A) $\dfrac{(10-8)^2}{8} + \dfrac{(80-64)^2}{64} + \dfrac{(110-128)^2}{128}$

(B) $\dfrac{(10-8)^2}{10} + \dfrac{(80-64)^2}{80} + \dfrac{(110-128)^2}{110}$

(C) $\dfrac{(10-10)^2}{10} + \dfrac{(80-30)^2}{30} + \dfrac{(110-160)^2}{160}$

(D) $\dfrac{(10-10)^2}{10} + \dfrac{(80-30)^2}{80} + \dfrac{(110-160)^2}{110}$

(E) $\dfrac{(10-66)^2}{10} + \dfrac{(80-67)^2}{80} + \dfrac{(110-67)^2}{110}$

20. A company has a choice of three investment schemes. Option I gives a sure $25,000 return on investment. Option II gives a 50% chance of returning $50,000 and a 50% chance of returning $10,000. Option III gives a 5% chance of returning $100,000 and a 95% chance of returning nothing. Which option should the company choose?

(A) Option I if the company needs at least $20,000 to pay off an overdue loan
(B) Option II if the company wants to maximize expected return
(C) Option III if the company needs at least $80,000 to pay off an overdue loan
(D) All of the above answers are correct.
(E) Because of chance, it really doesn't matter which option the company chooses.

21. Suppose $P(X) = 0.35$ and $P(Y) = 0.40$. If $P(X|Y) = 0.28$, what is $P(Y|X)$?

(A) $\dfrac{(0.28)(0.35)}{0.40}$

(B) $\dfrac{(0.28)(0.40)}{0.35}$

(C) $\dfrac{(0.35)(0.40)}{0.28}$

(D) $\dfrac{0.28}{0.40}$

(E) $\dfrac{0.28}{0.35}$

22. To test whether extensive exercise lowers the resting heart rate, a study is performed by randomly selecting half of a group of volunteers to exercise 1 hour each morning, while the rest are instructed to perform no exercise. Is this study an experiment or an observational study?

(A) An experiment with a control group and blinding
(B) An experiment with blocking
(C) An observational study with comparison and randomization
(D) An observational study with little if any bias
(E) None of the above statements are correct.

23. The waiting times for a new roller coaster ride are approximately normally distributed with a mean of 35 minutes and a standard deviation of 10 minutes. If there are 150,000 riders the first summer, which of the following is the shortest time interval associated with 100,000 riders?

(A) 0 to 31.7 minutes
(B) 31.7 to 39.3 minutes
(C) 25.3 to 44.7 minutes
(D) 25.3 to 35 minutes
(E) 39.3 to 95 minutes

PRACTICE TEST 4

24. A medical researcher, studying the effective durations of three over-the-counter pain relievers, obtains the following boxplots from three equal-sized groups of patients, one group using each of the pain relievers.

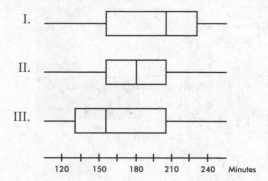

Which of the following is a correct statement with regard to comparing the effective durations in minutes of the three pain relievers?

(A) All three have the same interquartile range.
(B) More patients had over 210 minutes of pain relief in the (I) group than in either of the other two groups.
(C) More patients had over 240 minutes of pain relief in the (I) group than in either of the other two groups.
(D) More patients had less than 120 minutes of pain relief in the (III) group than in either of the other two groups.
(E) The durations of pain relief in the (II) group form a roughly normal distribution.

25. People at a high school were surveyed as to whether they were students or teachers and what their favorite ice cream flavor was, excluding vanilla and chocolate. The results are displayed in the following mosaic plot.

Based on this plot, which of the following is a true statement?

(A) The number of students answering "coffee" is greater than the number of teachers answering "coffee."
(B) More teachers answered either "cookie dough" or "strawberry" than students who answered "cookie dough."
(C) Of those who answered "strawberry," a greater proportion were teachers than students.
(D) There were more students than teachers in the survey.
(E) More people chose cookie dough than strawberry.

26. Suppose that 54% of the graduates from your high school go on to 4-year colleges, 20% go on to 2-year colleges, 19% find employment, and the remaining 7% search for a job. If a randomly selected student is not going on to a 2-year college, what is the probability she will be going on to a 4-year college?

(A) 0.460
(B) 0.540
(C) 0.630
(D) 0.675
(E) 0.730

27. A congressman is interested in the proportion p of registered voters in his district who favor legalizing medical marijuana. Forty-four percent of a simple random sample of 750 registered voters in his district favor legalizing medical marijuana. What is the midpoint for a 95% confidence interval estimate of p?

(A) 0.025
(B) 0.05
(C) 0.418
(D) 0.44
(E) p

28. Random samples of size n are drawn from a population. The mean of each sample is calculated, and the standard deviation of this set of sample means is found. Then the procedure is repeated, this time with samples of size $4n$. How does the standard deviation of the second group compare with the standard deviation of the first group?

(A) It will be the same.
(B) It will be twice as large.
(C) It will be four times as large.
(D) It will be half as large.
(E) It will be one-quarter as large.

29. Leech therapy is used in traditional medicine for treating localized pain. In a double-blind experiment on 50 patients with osteoarthritis of the knee, half are randomly selected to receive injections of leech saliva while the rest receive a placebo. Pain levels 7 days later among those receiving the saliva show a mean of 19.5, while pain levels among those receiving the placebo show a mean of 25.6 (higher numbers indicate more pain). Partial calculator output is shown below.

```
2-SampTTest
μ1<μ2
t=-3.939503313
df=43.43159286
x̄1=19.5
x̄2=25.6
Sx1=4.5
Sx2=6.3
n1=25
n2=25
```

Which of the following is a correct conclusion?

(A) After 7 days, the mean pain level with the leech treatment is significantly lower than the mean pain level with the placebo at the 0.01 significance level.
(B) After 7 days, the mean pain level with the leech treatment is significantly lower than the mean pain level with the placebo at the 0.05 significance level but not at the 0.01 level.
(C) After 7 days, the mean pain level with the leech treatment is significantly lower than the mean pain level with the placebo at the 0.10 significance level but not at the 0.05 level.
(D) After 7 days, the mean pain level with the leech treatment is not significantly lower than the mean pain level with the placebo at the 0.10 significance level.
(E) The proper test should be a one-sample t-test on a set of differences.

30. A survey was conducted to determine the percentage of parents who would support raising the legal driving age to 18. The results were stated as 67% with a margin of error of ±3%. What is meant by ±3%?

(A) Three percent of the population were not surveyed.
(B) In the sample, the percentage of parents who would support raising the driving age is between 64% and 70%.
(C) The percentage of the entire population of parents who would support raising the driving age is between 64% and 70%.
(D) It is unlikely that the given sample proportion result could be obtained unless the true percentage was between 64% and 70%.
(E) Between 64% and 70% of the population were surveyed.

31. It is estimated that 30% of all cars parked in a metered lot outside City Hall receive tickets for meter violations. In a random sample of 5 cars parked in this lot, what is the probability that at least one receives a parking ticket?

(A) $1 - (0.3)^5$
(B) $1 - (0.7)^5$
(C) $5(0.3)(0.7)^4$
(D) $5(0.3)^4 (0.7)$
(E) $5(0.3)^4(0.7) + 10(0.3)^3(0.7)^2 + 10(0.3)^2(0.7)^3 + 5(0.3)(0.7)^4 + (0.7)^5$

32. Data are collected on income levels x versus number of bank accounts y. Summary calculations give $\bar{x} = 32,000$, $s_x = 11,500$, $\bar{y} = 1.7$, $s_y = 0.4$, and $r = 0.42$. What is the slope of the least squares regression line of number of bank accounts on income level?

(A) $\dfrac{(0.42)(0.4)}{11,500}$

(B) $\dfrac{(0.42)(11,500)}{0.4}$

(C) $\dfrac{(1.7)(11,500)}{0.42}$

(D) $\dfrac{(1.7)(0.42)}{11,500}$

(E) $\dfrac{(0.42)(11,500)}{1.7}$

33. Given that the sample has a standard deviation of zero, which of the following is a true statement?

(A) The standard deviation of the population is also zero.
(B) The sample mean and sample median are equal.
(C) The sample may have outliers.
(D) The population has a symmetric distribution.
(E) All samples from the same population will also have a standard deviation of zero.

34. In one study, half of a class were instructed to watch exactly 1 hour of television per day, the other half were told to watch 5 hours per day, and then their class grades were compared. In a second study, students in a class responded to a questionnaire asking about their television usage and their class grades.

   (A) The first study was an experiment without a control group, and the second was an observational study.
   (B) The first study was an observational study, and the second was a controlled experiment.
   (C) Both studies were controlled experiments.
   (D) Both studies were observational studies.
   (E) Each study was part controlled experiment and part observational study.

35. All of the following statements are true for all discrete random variables *except* for which one?

   (A) The possible outcomes must all be numerical.
   (B) The possible outcomes must be mutually exclusive.
   (C) The mean (expected value) always equals the sum of the products obtained by multiplying each value by its corresponding probability.
   (D) The standard deviation of a random variable can never be negative.
   (E) Approximately 95% of the outcomes will be within two standard deviations of the mean.

36. In leaving for school on an overcast April morning, you make a judgment on the null hypothesis: the weather will remain dry. What would the results be of Type I and Type II errors?

   (A) Type I error: get drenched
       Type II error: needlessly carry around an umbrella
   (B) Type I error: needlessly carry around an umbrella
       Type II error: get drenched
   (C) Type I error: carry an umbrella, and it rains
       Type II error: carry no umbrella, and weather remains dry
   (D) Type I error: get drenched
       Type II error: carry no umbrella, and weather remains dry
   (E) Type I error: get drenched
       Type II error: carry an umbrella, and it rains

37. The mean thrust of a certain model jet engine is 9500 pounds. Concerned that a production process change might have lowered the thrust, an inspector tests a sample of units, calculating a mean of 9350 pounds with a z-score of -2.46 and a P-value of 0.0069. Which of the following is the most reasonable conclusion?

(A) 99.31% of the engines produced under the new process will have a thrust under 9350 pounds.

(B) 99.31% of the engines produced under the new process will have a thrust under 9500 pounds.

(C) 0.69% of the time, an engine produced under the new process will have a thrust over 9500 pounds.

(D) There is sufficient evidence to conclude that the new process is producing engines with a mean thrust under 9350 pounds.

(E) There is sufficient evidence to conclude that the new process is producing engines with a mean thrust under 9500 pounds.

38. In a group of 10 third graders, the mean height is 50 inches with a median of 47 inches, while in a group of 12 fourth graders, the mean height is 54 inches with a median of 49 inches. What is the median height of the combined group?

(A) 48 inches

(B) 52 inches

(C) $\dfrac{10(47)+12(49)}{22}$ inches

(D) $\dfrac{10(50)+12(54)}{22}$ inches

(E) With the information provided, a specific value for the median height of the combined group cannot be determined.

39. A reading specialist in a large public school system believes that the more time students spend reading, the better they will do in school. She plans a middle school experiment in which a simple random sample (SRS) of 30 eighth graders will be assigned four extra hours of reading per week, an SRS of 30 seventh graders will be assigned two extra hours of reading per week, and an SRS of 30 sixth graders with no extra assigned reading will be a control group. After one school year, the mean GPAs from each group will be compared. Is this a good experimental design?

(A) Yes, because simple random samples were used.

(B) No, because while this design may point out an association between reading and GPA, it cannot establish a cause-and-effect relationship.

(C) No, because without blinding, there is a strong chance of a placebo effect.

(D) No, because any conclusion would be flawed because of blocking bias.

(E) No, because grade level is a confounding variable.

40. A study at 35 large city high schools gives the following back-to-back stemplot of the percentages of students who say they have tried alcohol.

School Year 2016–17		School Year 2019–20
	0	
2	1	
9	2	
6 6 3	3	1 8
4 3 1 1	4	0 2 9
9 9 8 6 5 3 2 2 0	5	3 3 4 6
9 8 7 6 6 5 1 1	6	1 2 2 2 7
5 4 3 2 2	7	0 1 3 3 5 5 6 7 8 9
5 4 0	8	2 3 4 5 8 8 9
0	9	0 1 1 2

Key: 1|4|0 represents 41% for the 2016–17 school year and 40% for the 2019–20 school year

Which of the following does *not* follow from the above data?

(A) In general, the percentage of students trying alcohol seems to have increased from 2016–17 to 2019–20.

(B) The median alcohol percentage among the 35 schools increased from 2016–17 to 2019–20.

(C) The spread between the lowest and highest alcohol percentages decreased from 2016–17 to 2019–20.

(D) For both school years in most of the 35 schools, most of the students said they had tried alcohol.

(E) The percentage of students trying alcohol increased in each of the schools between 2016–17 and 2019–20.

STOP

If there is still time remaining, you may check your work on this section.

Part A

QUESTIONS 1–5

Spend about 65 minutes on this part of the exam.
Percentage of Section II grade—75

> **Directions:** You must show all work and indicate the methods you use. You will be graded on the correctness of your methods and on the accuracy of your results and explanations.

1. A guidance counselor at a large college is interested in interviewing students about their college experience. After talking with a local AP Statistics teacher, the guidance counselor plans the following three-stage sampling procedure.

   (1) The counselor will obtain separate lists of the four groups of students on the college campus: freshmen, sophomores, juniors, and seniors.

   (2) Each of the four groups takes a number of large classes together. The counselor will use a random number generator to pick three of these classes for each of the four groups.

   (3) In each of the chosen classes, the counselor will pick every fifth student entering the classroom.

   (a) The first stage above represents what kind of sampling procedure? Give an advantage in using it in this context.

   (b) The second stage above represents what kind of sampling procedure? Give an advantage in using it in this context.

   (c) The third stage above represents what kind of sampling procedure? Give a *disadvantage* in using it in this context.

2. (a) Suppose that in an election year, nationwide, 69% of all registered voters would answer "Yes" to the question, "Do you consider yourself highly focused on this year's presidential election?" Which of the following is more likely: a simple random sample (SRS) of 50 registered voters having over 75% answer "Yes" or an SRS of 100 registered voters having over 75% answer "Yes" to the given question? Explain.

   (b) A particular company with 95 employees wishes to survey their employees with regard to interest in the presidential election. They pick an SRS of 30 employees and ask, "Do you consider yourself highly focused on this year's presidential election?" Suppose that in fact 60% of all 95 employees would have answered "Yes." Explain why it is not reasonable to say that the distribution for the count in the sample who say "Yes" is a binomial with $n = 30$ and $p = 0.6$.

(c) Suppose that nationwide, 78% of all registered voters would answer "Yes" to the question, "Do you consider yourself highly focused on this year's presidential election?" You plan to interview an SRS of 20 registered voters. Explain why it is not reasonable to say that the distribution for the proportion in the sample who say "Yes" is approximately a normal distribution.

3. An instructor takes an anonymous survey and notes exam score, hours studied, and gender for the first exam in a large college statistics class ($n = 250$). A resulting regression model is

$$\widehat{Score} = 50.90 + 9.45(Hours) + 4.40(Gender)$$

where *Gender* takes the value 0 for men and 1 for women.

(a) Provide an interpretation in context for each of the three numbers appearing in the above model formula.

(b) Sketch the separate prediction lines for men and women resulting from using 0 or 1 in the above model.

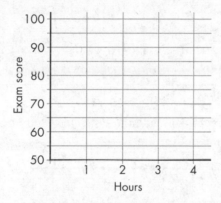

Looking at the data separately by gender results in the following two regression models:

For men: $\widehat{Score} = 51.8 + 8.5(Hours)$

For women: $\widehat{Score} = 54.4 + 10.4(Hours)$

(c) What comparative information do the slope coefficients from these two models give that does not show in the original model?

4. A laboratory is testing the concentration level in milligrams per milliliter for the active ingredient found in a pharmaceutical product. In a random sample of five vials of the product, the concentrations were measured at 2.46, 2.57, 2.70, 2.64, and 2.54 mg/mL.

(a) Determine a 95% confidence interval for the mean concentration level in milligrams per milliliter for the active ingredient found in this pharmaceutical product.

(b) Explain in words what effect an increase in confidence level would have on the width of the confidence interval.

(c) Suppose a concentration above 2.70 milligrams per milliliter is considered dangerous. What conclusion is justified by your answers to (a) and (b)?

5. A point is said to have *high leverage* if it is an outlier in the *x*-direction, and a point is said to be *influential* if its removal sharply changes the regression line.

(a) In the scatterplots below, compare points *A* and *B* with regard to having high leverage and with regard to being influential.

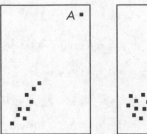

(b) In the scatterplots below, compare points *C* and *D* with regard to residuals and influence on β, the slope of the regression line.

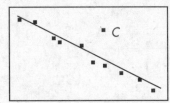

(c) In the scatterplot below, compare the effect of removing point *E* or point *F* to that of removing both *E* and *F*.

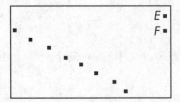

(d) In the scatterplot below, compare the effect of removing point *G* or point *H* to that of removing both *G* and *H*.

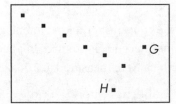

SECTION II

Part B

QUESTION 6

Spend about 25 minutes on this part of the exam.
Percentage of Section II grade—25

6. A college guidance counselor obtains a survey from a random sample of 40 students who live in the college dormitories. He concludes that students with higher GPAs (on a 4-point scale) appear to be less satisfied (on a 100-point scale) with dorm life. A scatterplot and linear regression analysis shows the following computer output:

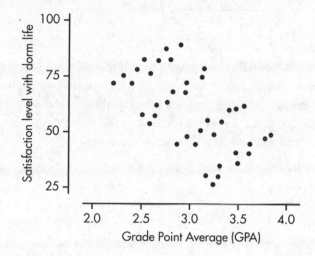

```
Predicted satisfaction level = 117.5 − 19.93(GPA)
S = 14.80    df = 38    t = −3.559    P = 0.001    R−Sq = 0.250
```

(a) Does the scatterplot and regression analysis support the guidance counselor's conclusion?
(b) What is the correlation, r?
(c) If one student's GPA is 0.5 greater than a second student's GPA, how would their predicted satisfaction levels compare?

A student who had taken AP Statistics in high school reads that the survey was obtained from a stratified random sample. She analyzes the data and produces the following revised scatterplot:

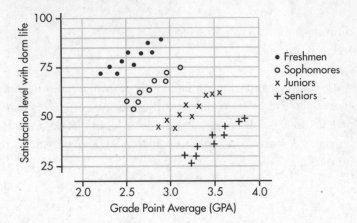

(d) If one student's GPA is 0.5 greater than a second student's GPA, how would their predicted satisfaction levels compare if the students are both sophomores?

(e) Based on the student's analysis, how should the counselor's conclusion be modified to give a better description of the relationship between GPA and satisfaction level for the students in the sample?

STOP

If there is still time remaining, you may check your work on this section.

Answer Key

1. **A**	9. **E**	17. **A**	25. **D**	33. **B**
2. **E**	10. **A**	18. **C**	26. **D**	34. **A**
3. **C**	11. **E**	19. **A**	27. **D**	35. **E**
4. **E**	12. **C**	20. **D**	28. **D**	36. **B**
5. **C**	13. **B**	21. **B**	29. **A**	37. **E**
6. **A**	14. **C**	22. **E**	30. **D**	38. **E**
7. **C**	15. **E**	23. **C**	31. **B**	39. **E**
8. **E**	16. **E**	24. **B**	32. **A**	40. **E**

Answers and Explanations

SECTION I

1. **(A)** The y-intercept, 12.1, gives the average aptitude test score for employees with 0 days training, and this can be considered to be a "base score." The slope, 6.2, gives the predicted average increase in the y-variable for each unit increase in the x-variable.

2. **(E)** For a simple random sample, every possible group of the given size has to be equally likely to be selected, and this is not true here. For example, with this procedure it will be impossible for all the early arrivals to be together in the final sample. This procedure is an example of systematic sampling (although the starting point should be randomly selected), but systematic sampling does not result in simple random samples.

3. **(C)** Power $= 1 - \beta$, and β is smallest when both α and n are greater.

4. **(E)** By the central limit theorem, the sampling distribution of $\bar{x}$ is approximately normal with mean equal to the population mean and with standard deviation equal to the population standard deviation divided by the square root of the sample size. In this example,

   $$\mu_{\bar{x}} = \mu = 435 \text{ and } \sigma_{\bar{x}} = \frac{\alpha}{\sqrt{n}} = \frac{85}{\sqrt{100}} = 8.5.$$

5. **(C)** We have H_0: $\mu = 3.5$ and H_a: $\mu < 3.5$. Then $\sigma_{\bar{x}} = \frac{0.9}{\sqrt{36}}$ and the z-score of 3.25 is $\frac{3.25 - 3.5}{\left(\frac{0.9}{\sqrt{36}}\right)}$.

6. **(A)** The cumulative proportions of 0.25 and 0.75 correspond to $Q_1 = 57$ and $Q_3 = 75$, respectively, and so the interquartile range is $75 - 57 = 18$.

7. **(C)** A placebo is a control treatment in which members of the control group do not realize whether or not they are receiving the experimental treatment.

8. **(E)** In experiments on people, subjects can be used as their own controls, with responses noted before and after the treatment. However, with such designs there is always the danger of a placebo effect. In this case, subjects might well have slower reaction times after drinking the alcohol because they think they should.

9. **(E)** Since $(-2, 4)$ is on the line $\hat{y} = 7x + b$, we have $4 = -14 + b$ and $b = 18$. Thus, the regression line is $\hat{y} = 7x + 18$. The point $(\bar{x}, \bar{y})$ is always on the regression line, and so we have $\bar{y} = 7\bar{x} + 18$.

10. **(A)** A simple random sample can be any size.

11. **(E)** With such small populations, censuses instead of samples are used, and there is no resulting probability statement about the difference.

12. **(C)** When the distribution is skewed to the right, the mean is usually greater than the median. Half the area is on either side of 23, so 23 is the median. With half the area to each side of 23, half the applicants' ages are to each side of 23. Histograms such as this show relative frequencies, not actual frequencies.

13. **(B)** While the procedure does use some element of chance, all possible groups of size 75 do not have the same chance of being picked, so the result is not a simple random sample. There is a real chance of selection bias. For example, a number of relatives with the same last name and all using the same long-distance carrier might be selected.

14. **(C)** While t-distributions do have mean 0, their standard deviations are greater than 1.

15. **(E)**

$$P(\text{smoker} \mid \text{cancer}) = \frac{P(\text{smoker} \cap \text{cancer})}{P(\text{cancer})} = \frac{0.014}{0.035}$$

16. **(E)** The critical z-scores are $\frac{80{,}000 - 75{,}000}{12{,}000} = 0.42$ and $\frac{100{,}000 - 75{,}000}{12{,}000} = 2.08$, with corresponding right tail probabilities of 0.3372 and 0.0188. The probability of being less than \$100,000 given that the starting salary is over \$80,000 is $\frac{0.3372 - 0.0188}{0.3372} = 0.94$. [Or

$$\frac{\text{normalcdf}(80000, 100000, 75000, 12000)}{\text{normalcdf}(80000, 1E99, 75000, 12000)} = 0.9450.]$$

17. **(A)**

$$\sigma_{\bar{x}_1 - \bar{x}_2} = \sqrt{\sigma_{\bar{x}_1}^2 + \sigma_{\bar{x}_2}^2} = \sqrt{\left(\frac{\sigma_1}{\sqrt{n}}\right)^2 + \left(\frac{\sigma_2}{\sqrt{n}}\right)^2} = \sqrt{\frac{0.8^2}{n} + \frac{0.5^2}{n}} = \frac{\sqrt{0.8^2 + 0.5^2}}{\sqrt{n}}$$

18. **(C)** The correlation coefficient is not changed by adding the same number to every value of one of the variables, by multiplying every value of one of the variables by the same positive number, or by interchanging which are the x- and y-variables.

19. **(A)** The binomial distribution with $n = 2$ and $p = 0.8$ is $P(0) = (0.2)^2 = 0.04$, $P(1) = 2(0.2)(0.8) = 0.32$, and $P(2) = (0.8)^2 = 0.64$, resulting in expected numbers of $0.04(200) = 8$, $0.32(200) = 64$, and $0.64(200) = 128$. Thus,

$$X^2 = \sum \frac{(\text{obs} - \text{exp})^2}{\text{exp}} = \frac{(10 - 8)^2}{8} + \frac{(80 - 64)^2}{64} + \frac{(110 - 128)^2}{128}$$

20. **(D)** Option I guarantees that the \$20,000 loan will be paid off. Option II gives the highest expected return: $(50{,}000)(0.5) + (10{,}000)(0.5) = 30{,}000$, which is greater than 25,000 and is also greater than $(100{,}000)(0.05) = 5000$. Option III provides the only chance of paying off the \$80,000 loan. The moral is that the highest expected value is not automatically the "best" answer.

21. **(B)**

$P(X|Y) = \dfrac{P(X \cap Y)}{P(Y)}$ gives that $P(X \cap Y) = P(X|Y)\,P(Y) = (0.28)(0.40)$.

Then $P(Y|X) = \dfrac{P(X \cap Y)}{P(X)} = \dfrac{(0.28)(0.40)}{0.35}$.

22. **(E)** This study is an experiment because a treatment (extensive exercise) is imposed. There is no blinding because subjects clearly know whether or not they are exercising. There is no blocking because subjects are not divided into blocks before random assignment to treatments. For example, blocking would have been used if subjects had been separated by gender or age before random assignment to exercise or not.

23. **(C)** From the shape of the normal curve, the answer is in the middle. We have $\dfrac{100,000}{150,000} = \dfrac{2}{3}$. The middle two-thirds (with $\dfrac{1}{6}$ in each tail) is between z-scores of $\pm\texttt{invNorm(5/6)} = \pm 0.97$, and $35 \pm 0.97(10)$ gives $(25.3, 44.7)$.

24. **(B)** More than 25% of the patients in the (I) group had over 210 minutes of pain relief, which is not the case for the other two groups. The interquartile range is the length of the box, so they are not all equal. There is no way to positively conclude a normal distribution from a boxplot.

25. **(D)** In mosaic plots, the area of a box is proportional to the count corresponding to that box. The vertical axis in this plot indicates the proportion of students versus teachers. The horizontal axis in this plot indicates the proportion of people choosing each of the three ice cream flavors for students and for teachers. Choice (D) is true because the three boxes corresponding to students have a total area greater than the three boxes corresponding to teachers, or noting that along the vertical axis, "Student" has a greater length than "Teacher."

26. **(D)**

$$\frac{0.54}{0.54 + 0.19 + 0.07} = 0.675$$

27. **(D)** The midpoint of the confidence interval is the sample proportion $\hat{p} = 0.44$.

28. **(D)** $\sigma_{\bar{x}} = \dfrac{\sigma}{\sqrt{n}}$, so the standard deviation of sample means is inversely related to the square root of the sample size. Thus, increasing the sample size by a multiple of d^2 divides the standard deviation of the set of sample means by d.

29. **(A)** With $df = 43.43$ and $t = -3.94$, tcdf gives that the P-value is $0.000146 < 0.01$. With a P-value this small, there is sufficient evidence to reject H_0; that is, there is sufficient evidence that the mean pain level with the leech treatment is lower than the mean pain level with the placebo.

30. **(D)** While the sample proportion is between 64% and 70% (more specifically, it is 67%), this is not the meaning of $\pm 3\%$. While the percentage of the entire population is likely to be between 64% and 70%, this is not known for certain.

31. **(B)** This is a binomial distribution with $n = 5$ and $p = 0.3$. The probability that a car does not receive a ticket is $1 - 0.3 = 0.7$, the probability that none of the five cars receives a ticket is $(0.7)^5$, and thus the probability that at least one receives a ticket is $1 - (0.7)^5$.

32. **(A)**

$$b = r\frac{s_y}{s_x} = 0.42\frac{0.4}{11,500} = \frac{(0.42)(0.4)}{11,500}$$

33. **(B)** If the standard deviation of a set is zero, all the values in the set are equal. The mean and median would both equal this common value and so would equal each other. If all the values are equal, there are no outliers. Just because the sample happens to have one common value, there is no reason for this to be true for the whole population. Statistics from one sample can be different from statistics from any other sample.

34. **(A)** The first study is an experiment with two treatment groups (1 hour and 5 hours of television per night) and no control group. The second study is observational; the researcher simply noted the students' self-reported responses.

35. **(E)** This refers only to very particular random variables, for example, random variables whose values are the number of successes in a binomial probability distribution with large n.

36. **(B)** A Type I error means that the null hypothesis is correct (the weather will remain dry) but you reject it (thus you needlessly carry around an umbrella). A Type II error means that the null hypothesis is wrong (it will rain) but you fail to reject it (thus you get drenched).

37. **(E)** If the sample statistic is far enough away from the claimed population parameter, we say that there is sufficient evidence to reject the null hypothesis. In this case the null hypothesis is that $\mu = 9500$. The P-value is the probability of obtaining a sample statistic as extreme as, or more extreme than, the one obtained if the null hypothesis is assumed to be true. The smaller the P-value, the more significant the difference between the null hypothesis and the sample results. With $P = 0.0069$, there is strong evidence to reject H_0.

38. **(E)** There are $10 + 12 = 22$ students in the combined group. In ascending order, where are the two middle scores? At least 5 third graders and 6 fourth graders have heights less than or equal to 49 inches, so at most 11 students have heights greater than or equal to 49. Thus, the median is less than or equal to 49. At least 5 third graders and 6 fourth graders have heights greater than or equal to 47 inches, so at most 11 students have heights less than or equal to 47. Thus, the median is greater than or equal to 47. All that can be said about the median of the combined group is that it is between 47 and 49 inches.

39. **(E)** Good experimental design aims to give each group the same experiences except for the treatment under consideration, Thus, all three SRSs should be picked from the same grade level.

40. **(E)** The stemplot does not indicate what happened for any individual school.

SECTION II: PART A

1. (a) This is an example of *stratified sampling*, where the groups (freshmen, sophomores, juniors, and seniors) are strata. The advantage is that the counselor will ensure that the final sample will represent all four groups of students, each of which might have different views of their college experience.

(b) This an example of *cluster sampling*, where each class (cluster) resembles the overall group from which it is selected. The advantage is that using these clusters is much more practical than trying to sample from among the large groups (freshman, sophomores, juniors, and seniors).

(c) This is an example of *systematic sampling*, which is quicker and easier than many other procedures. A possible disadvantage occurs if ordering is related to the variable under consideration. For example, in this study, if the order students enter the classroom is related to their views of their college experience, the counselor could end up with a nonrepresentative sample.

SCORING

Part (a) is essentially correct for identifying the procedure and giving a correct advantage in context. Part (a) is partially correct for one of these two elements.

Part (b) is essentially correct for identifying the procedure and giving a correct advantage in context. Part (b) is partially correct for one of these two elements.

Part (c) is essentially correct for identifying the procedure and giving a correct disadvantage in context. Part (c) is partially correct for one of these two elements.

4	**Complete Answer**	All three parts essentially correct.
3	**Substantial Answer**	Two parts essentially correct and one part partially correct.
2	**Developing Answer**	Two parts essentially correct OR one part essentially correct and one or two parts partially correct OR all three parts partially correct.
1	**Minimal Answer**	One part essentially correct OR two parts partially correct.

2. (a) The SRS with $n = 50$ is more likely to have a sample proportion greater than 75%. In each case, the sampling distribution of $\hat{p}$ is approximately normal with a mean of 0.69 and a standard deviation of $\sigma_{\hat{p}} = \sqrt{\dfrac{pq}{n}} = \sqrt{\dfrac{(0.69)(0.31)}{n}} = \dfrac{0.462}{\sqrt{n}}$. Thus, the sampling distribution with $n = 50$ will have more variability than the sampling distribution with $n = 100$. Thus, the tail area ($\hat{p} > 0.75$) will be larger for $n = 50$.

(b) The size of the sample, 30, is much too large compared to the size of the population, 95. With a sample size this close to the population size, the necessary assumption of independence does not follow. A commonly accepted condition is that the sample should be no more than 10% of the population.

(c) A normal distribution may be used to approximate a binomial distribution only if the sample size n is not too small. The commonly used condition check is that both np and nq are at least 10. In this case, $nq = (20)(1 - 0.78) = 4.4$.

SCORING

Part (a) is essentially correct for correctly giving $n = 50$ and linking in context to variability in the sampling distributions. Part (a) is partially correct for a correct answer missing comparison of variability in the sampling distributions.

Part (b) is essentially correct for a clear explanation in context. Part (b) is partially correct for saying the sample size is too close to the population size but not linking to this context (sample size 30 and population size 95).

Part (c) is essentially correct for a clear explanation in context. Part (c) is partially correct for saying the sample size is too small but not linking to this context ($nq = 4.4 < 10$).

4	Complete Answer	All three parts essentially correct.
3	Substantial Answer	Two parts essentially correct and one part partially correct.
2	Developing Answer	Two parts essentially correct OR one part essentially correct and one or two parts partially correct OR all three parts partially correct.
1	Minimal Answer	One part essentially correct OR two parts partially correct.

3. (a) The value 50.90 estimates the average score of men who spend 0 hours studying. The value 9.45 estimates the average increase in score for each additional hour of study time. For any fixed number of hours of study time, the value 4.40 estimates the average number of points that women score higher than men.

(b)

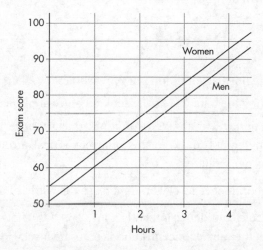

(c) For each additional hour of study time, the scores of men increase an average of 8.5 while those of women increase an average of 10.4. Additional hours of study time appear to benefit women more than men, something that does not show in the original model.

4. (a) *Parameter:* Let μ represent the mean concentration level for the active ingredient found in the population of vials of this pharmaceutical product.

Procedure: A one-sample t-interval for the mean.

Checks: We are given that this is a random sample, $n = 5$ is less than 10% of all vials of this product, and a dotplot

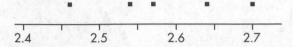

TIP

Use TInterval on the TI-84 or 1-Sample tInterval on the Casio Prizm.

makes the nearly normal condition not unreasonable.

Mechanics: Putting the data in a List, calculator software gives (2.4674, 2.6966).

Conclusion in context: We are 95% confident that the true mean concentration level for the active ingredient found in this pharmaceutical product is between 2.467 and 2.697 milligrams per milliliter.

(b) Raising the confidence level would increase the width of the confidence interval.

(c) Since the whole confidence interval (2.467, 2.697) is below the critical 2.7, at the 95% confidence level the mean concentration is at a safe level. However, with 2.697 so close to 2.7, based on the statement in (c), if the confidence level is raised we are no longer confident that the mean concentration is at a safe level.

SCORING

Section 1, Part (a), is essentially correct for correctly naming the procedure and checking the conditions, and is partially correct for one of these two.

Section 2, Part (a), is essentially correct for correct mechanics and a correct conclusion in context, and is partially correct for one of these two.

Section 3, Parts (b) and (c), is essentially correct for 1) a correct statement in Part (b), 2) a correct conclusion in context for the 95% interval, and 3) a correct conclusion in context for the case where the confidence is raised. Section 3 is partially correct for two of these three components correct.

4   Complete Answer   All three sections essentially correct.

3   Substantial Answer   Two sections essentially correct and one section partially correct.

2   Developing Answer   Two sections essentially correct OR one section essentially correct and one or two sections partially correct OR all three sections partially correct.

1   Minimal Answer   One section essentially correct OR two sections partially correct.

5. (a) Points A and B both have high leverage; that is, both their x-coordinates are outliers in the x-direction. However, point B is influential (its removal sharply changes the regression line), while point A is not influential (it appears to lie close to or directly on the regression line, so its removal will not change the line).

(b) Point C lies off the regression line. So, its residual is much greater than that of point D, whose residual is 0 (point D lies on the regression line). However, the removal of point C will very minimally affect the slope of the regression line, if at all, while the removal of point D dramatically affects the slope of the regression line.

(c) Removal of either point E or point F minimally affects the regression line, while removal of both has a dramatic effect.

(d) Removing either point G or point H will definitely affect the regression line (pulling the line toward the remaining of the two points). However, removing both will have little, if any, effect on the line.

SCORING

Each of Parts (a), (b), (c), and (d) has two components and is scored essentially correct for both components correct and partially correct for one component correct.

Count essentially correct answers as one point and partially correct answers as one-half point.

4	**Complete Answer**	4 points
3	**Substantial Answer**	3 points
2	**Developing Answer**	2 points
1	**Minimal Answer**	1 point

Use a holistic approach to decide a score totaling between two numbers, deciding whether to score up or down depending on the strength of the response and communication.

SECTION II: PART B

6. (a) The scatterplot is generally negative, and in fact the slope is given to be a negative number. So, yes, this supports the guidance counselor's conclusion.

(b) $r = -\sqrt{0.250} = -0.5$.

(c) $(-19.93)(0.5) = -9.965$. So, the student with the greater GPA is predicted to have a satisfaction level 9.965 lower than the other student.

(d) Two points on the sophomore regression line are approximately $(2.5, 55)$ and $(3.0, 75)$, so the sophomore student with the 0.5 higher GPA is predicted to have a $75 - 55 = 20$ higher satisfaction level.

(e) Overall, higher grade point averages are associated with lower satisfaction with dorm life. However, within each class level (freshmen, sophomores, juniors, and seniors), students with higher grade point averages tend to have higher satisfaction with dorm life.

Section 1 is essentially correct for answering yes in (a), justifying by noting the negative slope, and correctly calculating the correlation in (b). Section 1 is partially correct for two out of the three steps above.

Section 2 is essentially correct for calculating the "9.965 lower" in (c) and the "20 higher" in (d) and is partially correct for one out of these two steps correct.

Section 3 is essentially correct if the response states there is a positive association for each class AND the response notes the overall negative association. Section 3 is partially correct if the response states there is a positive association for each class but does not note the overall negative association.

4   **Complete Answer**   All three sections essentially correct.

3   **Substantial Answer**   Two sections essentially correct and one section partially correct.

2   **Developing Answer**   Two sections essentially correct OR one section essentially correct and one or two sections partially correct OR all three sections partially correct.

1   **Minimal Answer**   One section essentially correct OR two sections partially correct.

ANSWER SHEET
Practice Test 5

1. Ⓐ Ⓑ Ⓒ Ⓓ Ⓔ
2. Ⓐ Ⓑ Ⓒ Ⓓ Ⓔ
3. Ⓐ Ⓑ Ⓒ Ⓓ Ⓔ
4. Ⓐ Ⓑ Ⓒ Ⓓ Ⓔ
5. Ⓐ Ⓑ Ⓒ Ⓓ Ⓔ
6. Ⓐ Ⓑ Ⓒ Ⓓ Ⓔ
7. Ⓐ Ⓑ Ⓒ Ⓓ Ⓔ
8. Ⓐ Ⓑ Ⓒ Ⓓ Ⓔ
9. Ⓐ Ⓑ Ⓒ Ⓓ Ⓔ
10. Ⓐ Ⓑ Ⓒ Ⓓ Ⓔ

11. Ⓐ Ⓑ Ⓒ Ⓓ Ⓔ
12. Ⓐ Ⓑ Ⓒ Ⓓ Ⓔ
13. Ⓐ Ⓑ Ⓒ Ⓓ Ⓔ
14. Ⓐ Ⓑ Ⓒ Ⓓ Ⓔ
15. Ⓐ Ⓑ Ⓒ Ⓓ Ⓔ
16. Ⓐ Ⓑ Ⓒ Ⓓ Ⓔ
17. Ⓐ Ⓑ Ⓒ Ⓓ Ⓔ
18. Ⓐ Ⓑ Ⓒ Ⓓ Ⓔ
19. Ⓐ Ⓑ Ⓒ Ⓓ Ⓔ
20. Ⓐ Ⓑ Ⓒ Ⓓ Ⓔ

21. Ⓐ Ⓑ Ⓒ Ⓓ Ⓔ
22. Ⓐ Ⓑ Ⓒ Ⓓ Ⓔ
23. Ⓐ Ⓑ Ⓒ Ⓓ Ⓔ
24. Ⓐ Ⓑ Ⓒ Ⓓ Ⓔ
25. Ⓐ Ⓑ Ⓒ Ⓓ Ⓔ
26. Ⓐ Ⓑ Ⓒ Ⓓ Ⓔ
27. Ⓐ Ⓑ Ⓒ Ⓓ Ⓔ
28. Ⓐ Ⓑ Ⓒ Ⓓ Ⓔ
29. Ⓐ Ⓑ Ⓒ Ⓓ Ⓔ
30. Ⓐ Ⓑ Ⓒ Ⓓ Ⓔ

31. Ⓐ Ⓑ Ⓒ Ⓓ Ⓔ
32. Ⓐ Ⓑ Ⓒ Ⓓ Ⓔ
33. Ⓐ Ⓑ Ⓒ Ⓓ Ⓔ
34. Ⓐ Ⓑ Ⓒ Ⓓ Ⓔ
35. Ⓐ Ⓑ Ⓒ Ⓓ Ⓔ
36. Ⓐ Ⓑ Ⓒ Ⓓ Ⓔ
37. Ⓐ Ⓑ Ⓒ Ⓓ Ⓔ
38. Ⓐ Ⓑ Ⓒ Ⓓ Ⓔ
39. Ⓐ Ⓑ Ⓒ Ⓓ Ⓔ
40. Ⓐ Ⓑ Ⓒ Ⓓ Ⓔ

Practice Test 5

SECTION I

Questions 1–40

Spend 90 minutes on this part of the exam.

> **Directions:** The questions or incomplete statements that follow are each followed by five suggested answers or completions. Choose the response that best answers the question or completes the statement.

1. The mean and standard deviation of the population $\{1, 5, 8, 11, 15\}$ are $\mu = 8$ and $\sigma = 4.8$, respectively. Let S be the set of the 125 *ordered* triples (repeats allowed) of elements of the original population. Which of the following is a correct statement about the mean $\mu_{\bar{x}}$ and standard deviation $\sigma_{\bar{x}}$ of the means of the triples in S?

   (A) $\mu_{\bar{x}} = 8, \sigma_{\bar{x}} = 4.8$
   (B) $\mu_{\bar{x}} = 8, \sigma_{\bar{x}} < 4.8$
   (C) $\mu_{\bar{x}} = 8, \sigma_{\bar{x}} > 4.8$
   (D) $\mu_{\bar{x}} < 8, \sigma_{\bar{x}} = 4.8$
   (E) $\mu_{\bar{x}} > 8, \sigma_{\bar{x}} > 4.8$

2. Consider the following studies being run by three different AP Statistics instructors.

   I. One rewards students every day with lollipops for relaxation, encouragement, and motivation to learn the material.
   II. One promises that all students will receive A's as long as they give their best efforts to learn the material.
   III. One is available every day after school and on weekends so that students with questions can come in and learn the material.

   Which of the following statements is true about the three studies described in I, II, and III above?

   (A) None of these studies use randomization.
   (B) None of these studies use control groups.
   (C) None of these studies use blinding.
   (D) Important information can be found from all of these studies, but none can establish causal relationships.
   (E) All of the above.

3. A survey to measure job satisfaction of high school mathematics teachers was taken in 2015 and repeated 5 years later in 2020. Each year a random sample of 50 teachers rated their job satisfaction on a 1-to-100 scale with higher numbers indicating greater satisfaction. The results are given in the following back-to-back stemplot.

2015		2020	
	0		Key: $0\,|\,2\,|\,1$ represents
98775	1		a rating of 20 in
98530	2	1	2015 and a rating
65210	3	3589	of 21 in 2020
96430	4	01122233455667889999	
87421	5	035667899	
99877555322100	6	1344789	
976442	7	22689	
7511	8	138	
0	9	7	

What is the trend from 2015 to 2020 with regard to the standard deviation and range of the two samples?

(A) Both the standard deviation and range increased.
(B) The standard deviation increased, while the range decreased.
(C) The range increased, while the standard deviation decreased.
(D) Both the standard deviation and range decreased.
(E) Both the standard deviation and range remained unchanged.

4. The number of days it takes to build a new house has a variance of 386. A sample of 40 new homes shows an average building time of 83 days. With what confidence can we assert that the average building time for a new house is between 80 and 90 days?

(A) 15.4%
(B) 17.8%
(C) 20.0%
(D) 38.8%
(E) 82.1%

5. A shipment of resistors has an average resistance of 200 ohms with a standard deviation of 5 ohms, and the resistances are roughly normally distributed. Suppose a randomly chosen resistor has a resistance under 194 ohms. What is the probability that its resistance is greater than 188 ohms?

(A) 0.07
(B) 0.12
(C) 0.50
(D) 0.93
(E) 0.97

6. Suppose 4% of the population have a certain disease. A laboratory blood test gives a positive reading for 95% of people who have the disease and for 5% of people who do not have the disease. If a person tests positive, what is the probability the person has the disease?

   (A) 0.038
   (B) 0.086
   (C) 0.442
   (D) 0.558
   (E) 0.950

7. Consider the following three inference projects:

   I. Finding a 95% confidence interval of mean height of teachers in a small town
   II. Finding a 95% confidence interval of the proportion of students in a small town who are taking some AP class
   III. Performing a two-tailed hypothesis test where the null hypothesis was that the mean expenditure on entertainment by male students at a high school is the same as that of female students

   For which of the above projects is it appropriate to use a census?

   (A) A census is not appropriate for any of the above projects
   (B) Only I
   (C) Only II
   (D) Only I and II
   (E) I, II, and III

8. On the same test, Mary and Pam scored at the 64th and 56th percentiles, respectively. Which of the following is a true statement?

   (A) Mary scored eight more points than Pam.
   (B) Mary's score is 8% higher than Pam's.
   (C) Eight percent of those who took the test scored between Pam and Mary.
   (D) Thirty-six people scored higher than both Mary and Pam.
   (E) None of the above.

9. Which of the following is a true statement?

   (A) While observational studies gather information on an already existing condition, they still often involve intentionally forcing some treatment to note the response.
   (B) In an experiment, researchers decide on the treatment but typically allow the subjects to self-select into the control group.
   (C) If properly designed, either observational studies or controlled experiments can easily be used to establish cause and effect.
   (D) Wording to disguise hidden interests in observational studies is the same idea as blinding in experimental design.
   (E) Stratifying in sampling is a similar idea to blocking for experiments.

10. The random variable describing the number of minutes high school students spend in front of a computer daily has a mean of 200 minutes. Samples of two different sizes result in sampling distributions with the two graphs below.

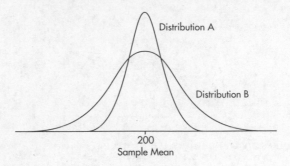

Which of the following is a true statement?

(A) Based on these graphs, no comparison between the two sample sizes is possible.
(B) More generally, sample sizes have no effect on sampling distributions.
(C) The sample size in A is the same as the sample size in B.
(D) The sample size in A is less than the sample size in B.
(E) The sample size in A is greater than the sample size in B.

11. To determine the mean cost of groceries in a certain city, an identical grocery basket of food is purchased at each store in a random sample of ten stores. If the average cost is $47.52 with a standard deviation of $1.59, find a 98% confidence interval estimate for the cost of these groceries in the city.

(A) $47.52 \pm 2.33\sqrt{1.59}$

(B) $47.52 \pm 2.33\left(\dfrac{1.59}{\sqrt{10}}\right)$

(C) $47.52 \pm 2.33\left(\sqrt{\dfrac{1.59}{10}}\right)$

(D) $47.52 \pm 2.821\left(\dfrac{1.59}{\sqrt{10}}\right)$

(E) $47.52 \pm 2.821\sqrt{\dfrac{1.59}{10}}$

12. A set consists of four numbers. The largest value is 200, and the range is 50. Which of the following statements is true?

(A) The mean is less than 185.
(B) The mean is greater than 165.
(C) The median is less than 195.
(D) The median is greater than 155.
(E) The median is the mean of the second and third numbers if the set is arranged in ascending order.

13. Many Americans believe that "fake news" is sowing confusion. In a random sample of 400 people, 92 say they have shared a made-up news story—either knowingly or not. How sure can we be that between 19% and 27% of adults have either knowingly or not shared a made-up news story?

    (A) 0.38%
    (B) 17.9%
    (C) 23.0%
    (D) 89.0%
    (E) 94.3%

14. Suppose we have a random variable X where the probability associated with the value k is $\binom{15}{k}(0.29)^k(0.71)^{15-k}$ for $k = 0,\dots,15$.

    What is the mean of X?

    (A) 0.29
    (B) 0.71
    (C) 4.35
    (D) 10.65
    (E) None of the above

15. The financial aid office at a state university conducts a study to determine the total student costs per semester. All students are charged $7500 for tuition. The mean cost for books is $350 with a standard deviation of $65. The mean outlay for room and board is $4800 with a standard deviation of $380. The mean personal expenditure is $1075 with a standard deviation of $125. Assuming independence among categories, what is the standard deviation of the total student costs?

    (A) $24
    (B) $91
    (C) $190
    (D) $405
    (E) $570

16. Suppose X and Y are random variables with $E(X) = 312$, $\text{var}(X) = 6$, $E(X) = 307$, and $\text{var}(Y) = 8$. What are the expected value and variance of the random variable $X + Y$?

    (A) $E(X + Y) = 619$, $\text{var}(X + Y) = 7$
    (B) $E(X + Y) = 619$, $\text{var}(X + Y) = 10$
    (C) $E(X + Y) = 619$, $\text{var}(X + Y) = 14$
    (D) $E(X + Y) = 309.5$, $\text{var}(X + Y) = 14$
    (E) There is insufficient information to answer this question.

17. In sample surveys, what is meant by *bias*?

    (A) A systematic error in a sampling method that tends to lead to unrepresentative samples

    (B) Prejudice, typically in ethnic- and gender-related studies

    (C) Natural variability seen between samples

    (D) Tendency for some distributions to be skewed

    (E) Tendency for some distributions to vary from normality

18. The following histogram gives the shoe sizes of people in an elementary school building one morning.

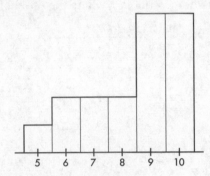

Which of the following is a true statement?

    (A) The distribution of shoe sizes is bimodal.

    (B) The median shoe size is $7\frac{1}{2}$.

    (C) The mean shoe size is probably less than the median shoe size.

    (D) The five-number summary is: $5\frac{1}{2}, 6\frac{1}{2}, 7\frac{1}{2}, 8\frac{1}{2}, 9\frac{1}{2}$.

    (E) Only 10% of the people had size 5 shoes.

19. When comparing the standard normal (z) distribution to the t-distribution with $df = 30$, which of the following is a false statement?

    (A) Both are symmetric.

    (B) Both are bell-shaped.

    (C) Both have center 0.

    (D) Both have standard deviation 1.

    (E) Both are unimodal.

20. Given a probability of 0.65 that interest rates will jump this year and a probability of 0.72 that if interest rates jump the stock market will decline, what is the probability that interest rates will jump and the stock market will decline?

    (A) $0.72 + 0.65 - (0.72)(0.65)$

    (B) $(0.72)(0.65)$

    (C) $1 - (0.72)(0.65)$

    (D) $\dfrac{0.65}{0.72}$

    (E) $1 - \dfrac{0.65}{0.72}$

21. Sampling error is

    (A) the mean of a sample statistic.

    (B) the standard deviation of a sample statistic.

    (C) the standard error of a sample statistic.

    (D) the result of bias.

    (E) the difference between a population parameter and an estimate of that parameter.

22. Which of the following statements about scatterplots and their associated regression lines is *false*?

    (A) A scatterplot gives an immediate indication of the shape (linear or not), strength (weak, moderate, or strong), direction (positive or negative), and unusual features (such as outliers or clusters) of a possible relationship between two variables.

    (B) Outliers are points whose residuals in absolute value are large when compared to those of other residuals.

    (C) Influential points are points whose removal would sharply change the regression line.

    (D) High-leverage points are those whose x-values are far from the mean of the x-values.

    (E) The best-fitting straight line, called the linear regression line, minimizes the sum of the residuals.

23. If the correlation coefficient $r = 0.78$, what percentage of variation in y is not accounted for by the linear regression model?

    (A) 22%

    (B) 39%

    (C) 56%

    (D) 61%

    (E) 78%

24. Consider the following scatterplot:

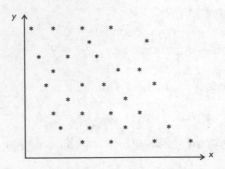

Which of the following is the best estimate of the correlation between *x* and *y*?

(A) −0.95
(B) −0.15
(C) 0
(D) 0.15
(E) 0.95

25. For one NBA playoff game, the actual percentage of the television viewing public who watched the game was 24%. If you had taken a survey of 50 television viewers that night and constructed a confidence interval estimate of the percentage watching the game, which of the following would have been true?

   I. The center of the interval would have been 24%.
  II. The interval would have contained 24%.
 III. A 99% confidence interval estimate would have contained 24%.

(A) I and II only
(B) I and III only
(C) II and III only
(D) All are true.
(E) None is true.

26. Which of the following is a true statement?

(A) In a well-designed, well-conducted sample survey, sampling error is effectively eliminated.
(B) In a well-designed observational study, responses are influenced through an orderly, carefully planned procedure during the collection of data.
(C) In a well-designed experiment, the treatments are carefully planned to result in responses that are as similar as possible.
(D) In a well-designed experiment, double-blinding is a useful matched pairs design.
(E) None of the above is a true statement.

27. Consider the following scatterplot showing the relationship between caffeine intake and job performance.

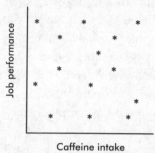

Which of the following is a reasonable conclusion?

(A) Low caffeine intake is associated with low job performance.
(B) Low caffeine intake is associated with high job performance.
(C) High caffeine intake is associated with low job performance.
(D) High caffeine intake is associated with high job performance.
(E) Job performance cannot be predicted from caffeine intake.

28. An author of a new book claims that his diet program will lead to an average weight loss of 2.8 pounds per week for anyone following the program. A researcher believes that the true figure will be lower and plans a test involving a random sample of 36 overweight people. She will reject the author's claim if the mean weight loss in the volunteer group is less than 2.5 pounds per week. Assume that the standard deviation among individuals is 1.2 pounds per week. If the true mean value is 2.4 pounds per week, what is the probability that the researcher will mistakenly fail to reject the author's false claim of 2.8 pounds?

(A) $P\left(z > \dfrac{2.5 - 2.4}{\left(\dfrac{1.2}{\sqrt{36}}\right)}\right)$

(B) $P\left(z < \dfrac{2.5 - 2.4}{\left(\dfrac{1.2}{\sqrt{36}}\right)}\right)$

(C) $P\left(z < \dfrac{2.8 - 2.5}{\left(\dfrac{1.2}{\sqrt{36}}\right)}\right)$

(D) $P\left(z > \dfrac{2.8 - 2.4}{\left(\dfrac{1.2}{\sqrt{36}}\right)}\right)$

(E) $P\left(z < \dfrac{2.8 - 2.4}{\left(\dfrac{1.2}{\sqrt{36}}\right)}\right)$

29. Which of the following is the central limit theorem?

   (A) No matter how the population is distributed, as the sample size increases, the mean of the sample means becomes closer to the mean of the population.
   (B) No matter how the population is distributed, as the sample size increases, the standard deviation of the sample means becomes closer to the standard deviation of the population divided by the square root of the sample size.
   (C) If the population is normally distributed, then as the sample size increases, the sampling distribution of the sample mean becomes closer to a normal distribution.
   (D) No matter how the original population is distributed, as the sample size increases, the sampling distribution of the sample mean becomes closer to a normal distribution.
   (E) No matter how the original population is distributed, as the sample size increases, the distribution of the sample becomes closer to the distribution of the population.

30. What is a sampling distribution?

   (A) A distribution of all the statistics that can be found in a given sample
   (B) A histogram, or other such visual representation, showing the distribution of a sample
   (C) A normal distribution of some statistic
   (D) A distribution of all the values taken by a statistic from all possible samples of a given size
   (E) All of the above

31. A judge chosen at random reaches a just decision roughly 80% of the time. What is the probability that in randomly chosen cases, at least two out of three judges reach a just decision?

   (A) $3(0.8)^2(0.2)$
   (B) $1 - 3(0.8)^2(0.2)$
   (C) $(0.8)^3$
   (D) $1 - (0.8)^3$
   (E) $3(0.8)^2(0.2) + (0.8)^3$

32. Miles per gallon versus speed (miles per hour) for a new model automobile is fitted with a least squares regression line. The following is computer output of the statistical analysis of the data.

```
Dependent variable: Miles per gallon

Variable    Coefficient      SE Coef         t-ratio        P
Constant     38.929          5.651            6.89        0.000
Speed        -0.2179         0.112           -1.95        0.099

R-Sq = 38.7%     R-Sq(adj) = 28.5%
s = 7.252 with 8 - 2 = 6 degrees of freedom
```

Which of the following gives a 99% confidence interval for the slope of the regression line of miles per gallon versus speed for the new-model automobile?

(A) $-0.2179 \pm 3.707(0.112)$

(B) $-0.2179 \pm 3.143\left(\dfrac{0.112}{\sqrt{8}}\right)$

(C) $-0.2179 \pm 3.707\left(\dfrac{0.112}{\sqrt{8}}\right)$

(D) $38.929 \pm 3.143\left(\dfrac{0.561}{\sqrt{8}}\right)$

(E) $38.929 \pm 3.707\left(\dfrac{5.651}{\sqrt{8}}\right)$

33. What fault do all these sampling designs have in common?

  I. The Parent Teacher Association (PTA), concerned about rising teenage pregnancy rates at a high school, randomly picks a sample of high school students and interviews them concerning unprotected sex they have engaged in during the past year.

  II. A radio talk show host asks people to phone in their views on whether the United States should keep troops in Afghanistan indefinitely to fight terrorist groups.

  III. The *Ladies Home Journal* plans to predict the winner of a national election based on a survey of its readers.

(A) All the designs make improper use of stratification.
(B) All the designs have errors that can lead to strong bias.
(C) All the designs confuse association with cause and effect.
(D) All the designs suffer from sampling error.
(E) None of the designs makes use of chance in selecting a sample.

34. Hospital administrators wish to determine the average length of stay for all surgical patients. A statistician determines that for a 95% confidence level estimate of the average length of stay to within ± 0.50 days, 100 surgical patients' records would have to be examined. How many records should be looked at for a 95% confidence level estimate to within ± 0.25 days?

   (A) $\dfrac{100}{4}$

   (B) $\dfrac{100}{2}$

   (C) $2(100)$

   (D) $4(100)$

   (E) There is not enough information given to determine the necessary sample size.

35. A chess master wins 80% of her games, loses 5%, and draws the rest. If she receives 1 point for a win, $\frac{1}{2}$ point for a draw, and no points for a loss, what is true about the sampling distribution X of the points scored in two independent games?

   (A) X takes on the values 0, 1, and 2 with respective probabilities 0.10, 0.26, and 0.64.

   (B) X takes on the values 0, $\frac{1}{2}$, 1, $1\frac{1}{2}$, and 2 with respective probabilities 0.0025, 0.015, 0.1025, 0.24, and 0.64.

   (C) X takes on values according to a binomial distribution with $n = 2$ and $p = 0.8$.

   (D) X takes on values according to a binomial distribution with mean $1(0.8) + \frac{1}{2}(0.15) + 0(0.05)$.

   (E) X takes on values according to a distribution with mean $(2)(0.8)$ and standard deviation $\sqrt{2(0.8)(0.2)}$.

36. Which of the following is a true statement?

   (A) The P-value is a conditional probability.

   (B) The P-value is usually chosen before an experiment is conducted.

   (C) The P-value is based on a specific test statistic and thus should not be used in a two-sided test.

   (D) P-values are more appropriately used with t-distributions than with z-distributions.

   (E) If the P-value is less than the level of significance, the null hypothesis is proved false.

37. An assembly line machine is supposed to turn out ball bearings with a diameter of 1.25 centimeters. Each morning the first 30 bearings produced are pulled and measured. If their mean diameter is under 1.23 centimeters or over 1.27 centimeters, the machinery is stopped and an engineer is called to make adjustments before production is resumed. The quality control procedure may be viewed as a hypothesis test with the null hypothesis $H_0: \mu = 1.25$ and the alternative hypothesis $H_a: \mu \neq 1.25$. The engineer is asked to make adjustments when the null hypothesis is rejected. In test terminology, what would a Type II error result in?

(A) A warranted halt in production to adjust the machinery
(B) An unnecessary stoppage of the production process
(C) Continued production of wrong size ball bearings
(D) Continued production of proper size ball bearings
(E) Continued production of ball bearings that randomly are the right or wrong size

38. Both over-the-counter niacin and the prescription drug Lipitor are known to lower blood cholesterol levels. In one double-blind study, Lipitor outperformed niacin. The 95% confidence interval estimate of the difference in mean cholesterol level lowering was (18, 41). Which of the following is a reasonable conclusion?

(A) Niacin lowers cholesterol an average of 18 points, while Lipitor lowers cholesterol an average of 41 points.
(B) There is a 0.95 probability that Lipitor will outperform niacin in lowering the cholesterol level of any given individual.
(C) There is a 0.95 probability that Lipitor will outperform niacin by at least 23 points in lowering the cholesterol level of any given individual.
(D) We are 95% confident that Lipitor will outperform niacin as a cholesterol-lowering drug.
(E) We are 95% confident the mean number of points by which Lipitor outperforms niacin is between 18 and 41.

39. The following parallel boxplots show the average daily hours of bright sunshine in Liberia, West Africa:

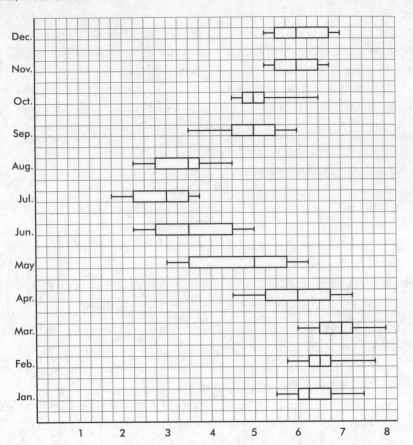

Average daily hours of bright sunshine

For how many months is the median below 4 hours?

(A) One

(B) Two

(C) Three

(D) Four

(E) Five

40. The following is a cumulative probability graph for the number of births per day in a city hospital.

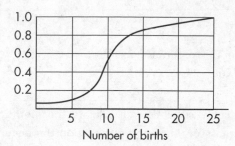

Number of births

Assuming that a birthing room can be used by only one woman per day, how many rooms must the hospital have available to be able to meet the demand at least 90 percent of the days?

(A) 5
(B) 10
(C) 15
(D) 20
(E) 25

STOP

If there is still time remaining, you may check your work on this section.

SECTION II

Part A

QUESTIONS 1–5

Spend about 65 minutes on this part of the exam.
Percentage of Section II grade—25

> **Directions:** You must show all work and indicate the methods you use. You will be graded on the correctness of your methods and on the accuracy of your results and explanations.

1. An experiment is being planned to study urban land use practices aimed at reviving and sustaining native bird populations. Vegetation types A and B are to be compared, and eight test sites are available. After planting, volunteer skilled birdwatchers will collect data on the abundance of bird species making each of the two habitat types their home. The east side of the city borders a river, while the south side borders an industrial park.

   (a) Suppose the decision is made to block using the scheme below (one block is white, one gray). How would you use randomization, and what is the purpose of the randomization?

   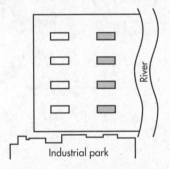

   (b) Comment on the strength and weakness of the above scheme as compared to the following blocking scheme (one block is white, one gray).

   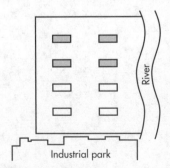

2. A world organization report gives the following percentages for primary-school age children enrolled in school in the 17 countries of each of two geographic regions.

Region A: 36, 45, 52, 56, 56, 58, 60, 63, 65, 66, 69, 71, 72, 74, 77, 82, 92
Region B: 35, 37, 41, 43, 43, 48, 50, 54, 65, 71, 78, 82, 83, 87, 89, 91, 92

(a) Draw a back-to-back stemplot of these data.

(b) The report describes both regions as having the same median percentage (for primary-school-age children enrolled in school) among their 17 countries and approximately the same range. What about the distributions is missed by the report?

(c) If the organization has education funds to help only one region, give an argument for which region should be helped.

(d) A researcher plans to run a two-sample *t*-test to study the difference in means between the percentages from each region. Comment on his plan.

3. Data were collected from a random sample of 100 student athletes at a large state university. The plot below shows grade point average (GPA) versus standard normal value (*z*-score) corresponding to the percentile of each GPA (when arranged in order). Also shown is the *normal line*, that is, a line passing through expected values for a normal distribution with the mean and SD of the given data.

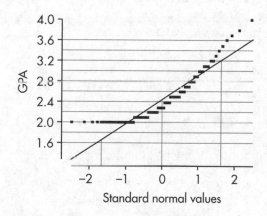

(a) What is the shape of this distribution? Explain.

(b) What is the 95th percentile of the data? Explain.

(c) In a normal distribution with the mean and SD of these data, what is the 95th percentile? Explain.

4. Although blood type frequencies in the United States are in the ratios of 9:8:2:1 for types O, A, B, and AB, respectively, local differences are often found depending upon a variety of demographic characteristics. Two researchers are independently assigned to determine if patients at a particular large city general hospital exhibit blood types supporting the above model. The table below gives the data results from what each researcher claims to be random samples of 500 patient lab results.

	O	A	B	AB
Researcher 1	253	194	38	15
Researcher 2	228	198	48	26

(a) Do the data reported by Researcher 1 support the 9:8:2:1 model for blood types of patients at the particular hospital? Justify your answer.

(b) The editorial board of a medical publication rejects the findings of Researcher 2, claiming that his data are suspicious in that they are "too good to be true." Give a statistical justification for the board's decision.

5. In a random sample of automobiles, the highway mileage (in mpg) and the engine size (in liters) are measured, and the following computer output for regression is obtained:

```
mpg: Mean = 28.5882 and SD = 5.88493
Mean liter = 2.52353 and SD = 0.898978
Dependent variable is: mpg
R-sq = 49.8%   R-sq(adj) = 46.5%
Variable    Coef       s.e.      t       p
Constant    40.2462    3.197     12.6    0.0001
liter       -4.61969   1.198     -3.86   0.0015
s = 4.306
```

Assume all conditions for regression inference are met.

(a) One of the points on the regression line corresponds to an auto with an engine size of 1.5 liters and highway mileage of 35 miles per gallon. What is the residual of this point? Interpret this residual in context.

Suppose another auto with an engine size of 5.0 liters and highway mileage of 17 mpg is added to the data set.

(b) Explain whether the new slope will be greater than, less than, or about the same as the slope given by the output above.

(c) Explain whether the new correlation, in absolute value, will be greater than, less than, or about the same as the correlation given by the output above.

SECTION II

Part B

QUESTION 6

Spend about 25 minutes on this part of the exam.
Percentage of Section II grade—25

SECTION II: PART B

6. An elite private tutoring service holds sessions around the country for students studying for the SAT exam and for students studying for the ACT exam. Each year, the service handles an equal number of students studying for each exam. The tutoring service has their own advertised "passing" scores for each exam. The proportion of their students passing the SAT exam and the proportion passing the ACT may be different. Their goal is for over 90 percent of all the students lumped together to pass whichever exam for which they were studying. With this in mind, the director of the service wants to estimate the average of the two proportions.

    A simple random sample of 250 of their students studying for the SAT exam is taken, and at year's end it is noted that 230 of them pass. Let p_{SAT} represent the proportion of all students using their service to study for the SAT exam who pass the exam.

    (a) Assume the students are chosen independently. Is it reasonable to assume that the sampling distribution of p_{SAT} is approximately normal? Explain.

    (b) Calculate the standard error of p_{SAT}.

    (c) How many standard errors is the observed value of p_{SAT} from 0.90?

    A simple random sample of 250 of their students studying for the ACT exam is taken, and at year's end it is noted that 235 of them pass. This gives a standard error of 0.0150. The parameter of interest is $p_{PASS} = \dfrac{p_{SAT} + p_{ACT}}{2}$.

    (d) Calculate $\hat{p}_{PASS}$ for the data from the two samples.

    (e) Calculate the standard error, $SE(\hat{p}_{PASS})$, of $\hat{p}_{PASS}$.

    Consider the hypotheses:

    H_0: The average proportion of students passing is 0.90. $(p_{PASS} = 0.90)$

    H_a: The average proportion of students passing is > 0.90. $(p_{PASS} > 0.90)$

    (f) Calculate the test statistic $\dfrac{\hat{p}_{PASS} - 0.90}{SE(\hat{p}_{PASS})}$.

    Chebyshev's inequality states that at least $1 - \dfrac{1}{k^2}$ of the values in any distribution are within k standard errors of the mean.

    (g) Using Chebyshev's inequality and the answer from (f), conclude whether or not there is sufficient evidence that p_{PASS}, the average proportion passing their respective exam, SAT or ACT, is greater than 0.90.

If there is still time remaining, you may check your work on this section.

Answer Key

SECTION I

1. **B**	9. **E**	17. **A**	25. **E**	33. **B**
2. **E**	10. **E**	18. **C**	26. **E**	34. **D**
3. **C**	11. **D**	19. **D**	27. **E**	35. **B**
4. **E**	12. **E**	20. **B**	28. **A**	36. **A**
5. **D**	13. **E**	21. **E**	29. **D**	37. **C**
6. **C**	14. **C**	22. **E**	30. **D**	38. **E**
7. **A**	15. **D**	23. **B**	31. **E**	39. **C**
8. **C**	16. **E**	24. **B**	32. **A**	40. **D**

Answers and Explanations

SECTION I

1. **(B)** $\mu_{\bar{x}} = \mu = 8$, and $\sigma_{\bar{x}} = \dfrac{\sigma}{\sqrt{n}} = \dfrac{4.8}{\sqrt{3}} < 4.8$.

2. **(E)** None of the studies has any controls such as randomization, a control group, or blinding, and so while they may give valuable information, they cannot establish cause and effect.

3. **(C)** The range increased slightly from $90 - 15 = 75$ to $97 - 21 = 76$, while the standard deviation decreased (note how the values are bunched together more closely in 2020).

4. **(E)** $\sigma_{\bar{x}} = \dfrac{\sqrt{386}}{\sqrt{40}} = 3.106$, and `normalcdf(80,90,83,3.106)` $= 0.8208$.

5. **(D)** The critical z-scores are $\dfrac{188 - 200}{5} = -2.4$ and $\dfrac{194 - 200}{5} = -1.2$, with corresponding left tail probabilities of 0.0082 and 0.1151, respectively. The probability of being greater than 188 given that it is less than 194 is $\dfrac{0.1151 - 0.0082}{0.1151} = 0.93$. [Or $\dfrac{\texttt{normalcdf}(188, 194, 200, 5)}{\texttt{normalcdf}(0, 194, 200, 5)} = 0.92876$.]

6. **(C)**
$$P(\text{pos test}) = P(\text{disease} \cap \text{pos}) + P(\text{healthy} \cap \text{pos})$$
$$= (0.04)(0.95) + (0.96)(0.05)$$
$$= 0.038 + 0.048 = 0.086$$
$$P(\text{disease} \mid \text{pos test}) = \frac{0.038}{0.086} = 0.442$$

7. **(A)** Given a census, the population parameter is known, and there is no need to use the techniques of inference.

8. **(C)** Sixty-four percent of the students scored the same as or below Mary's score, and 56% scored the same as or below Pam's score; so, 8% must have scored between them.

9. **(E)** *Stratification* is sampling that first divides the population into representative groups called strata, and *blocking* in experimental design first divides the subjects into representative groups called blocks. Intentionally forcing some treatment to note the response is associated with controlled experiments, not with observational studies. In experiments, the researchers decide how people are placed in different groups; self-selection is associated with observational studies. Results of observational studies may suggest cause-and-effect relationships; however, controlled studies are used to establish such relationships.

10. **(E)** With $\sigma_{\bar{x}} = \dfrac{\sigma}{\sqrt{n}}$, the greater the sample size n, the smaller the standard deviation $\sigma_{\bar{x}}$.

11. **(D)** $df = 9$, $SE(\bar{x}) = \dfrac{s}{\sqrt{n}} = \dfrac{1.59}{\sqrt{10}}$, and the critical t-value is `invT(0.99,9)` $= 2.821$. The confidence interval is $\bar{x} \pm t^* SE(\bar{x}) = 47.52 \pm 2.821 \left(\dfrac{1.59}{\sqrt{10}} \right)$.

12. **(E)** The set could be {150, 150, 150, 200} with mean 162.5 and median 150. It might also be {150, 200, 200, 200} with mean 187.5 and median 200. The median of a set of four elements is the mean of the two middle elements.

13. **(E)** $\hat{p} = \dfrac{92}{400} = 0.23$, and $SE(\hat{p}) = \sqrt{\dfrac{(0.23)(0.77)}{400}} = 0.0210$. Then `normalcdf(0.19,` `0.27,0.23,0.0210)` $= 0.943$.

14. **(C)** This is a binomial with $n = 15$ and $p = 0.29$, and so the mean is $np = 15(0.29) = 4.35$.

15. **(D)** With independence, variances add, so $\sqrt{65^2 + 380^2 + 125^2} = 405$.

16. **(E)** While $E(X + Y) = E(X) + E(Y)$, without *independence* we cannot determine $\text{var}(X + Y)$ from the information given.

17. **(A)** Bias is the tendency to favor the selection of certain members of a population. It has nothing to do with the shape of distributions. The natural variability between samples is called *sampling error* or *sampling variability*.

18. **(C)** The distribution is skewed to the left, and so the mean is probably less than the median. There are not two distinct peaks, so the distribution is not bimodal. Relative frequency is given by relative area.

19. **(D)** While the standard deviation of the z-distribution is 1, the standard deviation of every t-distribution is greater than 1.

20. **(B)**

$$P\left(\begin{array}{c} \text{market} \\ \text{decline} \end{array} \cap \begin{array}{c} \text{interest} \\ \text{jumps} \end{array} \right) = P\left(\begin{array}{c} \text{market} \\ \text{decline} \end{array} \middle| \begin{array}{c} \text{interest} \\ \text{jumps} \end{array} \right) P\left(\begin{array}{c} \text{interest} \\ \text{jumps} \end{array} \right) = (0.72)(0.65)$$

21. **(E)** Different samples give different sample statistics, all of which are estimates for the same population parameter. So error, called *sampling error* (also called *sampling variability*), is naturally present.

22. **(E)** The linear regression line minimizes the sum of the *squares* of the residuals. The sum of the residuals is always zero.

23. **(B)** The percentage of the variation in y accounted for by the variation in x is given by the coefficient of determination r^2. In this example, $(0.78)^2 = 0.61$. So, the variation that is not accounted for is $1 - 0.61 = 0.39$.

24. **(B)** There is a weak negative correlation, and so -0.15 is the only reasonable possibility among the choices given.

25. **(E)** There is no guarantee that 24 is anywhere near the interval, and so none of the statements is true.

26. **(E)** Sampling error (sampling variability) relates to natural variation between samples, and it can never be eliminated. In good observational studies, responses are not influenced during the collection of data. In good experiments, treatments are compared as to the differences in responses.

27. **(E)** The scatterplot suggests a zero correlation.

28. **(A)** $\sigma_{\bar{x}} = \dfrac{1.2}{\sqrt{36}}$. With a true mean of 2.4, the z-score of 2.5 is $\dfrac{2.5 - 2.4}{\left(\dfrac{1.2}{\sqrt{36}}\right)}$, and to the right is the probability of failing to reject the false claim.

29. **(D)** The central limit theorem says that no matter how the original population is distributed, as the sample size increases, the distribution of the sample means, that is, the sampling distribution of the sample mean, becomes closer to a normal distribution.

30. **(D)** A sampling distribution is the distribution of all the values taken by a statistic, such as sample mean or sample proportion, from all possible samples of a given size.

31. **(E)** B($n = 3$, $p = 0.8$). $P(\text{at least 2}) = P(\text{exactly 2}) + P(\text{exactly 3}) = 3(0.8)^2(0.2) + (0.8)^3$.

32. **(A)** The standard error of the slope is given to be 0.112. The critical t-scores for 99% confidence with $df = 6$ are $\pm\texttt{invT(0.995,6)} = \pm3.707$. The confidence interval is $b \pm t^* SE(b) = -0.2179 \pm 3.707(0.112)$.

33. **(B)** The PTA survey has strong *response bias* in that students may not give truthful responses to a parent or teacher about their engaging in unprotected sex. The talk show survey results in a *voluntary response* sample, which typically gives too much emphasis to persons with strong opinions. The *Ladies Home Journal* survey has strong *selection bias*; that is, people who read the *Journal* are not representative of the general population.

34. **(D)** To divide the interval width by d, the sample size must be increased by a multiple of d^2.

35. **(B)** $P(0 \text{ pts}) = (0.05)^2$, $P(\frac{1}{2} \text{ pt}) = 2(0.05)(0.15)$, $P(1 \text{ pt}) = (0.15)^2 + 2(0.05)(0.8)$, $P(1\frac{1}{2}\text{pts}) = 2(0.15)(0.8)$, and $P(2 \text{ pts}) = (0.8)^2$.

36. **(A)** The P-value is the probability of obtaining a result as extreme as or more extreme than the one seen *given that* the null hypothesis is true; thus, it is a conditional probability. The P-value depends on the sample chosen. The P-value in a two-sided test is calculated by doubling the indicated tail probability. P-values are not restricted to use with any particular distribution. With a small P-value, there is evidence to reject the null hypothesis, but we're not *proving* anything.

37. **(C)** A Type II error is a mistaken failure to reject a false null hypothesis or, in this case, a failure to realize that the machinery is turning out wrong size ball bearings.

38. **(E)** Using a measurement from a sample, we are never able to say *exactly* what a population mean is; rather, we always say we have a certain *confidence* that the population mean lies in a particular *interval*. In this case, we are 95% confident the mean number of points by which Lipitor outperforms niacin is between 18 and 41.

39. **(C)** The median number of hours is less than four for June, July, and August.

40. **(D)** A horizontal line drawn at the 0.9 probability level corresponds to roughly 19 rooms.

1. (a) In each block, two of the sites will be randomly assigned to receive Type A vegetation, while the remaining two sites in the block will receive Type B vegetation. Randomization of vegetation type to the sites within each block should reduce bias due to any confounding variables. In particular, the randomization in blocks in the first scheme should even out the effect of distance from the industrial park on vegetation in blocks.

(b) The first scheme creates homogeneous blocks with respect to distance from the river, while the second scheme creates homogeneous blocks with respect to the industrial park. Randomization of vegetation types to sites within blocks in the first scheme should even out effects of distance from the industrial park, while randomization of vegetation types to sites within blocks in the second scheme should even out effects of distance from the river.

SCORING

It is important to explain why randomization is important within the context of this problem, so use of terms like *bias* and *confounding* must be in context. One must explain the importance of homogeneous experimental units (sites, not vegetation types) within blocks.

4	**Complete Answer**	Correct explanation both of use of blocking and of use of randomization in the context of this problem.
3	**Substantial Answer**	Correct explanation of use of either blocking or randomization in context and a weak explanation of the other.
2	**Developing Answer**	Correct explanation of use of either blocking or randomization in context OR weak explanations of both.
1	**Minimal Answer**	Weak explanation of either blocking or randomization in context.

2. (a)

Region A		Region B	
6	3	5 7	Key: 6|3|5 means
5	4	1 3 3 8	36% in Region A
8 6 6 2	5	0 4	and 35% in
9 6 5 3 0	6	5	Region B
7 4 2 1	7	1 8	
2	8	2 3 7 9	
2	9	1 2	

(b) While the percentages in both regions have roughly symmetric distributions, the percentages from Region A form a distinctly unimodal pattern, while those from Region B are distinctly bimodal. That is, in Region B the countries tended to show either a very low or a very high percentage, while in Region A most countries showed a percentage near the middle one.

(c) Either an argument can be made for Region A because Region B has so many countries with high percentages (for primary-school-age children enrolled in school), or an argument can be made for Region B because Region B has so many countries with low percentages.

(d) Data are already given on *all* the countries in the two regions. In doing inference, one uses sample statistics to estimate population parameters. If the data are actually the whole population, there is no point of a *t*-test.

SCORING

Part (a) is essentially correct for a correct stemplot (numbers in each row do not have to be in order) with labeling as to which side refers to which region. Part (a) is partially correct for a correct stemplot missing the labeling.

Part (b) is essentially correct for noting that the percentages from Region A form a distinctly unimodal pattern, while those from Region B are distinctly bimodal. Part (b) is partially correct for noting either the unimodal pattern from the Region A data or the bimodal pattern from the Region B data.

Parts (c) and (d) together is essentially correct for 1) choosing either region and giving a reasonable argument for the choice and 2) explaining that inference is not proper when the data are the whole population. Parts (c) and (d) together is partially correct for one of the two parts correct.

4	**Complete Answer**	All three parts, (a), (b), and (c)–(d), essentially correct.
3	**Substantial Answer**	Two parts essentially correct and one part partially correct.
2	**Developing Answer**	Two parts essentially correct OR one part essentially correct and one or two parts partially correct OR all three parts partially correct.
1	**Minimal Answer**	One part essentially correct OR two parts partially correct.

3. (a) Note that there are a great number of values between 2.0 and 2.4, fewer values between 2.4 and 2.8, on down to the fewest values being between 3.6 and 4.0. Thus, the distribution is skewed right (skewed toward the higher values).

(b) Note that the greatest five values, out of the 100 values, are above a point roughly at 3.4 or 3.5, so the 95th percentile is roughly 3.4 or 3.5.

(c) In a normal distribution, the 95th percentile has a z-score of 1.645. Reading up from the x-axis to the normal line and across to the y-axis gives a GPA of approximately 3.2.

SCORING

Part (a) is essentially correct if the shape is correctly identified as skewed right and a correct explanation based on the given normal probability plot is given. Part (a) is partially correct for a correct identification with a weak explanation. Part (a) is incorrect if a correct identification is given with no explanation.

Part (b) is essentially correct for correctly noting 3.4 or 3.5 as the 95th percentile with a correct explanation. Part (b) is partially correct for giving 3.4 or 3.5 with a weak explanation or no explanation.

Part (c) is essentially correct for noting approximately 3.2 as the GPA and stating the method clearly. Part (c) is partially correct for giving 3.2 with a weak explanation or no explanation.

4	**Complete Answer**	All three parts essentially correct.
3	**Substantial Answer**	Two parts essentially correct and one part partially correct.
2	**Developing Answer**	Two parts essentially correct OR one part essentially correct and one or two parts partially correct OR all three parts partially correct.
1	**Minimal Answer**	One part essentially correct OR two parts partially correct.

---

4. (a) *Hypotheses:*

H_0: The distribution of blood types among patients at this hospital is in the ratios of 9:8:2:1 for types O, A, B, and AB, respectively.

H_a: The distribution of blood types among patients at this hospital is not in the ratios of 9:8:2:1 for types O, A, B, and AB, respectively.

Procedure: Chi-square goodness-of-fit test.

Checks:

1. The researcher claims that the data are from a random sample of patient records.

2. The ratios 9:8:2:1 give expected cell frequencies of $\frac{9}{20}(500) = 225$, $\frac{8}{20}(500) = 200$, $\frac{2}{20}(500) = 50$, $\frac{1}{20}(500) = 25$, each of which is at least 5.

Mechanics: χ^2GOF-Test gives $\chi^2 = 10.54$ and $P = 0.0145$ [or calculate $\chi^2 = \frac{(253 - 225)^2}{225} + \frac{(194 - 200)^2}{200} + \frac{(38 - 50)^2}{50} + \frac{(15 - 25)^2}{25} = 10.54$ and with $df = 4 - 1 = 3$, the P-value is 0.0145].

Conclusion in context with linkage to the P-value: With this small a *P*-value, 0.0145 < 0.05, there is sufficient evidence to reject H_0; that is, there is sufficient evidence that the distribution of blood types among patients at this hospital is not in the ratios of 9:8:2:1 for types O, A, B, and AB, respectively.

(b) In this case, the mechanics give:

$$\chi^2 = \frac{(228-225)^2}{225} + \frac{(198-200)^2}{200} + \frac{(48-50)^2}{50} + \frac{(26-25)^2}{25} = 0.18$$

With $df = 4 - 1 = 3$, the *P*-value is an incredibly high 0.9808. With such a large *P*-value, so close to 1, the probability of a χ^2-value of only 0.18 or less is very small, $1 - 0.9808 = 0.0192$. There is sufficient evidence that the data are "too good to be true"; that is, there is strong evidence that Researcher 2 made up the data to fit the 9:8:2:1 model.

SCORING

Part (a1) is essentially correct if the hypotheses are given, the test is identified, and the assumptions are checked. Part (a1) is partially correct for two of these three steps.

Part (a2) is essentially correct for correct mechanics and the conclusion given in context with linkage to the *P*-value. Part (a2) is partially correct if there is a minor error in mechanics OR the conclusion is not in context OR there is no linkage to the *P*-value.

Part (b) is essentially correct for noting that with a *P*-value of 0.9808, the probability of a χ^2-value of only 0.18 or less is very small, and thus the data are suspicious. Part (b) is partially correct for a correct idea but with a weak explanation.

4	**Complete Answer**	All three parts essentially correct.
3	**Substantial Answer**	Two parts essentially correct and one part partially correct.
2	**Developing Answer**	Two parts essentially correct OR one part essentially correct and one or two parts partially correct OR all three parts partially correct.
1	**Minimal Answer**	One part essentially correct OR two parts partially correct.

---

5. (a) The regression line is $\widehat{mpg} = 40.2462 - 4.61969 \, (liter)$.

Then $40.2462 - 4.61969(1.5) \approx 33.32$.

Residual = actual − predicted = 35 − 33.32 = 1.68 mpg

The actual mpg for this auto is 1.68 mpg greater than what was predicted by the regression model.

(b) The predicted value for mpg is $40.2462 - 4.61969(5.0) \approx 17.15$, which is very close to the actual value of 17. The added point is consistent with the linear pattern given by the computer output, so the new slope should be about the same as the old slope.

(c) Since the new point fits the old linear pattern so well and has an *x*-value (*liter*) much greater than $\bar{x} = 2.52$, the absolute value of the correlation will increase.

Part (a) is essentially correct for a correct calculation of the residual and a correct interpretation in context. Part (a) is partially correct for an error in calculating the residual but a correct interpretation using this error or a correct residual but an error in interpretation.

Part (b) is essentially correct for showing that the actual value is very close to the predicted value and concluding that the new slope should be about the same. Part (b) is partially correct if the predicted value is miscalculated but a proper conclusion based on the miscalculation is reached.

Part (c) is essentially correct for concluding that the absolute value of the correlation will increase because the new point fits the old linear pattern so well. Part (c) is partially correct for concluding that the absolute value of the correlation will increase but giving a weak explanation.

4	**Complete Answer**	All three parts essentially correct.
3	**Substantial Answer**	Two parts essentially correct and one part partially correct.
2	**Developing Answer**	Two parts essentially correct OR one part essentially correct and one or two parts partially correct OR all three parts partially correct.
1	**Minimal Answer**	One part essentially correct OR two parts partially correct.

SECTION II: PART B

6. (a) $n_{SAT}\hat{p}_{SAT} = 230 \geq 10$ and $n_{SAT}(1 - \hat{p}_{SAT}) = 20 \geq 10$. So, yes, it is reasonable to assume that the sampling distribution of $\hat{p}_{SAT}$ is approximately normal.

(b) $\hat{p}_{SAT} = \dfrac{230}{250} = 0.92$, and $SE(\hat{p}_{SAT}) = \sqrt{\dfrac{(0.92)(0.08)}{250}} - 0.0172$.

(c) $\dfrac{0.92 - 0.90}{0.0172} = 1.16$

(d) $\hat{p}_{SAT} = 0.92$, $\hat{p}_{ACT} = \dfrac{235}{250} = 0.94$, and $\hat{p}_{PASS} = \dfrac{0.92 + 0.94}{2} = 0.93$

(e) $SE(\hat{p}_{PASS}) = \sqrt{\dfrac{1}{4}(0.0172^2 + 0.0150^2)} = 0.0114$

(f) $\dfrac{\hat{p}_{PASS} - 0.90}{SE(\hat{p}_{PASS})} = \dfrac{0.93 - 0.90}{0.0114} = 2.63$

(g) Using Chebyshev, we calculate that at least $1 - \dfrac{1}{2.63^2} = 0.855$ of the values are within 2.63 standard errors of the mean. So, the probability of observing a value as far, or farther, from 0.90 as the one observed might be as high as $1 - 0.855 = 0.145$, which is greater than 0.05. So, there is not sufficient evidence to reject the null hypothesis. There is not sufficient evidence that the average proportion of students passing is > 0.90.

SCORING

Section 1 is essentially correct for answering yes in (a) with the appropriate checks to say the sampling distribution of $\hat{p}_{SAT}$ is approximately normal, the correct calculation of the standard error $SE(\hat{p}_{SAT})$, and the correct calculation of the number of standard errors from $\hat{p}_{SAT}$ to 0.90. Section 1 is partially correct for two out of the three steps above.

Section 2 is essentially correct for calculating the average $\hat{p}_{PASS}$ in (d) and the standard error $SE(\hat{p}_{PASS})$.

Section 3 is essentially correct for the correct calculation of the test statistic $\dfrac{\hat{p}_{PASS} - 0.90}{SE(\hat{p}_{PASS})}$ in (f), makes the correct calculation using Chebyshev's inequality, and gives a proper conclusion in context and is partially correct for two out of these three steps correct.

4 Complete Answer   All three sections essentially correct.

3 Substantial Answer   Two sections essentially correct and one section partially correct.

2 Developing Answer   Two sections essentially correct OR one section essentially correct and one or two sections partially correct OR all three sections partially correct.

1 Minimal Answer   One section essentially correct OR two sections partially correct.

Part Five: Appendices

Appendices

FORMULAS

I. Descriptive Statistics

$$\bar{x} = \frac{1}{n}\sum x_i = \frac{\sum x_i}{n}$$

$$s_x = \sqrt{\frac{1}{n-1}\sum (x_i - \bar{x})^2} = \sqrt{\frac{\sum (x_i - \bar{x})^2}{n-1}}$$

$$\hat{y} = a + bx \qquad\qquad \bar{y} = a + b\bar{x}$$

$$r = \frac{1}{n-1}\sum \left(\frac{x_i - \bar{x}}{s_x}\right)\left(\frac{y_i - \bar{y}}{s_y}\right) \qquad b = r\frac{s_y}{s_x}$$

II. Probability and Distributions

$$P(A \cup B) = P(A) + P(B) - P(A \cap B) \qquad P(A \mid B) = \frac{P(A \cap B)}{P(B)}$$

Probability Distribution	Mean	Standard Deviation
Discrete random variable X	$\mu_x = E(X) = \sum x_i \cdot P(x_i)$	$\sigma_x = \sqrt{\sum (x_i - \mu_x)^2 \cdot P(x_i)}$
If X has a **binomial** distribution with parameters n and p, then: $P(X = x) = \binom{n}{x} p^x (1-p)^{(n-x)}$, where $x = 0, 1, 2, 3, ..., n$	$\mu_x = np$	$\sigma_x = \sqrt{np(1-p)}$
If X has a **geometric** distribution with parameters p, then: $P(X = x) = (1-p)^{(x-1)} p$, where $x = 0, 1, 2, 3, ...$	$\mu_x = \dfrac{1}{p}$	$\sigma_x = \dfrac{\sqrt{1-p}}{p}$

III. Sampling Distributions and Inferential Statistics

Standardized test statistic: $\dfrac{\text{statistic} - \text{parameter}}{\text{standard error of the statistic}}$

Confidence interval: statistic $\pm$ (critical value)(standard error of statistic)

Chi-square statistic: $\chi^2 = \sum \dfrac{(\text{observed} - \text{expected})^2}{\text{expected}}$

Sampling distributions for proportions:

Random Variable	Parameters of Sampling Distribution		Standard Error * of Sample Statistic
For one proportion: $\hat{p}$	$\mu_{\hat{p}} = p$	$\sigma_{\hat{p}} = \sqrt{\dfrac{p(1-p)}{n}}$	$s_{\hat{p}} = \sqrt{\dfrac{\hat{p}(1-\hat{p})}{n}}$
For two proportions: $\hat{p}_1 - \hat{p}_2$	$\mu_{\hat{p}_1 - \hat{p}_2} = p_1 - p_2$	$\sigma_{\hat{p}_1 - \hat{p}_2} = \sqrt{\dfrac{p_1(1-p_1)}{n_1} + \dfrac{p_2(1-p_2)}{n_2}}$	$s_{\hat{p}_1 - \hat{p}_2} = \sqrt{\dfrac{\hat{p}_1(1-\hat{p}_1)}{n_1} + \dfrac{\hat{p}_2(1-\hat{p}_2)}{n_2}}$ When $p_1 = p_2$ is assumed: $s_{\hat{p}_1 - \hat{p}_2} = \sqrt{\hat{p}_c(1-\hat{p}_c)\left(\dfrac{1}{n_1} + \dfrac{1}{n_2}\right)}$ where $\hat{p}_c = \dfrac{x_1 - x_2}{n_1 - n_2}$

Sampling distributions for means:

Random Variable	Parameters of Sampling Distribution		Standard Error* of Sample Statistic
For one population: $\bar{X}$	$\mu_{\bar{x}} = \mu$	$\sigma_{\bar{x}} = \dfrac{\sigma}{\sqrt{n}}$	$s_{\bar{x}} = \dfrac{s}{\sqrt{n}}$
For two populations: $\bar{X}_1 - \bar{X}_2$	$\mu_{\bar{x}_1 - \bar{x}_2} = \mu_1 - \mu_2$	$\sigma_{\bar{x}_1 - \bar{x}_2} = \sqrt{\dfrac{\sigma_1^2}{n_1} + \dfrac{\sigma_2^2}{n_2}}$	$s_{\bar{x}_1 - \bar{x}_2} = \sqrt{\dfrac{s_1^2}{n_1} + \dfrac{s_2^2}{n_2}}$

Sampling distributions for simple linear regression:

Random Variable	Parameters of Sampling Distribution	Standard Error* of Sample Statistic
For slope: b	$\mu_b = \beta$ $\sigma_b = \dfrac{\sigma}{\sigma_x \sqrt{n}}$, where $\sigma_x = \sqrt{\dfrac{\sum (x_i - \bar{x})^2}{n}}$	$s_b = \dfrac{s}{s_x \sqrt{n-1}}$, where $s = \sqrt{\dfrac{\sum (y_i - \hat{y}_i)^2}{n-2}}$ and $s_x = \sqrt{\dfrac{\sum (x_i - \bar{x})^2}{n-1}}$

*Standard deviation is a measurement of variability from the theoretical population. Standard error is the estimate of the standard deviation. If the standard deviation of the statistic is assumed to be known, the standard deviation should be used instead of the standard error.

GRAPHICAL DISPLAYS

	Right-skewed	Symmetric	Left-skewed

Dotplots

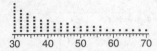

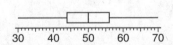

Stemplots

	Right-skewed
3	0000000002222222444444666668888
4	000022244466688
5	002244668
6	02468
7	0

	Symmetric
3	0246688
4	000222444466666888888
5	00000022222244444666688
6	000224468
7	0

	Left-skewed
3	02468
4	00224466888
5	000222444666688888
6	0000022222244444466666688888
7	0000

Histograms

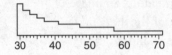

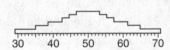

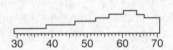

Cumulative Frequency Plots

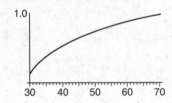

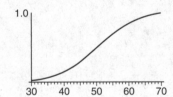

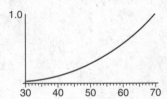

Boxplots

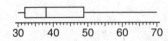

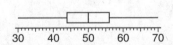

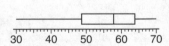

CHECKING ASSUMPTIONS FOR INFERENCE

Conditions for constructing confidence intervals and performing hypothesis tests

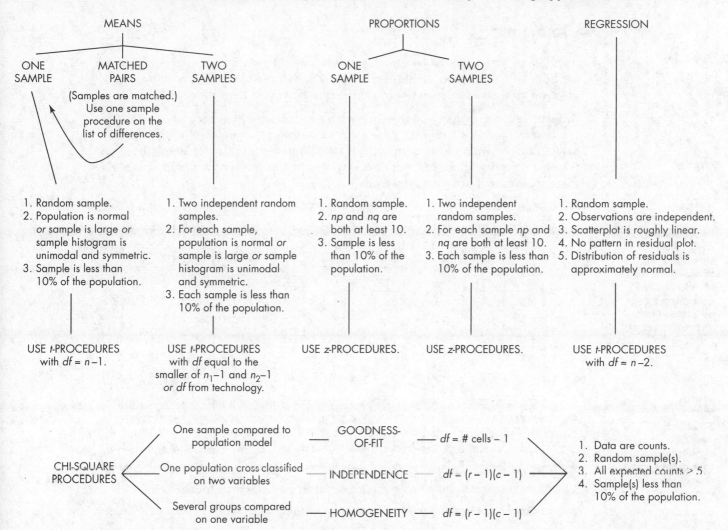

MEANS

ONE SAMPLE   MATCHED PAIRS   TWO SAMPLES

(Samples are matched.) Use one sample procedure on the list of differences.

PROPORTIONS

ONE SAMPLE   TWO SAMPLES

REGRESSION

1. Random sample.
2. Population is normal *or* sample is large *or* sample histogram is unimodal and symmetric.
3. Sample is less than 10% of the population.

1. Two independent random samples.
2. For each sample, population is normal *or* sample is large *or* sample histogram is unimodal and symmetric.
3. Each sample is less than 10% of the population.

1. Random sample.
2. np and nq are both at least 10.
3. Sample is less than 10% of the population.

1. Two independent random samples.
2. For each sample np and nq are both at least 10.
3. Each sample is less than 10% of the population.

1. Random sample.
2. Observations are independent.
3. Scatterplot is roughly linear.
4. No pattern in residual plot.
5. Distribution of residuals is approximately normal.

USE *t*-PROCEDURES with $df = n-1$.

USE *t*-PROCEDURES with df equal to the smaller of n_1-1 and n_2-1 *or df* from technology.

USE *z*-PROCEDURES.

USE *z*-PROCEDURES.

USE *t*-PROCEDURES with $df = n-2$.

CHI-SQUARE PROCEDURES

One sample compared to population model —— GOODNESS-OF-FIT —— $df = \#\text{ cells} - 1$

One population cross classified on two variables —— INDEPENDENCE —— $df = (r-1)(c-1)$

Several groups compared on one variable —— HOMOGENEITY —— $df = (r-1)(c-1)$

1. Data are counts.
2. Random sample(s).
3. All expected counts > 5
4. Sample(s) less than 10% of the population.

SIMULATION FOR PROBABILITY

Instead of algebraic calculations, sometimes we can use simulation to answer probability questions.

➡️ **EXAMPLE** _____

If left alone, 70% of birthmarks gradually fade away. If ten children, 5 boys and 5 girls, are born with birthmarks, what is the probability that the same number of boys and girls will lose their birthmarks? Answer the question using simulation.

Answer: Let the digits 1–7 represent having a birthmark that fades away, and 8, 9, and 0 represent having a birthmark that doesn't fade away. To simulate the 10 children, select 10 digits from the random number table, with the first 5 representing boys and the next 5 representing girls. Note the number of digits 1–7 in each group, and see if there is a match. Underlining the digits 1–7 gives

86961 94141	46633 32552	64452 34882	46266 23371	69214 70203
84873 03236	65618 34476	66335 55856	80396 76708	28303 11388
36392 37904	85601 08943	38324 38221	65492 52566	52057 25970
57161 68039	06748 19239	65683 21773	38557 66359	22908 88647
27941 43337	90190 59437	75107 40518	72389 78958	64819 40361
27492 40058	07181 31446	69345 47300	71402 60588	85664 55760
27827 61501	95435 91684	53217 66012	33702 26371	35333 32652
46511 30857	39655 92730	33859 00386	86309 36997	36997 91524
21567 49374	29800 76389	51752 72124	40673 49251	02103 80901
42505 78167	90045 72843	82718 74608	02434 32440	66180 82562

For example, in the first set of 10 digits, there are 3 boys and 4 girls whose birthmarks fade, so no match. In the second set of 10 digits, there are 5 boys and 5 girls whose birthmarks fade, so a match. We have 3-4, 5-5, 5-3, 5-5, 4-3; 3-4, 4-5, 5-4, 2-3, 3-3; 4-3, 3-2, 4-4, 4-5, 4-3; 5-2, 3-3, 4-5, 4-4, 2-3; 4-5, 1-4. 4-3, 3-2, 3-4; 4-2, 3-5, 4-3, 4-2, 4-4; 4-4, 4-3, 5-4, 4-5, 5-5; 5-3, 4-3, 3-2, 2-3, 3-4; 5-4, 1-3, 5-5, 4-4, 3-2; 4-3, 2-4, 3-3, 4-4, 3-4.

We count 13 matches out of a possible 50 and so estimate the probability that the same number of boys and girls will lose their birthmarks to be $\frac{13}{50} = 0.26$.

_____

➡️ **EXAMPLE** _____

The legendary baseball player Babe Ruth had a career batting average of 0.342—quite impressive for a home run hitter! Use simulation to estimate the probability that his first hit in a game is on the first at-bat. On the second at-bat. Not until the third at-bat. Fourth at-bat.

Answer: Using the random number table from the previous example, read off 3 digits at a time, with 001–342 representing a hit and anything else not a hit. Start all over again every time there is a hit. So, for example, reading off the first line gives

869-619-414-<u>146</u> is a first hit on the fourth at-bat,

633-<u>325</u> is a first hit on the second at-bat,

526-445-<u>234</u> is a first hit on the third at-bat,

882-462-662-<u>337</u> is a first hit on the fourth at-bat,

<u>169</u> is a first hit on the first at-bat, and

<u>214</u> is also a first hit on the first at-bat.

Continuing in this fashion and tabulating the results, we have

Number of the at-bat at which the first hit occurred	Frequency	Estimated probability
1	22	22/64 = 0.344
2	14	14/64 = 0.219
3	13	13/64 = 0.203
4	6	6/64 = 0.094
Over 4	9	9/64 = 0.141
Total	64	

The actual probabilities are 0.342, $(0.658)(0.342) = 0.225$, $(0.658)^2(0.342) = 0.148$, $(0.658)^3(0.342) = 0.097$, and $1 - (0.342 + 0.225 + 0.148 + 0.094) = 0.188$. [On the TI-84, one can calculate these *geometric probabilities*: `geometpdf(0.342,1),...,` `geometpdf(0.342,4)`, and $1 -$ `geometcdf(0.342,4)`.]

In performing a simulation, you must:

1. Set up a correspondence between outcomes and random numbers.
2. Give a procedure for choosing the random numbers (for example, pick three digits at a time from a designated row in a random number table).
3. Give a stopping rule.
4. Note what is to be counted (what is the purpose of the simulation), and give the count if requested.

INDEPENDENCE

Throughout this review book, the concept of *independence* has arisen many times, with different meanings in different contexts. A quick review is worthwhile.

1. Two events are *independent* if information about one variable conveys NO information about the other variable. In other words, two events are *independent* if the occurrence (or non-occurrence) of one is irrelevant to the occurrence (or non-occurrence) of the other.
2. If the events E and F are independent, then algebraically we have $P(E \mid F) = P(E)$, $P(F \mid E) = P(F)$, and $P(E \cap F) = P(E)P(F)$. If any one of these expressions is true, so are the other two. So, only one, any one, need be checked. The two conditional statements give insight into the meaning of independence, while the product rule is a very useful equivalent statement.
3. Starting with a 2×2 table plus row and column sums, two events are independent if all corresponding ratios are equal. Note that in a 2×2 table, any equality between two corresponding ratios will force all pairs of corresponding ratios to be equal.
4. When the distribution of one variable in a contingency table is the same for all categories of another variable, we say the two variables are *independent*.
5. When two random variables are *independent*, we can use the rule for adding variances with regard to the random variables $X + Y$ and $X - Y$.

6. In a binomial distribution, the trials must be *independent*; that is, the probability of success must be the same on every trial, irrespective of what happened on a previous trial. However, when we pick a sample from a population, we do change the size and makeup of the remaining population. We typically allow this as long as the sample is not too large—the usual rule of thumb is that the sample size should be smaller than 10% of the population.

7. In inference, we wish to make conclusions about a population parameter by analyzing a sample. A crucial assumption is always that the sampled values are *independent* of each other. This typically involves checking if there was proper randomization in the gathering of data. Also, to minimize the effect on independence from samples being drawn without replacement, we check that less than 10% of the population is sampled.

8. Furthermore, if inference involves the comparison of two groups, the two groups must be *independent* of each other.

9. We often need to check both independence of samples *and* independence of observations within each group.

10. With paired data, while the observations in each pair are not independent, the differences must be *independent* of each other.

11. With regression, the errors in the regression model must be *independent*, and one may check this by confirming there are no patterns in the residual plot.

MORE ON RESIDUALS

1. In one-variable quantitative analysis, we can calculate the sample mean $\bar{x} = \frac{\Sigma x_i}{n}$ and then ask how individual values differ from this sample mean. The deviations of values from the sample mean are called residuals; that is, $r_i = x_i - \bar{x}$ (*observed – mean*). A positive residual indicates a value greater than the mean, a negative residual indicates a value less than the mean, and a zero residual indicates a value equal to the mean. Both the sum and the mean of the residuals are equal to zero. That is, $\Sigma r_i = 0$ and $\bar{r} = 0$. The standard deviation $s = \sqrt{\frac{\Sigma(x_i - \bar{x})^2}{n-1}} = \sqrt{\frac{\Sigma r_i^2}{n-1}}$ is a "typical value" for the residuals. However, note that while s cannot be negative, a residual has both magnitude and direction and can be negative.

2. In linear regression, the difference between an observed value of the dependent variable y and the predicted value $\hat{y}$ is called the residual e, that is, $e_i = y_i - \hat{y}_i$ (*observed – predicted*). A positive residual indicates that the observed value is greater than what is predicted by the regression model, a negative residual indicates that the observed value is less than what is predicted by the regression model, and a zero residual indicates that the observed value is exactly what is predicted by the regression model. Both the sum and the mean of the residuals are equal to zero. That is, $\Sigma e_i = 0$ and $\bar{e} = 0$. Linear regression computer output typically gives the standard deviation $s = \sqrt{\frac{\Sigma(y_i - \hat{y}_i)^2}{n-2}} = \sqrt{\frac{\Sigma e_i^2}{n-2}}$ of the residuals, a "typical value" for the residuals, that is, a measure of how the points are spread around the regression line.

3. In chi-square analysis, the difference, *obs* − *exp* (*observed* − *expected*), between the observed value and the expected value for a given cell is called the residual. The standardized residual is defined by $c = \dfrac{obs - exp}{\sqrt{exp}}$. A positive residual indicates that there are more observations in the cell than what the null hypothesis predicts, a negative residual indicates that there are fewer observations in the cell than what the null hypothesis predicts, and a zero residual indicates that the number of observations in the cell are exactly what the null hypothesis predicts. Both the sum and the mean of the residuals are equal to zero. The standardized residuals have mean = 0 and standard deviation = 1. The most extreme standardized residuals, both positive and negative, contribute the most to the chi-square statistic,

$$\chi^2 = \sum \frac{(obs - exp)^2}{exp} = \sum c^2.$$

You should be able to explain what type of residuals are involved in a case at hand and what they mean in context. Note the role of squared residuals in the above three discussions.

ANSWERS AND EXPLANATIONS FOR QUIZZES 1–37

Quiz 1 (Pages 52–57)

MULTIPLE-CHOICE

1. **(D)** The minimum time is somewhere between 5 and 10 minutes but might not be exactly 5 minutes. Similarly, the maximum time is somewhere between 30 and 35 minutes. With 155 times, the middle time will be the 78th time if the times are arranged in order. There are 70 times less than 15 minutes and 40 times between 15 and 20 minutes, so the 78th time must be between 15 and 20 minutes.

2. **(C)** Stemplots are not used for categorical data sets, are too unwieldy to be used for very large data sets, and show every individual value. Stems should never be skipped over—gaps are important to see.

3. **(C)** A cumulative relative frequency plot that rises at a constant rate to start and then slowly at the end corresponds to a histogram that is horizontal to start and then has little area under the curve at the end.

4. **(A)** A cumulative relative frequency plot that rises slowly at first, then quickly in the middle, and finally slowly again at the end corresponds to a histogram with little area under the curve on the ends and much greater area in the middle.

5. **(D)** A cumulative relative frequency plot that rises slowly at first and then rises at a constant rate corresponds to a histogram with little area under the curve early on and then a horizontal section at the end.

6. **(B)** Removing a value that is greater than the average will always lower the average, and adding in a value that is less than the average will also always lower the average.

7. **(E)** The consumer group randomly picked 5 of the cars and found an average efficiency of only 32 mpg. The chance of picking 5 cars with an average of 32 mpg or lower from a fleet of cars of real 35 mpg efficiency is so small, only 4 out of 160, or 0.025, that it seems reasonable to conclude that the consumer group's set of 5 cars was not picked from a fleet of cars with 35 mpg efficiency and that the company's claim is wrong.

FREE-RESPONSE QUESTIONS

1. (a)

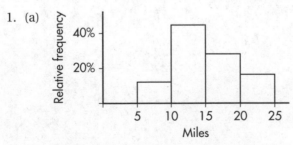

(b) $\dfrac{3+11}{25} = \dfrac{14}{25} = 56\%$

(c)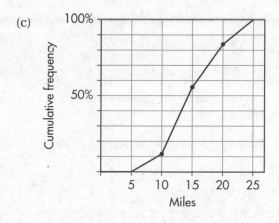

(d) The answer would still be 56%.

2. (a) A complete answer considers shape, center, and spread; unusual features like outliers; and mentions context.
   Shape: Unimodal, skewed right, probable outlier at 10.
   Center: Around 2 or 3.
   Spread: From 0 to 10 goals were scored by each team.

   (b) If the player scored six goals, his/her team must have scored either 7 or 10, but they lost, so they scored 7. The only possible final score is that they lost by a score of 10 to 7.

   (c) No, there were six teams that scored exactly two goals, but there were only five teams that scored less than two goals. So, not all the two-goal teams could have won.

3. (a) The lowest winning percentage over the past 22 years was 46.0%.

   (b) A complete answer considers shape, center, and spread; unusual features like clusters and gaps; and mentions context.
   Shape: Two clusters, each somewhat bell-shaped.
   Center: Around 50%.
   Spread: The winning percentages were between 46.0% and 55.6%.

   (c) The team had more losing seasons (13) than winning seasons (9).

   (d) The cluster of winning percentages is further above 50% than the cluster of losing percentages is below 50%.

Quiz 2 (Pages 58–62)

MULTIPLE-CHOICE

1. **(B)** There is no such thing as being skewed both left and right.

2. **(B)** Histograms give information about relative frequencies (relative areas correspond to relative frequencies) and may or may not have an axis with actual frequencies. Symmetric histograms can have any number of peaks. Choice of width and number of classes changes the appearance of a histogram. Stemplots clearly show outliers; however, in histograms outliers may be hidden in large class widths.

3. **(B)** A histogram with little area under the curve early and much greater area later results in a cumulative relative frequency plot that rises slowly at first and then at a much faster rate later.

4. **(E)** A histogram with little area under the curve in the middle and much greater area on both ends results in a cumulative relative frequency plot that rises quickly at first, then almost levels off, and finally rises quickly at the end.

5. **(A)** Uniform distributions result in cumulative relative frequency plots that rise at constant rates and thus are linear.

6. **(B)** There are 5 classes, and their average size is $\dfrac{150+25+25+25+25}{5} = \dfrac{250}{5} = 50$.

7. **(C)** Among the 250 students, there are 100 students in classes of size 25 and 150 students in a class of size 150. The average size of their history class is $\dfrac{100(25)+150(150)}{250} = \dfrac{25,000}{250} = 100$.

FREE-RESPONSE QUESTIONS

1. (a) 40% of the players averaged fewer than 20 points per game.

   (b) All the players averaged at least 3 points per game.

   (c) No players averaged between 5 and 7 points per game because the cumulative relative frequency was 10% for both 5 and 7 points.

   (d) Go over to the plot from 0.9 on the vertical axis and then down to the horizontal axis to result in 28 points per game.

   (e) Reading up to the plot and then over from 10 and from 20 shows that 0.25 of the players averaged under 10 points per game and 0.4 of the players averaged under 20 points per game. Thus, $0.4 - 0.25 = 0.15$ gives the proportion of players who averaged between 10 and 20 points per game.

2. (a) The center is roughly between 24.10 and 24.11, and the data are spread from 24.01 to 24.20.

   (b) There seems to be two "low" data points, 24.01 and 24.02, and one "high" data point, 24.20. These three data points are distinctly separated from the other points.

   (c) For this day's sample, $W = \dfrac{25.05 + 23.01}{2} = 24.03$. A value of 24.03 or less occurred only twice in the 100 samples. Thus, if the machinery was operating properly, a W measurement of 24.03 would be very unusual. The conclusion should be to recalibrate the machine.

3. (a) A complete answer considers shape, center, spread, and unusual features and mentions context.
   Shape: The distribution of months of dating before becoming engaged is skewed right.
   Center: Around 14.5 months.
   Spread: From 12.0 to 23.1 months of dating before becoming engaged.
   Unusual feature: There appears to be an outlier at 23.1 months.

   (b) The cumulative frequency plot of these data rises rapidly between 12 and 15 months and then rises at a much slower pace. There will be a horizontal piece between 20 and 23 months because with no data between these values, the proportion of the data less than 20 is the same as that less than 23.

Quiz 3 (Pages 79–84)

MULTIPLE-CHOICE

1. **(D)** The distribution is clearly skewed right, so the mean is greater than the median and the ratio is greater than one.

2. **(D)** Multiplying every value in a set by the same constant (in this case, by $\frac{1}{60}$) multiplies both the mean and the standard deviation by the same constant. Standardized scores (the number of standard deviations from the mean) are unchanged and without units.

3. **(E)** Outliers are any values below $Q_1 - 1.5(\text{IQR}) = 25 - 1.5(38 - 25) = 5.5$ or above $Q_3 + 1.5(\text{IQR}) = 38 + 1.5(38 - 25) = 57.5$. With a minimum of $10 > 5.5$, there are no outliers on the low end; however, with a maximum of $60 > 57.5$, the maximum is an outlier and so are any other values falling between 57.5 and 60.

4. **(A)** The value 50 seems to split the area under the histogram in two, so the median is about 50. Furthermore, the histogram is skewed to the left with a tail from 0 to 30.

5. **(B)** When looking at areas under the curve, Q_1 appears to be around 20, the median is around 30, and Q_3 is about 40.

6. **(C)** Adding the same constant to every value increases the mean by that same constant; however, the distances between the increased values and the increased mean stay the same, and so the standard deviation is unchanged. Graphically, you should picture the whole distribution as moving over by a constant; the mean moves, but the standard deviation (which measures spread) doesn't change.

7. **(E)** The median is somewhere between 20 and 30 but not necessarily at 25. Even a single very large score can result in a mean over 30 and a standard deviation over 10.

8. **(C)** The median corresponds to the 0.5 cumulative proportion. The 0.25 and 0.75 cumulative proportions correspond to $Q_1 = 1.8$ and $Q_3 = 2.8$, respectively, and so the interquartile range is $2.8 - 1.8 = 1.0$.

FREE-RESPONSE QUESTIONS

1. (a) Adding 10 to each value increases the mean by 10 but leaves measures of variability unchanged. So, the new mean is 340 hours while the range stays at 5835 hours, the standard deviation remains at 245 hours, and the variance remains at $245^2 = 60,025$ hr^2.

   (b) Increasing each value by 10% (multiplying by 1.10) will increase the mean to $1.1(330) = 363$ hours, the range to $1.1(5835) = 6418.5$ hours, the standard deviation to $1.1(245) = 269.5$ hours, and the variance to $(269.5)^2 = 72,630.25$ hr^2. (Note that the variance increases by a multiple of $(1.1)^2$ and not by a multiple of 1.1.)

2. A complete answer compares shape, center, and spread and mentions context in at least one of the responses.
   Shape: The distribution of times to complete all tasks by females is skewed right (toward the higher values), whereas the distributions of times to complete all tasks by males is roughly bell-shaped.
   Center: The center of the distribution of female times (at around $2\frac{1}{4}$ minutes) is less than the center of the distribution of male times (at around $3\frac{3}{4}$ minutes).

Spread: The spreads of the two distributions are roughly the same; the range of the female times $(5 - 1\frac{1}{2} = 3\frac{1}{2}$ minutes) equals the range of the male times $(5\frac{1}{2} - 2 = 3\frac{1}{2}$ minutes).

3.

Men		Women
6	1	
7 3	2	2
9 6 2	3	4 5 5 8 8
8 5 3 0	4	2 7 9 9
7 7 4 1	5	0 4 8
9 4 0	6	0 5 9
5 2	7	5 9
2	8	7
	9	8

2 | 3 | 4   means a
man's time of 32 min.
and a woman's of 34 min.

A complete answer compares shape, center, and spread and mentions context in at least one of the responses.

Shape: The men's distribution of hours grooming is roughly symmetric, whereas the women's distribution of hours grooming is skewed right (toward higher values).

Center: The center of the men's distribution is about the same as the center of the women's distribution, both about 50 min.

Spread: The spread of the men's distribution (with a range of $82 - 16 = 66$ min) is less than the spread of the women's distribution (with a range of $98 - 22 = 76$ min).

4. (a) For females, $Q_1 - 1.5(IQR) = 130 - 1.5(358 - 130) = -212$ and $Q_3 + 1.5(IQR) = 700$, so 1,098 is an outlier. For males, $Q_1 - 1.5(IQR) = 72 - 1.5(273 - 72) = -229.5$ and $Q_3 + 1.5(IQR) = 574.5$, so there are no outliers.

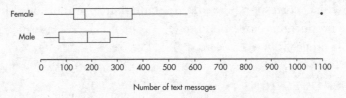

Number of text messages

(b) The medians are roughly equal. The male distribution appears roughly symmetric, so the mean is probably close to the median; however, the female distribution shows extreme right skewness, so the mean is much greater than the median. Thus, the females had a greater mean number of text messages than did the males.

Quiz 4 (Pages 85–90)

MULTIPLE-CHOICE

1. **(C)** The boxplot indicates that 25% of the data lie in each of the intervals 10–20, 20–35, 35–40, and 40–50. Counting boxes, only histogram C has this distribution.

2. **(E)** The boxplot indicates that 25% of the data lie in each of the intervals 10–20, 20–30, 30–40, and 40–50. Counting boxes, only histogram E has this distribution.

3. **(A)** Subtracting 10 from one value and adding 5 to two values leaves the sum of the values unchanged, so the mean will be unchanged. Exactly what values the outliers take will not change what value is in the middle, so the median will be unchanged.

4. **(C)** The high outlier is further from the bulk of values than is the low outlier, so removing both will decrease the mean. However, removing the lowest and highest values will not change what value is in the middle, so the median will be unchanged.

5. **(E)** When every value is multiplied by the same constant, both the mean and the standard deviation are multiplied by that constant. Graphically, increasing each value by 25% (multiplying by 1.25) both moves and spreads out the distribution.

6. **(A)** The sum of the scores in one class is $20 \times 92 = 1840$, while the sum in the other is $25 \times 83 = 2075$. The total sum is $1840 + 2075 = 3915$. There are $20 + 25 = 45$ students, and so the average score is $\frac{3915}{45} = 87$.

7. **(B)** Increasing every value by 5 gives 10% between 45 and 65, and then doubling gives 10% between 90 and 130.

8. **(E)** The minimum of the combined set of scores must be the minimum of the boys' scores since it is the lowest of any score. The maximum of the combined set of scores must be the maximum of the girls' scores since it is the highest of any score. The first quartile must be the same as the identical first quartiles of the two original distributions. There are no outliers (scores more than 1.5(IQR) from the first and third quartiles).

FREE-RESPONSE QUESTIONS

1. z-scores give the number of standard deviations from the mean, so

   $Q_1 = 300 - 0.7(25) = 282.5$ and $Q_3 = 300 + 0.7(25) = 317.5$.

   The interquartile range is $IQR = 317.5 - 282.5 = 35$, and $1.5(IQR) = 1.5(35) = 52.5$.

   The standard definition of outliers encompasses all values less than $Q_1 - 52.5 = 230$ and all values greater than $Q_3 + 52.5 = 370$.

2. (a) $1.6(9.5) = 15.2$ km/hr.

   (b) $1.6(10.8 - 7.9) = 4.64$ km/hr. (Note that subtracting 0.2 from every value does not change the IQR.)

   (c) $0.625(14) = 8.75$ mph. (Note that the runner in the 65th percentile when the units are km/hr will also be in the 65th percentile when the units are mph.)

3. (a) The Liberian median age lies in the 15–19 age interval because roughly 50% of the total bar lengths is above and below the 15–19 interval.

   (b) Numerical values are needed to calculate means. Histograms such as these give how many values are in specific intervals but do not give the actual values.

   (c) Canada has more children younger than 10 years of age. There are about 1.2 million children younger than age 10 in Liberia (boys and girls) and roughly 3.5 million in Canada. (Note the difference in scales!)

   (d) The population pyramid indicates that Canadian women live longer than men because all the higher age intervals show greater numbers of women than men.

   (e) In the Liberian graph, the smaller 15–19 age group shows a definite break with the overall pattern. A plausible explanation is that a great number of children died in the civil war.

4. A complete answer compares shape, center, and spread and mentions context in at least one of the responses.

Shape: Cruise A, for which the cumulative frequency plot rises steeply at first, has more younger passengers and thus a distribution skewed to the right (toward the higher ages). Cruise C, for which the cumulative frequency plot rises slowly at first and then steeply toward the end, has more older passengers and thus a distribution skewed to the left (toward the younger ages). Cruise B, for which the cumulative frequency plot rises slowly at each end and steeply in the middle, has a more bell-shaped distribution.

Center: Considering the center to be a value separating the area under the histogram roughly in half, the centers will correspond to a cumulative frequency of 0.5. Reading across from 0.5 to the intersection of each graph, and then down to the *x*-axis, shows centers of approximately 18, 40, and 61 years, respectively. Thus, the center of distribution A is the least, and the center of distribution C is the greatest.

Spread: The spreads of the age distributions of all three cruises are the same: from 10 to 70 years.

Quiz 5 (Pages 93–95)

MULTIPLE-CHOICE

1. **(B)** Curve *a* appears to have a mean of 6 and a standard deviation of 2, while curve *b* appears to have a mean of 18 and a standard deviation of 1.

2. **(D)** Point E appears to be one standard deviation above the mean. $74.3 + 9.7 = 84.0$. Note that point E (and point C) are points where the slope is steepest.

3. **(D)** 54.9 and 93.7 are two standard deviations below and above the mean, respectively. By the empirical rule, 95% of the data are in this interval.

4. **(A)** 64.6 is one standard deviation below the mean. By the empirical rule, 68% of the data is between one standard deviation below and above the mean. This leaves 34% outside this interval and 17% in each tail.

5. **(B)** With bell-shaped data, the empirical rule applies. Given that the spread from 92 to 98 is roughly 6 standard deviations, one standard deviation is about 1.

FREE-RESPONSE QUESTIONS

1. (a) $z = \dfrac{720 - 600}{120} = 1$, and for the normal model, the proportion of *z*-scores below 1 is $0.5 + 0.34 = 0.84$, or the 84th percentile.

   (b) Both $360 and $840 are $240 from the mean of $600, which makes $360 two standard deviations below the mean and $840 two standard deviations above. For the normal model, the proportion of *z*-scores between -2 and $+2$ is 0.95 or 95%.

   (c) Under a normal curve, 99.7% of the values are within ± 3 standard deviations of the mean: $600 - 3(120) = 240$ and $600 + 3(120) = 960$. So, 99.7% of American teens spend between $240 and $960 on food per year.

2. No. In a normal distribution the mean and median should be equal or nearly equal, but here the mean, 6.4, is significantly greater than the median, 5. Furthermore, in a normal distribution the distance from the minimum to the median should be about the same as the distance from the median to the maximum. In this data set, however, the distance median $-$ min $= 5 - 0 = 5$ is different from the distance max $-$ median $= 15 - 5 = 10$. Both these facts suggest the distribution is skewed right.

Quiz 6 (Pages 105–108)

MULTIPLE-CHOICE

1. **(E)** Of the 500 people surveyed, $50 + 150 + 50 = 250$ were students, and $\frac{250}{500} = 0.5$, or 50%.

2. **(A)** Of the 500 people surveyed, 125 both picked challenging as most important and were teachers: $\frac{125}{500} = 0.25$, or 25%.

3. **(E)** There were $15 + 10 + 25 = 50$ administrators, and 25 of them picked strict as most important: $\frac{25}{50} = 0.5$ or 50%.

4. **(E)** There were $150 + 50 + 10 = 210$ people picking enthusiastic as most important, and 150 of them were students: $\frac{150}{210} = 0.714$, or 71.4%.

5. **(C)** The percentages of students, teachers, and administrators picking strict as most important were 20%, 12.5%, and 50%, respectively.

6. **(E)** In mosaic plots, the area of a box is proportional to the count corresponding to that box. The horizontal axis in this plot indicates the proportions of humanities versus science students. The vertical axis in this plot indicates the proportions of living on campus versus living off campus for humanities students and for science students. Choice (E) is not true because of the two boxes corresponding to students living on campus, the humanities box has greater area than the science box.

7. **(D)** Relative frequencies must be equal. You can use either rows or columns: the rows give $\frac{20}{70} = \frac{30}{30 + n}$, and the columns give $\frac{20}{50} = \frac{50}{50 + n}$. Solving for n in either proportion works: $\frac{n}{30} = \frac{50}{20}$ or $\frac{n}{50} = \frac{30}{20}$. In both cases, $n = 75$.

FREE-RESPONSE QUESTIONS

1. (a) i. $\frac{1164}{1389} = 83.80\%$

    ii. $1 - \frac{20}{1389} = 98.56\%$

    iii. $\frac{52}{225} = 23.11\%$

(b) Calculate row or column totals, and then show either a side-by-side bar graph or a segmented bar graph, showing percentages, and conditioned on either career path (officer vs. enlisted) or military branch:

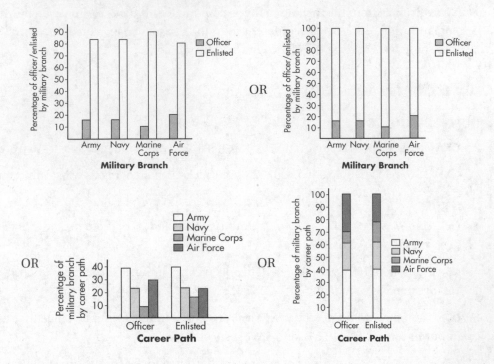

OR

(c) The Army and the Navy have about the same percentage of officers (16%), while the Air Force has a higher percentage of officers (20%), and the Marine Corps has a lower percentage of officers (10%).

OR

Among the officers and the enlisted career paths, there are about the same percentage Army (39%) and about the same percentage Navy (23%), while the officers have a lower percentage Marine Corps than the enlisted (9% vs. 16%) and the officers have a higher percentage Air Force than the enlisted (29% vs. 22%).

2. (a) Of the people identifying as liberal, 273 out of 420, or $\frac{273}{420}$, support a guaranteed income.

(b) Of the people who oppose a guaranteed income, 193 out of 315, or $\frac{193}{315} = 0.613 = 61.3\%$, identify as conservative.

(c) Based on the mosaic plot, there is an association between identification as liberal or conservative and support for a guaranteed income. Most adults identifying as liberal support a guaranteed income, while most adults identifying as conservative oppose a guaranteed income.

3. (a) Each employee racial group experienced an increase in mean salary between 2005 and 2015. White employees' mean salary rose from $37,000 to $40,000, Black employees' mean salary rose from $22,000 to $23,000, and Hispanic (non-White) employees' mean salary rose from $23,000 to $25,000.

(b) The mean company salary in 2005 was $\dfrac{150(37{,}000) + 50(22{,}000) + 50(23{,}000)}{250} =$ 31,200, while in 2015 the mean was $\dfrac{90(40{,}000) + 60(23{,}000) + 100(25{,}000)}{250} =$ 29,920. The company's mean employee salary dropped from \$31,200 to \$29,920 between 2005 and 2015.

(c) You should tell the prospective employee that while it is true that the company's mean employee salary has dropped between 2005 to 2015, the mean salary paid to each racial group of employees has risen. The reason behind this apparent paradox is related to the changing employee racial composition. (You might want to address if and how the company is addressing the apparent unequal salary scales based on race!)

Quiz 7 (Pages 109–113)

MULTIPLE-CHOICE

1. **(A)** In the bar corresponding to the Northeast, the segment corresponding to country music stretches from the 50% level to the 70% level, indicating a length of 20%.

2. **(B)** Based on lengths of indicated segments, the percentage from the West who prefer country is the greatest.

3. **(E)** The given bar chart shows percentages, not actual numbers.

4. **(B)** In a complete distribution, the probabilities sum to 1 and the relative frequencies total 100%.

5. **(A)** The different lengths of corresponding segments show that in different geographic regions, different percentages of people prefer each of the music categories.

6. **(D)** In mosaic plots, the area of a box is proportional to the count corresponding to that box. Choice (D) is true because the three boxes corresponding to students with GPAs under 3.0 have a total area greater than the three boxes corresponding to students with GPAs 3.0 or higher, or note that along the vertical axis, "GPA under 3.0" has a greater length than "GPA 3.0 or higher."

7. **(E)** It is possible for both to be correct, for example, if there were 11 secretaries (10 women, 3 of whom receive raises, and 1 man who receives a raise) and 11 executives (10 men, 1 of whom receives a raise, and 1 woman who does not receive a raise). Then 100% of the male secretaries receive raises while only 30% of the female secretaries do; and 10% of the male executives receive raises while 0% of the female executives do. At the same time, of all the employees, 3 out of 11 women receive raises while only 2 out of 11 men receive raises. This is an example of Simpson's paradox.

FREE-RESPONSE QUESTIONS

1. (a)

Program	Percentage of Men Accepted (%)	Percentage of Women Accepted (%)
A	62	82
E	28	24

   Women seem to be favored in program A, while men seem to be slightly favored in program E.

   (b) Overall, 564 out of 1016 male applicants were accepted, for a 55.5% acceptance rate, while 184 out of 501 female applicants were accepted, for a 36.7% acceptance rate. This appears to contradict the results from part (a).

   (c) You should tell the reporter that while it is true that the overall acceptance rate in these two programs for women is 36.7% compared to the 55.5% acceptance rate for men, in one program women have a much higher acceptance rate than men while in the other program women have only a slightly lower acceptance rate than men. The reason behind this apparent paradox is that between these two programs, most men applied to program A, which had a high acceptance rate. However, most women applied to program E, which had a low acceptance rate.

2. (a) Of the speeding pullovers who were given a ticket, 27 out of 93, or $\frac{27}{93}$, were Latino.

   (b) Of the speeding pullovers who were White, 84 out of 120, or $\frac{84}{120} = 0.70 = 70\%$, were given a warning.

   (c) Based upon the mosaic plot, there is an association between race and whether or not a ticket or warning is given to drivers pulled over for speeding at night. Whites are more likely to receive a warning, Blacks are equally as likely to receive a warning or a ticket, while Latinos are more likely to receive a ticket. Latinos were twice as likely to receive a ticket as Whites.

3. (a) $\bar{R}_A = \frac{(3+2+1)}{3} = 2, \bar{R}_B = \frac{(2+1+2)}{3} = \frac{5}{3}, \bar{R}_C = \frac{(1+3+3)}{3} = \frac{7}{3}$

   $Q = 3\left[(2-2)^2 + \left(\frac{5}{3} - 2\right)^2 + \left(\frac{7}{3} - 2\right)^2 \right] = \frac{2}{3}$

   (b) If there was a significant difference in the players' abilities, the average rankings would have varied significantly from 2, and thus the resulting Q-value would have been large. However, as seen in the simulations, $\frac{2}{3}$ is not large at all. Thus, there is not sufficient evidence of a significant difference in the players' abilities.

Quiz 8 (Pages 132–139)

MULTIPLE-CHOICE

1. **(D)** Residual = Measured − Predicted, so if the residual is negative, the predicted must be greater than the measured (observed).

2. **(E)** The correlation r cannot take a value greater than 1.

3. **(B)** The "Predictor" column indicates the independent variable with its coefficient to the right.

4. **(E)** $r = \sqrt{0.986} = 0.993$. (The sign is the same as the sign of the slope, which in this case is positive.)

5. **(C)** $\widehat{back} = 0.056 + 0.920(0.55) = 0.562$, and so the *residual* $= 0.59 - 0.562 = 0.028$.

6. **(C)** The coefficient of determination r^2 gives the proportion of the y-variance that is accountable from a knowledge of the variability of x. In this case $r^2 = (0.632)^2 = 0.399$ or 39.9%.

7. **(C)** The correlation is not changed by adding the same number to every value of one of the variables, by multiplying every value of one of the variables by the same positive number, or by interchanging the x- and y-variables.

8. **(B)** The point (35, 14) appears to follow the linear trend of the rest of the data. The slope wouldn't change, but with another point added to the trend, r^2 will increase, and s, the measure of typical deviation from the line, will decrease.

9. **(D)** The word "negative" in the phrase "strong negative linear association" means that generally as one variable increases, the other variable decreases. Thus, regions with a higher percentage of seat belt usage tend to have lower numbers of highway deaths due to failure to wear seat belts. Choice (A) is an interpretation of the y-intercept. Correlation is a measure of the strength of a linear relationship but does not by itself explain the meaning of "positive" or "negative." While the overall pattern is of a negative association, anything can be true about two points on the scatterplot. Choice (E) is an interpretation of the slope.

10. **(A)** Slope $= 0.15\left(\dfrac{42{,}000}{1.3}\right) \approx 4850$ and intercept $= 208{,}000 - 4850(6.2) \approx 178{,}000$.

11. **(E)** A scatterplot readily shows that while the first three points lie on a straight line, the fourth point does not lie on this line. Thus, no matter what the fifth point is, all the points cannot lie on a straight line, and so r cannot be 1.

12. **(C)** The point X has low or no leverage because its x-value appears to be close to the mean x-value. The point X has a large residual because it is far above the line of best fit.

FREE-RESPONSE QUESTIONS

1. (a) The correlation coefficient $r = \sqrt{0.819} = 0.905$. It is positive because the slope of the regression line is positive.

   (b) The slope is 8.5, signifying that each gram of medication lowers the pulse rate by 8.5 beats per minute, on average.

   (c) $\hat{y} = -1.68 + 8.5(2.25) = 17.4$ beats per minute.

(d) There is always danger in using a regression line to extrapolate beyond the values of *x* contained in the data. In this case, the 5 grams was an overdose, the patient died, and the regression line cannot be used for such values beyond the data set.

(e) Removing the 3-gram result from the data set will increase the correlation coefficient because the 3-gram result appears to be far off a regression line through the remaining points.

(f) Removing the 3-gram result from the data set will swing the regression line upward so that the slope will increase.

2. (a) There is a strong, positive, nonlinear relationship between MAUs and year.

(b) 97.9% of the variation in log(MAUs) can be explained by the linear relationship between log(MAUs) and log(year).

(c) The residual plot shows a definite pattern, indicating that while the linear model may be good, there is a better nonlinear model out there to describe the relationship between log(MAUs) and log(year).

3. (a) In scatterplot I, the points fall exactly on a downward sloping straight line, so $r = -1$. In scatterplot II, the isolated point is an influential point, and *r* is close to $+1$. In scatterplot III, the isolated point is also influential, and *r* is close to 0.

(b) Inserting a point that lies directly on a straight line drawn through the three points will not change the correlation of $r = -1$.

4. (a) The association between stress level and debt for this sample of 25 college graduates is linear, moderate, and positive.

(b) $19.376 + 0.42640(60) = 44.96$ and *Observed − Predicted* $= 25 - 44.96 = -19.96$. The residual value means that the stress level is 19.96 points less than what would be predicted for a student with a $60,000 debt.

(c) Health is a better choice than income for including with debt in a regression model for predicting stress. The relatively strong association between health and residuals from regression of stress on debt shows that health will help explain the unexplained variation (the residuals) between stress and debt.

Quiz 9 (Pages 140–145)

MULTIPLE-CHOICE

1. **(E)** The negative value of the slope (–2.84276) gives that, on average, the predicted combined SAT score of a school is 2.84 points lower for each one unit higher in the percentage of students taking the exam. Choices (A) through (D) are incorrect for the following reasons. The variable column indicates the independent (explanatory) variable. The sign of the correlation is the same as the sign of the slope (negative here). In this example, the *y*-intercept is meaningless (predicted SAT result if no students take the exam). There can be a strong linear relation with high r^2 value but still a distinct pattern in the residual plot indicating that a nonlinear fit may be even stronger.

2. **(E)** A negative correlation shows a tendency for higher values of one variable to be associated with lower values of the other; however, given any two points, anything is possible.

3. **(E)** Since $(2, 5)$ is on the line $y = 3x + b$, we have $5 = 6 + b$ and $b = -1$. Thus, the regression line is $y = 3x - 1$. The point $(\bar{x}, \bar{y})$ is always on the regression line, and so we have $\bar{y} = 3\bar{x} - 1$.

4. **(C)** If the points lie on a straight line, $r = \pm 1$. Correlation has the formula $r = \frac{1}{n-1} \sum \left(\frac{x_i - \bar{x}}{s_x} \right) \left(\frac{y_i - \bar{y}}{s_y} \right)$, so x and y are interchangeable, and r does not depend on which variable is called x or y. However, since means and standard deviations can be strongly influenced by outliers, r too can be strongly affected by extreme values. While $r = 0.75$ indicates a better fit with a linear model than $r = 0.25$ does, we cannot say that the linearity is threefold.

5. **(D)** The sum and thus the mean of the residuals are always zero. In a good straight-line fit, the residuals show a random pattern.

6. **(C)** Predicted winning percentage $= 44 + 0.0003(34,000) = 54.2$, and Residual $=$ Observed $-$ Predicted $= 55 - 54.2 = 0.8$.

7. **(B)** The slope and the correlation coefficient have the same sign. Multiplying every y-value by -1 changes this sign.

8. **(D)** A linear association means that as the explanatory variable (home size here) changes by a constant amount, the response variable (selling price here) also changes by a constant amount, on average. Unless there was perfect linear correlation, the points will not line up on a straight line. No distinct pattern in the residual plot just means that there is no obvious better model out there, but it doesn't necessarily say that the data are linear. The coefficient of determination indicates something about the strength of the relationship but does not define linearity. Choice (E) is an interpretation of the slope, but again, not a definition of linearity.

9. **(E)** "Strong" means that the points in the related scatterplot fall close to the least squares regression line. In context, this means that the actual incidence of skin cancer at a given latitude will be very close to what is predicted by the least squares model. Association does not imply causation. While (C) refers to the existence of an association, and (B) and (D) further refer to the direction of the association, none of these statements refer to the strength of the association.

10. **(E)** The least squares line passes through $(\bar{x}, \bar{y}) = (2, 4)$, and the slope b satisfies $b = r\frac{s_y}{s_x} = \frac{5r}{3}$. Since $-1 \le r \le 1$, we have $-\frac{5}{3} \le b \le \frac{5}{3}$.

11. **(A)** Using your calculator, find the regression line to be $\hat{y} = -8 + 9x$. The regression line, also called the least squares regression line, minimizes the sum of the squares of the vertical distances between the points and the line. In this case $(2, 10)$, $(3, 19)$, and $(4, 28)$ are on the line, and so the minimum sum is $(10 - 11)^2 + (19 - 17)^2 + (28 - 29)^2 = 6$.

12. **(B)** The point X has high leverage because its x-value is much greater than the mean x-value. Point X has a small residual because the regression line would pass close to it.

FREE-RESPONSE QUESTIONS

1. (a) The correlation coefficient is $r = \sqrt{0.746} = 0.864$. It is positive because the slope of the regression line is positive.

   (b) The slope is 1.106, signifying that each additional page raises a grade by 1.106, on average.

   (c) Including Mary's paper will lower the correlation coefficient because her result seems far off the regression line through the other points.

   (d) Including Mary's paper will swing the regression line down and lower the value of the slope.

2. (a) Yes. The residual graph is not curved, does not show fanning, and appears to be random or scattered.

   (b) The slope is 0.95893, indicating that the winning jump improves 0.95893 inches per year on average or about 3.8 inches every four years on average.

   (c) With $r^2 = 0.921$ and a positive slope, the correlation r is 0.96.

   (d) $0.95893(80) + 256.576 \approx 333.3$ inches.

   (e) The residual for 1980 is $+2$, and so the actual winning distance must have been $333.3 + 2 = 335.3$ inches.

3. (a)

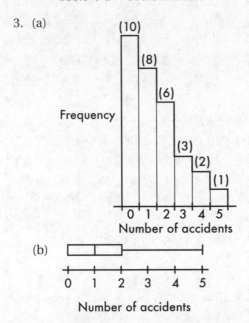

   (c) There is a roughly linear trend with daily accidents increasing during the month.

   (d) The daily number of accidents is strongly skewed to the right.

4. (a) The association between exam score and hours of sleep is *nonlinear* (or *curved*), weak (or moderate), and positive.

   (b) There appears to be very little, if any, association between exam score and hours of sleep for the 8 students who received at least 6 hours of sleep the night before the exam. (Because of the one very low score that also corresponds to hours of sleep just over 6, one could say that there is a very weak *positive* association.)

   (c) The student who scored 75 and slept for just over 2 hours the night before the exam has such a strong influence that that particular scatterplot point taken together with all the other points suggests a nonlinear form.

Quiz 10 (Pages 154–156)

MULTIPLE-CHOICE

1. **(C)** The four classes (freshmen, sophomores, juniors, and seniors) are strata.

2. **(D)** If there is bias, taking a larger sample just magnifies the bias on a larger scale. If there is enough bias, the sample can be worthless. Even when the subjects are chosen randomly, there can be bias due, for example, to nonresponse or to the wording of the questions. Convenience samples, like shopping mall surveys, are based on choosing individuals who are easy to reach, and they typically miss a large segment of the population. Voluntary response samples, like radio call-in surveys, are based on individuals who offer to participate, and they typically overrepresent persons with strong opinions.

3. **(B)** Different samples give different sample statistics, all of which are estimates of a population parameter. Sampling error (also called sampling variability) relates to natural variation between samples, can never be eliminated, can be described using probability, and is generally smaller if the sample size is larger. Furthermore, it is not an error or mistake on anyone's part!

4. **(D)** The dorms are clusters.

5. **(A)** This survey provides a good example of voluntary response bias, which often overrepresents negative opinions. The people who chose to respond were most likely parents who were very unhappy, and so there is very little chance that the 10,000 respondents were representative of the population. Knowing more about her readers, or taking a sample of the sample, would not have helped.

6. **(E)** In a simple random sample, every possible group of the given size has to be equally likely to be selected, and this is not true here. For example, with this procedure it would be impossible for all the players of one team to be together in the final sample. This procedure is an example of stratified sampling, but stratified sampling does not result in simple random samples.

FREE-RESPONSE QUESTIONS

1. (a) This is an example of *stratified sampling*, where the chapters are strata. The advantage is that the student is ensuring that the final sample will represent the 12 different authors, who may well use different average word lengths.

    (b) This is an example of *cluster sampling*, where for each chapter the three chosen pages are clusters. It is reasonable to assume that each page (cluster) resembles the author's overall pattern. The advantage is that using these clusters is much more practical than trying to sample from among all an author's words.

    (c) This is an example of *systematic sampling*, which is quicker and easier than many other procedures. A possible disadvantage is that if ordering is related to the variable under consideration, this procedure will likely result in an unrepresentative sample. For example, in this study if an author's word length is related to word order in sentences, the student could end up with words of particular lengths.

2. Version I probably resulted in 52% choosing domestic policy over foreign policy. The words used to define a category can make a significant difference. In this example, using the words "war on terrorism" instead of the more neutral "foreign policy" probably resonated with people and led to the dramatic difference in results.

3. (a) To obtain an SRS, you might use a random number table and note the first two different numbers between 1 and 5 that appear. Or you could use a calculator to generate numbers between 1 and 5, again noting the first two different numbers that result.

    (b) Time and cost considerations would be the benefit of substitution. However, substitution rather than returning to the same home later could lead to selection bias because certain types of people are not and will not be home at 9:00 a.m. With substitution, the sample would no longer be a simple random sample.

    (c) Corner lot homes like homes 1 and 5 might have different residents (perhaps with higher income levels) than other homes. With this in mind, a stratified sample might be more appropriate, randomly choosing one of the corner lot homes and one of the interior lot homes.

TIP

In (a), a complete answer addresses form, strength, direction, and context.

4. The direct telephone and mailing options will both suffer from undercoverage bias. For example, undocumented immigrants themselves and those housing them may be reluctant to put themselves on record by answering either a phone or mail survey. The pollster interviews will result in a convenience sample, which can be highly unrepresentative of the population. In this case, there might be a real question concerning which members of her constituency spend any time in the downtown area where her office is located. The radio appeal will lead to a voluntary response sample, which typically gives too much emphasis to persons with strong opinions.

Quiz 11 (Pages 157–160)

MULTIPLE-CHOICE

1. **(E)** With a random starting point and by picking every *k*th person for the sample, the method is called systematic sampling.

2. **(E)** The wording "creating a level playing field" and "a right to express their individuality" are nonneutral and clearly leading phrasings.

3. **(E)** Each of the 50 states, with its own longtime past state standards and different regional culture, is considered a homogeneous stratum.

4. **(C)** It is most likely that the apartments at which the interviewer had difficulty finding someone home were apartments with fewer students living in them. Replacing these with other randomly picked apartments most likely replaces smaller-occupancy apartments with larger-occupancy ones.

5. **(E)** In a simple random sample, every possible group of the given size has to be equally likely to be selected, and this is not true here. For example, with this procedure it will be impossible for all the early arrivals to be together in the final sample. This procedure is an example of systematic sampling, but systematic sampling does not result in simple random samples.

6. **(C)** Surveying people coming out of any church results in a very unrepresentative sample of the adult population, especially given the question under consideration. Using chance and obtaining a high response rate will not change the selection bias and make this into a well-designed survey.

FREE-RESPONSE QUESTIONS

1. (a) Method A is an example of *cluster sampling*, where the population is divided into heterogeneous groups called *clusters* and individuals from a random sample of the clusters are surveyed. It is often more practical to simply survey individuals from a random sample of clusters (in this case, a random sample of city blocks) than to try to randomly sample a whole population (in this case, the entire city population).

   (b) Method B is an example of *stratified sampling*, where the population is divided into homogeneous groups called *strata* and random individuals from each stratum are chosen. Stratified samples can often give useful information about each stratum (in this case, about each of the five neighborhoods) in addition to information about the whole population (the city population).

2. There are many possible examples, such as "Are you in favor of sending U.S. troops to participate in U.N. peacekeeping operations even though these operations usually have nothing to do with our national interests?" and "Many people in the world live in conflict zones and experience great difficulty finding adequate housing, food, and medical care. Are you in favor of sending U.S. troops to participate in U.N. peacekeeping operations?" Because of wording bias, the first sentence will result in a lower proportion of people in favor of sending U.S. troops than will the second sentence.

3. (a) Method I results in cluster sampling (the rows are clusters) and can be implemented by numbering the rows 1 through 10, using a random number generator to randomly pick a number between 1 and 10, and using all 15 plots in that row. Method II results in stratified sampling (the columns are strata) and can be implemented by numbering the rows 1 through 10, using a random number generator to pick a number between 1 and 10 to pick a plot in the first column, repeating this process to pick a random plot in each of the 15 columns, and using the 15 plots resulting from picking one plot from each column.

   (b) Method II, a stratified sample (with the columns as strata) would be preferable here as each column is somewhat different, so picking a plot from each is meaningful. Method I, an attempt at cluster sampling with the rows as clusters, would work only if each of the rows was heterogeneous (representative of the whole area); this is clearly not true as, for example, the top and bottom rows have far fewer nests than the middle rows.

4. (a) People who attend the school's basketball game are probably more interested in supporting the athletic program than those who do not attend. This study will probably show that a higher proportion of people favor the increased spending than what is the true proportion.

   (b) Subscribers to a health magazine are more likely to believe in the benefits of eating organic than the general population, so the resulting proportion of believers in this study will be greater than it is in the general population.

   (c) The distributer could put any moldy strawberries on the bottom, out of sight! The proportion of accepted boxes will be higher than it should be.

   (d) Many patients will be reluctant to admit to their doctor that they are not following their diet instructions. So, the proportion who say they are following the instructions will be higher than the true proportion following the instructions.

   (e) Parents who are satisfied with the school system are probably less likely to take the time to respond to this voluntary response survey, while those who are unhappy with the education their kids are receiving are more likely to respond. So, the proportion of "unhappy" responses will probably be higher than the true proportion of parents who are unhappy.

Quiz 12 (Pages 168–169)

MULTIPLE-CHOICE

1. **(A)** The main office at your school should be able to give you the class sizes of every math and English class. If need be, you can check with every math and English teacher.

2. **(C)** In the first study, the families were already in the housing units, while in the second study, one of two treatments was applied to each family.

3. **(D)** Using only a sample from the observations gives less information. It may well be that very bright students are the same ones who both choose to take AP Statistics and have high college GPAs. If students could be randomly assigned to take or not take AP Statistics, the results would be more meaningful. Of course, ethical considerations might make it impossible to isolate the confounding variable in this way.

4. **(C)** In experiments on people, the subjects can be used as their own controls, with responses noted before and after the treatment. However, with such designs there is always the danger of a placebo effect. Thus, the design of choice would involve a separate control group to be used for comparison.

5. **(B)** Blocking divides the subjects into groups, such as men and women, or political affiliations, and thus reduces variation. That is, when we group similar individuals together into blocks and then randomize within the blocks, much of the variability due to the differences between the blocks is accounted for and so comparison of the treatment groups is clearer.

FREE-RESPONSE QUESTIONS

1. (a) The explanatory variable is the frequency of substance abuse. The response variable is the age at death.

   (b) This is a prospective observational study. No treatments were assigned, and the researchers asked the rock and pop stars to report their frequency of substance abuse and later noted their ages at time of death.

   (c) No, this is an observational study, so no cause-and-effect conclusion is possible. It is possible that other variables are influencing the response. For example, rock and pop stars might engage in other forms of risk taking, such as poor eating and poor sleep patterns, which might negatively affect life expectancy much more than substance abuse.

2. Every day for some specified period of time, look at the next digit on a random number table. If it is odd, flash the subliminal message all day on the screen, while if it is even, don't flash the message that day (randomization). Don't let the customers know what is happening (blinding) and don't let the clerks selling the popcorn know what is happening (double-blinding). Compare the quantity of popcorn bought by the treatment group, that is, by the people who receive the subliminal message, to the quantity bought by the control group, the people who don't receive the message (comparison).

3. To achieve blocking by sunlight, first separate the sunlit and shaded plots. Label the 15 sunlit plots 1 through 15. Use a random integer generator on a calculator to pick integers between 1 and 15, throwing away repeats, until 5 unique integers have been selected. The sunlit plots corresponding to these 5 integers will receive the fertilizer at regular concentration. Continue using the calculator, throwing away repeats, until 5 more unique integers between 1 and 15 have been selected. The sunlit plots corresponding to these 5 new integers will receive the fertilizer at double concentration, while the 5 remaining sunlit plots will be a control group receiving no fertilizer. Now repeat the procedure, this time labeling the shaded plots 1 through 15. Assuming size is the pertinent outcome, weigh all vegetables at the end of the season, compare the average weights among the three sunlit groups, and compare the average weights among the three shaded groups to determine the effect of the fertilizer, if any, at different levels on sunlit plots and separately on shaded plots.

Quiz 13 (Pages 170–172)

MULTIPLE-CHOICE

1. **(A)** This study is an experiment because a treatment (periodic removal of a pint of blood) is imposed. There is no blinding because the subjects clearly know whether or not they are giving blood. There is no blocking because the subjects are not divided into blocks before random assignment to treatments. For example, blocking would have been used if the subjects had been separated by gender or age before random assignment to give or not give blood donations. There is a single factor—giving or not giving blood.

2. **(B)** The desire of the workers for the study to be successful led to a placebo effect. That is, they saw that the lighting was being changed, and they realized they were being observed. They then assumed that the lighting would be causing a change in production and responded by making this assumption come true.

3. **(D)** Unnecessary blocking detracts from accuracy because of smaller sample sizes. Blocking in experiment design first divides the subjects into representative groups called blocks, just as stratification in sampling design first divides the population into representative groups called strata. This procedure can control certain variables by bringing them directly into the picture, and thus conclusions are more specific. The paired comparison design is a special case of *blocking* in which each pair can be considered a block. In a block design, subjects within *each* block are randomly assigned treatments. One can think of blocking as running parallel experiments before combining the results.

4. **(D)** Octane is the only explanatory variable, and it is being tested at four levels. Miles per gallon is the single response variable.

5. **(A)** Blinding does have to do with whether or not the subjects know which treatment (color in this experiment) they are receiving. However, drinking out of solid black thermoses makes no sense when the beverages are identical except for color and the point of the experiment is the teenager's reaction to color. Blinding has nothing to do with blocking (sports team participation in this experiment).

FREE-RESPONSE QUESTIONS

1. (a) The explanatory variable is the number of minutes a student spends on power naps. The response variable is the GPA at graduation.

   (b) This is a retrospective observational study. No treatments were assigned, and the researchers gathered all the data after graduation.

   (c) No, this is an observational study, so there cannot be a cause-and-effect conclusion. It is possible that other variables are influencing the response. For example, the better students may happen to like to take naps, and the naps might have nothing to do with their grades.

> **NOTE**
>
> The only way to show causation is with a well-designed, well-controlled *experiment*!

2. For each new heart attack patient entering the hospital, look at the next digit from a random number table. If it is odd, give the name to a group of people who will pray for the patient throughout his or her hospitalization. If it is even, don't ask the group to pray (randomization). Don't let the patients know what is happening (blinding), and don't let the doctors know what is happening (double-blinding). Compare the lengths of hospitalization of patients who receive prayers with those of control group patients who don't receive prayers (comparison).

3. (a) Allowing the students to self-select which class to take leads to confounding, which could be significant. For example, perhaps the brighter students all want to learn a certain one of the three languages.

   (b) It is possible for the average score of all science majors to be lower than the average for all math majors even though the science majors averaged higher in each class. For example, suppose that the students taking Java scored much higher than the students in the other two classes. Furthermore, only one science major took Java, and she had the highest score in the class. Then the overall average of the math majors could well be higher than the overall average of the science majors. This is an example of Simpson's paradox, in which a comparison can be reversed when more than one group is combined to form a single group.

   (c) Number the students 1 through 300. Use a random number generator to pick integers between 1 and 300, ignoring repeats, until 100 unique such integers have been selected. The students with the corresponding selected integers should take the Python class. Keep generating integers, ignoring repeats, until 100 more unique new integers between 1 and 300 are selected. The students corresponding to these integers should take the C++ class, while the remaining 100 students should take the Java class.

   (d) Go through the list of students, flipping a die for each. If a 1 or a 2 shows, the student takes Python; if a 3 or a 4 shows, C++; and if a 5 or a 6 shows, Java.

   (e) Another possible variable is the teachers. For example, perhaps the better teachers teach Java.

Quiz 14 (Pages 181–185)

MULTIPLE-CHOICE

1. **(E)** Column total divided by table total $= \dfrac{475}{1000}$

2. **(C)** Cell value divided by table total $= \dfrac{80}{1000}$ (probability of an intersection)

3. **(E)** $\dfrac{110 + 255 - 80}{1,000}$ or $\dfrac{80 + 25 + 5 + 175}{1,000}$ (probability of a union)

4. **(A)** Cell value divided by row total $= \dfrac{80}{110}$ (conditional probability)

5. **(A)** $P(2.0 - 3.0\,\text{GPA}) = \dfrac{475}{1000} = 0.475$; however,

   $P(2.0 - 3.0\,\text{GPA} \mid \text{few skips}) = \dfrac{450}{890} = 0.506$. If independent, these would have been equal.

6. **(A)** Conditional probability: $P(over\ 70 \mid at\ least\ 20) = \dfrac{5,592,012}{9,664,994}$

7. **(B)** The probability of the next child being a girl is independent of the sex of the previous children. Before she had any children, if the question had been about the probability of having eight girls in a row, then the answer would have been $(0.5)^8$, or about 1 in 256.

8. **(D)** The probability of throwing heads is 0.5. By the law of large numbers, the more times you flip the coin, the more the relative frequency tends to become closer to this probability. With fewer tosses, there is a greater chance of wide swings in the relative frequency.

9. **(B)** If E and F are independent, then $P(E \cap F) = P(E)P(F)$; however, in this problem, $(0.4)(0.35) \neq 0.3$.

10. **(D)** $P(\text{famine} \mid \text{plague}) = \dfrac{P(\text{famine} \cap \text{plague})}{P(\text{plague})} = \dfrac{0.15}{0.39}$

11. **(E)**

$$P(\text{positive test}) = P(\text{cancer} \cap \text{positive test}) + P(\text{no cancer} \cap \text{positive test})$$
$$= (0.02)(0.98) + (0.98)(0.03)$$

$$P(\text{cancer} \mid \text{positive test}) = \frac{P(\text{cancer} \cap \text{positive test})}{P(\text{positive test})}$$

$$= \frac{(0.02)(0.98)}{(0.02)(0.98) + (0.98)(0.03)}$$

FREE-RESPONSE QUESTIONS

1. It's easiest to first sum the rows and columns:

	0–5	6–10	>10	
Less than 50 years old	80	125	20	225
More than 50 years old	10	75	50	135
	90	200	70	360

Years of experience

(a) $P(\text{age} < 50) = \dfrac{225}{360} = 0.625$

$P(\text{experience} > 10) = \dfrac{70}{360} = 0.194$

$P(\text{age} > 50 \cap \text{experience } 0\text{–}5) = \dfrac{10}{360} = 0.028$

(b) $P(\text{age} > 50 \mid \text{experience } 6\text{–}10) = \dfrac{125}{200} = 0.625$

(c) $P(\text{age} < 50) = \dfrac{225}{360} = 0.62$; however, $P(\text{age} < 50 \mid \text{experience} > 10) = \dfrac{20}{70} = 0.28$

and so they are not independent. $P(\text{age} > 50) = \dfrac{135}{360} = 0.375$ and also

$P(\text{age} > 50 \mid \text{experience } 6\text{–}10) = \dfrac{75}{200} = 0.375$, and so these two are independent.

2. (a) The probability of the complement is 1 minus the probability of the event, but $1 - 0.43 \neq 0.47$.

(b) Probabilities are never greater than 1, but $6(0.18) = 1.08$.

(c) The probability of an intersection cannot be greater than the probability of one of the separate events.

(d) The probability of a union cannot be less than the probability of one of the separate events.

(e) Probabilities are never negative.

Quiz 15 (Pages 186–189)

MULTIPLE-CHOICE

1. **(A)** 35 out of the entire population of 500 both have HIV and tested positive.

2. **(A)** Of the 460 healthy people, 25 tested positive.

3. **(E)** Of the 40 people with HIV, 35 tested positive.

4. **(E)** Of the 460 healthy people, 435 tested negative.

5. **(E)** Coins have no memory. So, the probability that the next toss will be heads is 0.5, and the probability that it will be tails is 0.5. The law of large numbers says that as the number of tosses becomes larger, the percentage of heads tends to become closer to 0.5.

6. **(E)** While the outcome of any single play on a roulette wheel or the age at death of any particular person is uncertain, the law of large numbers gives that the relative frequencies of specific outcomes in the long run tend to become closer to numbers called probabilities.

7. **(D)** $P(E \cap F) = P(E)P(F)$ only if the events are independent. In this case, it is well known that women, on average, live longer than men, and so the events are not independent.

8. **(E)** The probabilities of each pump not failing are $1 - 0.025$, $1 - 0.034$, and $1 - 0.02$, respectively. The probability of none failing is the product of $(1 - 0.025)(1 - 0.034)(1 - 0.02)$. So the probability of at least one failing is $1 - (1 - 0.025)(1 - 0.034)(1 - 0.02)$.

9. **(A)** The probabilities $\dfrac{7}{24}, \dfrac{8}{24}$, and $\dfrac{9}{24}$ are all nonnegative, and they sum to 1.

10. **(D)** $\dfrac{36 + 22}{36 + 22 + 5 + 2} = 0.892$

11. **(E)**

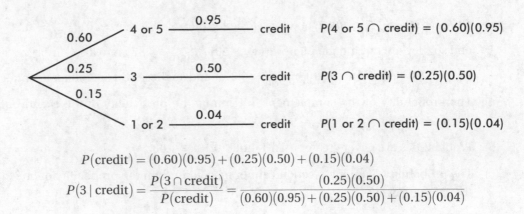

$$P(\text{credit}) = (0.60)(0.95) + (0.25)(0.50) + (0.15)(0.04)$$

$$P(3 \mid \text{credit}) = \frac{P(3 \cap \text{credit})}{P(\text{credit})} = \frac{(0.25)(0.50)}{(0.60)(0.95) + (0.25)(0.50) + (0.15)(0.04)}$$

FREE-RESPONSE QUESTIONS

1. (a) She is assuming that acceptances to the two colleges are independent. This does not seem reasonable as most colleges use similar acceptance criteria.

   (b) She is assuming that acceptances to the two colleges are mutually exclusive. This does not seem reasonable as students can simultaneously be accepted at both of the colleges.

2. (a) $P(E \mid F) = \dfrac{P(E \cap F)}{P(F)}$ so $P(E \cap F) = P(E \mid F)P(F)$, and thus,

   $P(\text{woman} \cap \text{glasses}) = (0.55)(0.56) = 0.308$.

   (b) $P(\text{woman} \cap \text{contacts}) = (0.63)(0.04) = 0.0252$

   (c) If 55% of those who wear glasses are women, 45% of those who wear glasses must be men. If 63% of those who wear contacts are women, 37% of those who wear contacts must be men. Thus, we have $P(\text{man} \cap \text{glasses}) = (0.45)(0.56) = 0.252$.

   (d) $P(\text{man} \cap \text{contacts}) = (0.37)(0.04) = 0.0148$

   (e) The probability that you will encounter a person not wearing glasses or contacts is $1 - (0.56 + 0.04) = 0.4$.

Quiz 16 (Pages 197–199)

MULTIPLE-CHOICE

1. **(B)** $E(X) = \mu_X = \Sigma x_i p_i = \$700(0.05) + \$540(0.25) + \$260(0.7) = \$352$

2. **(D)** Expected values and variances can be added. Thus,

$$E(\text{Total}) = E(X_1) + E(X_2) + \cdots + E(X_{50})$$
$$= 2{,}450 + 2{,}450 + \cdots + 2{,}450 = (2{,}450) = 122{,}500$$
$$SD(\text{Total}) = \sqrt{\text{var}(X_1) + \text{var}(X_2) + \cdots + \text{var}(X_{50})}$$
$$= \sqrt{575^2 + 575^2 + \cdots + 575^2} = \sqrt{50\left(575^2\right)} = 4065.86$$

3. **(C)** For a set of differences, means subtract, but variances add. Thus,

$$E(\text{Sci} - \text{Eng}) = E(\text{Sci}) - E(\text{Eng}) = 650 - 465 = 185$$
$$SD(\text{Sci} - \text{Eng}) = \sqrt{\text{var}(\text{Sci}) + \text{var}(\text{Eng})} = \sqrt{130^2 + 90^2} = 158.11$$

4. **(C)** Means and variances add. Thus,

$$E(D + C) = E(D) + E(C),\ 212 = 98 + E(C),\ E(C) = 114$$
$$\text{var}(D + C) = \text{var}(D) + \text{var}(C),\ 39^2 = 25^2 + \text{var}(C),$$
$$SD(C) = \sqrt{39^2 - 25^2} = 29.93$$

5. **(B)** Means and variances add. Thus,

$$E(\text{Total}) = E(\text{Box}) + 50E(\text{Hole}),\ 16.0 = 1.0 + 50E(\text{Hole}),\ E(\text{Hole}) = 0.3$$
$$\text{var}(\text{Total}) = \text{var}(\text{Box}) + 50\text{var}(\text{Hole})$$
$$0.245^2 = 0.2^2 + 50\left(SD(\text{Hole})\right)^2,\ SD(\text{Hole}) = 0.02$$

6. **(B)** Your expected winnings are only

$$\frac{1}{7{,}500{,}000}(1{,}000{,}000) + \frac{5}{7{,}500{,}000}(10{,}000) + \frac{20}{7{,}500{,}000}(100) = 0.14$$

FREE-RESPONSE QUESTIONS

1. (a) $E(S) = 0(0.2) + 1(0.5) + 2(0.3) = 1.1$

   (b) $\text{var}(S) = (0 - 1.1)^2(0.2) + (1 - 1.1)^2(0.5) + (2 - 1.1)^2(0.3) = 0.49$ and $SD(S) = \sqrt{0.49} = 0.7$

   (c) $E(S + L) = E(S) + E(L) = 1.1 + 1.4 = 2.5$

   (d) $\text{var}(S + L) = \text{var}(S) + \text{var}(L) = (0.7)^2 + (0.6)^2 = 0.85$ and $SD(S + L) = \sqrt{0.85} = 0.922$

2. (a) $F = 5C$ (the weight of coconut is being multiplied by a constant)

   $E(F) = E(5C) = 5E(C) = 5(3.2) = 16$ ounces, and

   $SD(F) = SD(5C) = 5SD(C) = 5(0.7) = 3.5$ ounces

   (b) $T = C + C + C + C + C$ (total weight is the sum of the weights of 5 individual coconuts)
   $E(T) = E(C) + E(C) + E(C) + E(C) + E(C) = 5(3.2) = 16$ pounds and $\text{var}(T) = \text{var}(C) + \text{var}(C) + \text{var}(C) + \text{var}(C) + \text{var}(C) = 5(0.7)^2 = 2.45$ with $SD(T) = \sqrt{2.45} = 1.56552$

Quiz 17 (Pages 200–202)

MULTIPLE-CHOICE

1. **(D)** $\bar{x} = E(X) = \sum xP(x) = 0(0.999) + 25{,}000(0.001) = 25$

$$\sigma = \sqrt{\sum (x - \bar{x})^2 P(x)} = \sqrt{(0 - 25)^2(0.999) + (25{,}000 - 25)^2(0.001)} = 790.17$$

[Or on the TI-84, put returns and probabilities into two lists and run `1-Var Stats L1,L2`.]

2. **(D)** $1000(0.2) + 5000(0.05) = 450$, and $800 - 450 = 350$

3. **(C)** Expected values and variances add. Thus,

$E(\text{Thefts}) = E(X_1) + E(X_2) + \cdots + E(X_{45}) = 361 + 361 + \cdots + 361 = 45 \times 361$ and
$\text{SD}(\text{Thefts}) = \sqrt{74^2 + 74^2 + \ldots + 74^2} = \sqrt{45 \times 74^2} = \sqrt{45} \times 74$

4. **(C)** For a set of differences, means subtract, but variances add. Thus,
$\text{SD}(X - Y) = \sqrt{\text{var}(X) + \text{var}(Y)} = \sqrt{3100^2 + 1200^2}$

5. **(B)** Expected values and variances add. Thus,

$$E(\text{Broken}) = E(X_1) + E(X_2) + \cdots + E(X_{100}) = 0.3 + 0.3 + \cdots + 0.3 = 100(0.3) = 30$$
$$\text{SD}(\text{Broken}) = \sqrt{\text{var}(X_1) + \text{var}(X_2) + \cdots + \text{var}(X_{100})}$$
$$= \sqrt{0.18^2 + 0.18^2 + \cdots + 0.18^2} = \sqrt{100(0.18)^2} = 1.8$$

6. **(A)** $E(Un + As + FT) = E(Un) + E(As) + E(FT)$,
$5.6 = 1.5 + 28 + E(FT), E(FT) = 1.3$

Means and variances add. Thus, $\text{var}(Un + As + FT) = \text{var}(Un) + \text{var}(As) + \text{var}(FT)$,
$$0.886^2 = 0.2^2 + 0.85^2 + \left[\text{SD}(FT)\right]^2,$$
$$\text{SD}(FT) = \sqrt{0.886^2 - 0.2^2 - 0.85^2} = 0.15$$

FREE-RESPONSE QUESTIONS

1. (a) $E(X) = 35(0.1) + 36(0.5) + 37(0.4) = 36.3$

   (b) $\text{var}(X) = (35 - 36.3)^2(0.1) + (36 - 36.3)^2(0.5) + (37 - 36.3)^2(0.4) = 0.41$ and $\text{SD}(X)$
   $= \sqrt{0.41} = 0.640$

   (c) $E(\text{Empty}) + 36E(\text{Fruit}) = E(\text{Full}), 1 + 36E(\text{Fruit}) = 10$

   $E(\text{Fruit}) = \dfrac{10 - 1}{36} = 0.25$ pounds

   (d) $\text{var}(\text{Empty}) + 36\,\text{var}(\text{Fruit}) = \text{var}(\text{Full}), 0.1^2 + 36\text{var}(\text{Fruit}) = 0.5^2$

   $\text{var}(\text{Fruit}) = \dfrac{0.25 - 0.01}{36} = 0.00667$

   $\text{SD}(\text{Fruit}) = \sqrt{0.0067} = 0.082$ pounds

2. (a) $F = 0.01P$ (the weight of pineapple is being multiplied by a constant)

$E(F) = E(0.01P) = 0.01E(P) = 0.01(900) = 9$ grams, and

$SD(F) = SD(0.01P) = 0.01SD(P) = 0.01(120) = 1.2$ grams

(b) $T = P + P + P + P$ (total weight is the sum of the weights of 4 individual pineapples)

$E(T) = E(P) + E(P) + E(P) + E(P) = 4(900) = 3600$ grams, and

$var(T) = var(P) + var(P) + var(P) + var(P) = 4(120)^2 = 57{,}600$ with

$SD(T) = \sqrt{57{,}600} = 240$ grams

Quiz 18 (Pages 209–211)

MULTIPLE-CHOICE

1. **(A)** This is a binomial with $n = 5$ and $p = 0.15$. Then, $P(X = 2) = \binom{5}{2}(0.15)^2(0.85)^3$.

2. **(C)** This is a binomial with $n = 10$ and $p = 0.37$, and so the mean is $np = 10(0.37) = 3.7$.

3. **(B)** The sample has five adults, so $n = 5$, and the probability any of these adults experienced cyberbullying is given to be $\frac{1}{6}$.

4. **(C)** This is a binomial distribution with $n = 5$ and $p = \frac{2}{5} = 0.4$. Then,

$$P(X = 2) = \binom{5}{2}(0.4)^2(0.6)^3.$$

5. **(D)** This is a geometric distribution with $p = 0.4$. Then, $P(X = 3) = (0.6)^2(0.4)$.

6. **(D)** This is a geometric distribution with $p = \frac{1}{20} = 0.05$. The mean $= \frac{1}{p} = \frac{1}{0.05} = 20$, and the standard deviation $= \sqrt{\frac{1-p}{p^2}} = \sqrt{\frac{1-0.05}{(0.05)^2}} = \sqrt{380}$.

FREE-RESPONSE QUESTIONS

1. (a) We have a binomial with $n = 20$ and $p = 0.15$, so $P(X \geq 10) = 1 -$ binomcdf$(20, 0.15, 9) = 0.00025$. [Or we can calculate $P(X \geq 10) =$ $\binom{20}{10}(0.15)^{10}(0.85)^{10} + \binom{20}{11}(0.15)^{11}(0.85)^9 + \cdots + (0.15) = 0.00025$.]

(b) If the probability of *Legionella* bacteria growing in an automatic faucet is 0.15, the probability of a result as extreme or more extreme than what was obtained in the Johns Hopkins study is only 0.00025. With such a low probability, there is strong evidence to conclude that the probability of *Legionella* bacteria growing in automatic faucets is greater than 0.15. That is, there is strong evidence that automatic faucets actually house more bacteria than the manual kind.

2. (a) $\dfrac{0.02}{0.11 + 0.02 + 0.014 + 0.006} = \dfrac{0.02}{0.15} = \dfrac{2}{15} = 0.133$

(b) P(at least 2 tattoos) $= 0.02 + 0.014 + 0.006 = 0.04$. The student with tattoos can be the first, second, or third student, so we have $3(0.04)(0.85)^2 = 0.0867$.

(c) $E(X) = 0(0.85) + 1(0.11) + 2(0.02) + 3(0.0014) + 4(0.006) = 0.216$

3. (a) This is a binomial distribution with $n = 4$ and $p = 0.35$. Then $P(X = 2) =$ binompdf(4, 0.35, 2) = 0.3105. [Or $P(X = 2) = \binom{4}{2}(0.35)^2(0.65)^2 = 0.3105.$]

(b) Binomial, $n = 4$, $p = 0.35$, and $P(X \geq 2) = 1 - P(X \leq 1) = 1 - \text{binomialcdf}(4, 0.35, 1)] = 0.4370$. [Or $1 - \left[(0.65)^4 + 4(0.35)(0.65)^3\right] = 0.4370.$]

(c) This is a geometric distribution with $p = 0.35$. $P(X = 2) = \text{geometpdf}(0.35, 2) = 0.2275$. [Or $(0.65)(0.35) = 0.2275.$]

(d) Geometric, $p = 0.35$. Mean $= \dfrac{1}{p} = \dfrac{1}{0.35} = 2.8571$ and standard deviation $= \sqrt{\dfrac{1-p}{p^2}} = \sqrt{\dfrac{1-0.35}{(0.35)^2}} = 2.3035.$

Quiz 19 (Pages 212–214)

MULTIPLE-CHOICE

1. **(E)** This is a binomial with $n = 3$ and $p = 0.90$. Then $P(X \geq 2) = (0.9)^3 + 3(0.9)^2(0.1)$.

2. **(D)** This is a binomial with $n = 4$ and $p = 0.65$.

$$P(X = 2) = \binom{4}{2}(0.65)^2(0.35)^2 = 6(0.65)^2(0.35)^2$$

3. **(B)** The sample has seven adults, so $n = 7$, and the probability any of these adults using recreational marijuana is given to be $\dfrac{1}{10}$.

4. **(E)** This is a binomial with $n = 4$, and $p = \dfrac{1}{4} = 0.25$. Then

$$P(X \geq 2) = 1 - P(X \leq 1) = 1 - \left[P(X = 0) + P(X = 1)\right] = 1 - \left[(0.75)^4 + \binom{4}{1}(0.25)(0.75)^3\right].$$

5. **(D)** This is a geometric distribution with $p = 0.25$. Then $P(X \leq 2) = 1 - P(X > 2) = 1 - (0.75)^2$. Note that the answer could also have been expressed as $P(X = 1) + P(X = 2) = (0.25) + (0.25)(0.75)$.

6. **(C)** This is a geometric distribution with $p = \dfrac{2}{3}$. Then the mean $= \dfrac{1}{p} = \dfrac{1}{\left(\frac{2}{3}\right)} = 1.5$, and the

standard deviation $= \sqrt{\dfrac{1-p}{p^2}} = \sqrt{\dfrac{1-\left(\frac{2}{3}\right)}{\left(\frac{2}{3}\right)^2}} = \sqrt{\dfrac{\left(\frac{1}{3}\right)}{\left(\frac{4}{9}\right)}} = \sqrt{\dfrac{3}{4}} = \dfrac{\sqrt{3}}{2}.$

FREE-RESPONSE QUESTIONS

1. With Option A, this is a binomial with $n = 6$ and $p = 0.75$. If in reality only 75% of the articles meet all specifications, the probability of rejecting the day's production is $P(X \leq 4) = \text{binomcdf}(6, .75, 4) = 0.466$. [Or we can calculate: $P(\text{rejection}) = 1 - P(\text{acceptance}) =$

$$1 - \left[\binom{6}{5}(0.75)^5(0.25) + (0.75)^6\right] = 0.466.]$$

With Option B, this is a binomial with $n = 12$ and $p = 0.75$. If in reality only 75% of the articles meet all specifications, the probability of rejecting the day's production is $P(X \leq 9) = \text{binomcdf}(12, .75, 9) = 0.609$. [Or we can calculate: $P(\text{rejection}) = 1 - P(\text{acceptance}) =$

$$1 - \left[\binom{12}{10}(0.75)^{10}(0.25)^2 + \binom{12}{11}(0.75)^{11}(0.75)^{12} \right] = 0.609 .]$$

For the greatest probability of rejecting the day's production if only 75% of the articles meet all specifications, the buyer should request the manufacturer to use Option B with a probability of rejection of 0.609 as opposed to Option A with a probability of rejection of only 0.466.

2. This is a binomial with $n = 7$ and $p = 0.2$. If USAir accounted for 20% of the major disasters, the chance that it would be involved in at least four of seven such disasters is $P(X \geq 4) = 1 - \text{binomcdf}(7, .2, 3) = 0.033$. [Or we can calculate

$$\binom{7}{4}(0.2)^4(0.8)^3 + \binom{7}{5}(0.2)^5(0.8)^2 + \binom{7}{6}(0.2)^6(0.8)^1 + (0.2)^7$$
$$= 0.029 + 0.004 + 0.000 + 000 = 0.033.]$$

Mathematically, if USAir accounted for only 20% of the major disasters, there is only a 3.3% chance of it being involved in four of seven such disasters. This seems more than enough evidence to be suspicious! There is sufficient evidence that USAir accounts for more than 20% of major disasters in the years leading up to 1994.

3. (a) This is a geometric distribution with $p = 0.14$. $P(X = 3) = \text{geometpdf}(.14, 3) = 0.1035$. [Or we can calculate $(0.86)^2(0.14) = 0.1035$.]

   (b) This is a geometric distribution with $p = 0.14$. $P(X < 10) = \text{geometcdf}(.14, 9) = 0.7427$. [Or we could calculate $1 - (0.86)^9 = 0.7427$.]

   (c) This is a geometric distribution with $p = 0.14$. $\mu = \dfrac{1}{p} = \dfrac{1}{0.14} = 7.143$ and $\sigma = \sqrt{\dfrac{1-p}{p^2}} = \sqrt{\dfrac{1-0.14}{0.14^2}} = 6.624$.

Quiz 20 (Pages 227–229)

MULTIPLE-CHOICE

1. **(A)** The area under any probability distribution is equal to 1. Many bell-shaped curves are *not* normal curves. The smaller the standard deviation of a normal curve, the higher and narrower the graph is. The mean determines the value around which the curve is centered; different means give different centers. Because of symmetry, the mean and median are identical for normal distributions.

2. **(E)** All normal distributions have the same percentage, about 95%, of their observations within two standard deviations of the mean.

3. **(B)** The z-score of 10 is $\dfrac{10-12.4}{1.2} = -2$. Then $P(X < 10) = P(z < -2)$.

4. **(E)** The z-scores of 3 and 4 are $\dfrac{3-3.4}{0.5}$ and $\dfrac{4-3.4}{0.5}$, respectively. Then

$$P(3 < X < 4) = P\left(\dfrac{3-3.4}{0.5} < z < \dfrac{4-3.4}{0.5}\right).$$

5. **(E)** The critical z-score associated with 99% to the left is invNorm(0.99) = 2.326.

6. **(B)** $\mu_{X+Y} = 150 + 120 = 270$ and $\sigma_{X+Y} = \sqrt{40^2 + 30^2} = 50$

7. **(C)** The critical z-score is invNorm(0.85) = 1.036. Then $\mu + 1.036(2) = 16$ gives $\mu = 16 - 1.036(2)$.

FREE-RESPONSE QUESTIONS

1. (a) $P\left(z > \dfrac{x-150}{25}\right) = 0.90$, so $\dfrac{x-150}{25} = -1.282$ and $x = 117.95$ for a cutoff score of 118.

   (b) $P(x > 118 \mid x > 100) = \dfrac{P(x > 118)}{P(x > 100)} = \dfrac{0.900}{0.9772} = 0.921$

   (c) $1 - (0.921)^3 = 0.219$

2. (a) Let the random variable D be the difference in walking times (Steve – Jan). Then $\mu_D = 30 - 25 = 5$, and $\sigma_D = \sqrt{5^2 + 4^2} = 6.403$. For Steve to arrive before Jan, the difference in walking times must be < 0.

   $$P(D < 0) = P\left(z < \frac{0-5}{6.403}\right) = 0.217$$

   (b) If m is the number of minutes he should leave early, the new mean of the differences is $5 - m$, and we want $P\left(z < \dfrac{0 - (5-m)}{6.403}\right) = 0.9$, which gives $\dfrac{-5+m}{6.403} = 1.282$ and $m = 13.2$ minutes.

Quiz 21 (Pages 230–233)

MULTIPLE-CHOICE

1. **(B)** The normal curve is symmetric around its center, the mean. The function normalcdf(0, 2) = 0.4772 is not twice normalcdf(0, 1) = 0.3413. The function invNorm(0.75) − invNorm(0.25) = 0.67 − (−0.67) is not 3. The range is not finite. And, $P(z < 0.1)$ is more than 0.5, while $P(z > 0.9)$ is less than 0.5.

2. **(C)** Curve a has mean 6 and standard deviation 2, while curve b has mean 18 and standard deviation 1.

3. **(A)** The z-score of 3 is $\dfrac{3 - \mu}{\sigma} = \dfrac{3 - 2.43}{0.88}$.

4. **(B)** If 95% of the area is to the right of a score, then 5% is to the left. The critical z-score is invNorm(0.05) = −1.645. Converting this to a raw score gives $11 - 1.645(1.5)$.

5. **(B)** The critical z-scores associated with the middle 95% are $\pm$invNorm(0.975) = ±1.96. Converting to raw scores gives $9500 \pm 1.96(1750)$.

6. **(A)** The critical z-score associated with 18% to the right or 82% to the left is invNorm(0.82) = 0.915. Then $100 + 0.915\sigma = 120$ gives $\sigma = \dfrac{120 - 100}{0.915}$.

7. **(E)** The critical z-score associated with 0.5% to the right (99.5% to the left) is invNorm(0.995) = 2.576. Then $c + 2.576(0.4) = 8$, which gives $c = 8 - 2.576(0.4)$.

FREE-RESPONSE QUESTIONS

1. (a) $P(2740 \mid 2500) = \dfrac{P\left(z > \dfrac{2740 - 3000}{400}\right)}{P\left(z > \dfrac{2500 - 3000}{400}\right)} = \dfrac{0.742154}{0.894350} = 0.829825$ and

   $P(3040 \mid 2800) = \dfrac{P\left(z > \dfrac{3040 - 3000}{400}\right)}{P\left(z > \dfrac{2800 - 3000}{400}\right)} = \dfrac{0.460172}{0.691462} = 0.665506$

   Given independence, the probability that both components last 240 more hours is $(0.829825)(0.665506) = 0.552$.

   (b) The probability that both fail is $(1 - 0.829825)(1 - 0.665506) = 0.056923$. The probability that at least one doesn't fail is $1 - 0.056923 \approx 0.943$.

2. (a) $P\left(z < \dfrac{8500 - 9000}{590}\right) \approx 0.2$

   (b) The probability of a caplet having more than 9000 units is 0.5, and thus one would expect to find two caplets with more than 9000 units much more quickly than finding two caplets with less than 8500 units. Thus, the first histogram results from looking for two caplets with more than 9000 units and the second histogram results from looking for two caplets with less than 8500 units.

Quiz 22 (Pages 247–252)

MULTIPLE-CHOICE

1. **(D)** The relevant condition is that both np and $nq \geq 10$, and only Choice (D) with $np = (75)(0.15) = 11.25$ and $nq = (75)(0.85) = 63.75$ satisfies this.

2. **(C)** The spread of the sampling distribution depends upon $\sigma_{\hat{p}} - \sqrt{\dfrac{p(1-p)}{n}}\sqrt{1 - \dfrac{n}{N}}$. As long as $n < 10\%(N)$, the finite population factor, $\sqrt{1 - \dfrac{n}{N}}$, is close to 1, and so $\sigma_{\hat{p}} \approx \sqrt{\dfrac{p(1-p)}{n}}$ and does not depend on the population size N. The larger the sample, the smaller the spread in the sampling distribution is. Bias relates to the center, not the spread, of a sampling distribution. Sample statistics, not sample parameters, are used to make inferences about population proportions. Statistics from smaller samples have more variability, not less.

3. **(E)** $\dfrac{\sigma}{\sqrt{n}}$ and σ are not equal unless $n = 1$. It is always true that the sampling distribution of $\bar{x}$ has mean μ and standard deviation $\dfrac{\sigma}{\sqrt{n}}$. In addition, the sampling distribution will be normal if the population is normal and will be approximately normal if n is large, even if the population is not normal.

4. **(A)** The maximum of a sample is never larger than the maximum of the population, so the mean of the sample maximums will not be equal to the population maximum. (A sampling distribution is unbiased if its mean is equal to the population parameter.)

5. **(D)** The sample is given to be random, both $np = (400)(0.34) = 136 \geq 10$ and $n(1 - p) = (400)(0.66) = 264 \geq 10$, and our sample is clearly less than 10% of all people. So, the sampling distribution of $\hat{p}$ is approximately normal with mean 0.34 and standard deviation $\sigma_{\hat{p}} = \sqrt{\dfrac{(0.34)(0.66)}{400}}$. With z-scores of $\dfrac{0.30 - 0.34}{\sqrt{\dfrac{(0.34)(0.66)}{400}}}$ and $\dfrac{0.35 - 0.34}{\sqrt{\dfrac{(0.34)(0.66)}{400}}}$, the probability that the sample proportion is between 0.30 and 0.35 is

$$P(0.30 < X < 0.35) = P\left(\frac{0.30 - 0.34}{\sqrt{\dfrac{(0.34)(0.66)}{400}}} < z < \frac{0.35 - 0.34}{\sqrt{\dfrac{(0.34)(0.66)}{400}}} \right)$$

6. **(A)** We have a random sample that is less than 10% of all schools. With a sample size of 30, the central limit theorem applies, and the sampling distribution of $\bar{x}$ is approximately normal with mean $\mu_{\bar{x}} = 1200$ and standard deviation $\sigma_{\bar{x}} = \dfrac{300}{\sqrt{30}}$. The z-score of 1000 is $\dfrac{1000 - 1200}{\dfrac{300}{\sqrt{30}}}$, and the probability of a sample mean over 1000 is

$$P\left(z > \frac{1000 - 1200}{\dfrac{300}{\sqrt{30}}} \right).$$

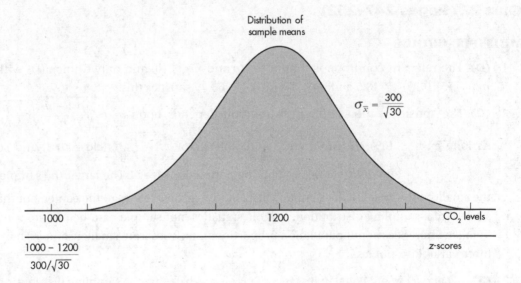

Distribution of
sample means

$\sigma_{\bar{x}} = \dfrac{300}{\sqrt{30}}$

1000   1200 CO_2 levels

$\dfrac{1000 - 1200}{300/\sqrt{30}}$ z-scores

7. **(C)** We have two independent random samples, each less than 10% of their respective populations, and we note that $n_1 p_1 = 75(0.43) = 32.25$, $n_1(1 - p_1) = 75(0.57) = 42.75$, $n_2 p_2 = 80(0.37) = 29.6$, and $n_2(1 - p_2) = 80(0.63) = 50.4$ are all ≥ 10. Thus, the sampling distribution of $\hat{p}_1 - \hat{p}_2$ is roughly normal with mean $\mu_{\hat{p}_1 - \hat{p}_2} = 0.43 - 0.37 = 0.06$ and standard deviation $\sigma_{\hat{p}_1 - \hat{p}_2} = \sqrt{\dfrac{(0.43)(0.57)}{75} + \dfrac{(0.37)(0.63)}{80}}$. The z-score of 0.05 is $\dfrac{0.05 - 0.06}{\sqrt{\dfrac{(0.43)(0.57)}{75} + \dfrac{(0.37)(0.63)}{80}}}$, the z-score of 0.10 is $\dfrac{0.10 - 0.06}{\sqrt{\dfrac{(0.43)(0.57)}{75} + \dfrac{(0.37)(0.63)}{80}}}$, and

$$P(0.05 < X < 0.10) = P\left(\dfrac{0.05 - 0.06}{\sqrt{\dfrac{(0.43)(0.57)}{75} + \dfrac{(0.37)(0.63)}{80}}} < z < \dfrac{0.10 - 0.06}{\sqrt{\dfrac{(0.43)(0.57)}{75} + \dfrac{(0.37)(0.63)}{80}}} \right).$$

8. **(D)** The variance for the sampling distribution of $\hat{p}$ equals $\dfrac{p(1-p)}{n}$. A larger n in the denominator results in a smaller quotient, and $(0.1)(0.9) < (0.5)(0.5)$.

9. **(C)** The sampling distribution is roughly normal with a mean of $\mu = 0.32 - 0.29 = 0.03$ and a standard deviation of $\sigma = \sqrt{\dfrac{(0.32)(0.68)}{400} + \dfrac{(0.29)(0.71)}{400}} = 0.0325$.

FREE-RESPONSE QUESTIONS

1. (a) $\sigma_{\hat{p}} = \sqrt{\dfrac{p(1-p)}{n}} = \sqrt{\dfrac{(0.16)(0.84)}{80}} = 0.041$

   (b) Yes, the conditions for inference are met. First, it is given that a random sample is planned. Second, the sample size, $n = 80$, is clearly less than 10% of all game players. Third, $np = (80)(0.16) = 12.8 \geq 10$ and $nq = (80)(0.84) = 67.2 \geq 10$.

   (c) The sample size would be larger than 80, because $0.037 < 0.041$. In order for $\sigma_{\hat{p}}$ to be smaller, the denominator in $\sigma_{\hat{p}} = \sqrt{\dfrac{p(1-p)}{n}}$ must be larger.

2. (a) 0.58

   (b) This is a binomial with $n = 3$ and $p = 0.58$.

   $$P(X \geq 3) = \binom{5}{3}(0.58)^3(0.42)^2 + \binom{5}{4}(0.58)^4(0.42) + (0.58)^5 = 0.647$$

   or $1 - \text{binomcdf}(5, .58, 2) = 0.647$.

   (c) The sample is given to be random, both $np = (350)(0.58) = 203 \geq 10$ and $n(1 - p) = (350)(0.42) = 147 \geq 10$, and our sample is clearly less than 10% of all Americans. So, the sampling distribution of $\hat{p}$ is approximately normal with mean 0.58 and standard deviation $\sigma_{\hat{p}} = \sqrt{\dfrac{(0.58)(0.42)}{350}} = 0.0264$. With a z-score of $\dfrac{0.5 - 0.58}{0.0264} = -3.030$, the probability that the sample proportion is greater than 0.5 is $\text{normalcdf}(-3.030, 1000) = 0.9988$.
   [Or $\text{normalcdf}(.5, 1, .58, .0264) = 0.9988$.]

> **TIP**
>
> On the exam, you can use expressions like binomcdf only if you note the parameters: binomcdf $(n = 5, p = 0.58, 2)$.

> **NOTE**
>
> You can also do a direct binomial calculation:
> $1 - \text{binomcdf}(350, 0.58, 174) = 0.9989$.

3. (a) $P(X > 4250) = P\left(z > \dfrac{4250 - 4000}{125}\right) = P(z > 2) = 0.0228.$

   [Or `normalcdf(4250,10000,4000,125) = 0.0228`.]

   (b) The original population is approximately normal, so the sampling distribution of $\bar{x}$ is approximately normal with mean $\mu_{\bar{x}} = \mu = 4000$ and standard deviation $\sigma_{\bar{x}} = \dfrac{\sigma}{\sqrt{n}} = \dfrac{125}{\sqrt{40}} = 19.764.$

   (c) $P(X < 3950) = P\left(z < \dfrac{3950 - 4000}{19.764}\right) = P(z < -2.530) = 0.0057.$

   [Or `normalcdf(0,3950,4000,19.764) = 0.0057`.]

   (d) The answer to part (a) would be affected because it assumes an approximately normal population. The other answers would not be affected because for large enough n, the central limit theorem gives that the sampling distribution of $\bar{x}$ is roughly normal regardless of the distribution of the original population.

4. (a) Using a calculator, find $\mu = 12$ and $\sigma = 5.0596.$

   (b)

Sample	{6, 9}	{6, 11}	{6, 13}	{6, 21}	{9, 11}	{9, 13}	{9, 21}	{11, 13}	{11, 21}	{13, 21}
Mean	7.5	8.5	9.5	13.5	10	11	15	12	16	17

   (c) $\mu_{\bar{x}} = 12$ and $\sigma_{\bar{x}} = 3.0984.$

   (d) $\mu_{\bar{x}} = \mu = 12$ and $\dfrac{\sigma}{\sqrt{n}}\sqrt{\dfrac{N-n}{N-1}} = \dfrac{5.0596}{\sqrt{2}}\sqrt{\dfrac{5-2}{5-1}} = 3.0984 = \sigma_{\bar{x}}.$

   (e) If N is very large, $\dfrac{N-n}{N-1}$ is approximately equal to 1, and so the expression is approximately equal to $\dfrac{\sigma}{\sqrt{n}}.$

Quiz 23 (Pages 253–257)

MULTIPLE-CHOICE

1. **(D)** The central limit theorem gives that the larger the sample size, the closer the sampling distribution will be to normal.

2. **(A)** The sampling distribution of $\hat{p}$ is *unbiased*; that is, its mean is equal to the population proportion p. Both (B) and (C) are incorrect because the sampling distribution of $\hat{p}$ has a standard deviation of $\sqrt{\dfrac{p(1-p)}{n}}$, which is smaller with larger n. Both (D) and (E) are incorrect because, while the sampling distribution of $\hat{p}$ is never exactly normal, it is considered close to normal *provided that* both np and $n(1-p)$ are large enough (greater than 10 is a standard guide).

3. **(B)** Sample proportions are an unbiased estimator for the population proportion, and larger sample sizes lead to reduced variability.

4. **(E)** The sampling distribution of $\bar{x}$ has mean μ, standard deviation $\dfrac{\sigma}{\sqrt{n}}$, and shape that becomes closer to normal with larger n.

5. **(E)** All are unbiased estimators for the corresponding population parameters; that is, the means of their sampling distributions are equal to the population parameters.

6. **(B)** The sample is given to be random, both $np = (300)(0.12) = 36 \geq 10$ and $n(1 - p) = (300)(0.88) = 264 \geq 10$, and our sample is clearly less than 10% of all butterfly larvae. So, the sampling distribution of $\hat{p}$ is approximately normal with mean $\mu_{\hat{p}} = 0.12$ and standard deviation $\sigma_{\hat{p}} = \sqrt{\dfrac{(0.12)(0.88)}{300}}$. With a z-score of $\dfrac{0.15 - 0.12}{\sqrt{\dfrac{(0.12)(0.88)}{300}}}$, the probability that

the sample proportion is more than 15% is $P\left(z > \dfrac{0.15 - 0.12}{\sqrt{\dfrac{(0.12)(0.88)}{300}}} \right)$.

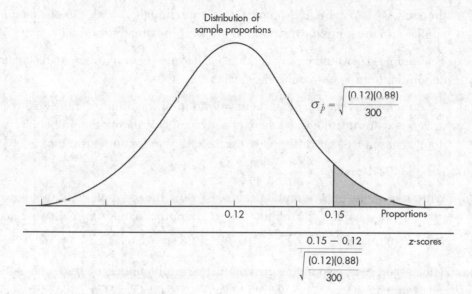

Distribution of sample proportions

$\sigma_{\hat{p}} = \sqrt{\dfrac{(0.12)(0.88)}{300}}$

0.12   0.15   Proportions

$\dfrac{0.15 - 0.12}{\sqrt{\dfrac{(0.12)(0.88)}{300}}}$ z-scores

7. **(E)** We have a random sample that is less than 10% of the high school football population. With a sample size of 48, the central limit theorem applies, and the sampling distribution of $\bar{x}$ is approximately normal with mean $\mu_{\bar{x}} = 355$ and standard deviation $\sigma_{\bar{x}} = \dfrac{80}{\sqrt{48}}$. The z-scores of 340 and 360 are $\dfrac{340 - 355}{\dfrac{80}{\sqrt{48}}}$ and $\dfrac{360 - 355}{\dfrac{80}{\sqrt{48}}}$, respectively, and the probability of a sample mean between 340 and 360 is

$$P(340 < X < 360) = P\left(\dfrac{340 - 355}{\dfrac{80}{\sqrt{48}}} < z < \dfrac{360 - 355}{\dfrac{80}{\sqrt{48}}} \right).$$

8. **(B)** We have independent random samples, each less than 10% of babies, and both sample sizes are $40 \geq 30$. So, the sampling distribution of $\bar{x}_1 - \bar{x}_2$ is roughly normal with mean $\mu_{\bar{x}_1 - \bar{x}_2} = 1.8 - 1.2 = 0.6$ and standard deviation $\sigma_{\bar{x}_1 - \bar{x}_2} = \sqrt{\dfrac{(0.5)^2}{40} + \dfrac{(0.3)^2}{40}}$. The

z-score of 0.75 is $\dfrac{0.75 - 0.6}{\sqrt{\dfrac{(0.5)^2}{40} + \dfrac{(0.3)^2}{40}}}$, and $P(X > 0.75) = P\left(z > \dfrac{0.75 - 0.6}{\sqrt{\dfrac{(0.5)^2}{40} + \dfrac{(0.3)^2}{40}}} \right)$.

9. **(D)** Variances add, and so $\sqrt{\left(\dfrac{68}{\sqrt{50}}\right)^2 + \left(\dfrac{45}{\sqrt{50}}\right)^2} = \sqrt{\dfrac{68^2}{50} + \dfrac{45^2}{50}} = \sqrt{\dfrac{68^2 + 45^2}{50}}$.

FREE-RESPONSE QUESTIONS

1. (a) When the mean is much greater than the median, the distribution is probably skewed right (toward the greater values).

   (b) Yes, the conditions for inference are met. First, it is given that a random sample was planned. Second, the sample size, $n = 90$, is clearly less than 10% of all men working on Wall Street. Third, while the population distribution is probably not normal (it's probably skewed right), the sample size is large enough, $n = 90 \geq 30$, so that the CLT applies and the sampling distribution of $\bar{x}$ is approximately normal.

   (c) The sample size would be smaller than 90, because in order for $\sigma_{\bar{x}}$ to be larger, the denominator in $\sigma_{\bar{x}} = \dfrac{\sigma}{\sqrt{n}}$ must be smaller.

2. (a) We do not know the shape of the distribution of the amount individual teenage drivers pay for insurance, so there is no way of calculating the probability a randomly chosen teenage driver pays over $2400 a year for auto insurance.

   (b) $\mu_{\bar{x}} = \mu = 2275$ and $\sigma_{\bar{x}} = \dfrac{\sigma}{\sqrt{n}} = \dfrac{650}{\sqrt{90}} = 68.52$

   (c) With this large of a sample size $(90 \geq 30)$, the central limit theorem tells us that the sampling distribution of $\bar{x}$ is approximately normal. We calculate $P(\bar{x} > 2400) = P\left(z > \dfrac{2400 - 2275}{68.52}\right) = P(z > 1.824) = 0.0341$.

3. (a) The z-scores corresponding to cumulative probabilities of 0.25 and 0.75 are ± 0.6745. Thus, $Q_1 = 374 - (0.6745)(38.55) = 348.00$, $Q_3 = 374 + (0.6745)(38.55) = 400.00$, and IQR $= Q_3 - Q_1 = 400 - 348 = 52$ hours.

   (b) From part (a), we have that the probability of less than 400 deprived hours is 0.75, and thus the probability of being more than 400 deprived hours is 0.25. Now we have a binomial with $n = 3$ and $p = 0.25$. $P(\text{majority} > 400) = P(2 \text{ or } 3 \text{ are } > 400) = 3(0.25)^2(0.75) + (0.25)^3 = 0.15625$.

   (c) The sampling distribution of the sample means is approximately normal (because the original population is normal) with mean $\mu_{\bar{x}} = \mu = 374$ and standard deviation $\sigma_{\bar{x}} = \dfrac{\sigma}{\sqrt{n}} = \dfrac{38.55}{\sqrt{3}} = 22.257$. The z-score of 400 is $\dfrac{400 - 374}{22.257} = 1.168$, and the probability of a sample mean over 400 is `normalcdf(1.168,1000)` $= 0.121$. [Or `normalcdf(400,1000,374,22.257)` $= 0.121$.]

4. (a) A calculator quickly gives $\bar{x} = 72$ and $s^2 = 4$.

   (b) $\dfrac{(n-1)s^2}{\sigma^2} = \dfrac{(5-1)(4)}{3} = 5.33$

   (c) Even though the sample variance of 4 is greater than the old variance of 3, by looking at where 5.33 is found in the dotplot (the dotplot generated from a population with variance 3), it is very plausible that the given sample came from a population with variance 3. It does not appear that there is evidence of a significant loss of consistency with regard to the graders being used.

Quiz 24 (Pages 274–278)

MULTIPLE-CHOICE

1. **(D)** The relevant condition is that both $n\hat{p}$ and $n\hat{q} \geq 10$, and only Choice (D) with $n\hat{p} = (75)(0.15) = 11.25$ and $n\hat{q} = (75)(0.85) = 63.75$ satisfies this.

2. **(B)** The critical z-scores with 0.01 in the tails are $\pm\texttt{invNorm(0.99)} = \pm 2.326$ and $SE(\hat{p}) = \sqrt{\dfrac{\hat{p}(1-\hat{p})}{n}} = \sqrt{\dfrac{(2/3)(1/3)}{60}}$. The confidence interval is $\hat{p} \pm z^{*}SE(\hat{p}) = \dfrac{2}{3} \pm 2.326\sqrt{\dfrac{(2/3)(1/3)}{60}}$.

3. **(E)** The margin of error is $\pm z^{*}SE(\hat{p})$, where $SE(\hat{p}) = \sqrt{\dfrac{(0.29)(0.71)}{1470}} = 0.011835$. Then $z^{*}(0.011835) = 0.03$, $z^{*} = 2.535$, and $\texttt{normalcdf(-2.535, 2.535)} = 0.9888$, or about 99%.

4. **(E)** The 98% confidence gives 0.01 is in each tail. So, $\texttt{invNorm(0.99)} = 2.326$, $\sqrt{p(1-p)} \leq 0.5$, and so $z^{*}\sqrt{\dfrac{p(1-p)}{n}} = z^{*}\dfrac{\sqrt{p(1-p)}}{\sqrt{n}} \leq \dfrac{2.326 \times 0.5}{\sqrt{n}}$, which we require to be ≤ 0.025. This gives $\sqrt{n} \geq \dfrac{2.326 \times 0.5}{0.025}$ and $n \geq \left(\dfrac{2.236 \times 0.5}{0.025}\right)^{2}$.

5. **(A)** $\hat{p}_1 - \dfrac{210}{361}$, $\hat{p}_2 - \dfrac{34}{86}$, $SE(\hat{p}_1 - \hat{p}_2) = \sqrt{\dfrac{\hat{p}_1(1-\hat{p}_1)}{n_1} + \dfrac{\hat{p}_2(1-\hat{p}_2)}{n_2}}$, and the 95% confidence interval is $(\hat{p}_1 - \hat{p}_2) \pm 1.96SE(\hat{p}_1 - \hat{p}_2) = \left(\dfrac{210}{361} - \dfrac{34}{86}\right) \pm 1.96\sqrt{\dfrac{\left(\dfrac{210}{361}\right)\left(1 - \dfrac{210}{361}\right)}{361} + \dfrac{\left(\dfrac{34}{86}\right)\left(1 - \dfrac{34}{86}\right)}{86}}$.

6. **(E)** We attempt to show that the null hypothesis is unacceptable by showing that it is improbable; that is, we are measuring the strength of evidence against the null hypothesis. Choices (A) and (D) are incorrect as we cannot show that the null hypothesis is definitely true or false. Choice (B) is incorrect as the null and alternative hypotheses are stated in terms of a population parameter, not a sample statistic. Choice (C) is incorrect as these hypothesis tests always assume simple random samples.

7. **(C)** A Type II error is a mistaken failure to reject a false null hypothesis or, in this case, a failure to realize that a person really does have ESP.

8. **(A)** This is a one-sided z-test, H_0: $p = 0.5$, H_a: $p > 0.5$, with $\sigma_{\hat{p}} = \sqrt{(0.5)(0.5)/565}$ and $\hat{p} = \dfrac{305}{565} = 0.540$. The P-value equals $P\left(z > \dfrac{\hat{p} - p_0}{\sigma_{\hat{p}}}\right) = P\left(z > \dfrac{0.54 - 0.50}{\sqrt{(0.5)(0.5)/565}}\right)$.

9. **(D)** $\hat{p}_1 = \dfrac{151}{500} = 0.302$, $\hat{p}_2 = \dfrac{345}{1000} = 0.345$, $\hat{p}_c = \dfrac{151 + 345}{500 + 1000} = 0.331$. This is a two-sided test $(H_0: p_1 - p_2 = 0, H_a: p_1 - p_2 \neq 0)$ with $SE(\hat{p}_1 - \hat{p}_2) = \sqrt{\hat{p}_c(1 - \hat{p}_c)\left(\dfrac{1}{n_1} + \dfrac{1}{n_2}\right)} = \sqrt{(0.331)(0.669)\left(\dfrac{1}{500} + \dfrac{1}{1000}\right)}$. The P-value equals twice the tail probability (because this is a two-sided test), so the P-value equals $2P\left(z < \dfrac{\hat{p}_1 - \hat{p}_2}{SE(\hat{p}_1 - \hat{p}_2)}\right) = 2P\left(z < \dfrac{0.302 - 0.345}{\sqrt{(0.331)(0.669)\left(\dfrac{1}{500} + \dfrac{1}{1000}\right)}}\right)$.

10. **(B)** If the null hypothesis is far off from the true parameter value, there is a greater chance of rejecting the false null hypothesis and thus a smaller risk of a Type II error. Power is the probability that a Type II error is not committed, so a lower Type II error results in higher power.

FREE-RESPONSE QUESTIONS

1. (a) The figure 93% comes from a sample, so it is a statistic.

   (b) The margin of error is proportional to $\frac{1}{\sqrt{n}}$. With such a very large sample size n, the margin of error will be very small.

   (c) This is a voluntary response survey, and so we should have very little confidence in the conclusion.

2. (a) *Parameters:* Let p_{US} represent the proportion of the population of new births in the United States that are to unmarried women. Let p_{UK} represent the proportion of the population of new births in the United Kingdom are to unmarried women.

   Procedure: Two-sample z-interval for a difference between population proportions, $p_{US} - p_{UK}$, the difference in population proportions of new births to unmarried women in the United States and the United Kingdom.

   Checks: It is given that these are random samples, they are clearly independent, $n_{US}\hat{p}_{US} = (500)(0.412) = 206 \geq 10$, $n_{US}(1 - \hat{p}_{US}) = (500)(0.588) = 294 \geq 10$, $n_{UK}\hat{p}_{UK} = (400)(0.465) = 186 \geq 10$, and $n_{UK}(1 - \hat{p}_{UK}) = (400)(0.535) = 214 \geq 10$, and the sample sizes, $n_{US} = 500$ and $n_{UK} = 400$, are less than 10% of the numbers of new births in the United States and the United Kingdom, respectively.

   Mechanics: Calculator software (such as `2-PropZInt` on the TI-84) gives $(-0.1182, 0.01219)$.

   [For instructional purposes, we note that:

   $$(0.412 - 0.465) \pm 1.96\sqrt{\frac{(0.412)(0.588)}{500} + \frac{(0.465)(0.535)}{400}} =$$
   $$-0.053 \pm 0.065 \text{ or } (-0.118, 0.012).]$$

   Conclusion in context: We are 95% confident that the true difference in proportions, $p_{US} - p_{UK}$, of new births to unmarried women in the United States and the United Kingdom is between -0.118 and 0.012.

   (b) Because 0 is in the interval of plausible values for the difference of population proportions, this confidence interval does not support the belief by the U.N. health-care statistician that the true proportions of new births to unmarried women are different in the United States and the United Kingdom.

3. (a) $z^* SE(\hat{p}) = 1.96\sqrt{\dfrac{(0.74)(0.26)}{1050}} = 0.0265$. So, the margin of error is $\pm 2.65\%$.

(b) $0.74 \pm 0.0265 = (0.7135, 0.7265)$. We are 95% confident that between 71.35% and 76.65% of all smokers would like to give up smoking.

(c) If this survey were conducted many times, we would expect about 95% of the resulting confidence intervals to contain the true proportion of all smokers who would like to give up smoking.

(d) Smoking is becoming more undesirable in society as a whole, so some smokers may untruthfully say they would like to stop.

(e) To be more confident, we must accept a *wider* interval (the critical *z*-score would go from 1.96 to 2.576).

(f) All other things being equal, the greater the sample size, the *smaller* the margin of error is (the standard error is inversely proportional to $\sqrt{n}$).

4. *Parameter:* Let *p* represent the proportion of the population of AAUP members who own a Roth IRA.

Hypotheses: H_0: $p = 0.174$ and H_a: $p > 0.174$, where *p* is the proportion of all AAUP members who own Roth IRAs.

Procedure: A one-sample *z*-test for a population proportion.

Checks: Random sample (given), $np_0 = 750(0.174) = 130.5 \geq 10$ and $n(1 - p_0) = 750(0.826) = 619.5 \geq 10$ (where p_0 is the claimed 0.174) and the sample size, $n = 750$, is assumed to be less than 10% of all AAUP members.

Mechanics: Calculator software gives $z = 1.878$ and $P = 0.030$.

[For instructional purposes, we note that:

$\hat{p} = \dfrac{150}{750} = 0.2$, $z = \dfrac{0.2 - 0.174}{\sqrt{\dfrac{(0.174)(0.826)}{750}}} = 1.878$, and the *P*-value $= P(z > 1.878) = 0.030$.]

Conclusion in context with linkage to the P-value: With a *P*-value this small, $0.030 < 0.05$, there is sufficient evidence to reject H_0; that is, there is sufficient evidence that more than 17.4% of all AAUP members own Roth IRAs.

TIP

Use 1–PropZTest on the TI-84 or Z-1-PROP on the Casio Prizm.

Quiz 25 (Pages 279–283)

MULTIPLE-CHOICE

1. **(D)** Increasing the sample size by a multiple of *d* divides the interval estimate by $\sqrt{d}$.

2. **(E)** There is no guarantee that 13.4 is anywhere near the interval, so none of the statements are true.

3. **(A)** The margin of error has to do with measuring chance variation but has nothing to do with faulty survey design. Sampling error (also called sampling variability) refers to the natural variation among samples and is quantified by the margin of error. As long as *n* is large, *s* is a reasonable estimate of σ; however, again this is not measured by the margin of error. (As seen in Unit 7, with *t*-scores there is a correction for using *s* as an estimate of σ.)

4. **(D)** `invNorm(0.975) = 1.96` and

$$SE(\hat{p}) = \sqrt{\frac{\hat{p}(1-\hat{p})}{n}} = \sqrt{\frac{(0.750)(0.250)}{1000}}.$$ The confidence interval is

$$\hat{p} \pm z^* SE(\hat{p}) = 0.750 \pm 1.96\sqrt{\frac{(0.750)(0.250)}{1000}}.$$

5. **(C)** `invNorm(0.95) = 1.645`, $1.645\left(\dfrac{0.5}{\sqrt{n}}\right) \le 0.04$, $\sqrt{n} \ge 20.563$, $n \ge 422.8$, so choose $n = 423$.

6. **(B)** The critical z-scores are $\pm$`invNorm(0.995)` $= \pm 2.576$.

$$n_1 = 300 \quad n_2 = 400$$
$$\hat{p}_1 = 0.65 \quad \hat{p}_2 = 0.48$$
$$SE(\hat{p}_1 - \hat{p}_2) = \sqrt{\frac{\hat{p}_1(1-\hat{p}_1)}{n_1} + \frac{\hat{p}_2(1-\hat{p}_2)}{n_2}} = \sqrt{\frac{(0.65)(0.35)}{300} + \frac{(0.48)(0.52)}{400}}$$
$$(\hat{p}_1 - \hat{p}_2) \pm z^* SE(\hat{p}_1 - \hat{p}_2) = (0.65 - 0.48) \pm 2.576\sqrt{\frac{(0.65)(0.35)}{300} + \frac{(0.48)(0.52)}{400}}$$

7. **(E)** The P-value is a conditional probability; in this case, there is a 0.032 probability of an observed difference in sample proportions as extreme as (or more extreme than) the one obtained if the null hypothesis is assumed to be true.

8. **(B)** The level of significance is defined to be the probability of committing a Type I error, that is, of mistakenly rejecting a true null hypothesis. Here it is 0.05, no matter what the sample size.

9. **(E)** `1PropZTest` gives $P = 0.140956$. [For instructional purposes, we note that with $\hat{p} = \dfrac{198}{2300} = 0.086087$ and $p_0 = 0.08$, we get $z = \dfrac{0.086087 - 0.08}{\sqrt{(0.08)(0.92)/2300}} = 1.076$ and $P = P(z >$ $1.076) =$ `normalcdf(1.076, 100)` $= 0.141$.]

With a P-value this high, $0.141 > 0.10$, the government does not have sufficient evidence to reject the company's claim.

10. **(C)** With a smaller α, that is, with a tougher standard to reject H_0, there is a greater chance of failing to reject a false null hypothesis; that is, there is a greater chance of committing a Type II error. Power is the probability that a Type II error is not committed, so a higher Type II error results in lower power.

FREE-RESPONSE QUESTIONS

1. (a) *Parameter:* Let p represent the proportion of the population of family members who contract H1N1 after an initial family member does in this state.

   Procedure: One-sample z-interval for a proportion.

   Checks: It is given that this is a random sample. We calculate $n\hat{p} = 129 \ge 10$ and $n(1-\hat{p}) = 747 \ge 10$, and $n = 876$ is assumed to be less than 10% of the number of all family members of H1N1 patients in the state.

   Mechanics: Calculator software gives (0.12757, 0.16695).

[For instructional purposes, we note that $\hat{p} = \dfrac{129}{876} = 0.147$

and $0.147 \pm 1.645\sqrt{\dfrac{(0.147)(0.853)}{876}} = 0.147 \pm 0.020$ or $(0.127, 0.167)$.]

Conclusion in context: We are 90% confident that the true proportion of all family members who contract H1N1 after an initial family member does in this state is between 0.127 and 0.167.

(b) Because $\dfrac{1}{8} = 0.125$ is not in the interval of plausible values for the population proportion, there *is* sufficient evidence that the true proportion of all family members who contract H1N1 after an initial family member does in this state is different from the 1 in 8 chance concluded in the published study.

(c) $0.147 \pm 2.576\sqrt{\dfrac{(0.147)(0.853)}{876}} = 0.147 \pm 0.031$ or $(0.116, 0.178)$, which does include 0.125. So, using a 99% confidence interval, there is *not* sufficient evidence that the true proportion of all family members who contract H1N1 after an initial family member does in this state is different from the 1 in 8 chance concluded in the published study.

2. (a) The *P*-value of 0.138 gives the probability of observing a sample proportion of GBS complications as great as (or greater than) the proportion found in the study if, in fact, the proportion of GBS complications is 0.000001.

   (b) Since $0.138 > 0.10$, there is not sufficient evidence to reject H_0; that is, there is not sufficient evidence that under the new vaccine, the true proportion of GBS complications is greater than 0.000001 (one in a million).

   (c) The null hypothesis is not rejected, so there is the possibility of a Type II error, that is, of mistakenly failing to reject a false null hypothesis. A possible consequence is continued use of the vaccine with a higher rate of GBS complications than is acceptable.

3. (a) This was an observational study as no treatments were imposed. It would have been highly unethical to impose treatments, that is, to instruct randomly chosen volunteers to smoke, drink, skip exercise, and eat poorly.

   (b) H_0: $p_4 = p_0$, H_a: $p_4 > p_0$, where p_4 is the proportion of the population of adults with all four bad habits who die during a 20-year period, and p_0 is the proportion of the population of adults with none of the four bad habits who die during a 20-year period. (Note that the hypotheses are about the population of all adults with and with none of the four bad habits, not about the volunteers who took part in the study.)

   (c) A Type I error, that is, a mistaken rejection of a true null hypothesis, would result in people being encouraged to not smoke, not drink, exercise, and eat well when these actions actually will not help decrease 20-year death rates.

   (d) A Type II error, that is, a mistaken failure to reject a false null hypothesis, would result in people thinking that smoking, drinking, inactivity, and poor diet don't increase 20-year death rates when actually they do contribute to higher 20-year death rates.

   (e) Calculator software (such as `2-PropZTest` on the TI-84 or Z-2-PROP on the Casio Prizm) gives $P = 0.000$.

[For instructional purposes, we note that:

$$\hat{p}_1 = \frac{91}{314} = 0.290, \ \hat{p}_2 = \frac{32}{387} = 0.083, \ \hat{p}_c = \frac{91 + 32}{314 + 387},$$

and $= 0.175$, and $P\left(z > \dfrac{0.290 - 0.083}{\sqrt{(0.175)(0.825)\left(\dfrac{1}{314} + \dfrac{1}{387}\right)}}\right) = 0.000.$]

If the null hypothesis were true, that is, if there was no difference in the 20-year death rates between people with all four bad habits and people with none of the bad habits, the probability of sample proportions with a difference as extreme as (or more extreme than) observed is 0.000 (to three decimals).

4. *Parameters:* Let p_B represent the proportion of the population of high school boys who meet the recommended level of physical activity. Let p_G represent the proportion of the population of high school girls who meet the recommended level of physical activity.

Hypotheses: H_0: $p_B - p_G = 0$ (or $p_B = p_G$) and H_a: $p_B - p_G > 0$ (or $p_B > p_G$).

Procedure: Two-sample z-test for a difference of two proportions.

Checks: Independent random samples (given), and with

$\hat{p}_c = \dfrac{370 + 218}{850 + 580} = \dfrac{588}{1430} = 0.411$, $n_B\hat{p}_c = (850)(0.411) = 349 \geq 10$,

$n_B(1 - \hat{p}_c) = (850)(0.589) = 501 \geq 10$, $n_G\hat{p}_c = (580)(0.411) = 238 \geq 10$, and

$n_G(1 - \hat{p}_c) = (580)(0.589) = 342 \geq 10$; and the sample sizes, 850 and 580, are clearly less than 10% of the respective populations.

Mechanics: Calculator software (such as `2-PropZTest` on the TI-84) gives $z = 2.2427$ and $P = 0.012$. [For instructional purposes, we note that

$$\hat{p}_B = \frac{370}{850} = 0.4353, \ \hat{p}_G = \frac{218}{580} = 0.3759, \ \hat{p}_c = \frac{370 + 218}{850 + 580} = \frac{588}{1430} = 0.4112,$$

$$z = \frac{0.4353 - 0.3759}{\sqrt{(0.4112)(0.5888)\left(\dfrac{1}{850} + \dfrac{1}{580}\right)}} = 2.2427, \text{ and}$$

$$P = P(z > 2.2427) = 0.012.]$$

Conclusion in context with linkage to the P-value: With a *P*-value this small ($0.012 < 0.05$), there is sufficient evidence to reject H_0; that is, there is sufficient evidence that the proportion of all high school boys who meet the recommended level of physical activity is greater than the proportion of all high school girls who meet the recommended level of physical activity.

Quiz 26 (Pages 301–306)

MULTIPLE-CHOICE

1. **(E)** The 95% refers to the method: 95% of all intervals obtained by this method will capture the true population parameter. Nothing is certain about any particular set of 20 intervals. For any particular interval, the probability that it captures the true parameter is 1 or 0 depending upon whether the parameter is or isn't in it.

2. **(A)** If you are willing to accept less confidence, that is, a smaller confidence level, you will have a smaller margin of error and a smaller confidence interval. The margin of error varies directly with the critical z-value and directly with the standard deviation of the sample but inversely with the square root of the sample size. The value of the sample mean and the population size do not affect the margin of error.

3. **(C)** $df = n - 1 = 30 - 1 = 29$, $\texttt{invT(0.975, 29)} = 2.045$, and

   $SE(\bar{x}) = \dfrac{s}{\sqrt{n}} = \dfrac{1.2}{\sqrt{30}}$, so the confidence interval is $\bar{x} \pm t_n^*{}_{-1} SE(\bar{x}) = 28.5 \pm 2.045 \left(\dfrac{1.2}{\sqrt{30}} \right)$.

4. **(E)** $1.96 \left(\dfrac{1.1}{\sqrt{n}} \right) \leq 0.2$ gives $\sqrt{n} \geq \dfrac{1.96 \times 1.1}{0.2}$ and $n \geq \left(\dfrac{1.96 \times 1.1}{0.2} \right)^2$.

5. **(A)** $SE(\bar{x}_1 - \bar{x}_2) = \sqrt{\dfrac{(s_1)^2}{n_1} + \dfrac{(s_2)^2}{n_2}} = \sqrt{\dfrac{(19.1)^2}{347} + \dfrac{(19.9)^2}{561}}$ and with such large sample sizes, the critical t-scores should be just slightly greater than the critical z-scores of ± 1.96 associated with 95% confidence. The confidence interval is $(\bar{x}_1 - \bar{x}_2) \pm t^* SE(\bar{x}_1 - \bar{x}_2) = (93.5 - 84.5) \pm 1.97 \sqrt{\dfrac{(19.1)^2}{347} + \dfrac{(19.9)^2}{561}}$.

6. **(C)** The data come in pairs, so we calculate a one-sample confidence interval on the set of differences. With $df = n - 1 = 9 - 1$ and 95% confidence, $\texttt{invT(0.975, 8)} = 2.306$. Then, the interval is: $\bar{x} \pm t^* \dfrac{s}{\sqrt{n}} = 11.6 \pm 2.306 \left(\dfrac{4.891}{\sqrt{9}} \right) = 11.6 \pm 3.76$.

7. **(C)** This is a hypothesis test with H_0: breaking strength is within specifications, and H_a: breaking strength is below specifications. A Type I error is committed when a true null hypothesis is mistakenly rejected. In this context, a Type I error would be committed if the breaking strength is within specifications, but because the sample wrongly indicates a drop below the specified level, the production process is unnecessarily halted.

8. **(D)** This is a one-sided test, H_0: $\mu = 12$, H_a: $\mu < 12$, and because the population standard deviation is known, we use z instead of t.

9. **(D)** $\texttt{T-Test}$ gives $P = 0.112423$. [Or: $t = \dfrac{145.8 - 150}{12.81/\sqrt{15}} = -1.270$, and with $df = 14$, $P = \texttt{tcdf}(-100, -1.270, 14) = 0.112$.] With this high of a P-value, $0.112 > 0.10$, the students do not have sufficient evidence to reject the fast food chain's claim.

10. **(C)** A larger sample size n reduces the standard deviation of the sampling distribution, resulting in a narrower sampling distribution. So, for the given sample statistic, the P-value is smaller, and the probabilities of mistakenly rejecting a true null hypothesis or mistakenly failing to reject a false null hypothesis are both decreased. Furthermore, a lower Type II error results in higher power.

11. **(E)** The two-sample hypothesis test is not the proper one and can be used only when the two sets are independent. In this case, there is a clear relationship between the data, in pairs, one pair for each student, and this relationship is completely lost in the procedure for the two-sample test. The proper procedure is to run a one-sample test on the single variable consisting of the differences from the paired data.

FREE-RESPONSE QUESTIONS

1. *Parameter:* Let μ represent the mean calories per serving for the population of breakfast cereals.

   Procedure: A one-sample t-interval for the mean number of calories of all breakfast cereals.

   Checks: We are given that we have an SRS, we assume the sample is less than 10% of all brands of breakfast cereals, and the nearly normal condition seems reasonable (a stemplot of the sample data is roughly unimodal and symmetric):

   $$\begin{array}{r|l} 21 & 7 \\ 22 & 0\ 4\ 8 \\ 23 & 3 \end{array} \qquad \begin{array}{l} 21\,|\,7 \\ \text{means } 217 \text{ cal} \end{array}$$

   Under these conditions, the mean calories can be modeled by a t-distribution with $n - 1 = 5 - 1 = 4$ degrees of freedom.

   Mechanics: Calculator software (such as `TInterval` using `Data` on the TI-84) gives (216.5, 232.3).

   [Note that a one-sample t-interval for the mean can be directly calculated: $\bar{x} \pm t\dfrac{s}{\sqrt{n}} = 224.4 \pm 2.776\dfrac{6.348}{\sqrt{5}} = 224.4 \pm 7.88.$]

   Conclusion in context: We are 95% confident that the true mean number of calories per serving of all breakfast cereals is between 216.5 and 232.3.

2. The parameter of interest is $\mu =$ the mean tire pressure in the front right tires of cars with recommended tire pressure of 35 psi. H_0: $\mu = 35$ and H_a: $\mu \neq 35$.

3. (a) We are calculating one-sample t-intervals for the mean price of gas in inner-city stations and for the mean price of gas in suburban stations. In each case, we are given simple random samples, and the sample sizes of $40 \geq 30$ and $120 \geq 30$ are large enough so that the CLT applies. We assume that the sample sizes are less than 10% of the populations of all gas stations. The population standard deviations are unknown, and so we find t-intervals. Calculator software gives (3.434, 3.466) for inner-city stations and (3.3655, 3.3945) for suburban stations. That is, we are 95% confident that the true mean price for gas in all inner-city stations is between \$3.434 and \$3.466, and we are 95% confident that the true mean price for gas in all suburban stations is between \$3.3655 and \$3.3945.

   (b) Because the sample size of inner-city stations is smaller than the sample size of suburban stations, and $SE(\bar{x}) = \dfrac{s}{\sqrt{n}}$.

(c) The value $3.50 is greater than all values in the interval (3.434, 3.466). So, based on this confidence interval, we are confident that the mean price of inner-city gasoline is less than $3.50. Or, you could reason as follows:

Yes, because the standard error of the set of sample means is $\dfrac{0.05}{\sqrt{40}} = 0.0079$, and so $3.50 is more than three standard deviations away from $3.45. Thus, the probability that the true mean is this far from the sample mean is extremely small.

4. (a) The P-value of 0.0197 gives the probability of observing a sample mean of 1002.4 or greater if, in fact, this system results in a mean *E. coli* concentration of 1000 MPN per 100 mL.

(b) Since $0.0197 < 0.05$, there is sufficient evidence to reject H_0; that is, there is sufficient evidence that the mean *E. coli* concentration is greater than 1000 MPN per 100 mL.

(c) With rejection of the null hypothesis, there is the possibility of a Type I error, that is, of mistakenly rejecting a true null hypothesis. Possible consequences are that sales of the system drop even though the system is doing what it claims or that the company performs an overhaul to fix the system even though the system is properly reducing the mean concentration of *E. coli* to 1000 MPN per 100 mL.

5. The data come in pairs, and the two-sample test does not apply the knowledge of what happened to each individual driver (the condition of independence of the two samples is violated). The appropriate test is a one-population, small-sample hypothesis test on the set of differences: $\{0, -3, -2, -2, -1, -1\}$. We proceed as follows:

Parameter: Let μ_D represent the mean difference, DWI − DWT, in reaction times between DWI and DWT in the population of adults.

Hypotheses: H_0: $\mu_D = 0$, H_a: $\mu_D \neq 0$

Procedure: A paired t-test, that is, a one-sample t-test on the set of differences.

Checks: The reaction times of any individual are assumed independent of the reaction times of the others, so the *differences are independent*. We must assume that the volunteers are a representative sample; the sample size, $n = 6$, is less than 10% of all possible drivers; and a dotplot of the sample data is roughly unimodal and symmetric. So, it is not unreasonable to assume the sample comes from a roughly normal population:

Differences

Mechanics: Calculator software (such as T-Test and Data on the TI-84) gives $t = -3.50$ and $P = 0.017$.

[For instructional purposes, we note that $\bar{x} = -1.5$ and $s = 1.049$ giving $t = \dfrac{-1.5 - 0}{1.049/\sqrt{6}} = -3.50$, and with $df = 6 - 1 = 5$, $P(t < -3.50) = \texttt{tcdf(-100,-3.50,5)} = 0.0086$. Since this is a two-sided test, we double this value to find the P-value to be $P = 0.017$.]

Conclusion in context with linkage to the P-value: With a P-value this small, $0.017 < 0.05$, there is sufficient evidence to reject H_0; that is, there is sufficient evidence of a difference between the true mean effect on reaction time between DWI and DWT.

Quiz 27 (Pages 307–312)

MULTIPLE-CHOICE

1. **(C)** The CLT gives that no matter what the population distribution is, the larger the sample size, the closer the sampling distribution of $\bar{x}$ will be to normal.

2. **(E)** The 90% refers to the method; 90% of all intervals obtained by this method will capture μ. Nothing is sure about any particular set of 100 intervals. For any particular interval, the probability that it captures μ is either 1 or 0 depending on whether μ is or isn't in it.

3. **(E)** The critical t-scores with 0.02 in the tails are $\pm\text{invT}(0.98, 79) = \pm2.088$ and $SE(\bar{x}) = \dfrac{s}{\sqrt{n}} = \dfrac{15}{\sqrt{80}}$.

4. **(D)** $SE(\bar{x}) = \dfrac{1.52}{\sqrt{64}} = 0.19$, $\dfrac{3.96 - 3.58}{0.19} = 2$, $\dfrac{3.20 - 3.58}{0.19} = -2$, $df = 64 - 1 = 63$, and $\text{tcdf}(-2, 2, 63) = 0.950 = 95\%$.

5. **(E)** To divide the interval estimate by d without affecting the confidence level, multiply the sample size by a multiple of d^2. In this case, $4(50) = 200$.

6. **(B)** $\bar{x}_{high} = 33.0$, $s_{high} = 6.3$, $n_{high} = 274$, $\bar{x}_{low} = 28.6$, $s_{low} = 6.3$, $n_{low} = 90$, and 2-SampTInt gives $(3.1333, 5.6667)$, or 4.4 ± 1.2667.

7. **(C)** The population standard deviation is unknown, so this is a t-test. Medications having an effect shorter or longer than claimed should be of concern, so this is a two-sided t-test: $H_a: \mu \neq 58.4$, and $df = n - 1 = 40 - 1 = 39$.

8. **(A)** With unknown population standard deviations, the t-distribution must be used, and $SE(\bar{x}_1 - \bar{x}_2) = \sqrt{\dfrac{s_1^2}{n_1} + \dfrac{s_2^2}{n_2}} = \sqrt{\dfrac{(0.3)^2}{10} + \dfrac{(0.2)^2}{10}}$. The critical t-score is $t = \dfrac{\bar{x}_1 - \bar{x}_2}{SE(\bar{x}_1 - \bar{x}_2)}$, and the P-value is $P\left(t < \dfrac{\bar{x}_1 - \bar{x}_2}{SE(\bar{x}_1 - \bar{x}_2)}\right) = P\left(t < \dfrac{4 - 4.8}{\sqrt{\dfrac{(0.3)^2}{10} + \dfrac{(0.2)^2}{10}}}\right)$.

9. **(B)** $P(\text{at least one Type I error}) = 1 - P(\text{no Type I errors}) = 1 - (0.99)^5 = 0.049$

10. **(B)** $N(385, 150/\sqrt{100})$. The test calls for rejection of the manager's claim if $X > 375$, and thus failure to reject if $X < 375$. $P(X < 375) = P\left(z < \dfrac{375 - 385}{150/\sqrt{100}}\right) = 0.2525$ [or $\text{normalcdf}(0, 375, 385, 150/\sqrt{100}) = 0.2525$].

11. **(B)**

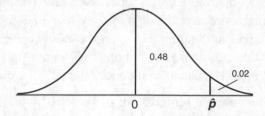

With 0.02 in the tail, a confidence interval over $0.48 + 0.48 = 0.96 = 96\%$ will contain 0.

FREE-RESPONSE QUESTIONS

1. (a) *Parameter:* Let μ represent the mean number of riders in the population of cars during rush hour.

    Procedure: A one-sample t-interval.

    Checks: We are given a simple random sample, the sample size of 30 is assumed to be less than 10% of all subway cars, and the sample size of 30 is large enough for the CLT to apply. (OR: the sample data are unimodal, reasonably symmetric with no extreme values or skewness, so it is not unreasonable that the data come from a roughly normal population.)

    Mechanics: The population SD is unknown, so we use a t-distribution. Calculator software gives (81.67, 85.33).

    [For instructional purposes, we note that with $df = n - 1 = 29$, the critical t-scores are $\pm\texttt{invT(0.95, 29)} = \pm 1.699$, and $\bar{x} \pm t^* \left(\dfrac{s}{\sqrt{n}} \right) = 83.5 \pm 1.699 \left(\dfrac{5.9}{\sqrt{30}} \right) = 83.5 \pm 1.83.$]

    Conclusion in context: We are 90% confident that the true mean number of riders per car during rush hour is between 81.67 and 85.33.

   (b) $1.645 \left(\dfrac{5.9}{\sqrt{n}} \right) \leq 1$, $\sqrt{n} \geq 9.7055$, and $n \geq 94.2$, so you must choose a random sample of at least 95 subway cars.

2. (a) *Parameter:* Let μ represent the mean income of the population of basketball players.

    Procedure: A one-sample t-interval.

    Checks: We are told to assume that all necessary assumptions are met.

    Mechanics: Calculator software gives $(-29{,}337, 421{,}337)$.

    [For instructional purposes, we note that with $df = 9$,

    $\bar{x} \pm t \dfrac{s}{\sqrt{n}} = 196{,}000 \pm 2.262 \dfrac{315{,}000}{\sqrt{10}} = 196{,}000 \pm 225{,}000.$]

    Conclusion in context: We are 95% confident that the mean salary of all basketball salaries from the population from which this sample was taken is between $-\$29{,}337$ and $\$421{,}337$.

(b) We must assume the sample is an SRS and that basketball salaries are approximately normally distributed. This does *not* seem reasonable—the salaries are probably strongly skewed to the right by a few high ones. With the given mean and standard deviation, and noting that salaries cannot be negative, clearly we do not have a normal distribution. Furthermore, the sample size, $n = 10$, is too small to invoke the CLT. The calculated confidence interval is not meaningful.

3. (a) *Procedure:* A two-sample t-interval for $\mu_{NRW} - \mu_{RW}$, the difference in population means of BPA body concentrations in nonretail workers and retail workers.

   Checks: It is given that these are random samples. It is reasonable to assume the samples are independent, both samples sizes ($528 \geq 30$ and $197 \geq 30$) are large enough so that the CLT applies, and we assume the sample sizes are less than 10% of the populations.

   Mechanics: Calculator software gives $(-0.9521, -0.7479)$ with $df = 332.3$.

   Conclusion in context: We are 99% confident that the difference in true means of BPA body concentrations in all nonretail and retail workers (nonretail mean minus retail mean) is between -0.75 and -0.95 μg/L.

   (b) Because 0 is not in the interval of plausible values for the difference of population means and the entire interval is negative, the interval does support the belief that retail workers carry higher amounts of BPA in their bodies than nonretail workers.

4. *Parameters:* Let μ_{NFL} represent the mean attendance of the population of NFL games. Let μ_{10} represent the mean attendance of the population of Big Ten football games.

   Hypotheses: H_0: $\mu_{NFL} - \mu_{10} = 0$ (or $\mu_{NFL} = \mu_{10}$) and H_a: $\mu_{NFL} - \mu_{10} < 0$ (or $\mu_{NFL} < \mu_{10}$)

   Procedure: A two-sample t-test for means.

   Checks: Independent random samples (given), both samples sizes, $n_{NFL} = 35 \geq 30$ and $n_{10} = 30 \geq 30$, are large enough for the CLT to apply, and the sample sizes, 35 and 30, are less than 10% of all NFL and Big Ten football games, respectively.

   Mechanics: The population SDs are unknown, so we use a t-distribution. Calculator software (such as `2-SampTTest` on the TI-84 or 2-Sample tTest on the Casio Prizm) gives $t = -0.8301$, $df = 49.3$, and $P = 0.2052$.

   Conclusion in context with linkage to the P-value: With a P-value this large, $0.2052 > 0.05$, there is not sufficient evidence to reject H_0; that is, there is not sufficient evidence that the true mean attendance at Big Ten Conference football games is greater than that at NFL games.

5. (a) Different schemes are possible. For example, assign each material a single-digit number, such as A-0, B-1, C-2, D-3, E-4, F-5, G-6, H-7, I-8, J-9. Then read off the digits from the random number list, one at a time, throwing away any repeats, until all of the materials have been picked (or nine have been picked, as the last one left will go last). The order of picking then gives the order of being tested. By using this scheme, we would get the following order:

```
FBIE    HD C      A       J                        G
51844  73424 84380 82259 28273 58102 18727 69708
```

(b) The mean drilling times in the ten materials are 4.69 seconds for Drill 1 and 4.77 seconds for Drill 2.

The proper hypothesis test is a matched pairs *t*-test on the set of differences, {0, −0.3, 0.1, −0.2, −0.1, −0.4, 0, 0.3, −0.1, −0.1}.

Parameter: Let μ_d represent the mean difference in the population of drilling times of the two drills through different materials.

Hypotheses: H_0: $\mu_d = 0$, H_a: $\mu_d \neq 0$

Procedure: A one-sample *t*-test for the mean of paired differences.

Checks: Random samples (given), $n = 10$ is less than 10% of all possible materials, hardnesses and thicknesses, and a dotplot of the difference sample is roughly uni-modal and symmetric (OR: the normal probability plot is roughly linear), so it is not unreasonable to assume that the sample comes from a roughly normal population.

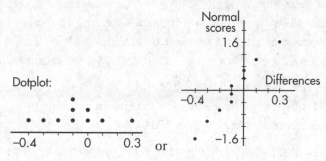

Normal probability plot:

Dotplot:

or

Mechanics: Calculator software (such as T-Test with Data) gives $t = -1.272$ and $P = 0.2353$.

[For instructional purposes, we note that $\bar{x} = -0.08$ and $s = 0.1989$, which gives $t = \dfrac{-0.08 - 0}{0.1989/\sqrt{10}} = -1.272$, and with $df = 10 - 1 = 9$, $P(t < -1.272) = 0.1176$. Since this is a two-sided test, we double this value to find the *P*-value to be $P = 0.2352$.]

Conclusion in context with linkage to the P-value: With a *P*-value this large, $0.2353 > 0.05$, there is not sufficient evidence to reject H_0; that is, there is not sufficient evidence of a difference in true mean drilling times of the two drills through different materials.

Quiz 28 (Pages 315–318)

MULTIPLE-CHOICE

1. **(B)** For this day's sample, $G = \dfrac{Max + Min}{2} = \dfrac{35.55 + 33.51}{2} = 34.53$. A value of 34.53 or less occurred only twice in the 100 samples. Thus, the estimated *P*-value is $\dfrac{2}{100} = 0.02$. If the machinery was operating properly, a *G* measurement of 34.53 would be very unusual. The conclusion should be to recalibrate the machine.

2. **(E)** The P-value is a conditional probability, the probability of as extreme as (or more extreme) a result as the one obtained given that the null hypothesis is true. This is a two-tailed test. In the simulation, there were $12 + 7 + 1 = 20$ values on the lower end and $5 + 2 = 7$ values on the upper end that are at least as extreme as the $W = -0.7$ value. Thus, the estimated P-value is $\frac{20 + 7}{1000} = 0.027$.

FREE-RESPONSE QUESTIONS

1. (a) $H_0: \mu = 350$ and $H_a: \mu < 350$

   (b) The Normal/Large Sample condition is not met. There is no indication that the population of all battery cranking powers have a roughly normal distribution; the sample size, $n = 5$, is not large enough for the central limit theorem to apply; and since the sample data is not given, there is no way to determine if it is reasonable to assume that the sample data come from a roughly normal distribution.

   (c) The consumer group's random sample of 5 batteries had an average cranking power of only 320 CCA. The simulation indicates that the chance of picking 5 batteries with an average of 320 CCA or lower from a group of new batteries manufactured to have 350 CCA power is 4 out of 160 or 0.025. So, the estimated P-value is 0.025.

   (d) With a P-value this small, $0.025 < 0.05$, there is sufficient evidence to reject H_0; that is, there is sufficient evidence that the true mean cranking power of these batteries is less than the company's claim of 350 CCA.

2. (a) $H_0: p_1 = p_2$ and $H_a: p_1 > p_2$, where p_1 is the proportion of all addicts taking desipramine who would not have relapsed after one year and p_2 is the proportion of all addicts taking lithium who would not have relapsed after one year.

   (b) $\frac{14}{25} - \frac{6}{25} = 0.32$

   (c) There are 3 values out of 500 that are 0.32 or greater, so the estimated P-value is $\frac{3}{500} = 0.006$.

   (d) With this small of a P-value, $0.006 < 0.05$, there is sufficient evidence to reject the null hypothesis; that is, there is sufficient evidence that the antidepressant desipramine is more effective than the mood stabilizer lithium in treating cocaine addiction.

Quiz 29 (Pages 333–337)

MULTIPLE-CHOICE

1. **(B)** The first quarter and full moon observed are 9 less and 11 more, respectively, than expected and contribute much more to the χ^2 test statistic than do the new moon and last quarter observed, which are only 4 less and 2 more than expected.

2. **(C)** Picking separate samples from each of the schools and classifying according to one variable (perception of quality education) is a survey design that is most appropriately analyzed using a chi-square test of homogeneity of proportions.

3. **(E)** With a P-value this small (less than 0.05), there is sufficient evidence in support of the alternative hypothesis H_a: the distributions of music preferences are different; that is, they differ for at least one of the proportions.

4. **(E)** Tests for independence involve a single sample cross-classified on two variables, while tests for homogeneity involve samples from each of two or more populations to compare the distribution of some variable.

5. **(E)** The proper test is a chi-square test of independence. The null hypothesis is always "independence" or "no association," and the hypotheses are always about the population, not the sample. Choice (C) refers to a chi-square test of homogeneity.

6. **(C)** Expected count $= \dfrac{(\text{row total}) \times (\text{column total})}{\text{table total}} = \dfrac{87 \times 141}{525} = 23.4$

7. **(C)** $df = (\text{rows} - 1)(\text{columns} - 1) = (5-1)(5-1) = 16$

8. **(B)** With this large a P-value, $0.290 > 0.05$, there is not sufficient evidence to reject the null hypothesis. If the null hypothesis is false, we will make a mistake. A Type II error is when we mistakenly fail to reject a false null hypothesis.

FREE-RESPONSE QUESTIONS

1. *Hypotheses:* H_0: The colors of the sugar shells are distributed according to 35% cherry red, 10% vibrant orange, 10% daffodil yellow, 25% emerald green, and 20% royal purple, and H_a: The colors of the sugar shells are not distributed as claimed by the manufacturer. [Or H_0: $P_{CR} = 0.35$, $P_{VO} = 0.10$, $P_{DY} = 0.10$, $P_{EG} = 0.25$, $P_{RP} = 0.20$, and H_a: at least one proportion is different from this distribution.]

   Procedure: χ^2 goodness-of-fit test.

   Checks: We are given a random sample, the data are measured as "counts," and all expected cells are at least 5: 35% of 300 = 105, 10% of 300 = 30, 25% of 300 = 75, and 20% of 300 = 60. It can be assumed that the sample size of 300 is less than 10% of all the sweets produced by the manufacturer.

   Mechanics: A calculator gives $\chi^2 = \sum \dfrac{(\text{obs} - \text{exp})^2}{\text{exp}} = 6.689$, and with $df = 5 - 1 = 4$, $P = 0.153$.

   Conclusion in context with linkage to the P-value: With a P-value this large, $0.153 > 0.10$, there is not sufficient evidence to reject H_0; that is, there is not sufficient evidence that the true distribution of colors is different from what is claimed by the manufacturer.

2. (a) Design I, with a single sample from one population classified on two variables (smoking and fitness), will result in a test of independence. Design II, with independent samples from two populations each with the single variable (fitness), will result in a test of homogeneity.

   (b) Design II, with its test of homogeneity, and using an equal sample size from each of the two populations (smokers and nonsmokers), is best for comparing proportions of smokers who have different fitness levels with proportions of nonsmokers who have different fitness levels.

> **NOTE**
>
> χ^2 GOF–Test on the TI-84; CHI-GOF on the Casio Prizm; or χ^2 Test: Goodness-of-fit on the HP Prime.

(c) Design I, which classifies one population on the two variables, smoking and fitness, is the only one of these two designs which will give data on the conditional distribution of people with given fitness levels who are smokers or are not smokers.

3. (a) *Hypotheses:* H_0: Happiness level is independent of busy/idle choice for high school students; and H_a: Happiness level is not independent of busy/idle choice for high school students.

Procedure: A chi-square test of independence.

NOTE

χ^2–Test on the TI-84; CHI-2WAY on the Casio Prizm; or χ^2-Test: 2-way test on the HP Prime.

Checks: It is given that there is a random sample, the sample of size 175 is less than 10% of all high school students, the data are measured as "counts," and the expected counts are all at least 5 (put the observed counts in a matrix; then χ^2-Test on the TI-84 gives expected counts of:

14.6	19.6	21.8	20.2	21.8
11.4	15.4	17.2	15.8	17.2

Mechanics: Calculator software gives $\chi^2 = 14.54$ with $P = 0.0058$ and $df = 4$.

Conclusion in context with linkage to the P-value: With a P-value this small, $0.0058 < 0.01$, there is strong evidence to reject H_0; that is, there is strong evidence of a relationship between happiness level and the busy/idle choice for high school students.

(b) No, it is not reasonable to conclude that encouraging high school students to keep busy will lead to higher happiness levels. This was not an experiment with students randomly chosen to sit or walk. The students themselves chose whether or not to sit or walk, so no cause-and-effect conclusion is possible. For example, it could well be that the happier students choose to walk, whereas the less happy students choose to sit.

Quiz 30 (Pages 338–341)

MULTIPLE-CHOICE

1. **(D)** With $df = (3 - 1)(5 - 1) = 8$, $P(\chi^2 > 13.95) = \chi^2\text{cdf}(13.95, 1000, 8) = 0.083$. Since $0.05 < 0.083 < 0.10$, there is sufficient evidence at the 10% significance level, but not at the 5% significance level, of a relationship between education level and sports interest.

2. **(D)** With $1 + 3 + 3 + 9 = 16$ and $n = 2000$, the expected number of fruit flies for the first species is $\frac{3}{16}(2000) = 375$.

3. **(E)** With $df = 3$, the P-value is $\chi^2\text{cdf}(5.998, 1000, 3) = 0.1117$. With a P-value this large, $0.1117 > 0.10$, there is not sufficient evidence of a difference in cafeteria food satisfaction among the class levels.

4. **(A)** The degrees of freedom are $df = (\text{rows} - 1)(\text{columns} - 1) = (3 - 1)(4 - 1) = 6$, and the P-value is $\chi^2\text{cdf}(12.7, 1000, 6) = 0.048$. With this small of a P-value, $0.048 < 0.05$, there is sufficient evidence to reject the null hypothesis. When the null hypothesis is rejected, there is the possibility of a Type I error, that is, the possibility of mistakenly rejecting a true null hypothesis.

5. **(B)** The proper inference procedure is a chi-square goodness-of-fit test. The null hypothesis is that the racial/ethnic distribution of harsh sentences to nonviolent offenders is the same as the racial/ethnic distribution of people charged with nonviolent crimes. That is, the null hypothesis is that the racial/ethnic distribution of harsh sentences to nonviolent offenders is 58% White, 23% Black, 12% Hispanic, and 7% Other.

6. **(B)** The expected number if H_0 were true is 23% of 80, that is, $(0.23)(80) = 18.4$.

7. **(B)** We calculate χ^2`cdf(9.96,1000,3)` $= 0.0189$.

8. **(A)** With the observed values 35, 29, 12, and 4, the White and Black numbers observed are 11.4 less and 10.6 more, respectively, than expected and contribute much more to the χ^2 test statistic than do the Hispanic and Other numbers observed, which are only 2.4 more and 1.6 less, respectively, than expected.

FREE-RESPONSE QUESTIONS

1. (a) *Hypotheses: H_0*: Eating breakfast and morning energy level are independent; and *H_a*: Eating breakfast and morning energy level are not independent.

   Procedure: A χ^2 test of independence on:

110	120	120
60	50	40

   Checks: We are given a random sample, the sample of size 500 is less than 10% of all adults, and a calculator gives that all expected cells are at least 5:

119	119	112
51	51	48

   Mechanics: Calculator software gives $\chi^2 = 4.202$ and $P = 0.1224$.

   Conclusion in context with linkage to the P-value: With this large of a P-value, $0.1224 > 0.05$, there is not sufficient evidence to reject H_0; that is, there is not sufficient evidence of a relationship between eating breakfast and morning energy level.

   (b) Yes, the conclusion changes. With $n = 1000$, the observed numbers are

220	240	240
120	100	80

   with $\chi^2 = 8.403$ and $P = 0.0150$. With a P-value this small, $0.0150 < 0.05$, now there is sufficient evidence of a relationship between eating breakfast and morning energy level.

2. (a) *Hypotheses:* H_0: The different treatments lead to the same satisfaction levels; and H_a: The different treatments lead to different satisfaction levels.

Procedure: χ^2 test of homogeneity.

Checks: We are given independent random samples, it is reasonable to assume each sample size is less than 10% of the corresponding population, and a calculator gives that all expected cells are at least 5:

55.6	55.6	27.8
23.6	23.6	11.8
20.8	20.8	10.4

Mechanics: A calculator gives $\chi^2 = 10.9521$, and with $df = 4$, $P = 0.0271$.

Conclusion in context with linkage to the P-value: With this small of a P-value, $0.0271 < 0.05$, there is sufficient evidence to reject H_0; that is, there is sufficient evidence that the different acne treatments do lead to different satisfaction levels.

(b) An example of a possible confounding variable is severity of the acne outbreak. It could be that those with more severe cases have less satisfaction no matter what the treatment and are also the ones who are encouraged to use oral medications or laser therapy. So, it would be wrong to conclude that oral medications or laser therapy are the causes of less satisfaction.

3. (a) The z-scores for 22.5 and 27.5 are, respectively, $\frac{22.5 - 20}{5} = 0.5$ and $\frac{27.5 - 20}{5} = 1.5$. Similarly, 17.5 and 12.5 have z-scores of -0.5 and -1.5, respectively. Using the normal probability table or a calculator gives the probabilities $P(z < -1.5) = 0.0668$, $P(-1.5 < z < -0.5) = 0.2417$, $P(-0.5 < z < 0.5) = 0.3830$, $P(0.5 < z < 1.5) = 0.2417$, and $P(z > 1.5) = 0.0668$.

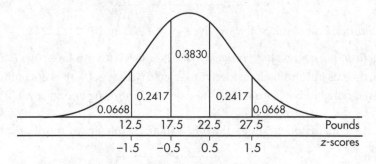

(b) Multiplying each probability by 500: $0.0668(500) = 33.4$, $0.2417(500) = 120.85$, $0.3830(500) = 191.5$, $0.2417(500) = 120.85$, and $0.0668(500) = 33.4$.

Weight (lb)	Below 12.5	12.5–17.5	17.5–22.5	22.5–27.5	Above 27.5
Expected #	33.4	120.85	191.5	120.85	33.4

(c) *Hypotheses:* H_0: The weights of student backpacks follow a normal distribution with $\mu = 20$ and $\sigma = 5$; and H_a: The weights of student backpacks do not follow a normal distribution with $\mu = 20$ and $\sigma = 5$.

Procedure: A chi-square test of goodness-of-fit.

Checks: We are given that the weights are from a random sample of students, and the sample of 500 students is less than 10% of all high school students. We note from above that all expected cells are ≥ 5.

Mechanics: A calculator gives $\chi^2 = \sum \dfrac{(\text{obs} - \text{exp})^2}{\text{exp}} = 6.7874$, and with $df = 5 - 1 = 4$, $P = 0.1476$.

Conclusion in context with linkage to the P-value: With a P-value this large, $0.1476 > 0.05$, there is not sufficient evidence to reject H_0; that is, there is not sufficient evidence that the data do not follow a normal distribution with $\mu = 20$ and $\sigma = 5$.

Quiz 31 (Pages 351–354)

MULTIPLE-CHOICE

1. **(E)** The sampling distribution of b is a t-distribution with $df = n - 2$. An explanation of why $df = n - 2$ is beyond this course.

2. **(C)** The spread around the regression line, that is, the differences between the predicted scores (the estimates) and the actual scores, are measured with the residual standard deviation. Usually, this value is labeled "S" in computer output.

3. **(E)** The relevant P-value is 0.065, which is less than 0.10 but greater than 0.05.

4. **(E)** The slope of the regression line is -0.163, which is the middle of the confidence interval. The correlation coefficient has the same sign as the slope.

5. **(C)** In the second column, headed by "Coef," are the y-intercept, 0.2336, which is usually labeled "Constant" or "Intercept," and the slope, 0.008051, which is labeled with the independent x-variable, the predictor, Salary. The dependent variable is a "predicted" value, signified with a "hat" as in $\hat{y}$ or written out as in "Predicted Batting Average."

6. **(D)** The confidence interval for the slope takes the form $b \pm t^*_{n-2} \times SE(b)$. The sample slope is $b = 0.008051$, the standard error of the sample slope is $SE(b) = 0.002825$, and the critical t-score is $t^*_{25-2} = \text{invT}(0.975, 23) = 2.0687$.

7. **(A)** The t-statistic for inference on the slope is given in the T-column in the row that also gives the sample slope and the standard error of the sample slope. Note that this can also be calculated by $t = \dfrac{0.008051 - 0}{0.002825} = 2.850$.

8. **(A)** The P-value for inference on the slope is given in the P-column in the row that also gives the sample slope and the standard error of the sample slope. However, unless otherwise stated, computer output gives the P-value for a two-tailed test, H_a: $\beta \neq 0$. With our one-tailed test, H_a: $\beta > 0$, the P-value is half this value. Thus, $P = (0.5)(0.0091) \approx 0.0045$.

FREE-RESPONSE QUESTIONS

1. (a) Assuming that all conditions for regression are met, the *y*-intercept and slope of the equation are found in the Coeff column of the computer printout.

$$\widehat{Selling\ price} = 0.890 + 1.029\,(Assessed\,value)$$

Both the selling price and the assessed value are in $1000s.

(b) *Procedure:* *t*-interval for the slope of a regression model.

Checks: It is given that all conditions are met.

Mechanics: From the printout, the standard error of the slope is $s_b = 0.08192$. With $df = 18$ and 0.005 in each tail, the critical *t*-values are ± 2.878. The 99% confidence interval of the true slope is

$$b \pm t\,s_b = 1.029 \pm 2.878(0.08192) = 1.029 \pm 0.236$$

Conclusion in context: We are 99% confident that the true slope of the regression line of selling price as a function of assessed value is between 0.79 and 1.27. In the context of the data, we are 99% confident that for every $1 increase in assessed value, the average increase in selling price is between $0.79 and $1.27 (or for every $1000s increase in assessed value, the average increase in selling price is between $790 and $1270).

(c) The entire interval, (0.79, 1.27), is positive. This shows evidence of a positive association between assessed value and selling price.

2. (a) Each additional year in age of teenagers is associated with an average of 0.4577 more texts per waking hour.

(b) The scatterplot of texts per hour versus age should be roughly linear, there should be no apparent pattern in the residual plot, and a histogram of the residuals should be approximately normal.

(c) *Hypotheses:* H_0: $\beta = 0$, H_a: $\beta \neq 0$, where β is the slope of the regression line that relates average texts per hour to age.

Procedure: A test of significance for the slope of the regression line.

Checks: We are given that all conditions for inference are met.

Mechanics: The computer printout gives that $t = 2.45$ and $P = 0.016$.

Conclusion in context with linkage to the P-value: With this small of a *P*-value, $0.016 < 0.05$, there is sufficient evidence to reject H_0; that is, there is sufficient evidence of a linear relationship between average texts per hour and age for teenagers ages 13–17.

(d) R-Sq $= 5.8\%$, so even though there is evidence of a linear relationship between average texts per hour and age for teenagers ages 13–17, only 5.8% of the variability in average texts per hour is explained by this regression model (or "is accounted for by the variation in age").

Quiz 32 (Pages 355–358)

MULTIPLE-CHOICE

1. **(C)** A confidence interval with all positive values gives evidence for a positive association, a confidence interval with all negative values gives evidence for a negative association, and a confidence interval with both positive and negative values gives no evidence of an association.

2. **(E)** We need to check that the distribution of the residuals is approximately normal, or at least is unimodal and roughly symmetric with no strong skewness or outliers. Ideally, this check should be made for the residuals at each possible value of x, but we very rarely have enough observations to check for normality at each x-value. So, we look at a histogram, dotplot, stemplot, or normal probability plot of all the residuals.

3. **(D)** With $H_0: \beta = 0$ and $H_a: \beta > 0$, $t = \dfrac{b - 0}{SE(b)} = \dfrac{0.6013 - 0}{0.2488}$.

4. **(D)** In the second column, headed by Coef, are the y-intercept, 0.7737, which is usually labeled "Constant" or "Intercept," and the slope, 0.4341, which is labeled with the independent x-variable, the predictor, Packs. The dependent variable is a "predicted" value, signified with a "hat" as in $\hat{y}$ or written out as in "Predicted Cox Ratio."

5. **(C)** The spread around the regression line, that is, the differences between the predicted scores (the estimates) and the actual scores, is measured with the residual standard deviation. This is usually simply labeled "S" in computer output.

6. **(E)** The confidence interval for the slope takes the form $b \pm t^{*}_{n-2} \times SE(b)$. The sample slope is $b = 0.4341$, the standard error of the sample slope is $SE(b)=0.07675$, and the critical t-score is $t^{*}_{7-2} = \text{invT}(0.95, 5) = 2.015$.

7. **(A)** The P-value for inference on the slope is given in the P-column in the row that also gives the sample slope and the standard error of the sample slope. However, unless otherwise stated, computer output gives the P value for a two-tailed test, $H_a: \beta \neq 0$. With our one-tailed test, $H_a: \beta > 0$, the P-value is half this value. Thus, $P = (0.5)(0.0012) = 0.0006$.

8. **(E)** The P-value is a conditional probability. It is the probability of a sample with as extreme as, or more extreme than, the data obtained given that the null hypothesis is true. In this case, the P-value is the probability of getting a random sample of 7 smokers that yields a least squares regression line with a slope of 0.4341 or greater if there is no linear relationship between smoking levels (packs per day) and risk of dementia (Cox hazard ratio).

FREE-RESPONSE QUESTIONS

1. (a) Predicted self-reported life satisfaction = 1.415 + 0.054 (Household income in $1000)

   (b) 82.6% of the variation in self-reported life satisfaction is accountable by this linear model (or is accountable by the variation in household income).

   (c) The standard deviation of the residuals is 0.977. That is, 0.977 gives a measure of how the points are spread around the regression line.

(d)   With H_0: $\beta = 0$ and H_a: $\beta > 0$, $t = \dfrac{b-0}{SE(b)} = \dfrac{0.054-0}{0.00358} = 15.1$.

(e)   *Procedure:* Confidence interval for a population slope.

Checks: We are given that all conditions for inference are met.

Calculation: With $df = 50 - 2 = 48$, the critical *t*-scores are $\pm$`invT(0.95, 48)` $= \pm1.677$. Then $0.054 \pm 1.677(0.00358) = 0.054 \pm 0.006$ gives $(0.048, 0.060)$.

Conclusion in context: We are 90 percent confident that the true slope of the least squares regression line linking self-reported life satisfaction and household income in \$1000 is between 0.048 and 0.060. That is, we are 90 percent confident that for each additional thousand dollars in household income, the self-reported life satisfaction goes up between 0.048 and 0.060, on average.

2.   *Hypotheses:* H_0: $\beta = 0$, H_a: $\beta \neq 0$, where β is the slope of the regression line that relates average fertility rate to women's life expectancy.

Procedure: *t*-test for the slope of a regression line.

Checks: We are given that the data come from a random sample of countries, and the sample size of 11 is less than 10% of all countries.

Scatterplot is approximately linear

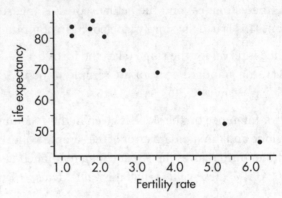

No apparent pattern in residual plot

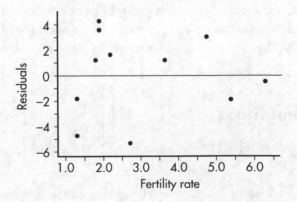

Distribution of residuals is approximately normal
(Because normal probability plot is roughly linear)

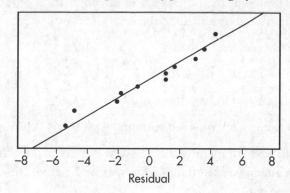

Residual

Mechanics: Using a calculator gives that $t = -12.53$ and $P = 0.0000$.

Conclusion in context with linkage to the P-value: With a P-value this small, $0.0000 < 0.05$, there is sufficient evidence to reject H_0; that is, there is sufficient evidence of a linear relationship between average fertility rate (children per woman) and life expectancy (women).

Quiz 33 (Pages 364–365)

1. *Procedure:* Chi-square test for homogeneity ("homogeneity" rather than "independence" because there are two independent samples)

   Checks: Independent random samples; all expected counts at least 5; $n \leq 0.10N$

   Hypotheses: H_0: There are no differences between the true distributions of reasons for trying out for varsity sports (social, health, or status) for men and women.

   H_a: There is at least one difference between the true distributions of reasons for trying out for varsity sports (social, health, or status) for men and women.

2. *Procedure:* One-sample z-interval for population proportion p

   Parameter: $p =$ the proportion of the population of AP Statistics students who graduate in the top ten percent of their senior class

   Checks: Random sample; at least 10 successes and 10 failures, that is, $n\hat{p} \geq 10$ and $n(1 - \hat{p}) \geq 10$; $n \leq 0.10N$

3. *Procedure:* One-sample t-interval for population mean μ

   Parameter: $\mu =$ the mean number of books read by all high school students during their four years in secondary school

   Checks: Random sample; population is normal or sample is large, at least 30, or sample distribution is unimodal and symmetric; $n \leq 0.10N$

4. *Procedure:* One-sample z-test for a population proportion p

   Parameter: Let p represent the proportion of the population of racially motivated hate crimes that are motivated by anti-Black bias

   Checks: Random sample; $np_0 \geq 10$ and $n(1 - p_0) \geq 10$; $n \leq 0.10N$

   Hypotheses: H_0: $p = 0.70$, H_a: $p > 0.70$

5. *Procedure:* Chi-square test for goodness-of-fit

   Checks: Random sample; all expected counts at least 5; $n \leq 0.10N$

   Hypotheses: H_0: In cases of identity theft, 30% of the victims use their mother's maiden name for their banking password, 25% use their pet's name, 20% use "password," and 25% use something else.

   H_a: In cases of identity theft, the overall distribution of banking passwords used is different from 30% mother's maiden name, 25% pet's name, 20% "password," and 25% something else.

6. *Procedure:* One-sample t-test for population mean μ

   Parameter: Let μ represent the mean speed of all fastballs thrown by a particular pitcher when participating in 10-minute yoga sessions

   Checks: Random sample; population is normal or sample is large, at least 30, or sample distribution is unimodal and symmetric

   Hypotheses: H_0: $\mu = 95$, H_a: $\mu > 95$

7. *Procedure:* Linear regression t-test for slope β

   Parameter: Let β represent the slope of the regression line relating school performance (GPA) to family income

   Checks: Random sample; scatterplot is roughly linear; no pattern in the residual plot; distribution of residuals is approximately normal; $n \leq 0.10N$

   Hypotheses: H_0: $\beta = 0$, H_a: $\beta \neq 0$. (The impact could be positive or negative.)

8. *Procedure:* Two-sample z-interval for difference in population proportions, $p_1 - p_2$

   Parameters: Let p_1 represent the proportion of the population of patients with warts who are cured with a treatment of cryotherapy. Let p_2 represent the proportion of the population of patients with warts who are cured with a treatment of duct tape occlusion.

   Checks: Independent random samples; at least 10 successes and 10 failures in each sample, that is, $n_1\hat{p}_1 \geq 10$, $n_1(1 - \hat{p}_1) \geq 10$, $n_2\hat{p}_2 \geq 10$, $n_2(1 - \hat{p}_2) \geq 10$; $n_1 \leq 0.10N_1$ and $n_2 \leq 0.10N_2$

9. *Procedure:* Two-sample t-test for difference in population means, $\mu_1 - \mu_2$

   Parameters: Let μ_1 represent the mean cholesterol level (measured in milligrams per deciliter) for all people living in "Western" countries. Let μ_2 represent the mean cholesterol level (measured in milligrams per deciliter) for all people living in "non-Western" countries.

   Checks: Independent random samples; for each sample, population is normal or sample is large, at least 30, or sample distribution is unimodal and symmetric; and for each sample $n \leq 0.10N$

   Hypotheses: H_0: $\mu_1 - \mu_2 = 0$, H_a: $\mu_1 - \mu_2 > 0$

10. *Procedure:* Linear regression t-interval for slope β

    Parameter: Let β represent the slope of the regression line relating average increase in CO_2 emissions (in tons per person) to income level (in GDP per capita) among all countries of the world

    Checks: Random sample; scatterplot is roughly linear; no pattern in the residual plot; distribution of residuals is approximately normal; $n \leq 0.10N$

11. *Procedure:* Chi-square test for independence

    Checks: Random sample; all expected counts at least 5; $n \leq 0.10N$

    Hypotheses: H_0: There is no association between race and whether or not an adult is unbanked

    H_a: There is an association between race and whether or not an adult is unbanked

12. *Procedure:* Two-sample z-test for difference in population proportions, $p_1 - p_2$

    Parameters: Let p_1 represent the proportion of the population of at-risk patients receiving rivaroxaban who still have strokes. Let p_2 represent the proportion of the population of at-risk patients receiving warfarin who still have strokes.

    Checks: Independent random samples; $n_1 \hat{p}_C \geq 10$, $n_1(1 - \hat{p}_C) \geq 10$, $n_2 \hat{p}_C \geq 10$, and $n_2(1 - \hat{p}_C) \geq 10$, where $\hat{p}_C = \dfrac{x_1 + x_2}{n_1 + n_2}$; $n_1 \leq 0.10N_1$ and $n_2 \leq 0.10N_2$

    Hypotheses: H_0: $p_1 - p_2 = 0$, H_a: $p_1 - p_2 \neq 0$

13. *Procedure:* Matched pair t-test for mean difference in population means, μ_{diff}

    Parameter: Let μ_{diff} represent the true mean difference (With pill − With placebo) in malaria episodes for the population of adults when they take the new pill and when they take the placebo

    Checks: Random sample; set of differences is from a normal population or set of differences is large, at least 30, or distribution of set of differences is unimodal and symmetric

    Hypotheses: H_0: $\mu_{\text{diff}} = 0$, H_a: $\mu_{\text{diff}} < 0$

14. *Procedure:* Two-sample *t*-interval for difference in population means, $\mu_1 - \mu_2$

Parameters: Let μ_1 represent the mean infant mortality rate in the population of live births in the U.S. (mean infant deaths for every 1000 live births)

Let μ_2 represent the mean infant mortality rate in the population of live births in Japan (mean infant deaths for every 1000 live births)

Checks: Independent random samples; for each sample, population is normal or sample is large, at least 30, or sample distribution is unimodal and symmetric; for each sample, $n \leq 0.10N$

15. *Procedure:* Matched pair *t*-interval for mean difference in population means, μ_{diff}

Parameter: Let μ_{diff} represent the mean difference (second score − first score) for the population of students taking a national standardized exam twice

Checks: Random sample; set of differences is from a normal population or set of differences is large, at least 30, or distribution of set of differences is unimodal and symmetric; $n \leq 0.10N$

16. *Procedure:* One-sample *z*-interval for population mean μ. (Note that we use a normal distribution rather than a *t*-distribution because the population standard deviation is given.)

Parameter: Let μ represent the mean waist size of the population of male high school math teachers.

Checks: Random sample; population is normal or sample is large, at least 30, or sample distribution is unimodal and symmetric; $n \leq 0.10N$

Quiz 34 (Pages 366–367)

1. *Procedure:* Two-sample *t*-test for difference in population means, $\mu_1 - \mu_2$

Parameters: Let μ_1 represent the true mean number of hours per week that first-year students study. Let μ_2 represent the true mean number of hours per week that sophomores study.

Checks: Independent random samples; for each sample, population is normal or sample is large, at least 30, or sample distribution is unimodal and symmetric; for each sample, $n \leq 0.10N$

Hypotheses: $H_0: \mu_1 - \mu_2 = 0$, $H_a: \mu_1 - \mu_2 > 0$

2. *Procedure:* One-sample *z*-interval for population proportion p

Parameter: Let p represent the proportion of the population of students who would report cheating by other students.

Checks: Random sample; at least 10 successes and 10 failures, that is, $n\hat{p} \geq 10$ and $n(1 - \hat{p}) \geq 10$; $n \leq 0.10N$

3. *Procedure:* Chi-square test for independence

   Checks: Random sample; all expected counts at least 5; $n \leq 0.10N$

   Hypotheses: H_0: There is no association between blood pressure level (low, average, or high) and personality type (high-strung or easy-going)

   H_a: There is an association between blood pressure level (low, average, or high) and personality type (high-strung or easy-going)

4. *Procedure:* One-sample z-test for a population proportion p

   Parameter: Let p represent the proportion of 3-point shot attempts successfully made by the population of Division 1 college basketball players this year.

   Checks: Random sample; $np_0 \geq 10$ and $n(1 - p_0) \geq 10$; $n \leq 0.10N$

   Hypotheses: H_0: $p = 0.34$, H_a: $p < 0.34$

5. *Procedure:* Matched pair t-test for mean difference in population means μ_{diff}

   Parameter: Let μ_{diff} represent the true mean difference (after program − before program) in the number of push-ups a clinic member can complete in 90 seconds.

   Checks: Random sample; set of differences is from a normal population or set of differences is large, at least 30, or distribution of set of differences is unimodal and symmetric

   Hypotheses: H_0: $\mu_{\text{diff}} = 0$, H_a: $\mu_{\text{diff}} > 0$

6. *Procedure:* One-sample t-test for population mean μ

   Parameter: Let μ represent the mean body temperature in the population.

   Checks: Random sample; population is normal or sample is large, at least 30, or sample distribution is unimodal and symmetric; $n \leq 0.10N$

   Hypotheses: H_0: $\mu = 98.6$, H_a: $\mu < 98.6$

7. *Procedure:* Matched pair t-interval for mean difference in population means μ_{diff}

   Parameter: Let μ_{diff} represent the true mean difference in resting heart rates before and after an extensive exercise program for teenagers.

   Checks: Random sample; set of differences is from a normal population or set of differences is large, at least 30, or distribution of set of differences is unimodal and symmetric

8. *Procedure:* Two-sample z-interval for difference in population proportions, $p_1 - p_2$

   Parameters: Let p_1 represent the proportion of the population of job applications with stereotypically "White" names who receive callbacks. Let p_2 represent the proportion of the population of job applications with stereotypically "Black" names who receive callbacks.

   Checks: Independent random samples; at least 10 successes and 10 failures in each sample, that is, $n_1\hat{p}_1 \geq 10$, $n_1(1 - \hat{p}_1) \geq 10$, $n_2\hat{p}_2 \geq 10$, $n_2(1 - \hat{p}_2) \geq 10$

9. *Procedure:* Chi-square test for homogeneity

   Checks: Independent random samples; all expected counts at least 5; $n \leq 0.10N$

   Hypotheses: H_0: There is no difference in the true distributions of highest school levels attained between Whites, Blacks, Asians, and Hispanics (non-White)

   H_a: There is a difference in the true distributions of highest school level attained between Whites, Blacks, Asians, and Hispanics (non-White)

10. *Procedure:* Linear regression *t*-interval for slope β

    Parameter: Let β represent the slope of the regression line relating selling price to miles noted on the odometer for all used cars of a particular model.

    Checks: Random sample; scatterplot is roughly linear; no pattern in the residual plot; distribution of residuals is approximately normal; $n \leq 0.10N$

11. *Procedure:* One-sample *t*-interval for population mean μ

    Parameter: Let μ represent the mean pause length in the population of telemarketing calls.

    Checks: Random sample; population is normal or sample is large, at least 30, or sample distribution is unimodal and symmetric; $n \leq 0.10N$

12. *Procedure:* Two-sample *z*-test for difference in population proportions, $p_1 - p_2$

    Parameters: Let p_1 represent the proportion of the population of White students who report that, at times, they feel unsafe to go to school. Let p_2 represent the proportion of the population of non-White students who report that, at times, they feel unsafe to go to school.

    Checks: Independent random samples; $n_1\hat{p}_C \geq 10$, $n_1(1 - \hat{p}_C) \geq 10$, $n_2\hat{p}_C \geq 10$, and $n_2(1 - \hat{p}_C) \geq 10$, where $\hat{p}_C = \dfrac{x_1 + x_2}{n_1 + n_2}$; $n_1 \leq 0.10N_1$ and $n_2 \leq 0.10N_2$

    Hypotheses: H_0: $p_1 - p_2 = 0$, H_a: $p_1 - p_2 < 0$

13. *Procedure:* Chi-square test for goodness-of-fit

    Checks: Random sample; all expected counts at least 5; $n \leq 0.10N$

    Hypotheses: H_0: The US incarcerated adult population by race is 39 percent White (non-Hispanic), 19 percent Hispanic, 40 percent Black, and 2 percent other

    H_a: The U.S. incarcerated adult population by race is something different than 39 percent White (non-Hispanic), 19 percent Hispanic, 40 percent Black, and 2 percent other

14. *Procedure:* Two-sample t-interval for difference in population means, $\mu_1 - \mu_2$

    Parameters: Let μ_1 represent the mean weight of the population of elephants living in captivity. Let μ_2 represent the mean weight of the population of elephants living in the wild.

    Checks: Independent random samples; for each sample, population is normal or sample is large, at least 30, or sample distribution is unimodal and symmetric; for each sample, $n \leq 0.10N$

15. *Procedure:* Linear regression t-test for slope β

    Parameter: Let β represent the slope of the regression line relating pulse rate to quantity of caffeine consumed in a sitting

    Checks: Random sample; scatterplot is roughly linear; no pattern in the residual plot; distribution of residuals is approximately normal; $n \leq 0.10N$

    Hypotheses: H_0: $\beta = 0$, H_a: $\beta > 0$. (Interest is in an increase in pulse rate.)

16. *Procedure:* One-sample z-test for population mean μ. (Note that we use a normal distribution rather than a t-distribution because the population standard deviation is given.)

    Parameter: Let μ represent the mean hourly wage paid for summer work to the population of high school students.

    Checks: Random sample; population is normal or sample is large, at least 30, or sample distribution is unimodal and symmetric; $n \leq 0.10N$

    Hypotheses: H_0: $\mu = 10.50$, H_a: $\mu < 10.50$

Quiz 35 (Pages 370–376)

1. (a)

Region	Population (in millions)	% of World Population	Wealth (GDP) (in billions)	% of World GDP
Africa	1111	15.5	2600	3.6
Asia	4299	60.0	18,500	25.8
Oceania	38	0.5	1800	2.5
Europe	742	10.4	24,400	34.0
U.S./Canada	355	5.0	20,300	28.3
Latin America	617	8.6	4200	5.8
World Total	7162	100.0	71,800	100.0

(b) The segmented bar chart treats each bar as a whole and divides each bar proportionally into segments corresponding to the percentage in each group.

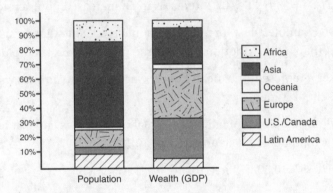

The segmented bar chart shows that both Europe and the U.S./Canada have a hugely disproportionate share of the wealth of the world when compared to their populations, while both Asia and Africa have a low share of the wealth of the world when compared to their populations.

(c) The slopes give each region's wealth in dollars per person. The U.S./Canada region has the steepest slope and thus the highest $/person, followed by Oceania, Europe, Latin America, Asia, and Africa.

(d) There are many possible answers. For example, switching to GPI takes into account industrial pollution, which is likely much higher in the U.S./Canada than in the islands of the South Pacific. With GDP per capita as the index, the U.S./Canada ranked slightly higher than Oceania. With CPI per capita as the index, the value for Oceania might go up and the value for U.S./Canada might go down sufficiently for the ranking to be reversed.

2. (a) Fifteen percent of all U.S. Black men are likely to go to prison for the first time by age 25.

   (b) No U.S. White males are likely to go to prison for the first time between ages 60 and 70.

   (c) Age 30. The graph indicates that 16% of all U.S. Hispanic males are likely to go to prison, and half of them (corresponding to 8% on the vertical axis) will go to prison for the first time by age 30.

   (d) Answers should explain how addressing the disproportionate suspensions among students of color should lower the initial slopes of the Black and Hispanic graphs. An answer might also conclude that dismantling the school-to-prison pipeline would move the x-intercept for all 3 curves to the right, that is, raise the ages for first admissions to prison for all young people.

3. (a) Predicted total SAT $= 903.36 + 0.00116$(Income)

   (b) Every \$1000 of additional family income is associated with an average of 1.16 points higher on a student's total SAT score.

   (c) The coefficient of determination $R^2 = 94.0\%$. That is, 94.0% of the variation in total SAT scores is explained by the linear regression model of total SAT on Family Income.

   (d) Predicted Total SAT $= 903.36 + 0.00116(63,500) = 977.02$

   Residual $=$ Actual $-$ Predicted $= 970 - 977.02 = -7.02$; that is, the actual total SAT score of this student is 7.02 *less* than what was predicted by the family income.

   (e) $S = 20.936$ is the standard deviation of the residuals. In other words, it is a measure of how the points are spread around the regression line.

   (f) Correlation does not imply causation. A possible confounding variable, for example, could be instructional money spent per student. Leveling per student expenditures across school districts might lead to higher SAT scores in economically disadvantaged areas.

4. (a)

SOHR

		Reported Deaths	Unreported Deaths	*Totals*
UN	Reported Deaths	2635	992	3627
	Unreported Deaths	425	$D - 4052$	$D - 3627$
	Totals	3060	$D - 3060$	D

   (b) Relative frequencies should be approximately equal if there is independence. The expected cell formula $\text{Exp} = \dfrac{(\text{Row sum}) \times (\text{Column sum})}{N}$ gives $2635 = \dfrac{(3627)(3060)}{D}$. (If you look at rows, you get instead $\dfrac{2635}{3627} = \dfrac{425}{D - 3627}$. If you look at columns, you get $\dfrac{2635}{3060} = \dfrac{992}{D - 3060}$.) Solving any of these equations gives $D - 4212$.

   (c) *Convenience samples* are based on choosing individuals who are easy to reach and tend to produce data highly unrepresentative of the entire population. In this case, only very readily obtainable reports of deaths are included, violating the assumption of a random sample of deaths for that month.

(d) Both samples will tend to report the same large casualty events, thus violating the assumption of independence.

(e) The calculated value of D is almost certainly an underestimate. As shown in parts (c) and (d), the reported deaths tend to be convenience samples of more widely reported deaths and with different samples tending to report the same deaths.

5. (a) This is an example of cluster sampling, which allows for a quicker, less expensive method of sampling than an SRS. It assumes that each cluster, in this case each county, is representative of the whole state.

(b) Number the counties 1 through 67. Use a random number generator to pick two unique integers between 1 and 67 (ignoring repeats). The two counties to be used will consist of the two counties with numbers corresponding to the two unique numbers.

(c) The purpose of randomization in picking the 90 jurors is to obtain a sample that is representative of all eligible White jurors in the state so that any conclusions can be generalized to all eligible White jurors in the state.

(d) Random assignment here is used to minimize the effect of possible confounding variables. For example, if self-selection is allowed, perhaps some White jurors are uncomfortable with people of color and would choose to be on the all-White juries. Because it is likely that these same jurors are more likely biased against a defendant of color, it might be impossible to say whether any conclusions can be made based on the racial makeup of a jury.

6. (a) $P(\text{Blast zone}) = \dfrac{5.5}{38.8} = 0.14175$ and $P(\text{Latino} \cap \text{Blast zone}) = P(\text{Blast zone})P(\text{Latino} \mid \text{Blast zone}) = (0.14175)(0.75) = 0.10631$

(b) $P(\text{Blast zone} \mid \text{Latino}) = \dfrac{P(\text{Latino} \cap \text{Blast zone})}{P(\text{Latino})} = \dfrac{0.10631}{0.390} = 0.27259$

(c) Is $P(\text{Blast zone} \mid \text{Latino}) = P(\text{Blast zone})$? No, $0.27259 \neq 0.14175$. So, in California, being Latino and living within a blast zone are not independent events.

(d) This is a binomial with $n = 5$ and $p = 0.14175$. Then $P(X \geq 2) = 1 - [(0.85825)^5 + 5(0.14175)(0.85825)^4] = 0.14979$. Calculator software gives $1 - \texttt{binomcdf(5, 0.14175, 1)} = 0.14979$.

(e) Another possible explanatory variable is poverty. It could be that blast zones tend to be associated with low-income housing and that a higher percentage of people of color than White people are low income.

7. (a) This is a binomial distribution with $n = 2500$ and $p = \dfrac{257}{268} = 0.959$.

Thus, $E(X) = np = 2500(0.959) = 2397.4$.

The standard deviation $= \sqrt{np(1-p)} = \sqrt{2500(0.959)(0.041)} = 9.915$.

(b) The probability of a case resulting in the death penalty, given it involves flawed forensic testimony, is $\dfrac{32}{257} = 0.1245$. Thus, the probability of the case not resulting in the death penalty is $1 - 0.1245 = 0.8755$. The probabilities of a death penalty occurring on the first or second case examined, respectively, are 0.1245 and $(0.8755)(0.1245) = 0.1090$. So, the probability that the first case with a resulting death penalty doesn't occur until at least the third case examined is

$$1 - (0.1245 + 0.1090) = 0.7665$$

(c) "The misapplication of forensic science" and "eyewitness misidentification" are not mutually exclusive because, if so, their intersection would be the empty set. Then the probability of their union would be $0.50 + 0.75 - 0 = 1.25 > 1$. It is possible that they are independent because, if so, the probability of their intersection would be $(0.50)(0.75)$ and the probability of their union would be

$$0.50 + 0.75 - (0.5)(0.75) = 0.875 \leq 1$$

8. (a) We have a normal distribution with mean 3287 and standard deviation 250. Then
$$P(X \geq 3500) = P\left(z \geq \frac{3500 - 3287}{250}\right) = 0.1971.$$

(b) We have a binomial distribution with $n = 5$ and $p = 0.1971$. Then $P(X \geq 3) = 1 - P(X \leq 2) = 1 - \text{binomcdf}(5, 0.1971, 2) = 0.05572$.

(c) We have a normal distribution with mean 3287 and standard deviation $\frac{250}{\sqrt{5}} = 111.80$. Then $P(\bar{x} \geq 3,500) = P\left(\frac{3,500 - 3,287}{111.80}\right) = 0.0284$.

(d) Possible answers might range from average yearly proceeds (about \$35 billion) to average number of slaves in the world on any given day (21 to 29 million), from numbers of child slaves to numbers of women in sexual slavery.

9. (a) A complete answer addresses center, spread, and shape and mentions context. The median age of hate site members is approximately 23. The ages go from 14 to 65 with a range of $65 - 14 = 51$ and an IQR $= 30 - 17 = 13$.

There are 6 outliers between 51 and 65. Although it is not usually possible to describe distribution shape from just a boxplot, in this case the distribution of ages does appear to be skewed to the right (to the larger values).

(b) In a skewed right distribution, the mean is usually larger than the median. So, we would conclude that the quotient $\frac{\text{Mean age}}{\text{Median age}}$ is greater than 1.

(c) If mostly the younger, computer savvy members of the hate groups are the ones who are online, the median age of all hate group members is probably greater than the median age of those self-reporting on the site, 23.

(d) This is a binomial with $n = 3$ and $p = 0.186$. Then $P(X \geq 2) = 0.0909$ or

$$3(0.186)^2(0.814) + (0.186)^3 = 0.0909$$

Quiz 36 (Pages 377–383)

1. (a) $0.19 + 0.17 - 0.06 = 0.30$

   (b) *Parameter:* Let p represent the proportion of the population of students in grades six through ten who are victims of moderate to frequent bullying.

   Procedure: One-sample z-interval for a population proportion.

   Checks: We must assume this was a random sample, $n\hat{p} = (15{,}686)(0.17) = 2667 \geq 10$, $n\hat{q} = (15{,}686)(0.83) = 13{,}019 \geq 10$, and $n = 15{,}686$ is less than 10% of all students in grades six through ten.

   Mechanics: Calculator software gives $(0.16415, 0.1759)$.

   Conclusion in context: We are 95% confident that the proportion of all students in grades six through ten who are victims of moderate to frequent bullying is between 0.16415 and 0.1759.

   (c) Because 0.22 is not in the interval of plausible values for the population proportion, there *is* evidence that the proportion of all students who are victims of moderate to frequent bullying is different from the 22% claimed in the one study.

   (d) Issues that might be addressed in planning a survey include stratification among the grade levels, anonymity, and timing (early enough in the school year to allow implementation of a plan and again later in the year to assess the effectiveness of any prevention and intervention program).

2. (a) *Parameter:* Let μ represent the mean number of firearm suicides in the U.S. among the population of all years.

   Procedure: A one-sample t-interval.

   Checks: We are given a random sample. The nearly normal condition seems reasonable because the sample data are very roughly unimodal and symmetric. Under these conditions, the mean firearm suicides per year can be modeled by a t-distribution with $n - 1 = 7 - 1 = 6$ degrees of freedom.

   Mechanics: Calculator software gives the interval $(20408, 21850)$. [We can instead calculate $22{,}128.6 \pm 2.447\dfrac{779.6}{\sqrt{7}} = 22{,}128.6 \pm 721.0$.]

   Conclusion in context: We are 95% confident that the mean number of firearm suicides in the U.S. each year is between 21,408 and 22,850.

   (b) It could well be true (actually, it is true) that most guns in Israel are technically owned by the government (by the IDF—the Israel Defense Forces). The 7.3 guns per 100 people statistic is a significant underestimate of guns in the population.

   (c) The conversation would change from number of guns in the population to strictness of gun laws pertaining to gun ownership.

   (d) The U.S. has 88.8 guns per 100 people but only $\dfrac{88.8}{3} = 29.6$ gun owners per 100 people. Thus, the "firearm suicides per gun owner" rate in the U.S. will be three times as great as the "firearm suicides per gun" rate. How "rates" are calculated can lead to different conclusions.

3. (a) A cumulative frequency plot that rises quickly at first and then rises at a much slower rate later corresponds to a histogram with a large area under the curve early and much less area later. Thus, it is *skewed right* (skewed to the higher values of lead). In skewed right distributions, we expect the *mean* to be greater than the median.

   (b) These figures indicate an extremely serious lead problem in the Flint water supply. The research team found that 25% of the homes had a lead water supply level of at least 15 ppb. Because 25% > 10% and 15 ppb > 5 ppb, the percent of homes with at least 5 ppb is larger than 10%.

   (c) With H_0: $\mu = 10$ and H_a: $\mu > 10$, a Type I error (mistaken rejection of a true null hypothesis) means that we think the lead level is greater than 10 ppb when, in fact, it is 10 ppb. So, we might institute costly and unnecessary cleanup measures. A Type II error (mistaken failure to reject a false null hypothesis) means that the mean lead level is greater than 10 ppb but we don't have enough evidence to think so. So, we might fail to institute a necessary cleanup. Clearly, with people's, and especially children's, health at stake, a Type II error is more of a concern.

4. (a) *Parameters:* Let μ_{LEAD} represent the mean yearly criminal and legal system costs of the population of low-level drug and prostitution offenders who are participating in the LEAD program. Let $\mu_{control}$ represent the mean yearly criminal and legal system costs of the population of low-level drug and prostitution offenders who are not participating in the LEAD program.

   Hypotheses: H_0: $\mu_{LEAD} = \mu_{control}$ and H_a: $\mu_{LEAD} < \mu_{control}$.

   Procedure: This is a two-sample t-test for means.

   Checks: We are given samples that, by design, are assumed representative and independent. The sample sizes, 203 and 115, are large enough so that by the CLT, the distribution of sample means is approximately normal and a t-test may be run. The sample sizes are assumed to be less than 10% of the population of all low-level drug and prostitution offenders.

   Mechanics: Calculator software gives $t = -33.41$ with a P-value of 0.000.

   Conclusion in context with linkage to the P-value: With this small of a P-value, 0.000 < 0.05, there is very strong evidence to reject H_0. In other words, there is very strong evidence that for those adults eligible for the LEAD program, the mean yearly criminal and legal system costs of those participating in the program are less than that for those not participating.

   (b) The null hypothesis was rejected. So, the possibility is of a Type I error, that is, of rejecting a true null hypothesis. A Type I error would mean that in reality, after participating in the program, the LEAD group would have the same mean days per year in jail as the control group. However, in error, we think the LEAD group has significantly fewer mean days per year in jail than the control group. A possible consequence is that the LEAD program is expanded, thinking it will reduce the future mean days per year in jail when, in reality, it won't.

(c) It is very difficult to "choose" representative samples. Using randomization has been shown to be the most effective technique to minimize the effect of possible confounding variables.

(d) "We ran a large sample statistical study and noted that for our sample, the average yearly criminal and legal costs spent on those offenders participating in the LEAD program were significantly lower than the average spent on those not participating in the program. The difference was so great that it is highly unlikely that it could have occurred by chance."

5. (a) $\frac{1710}{4658} = 36.7\%$, $\frac{1423}{4658} = 30.5\%$, and $\frac{543}{4658} = 11.7\%$

$\frac{64,412}{116,774} = 55.2\%$, $\frac{10,076}{116,774} = 8.6\%$, and $\frac{11,600}{116,774} = 9.9\%$

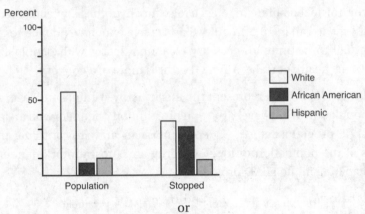

or

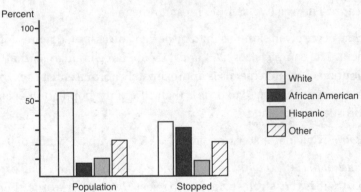

(b) *Parameter:* Let p represent the proportion of the population of all stopped drivers who are African American.

Hypotheses: H_0: $p = 0.086$ and H_a: $p > 0.086$

Procedure: One proportion z-test.

Checks: A representative (random) sample is assumed; $np_0 = 4658(0.086) = 400.6 \geq 10$ and $n(1 - p_0) = 4658(0.914) = 4257.4 \geq 10$; the sample size is < 10% of the population.

Mechanics: Software such as 1-PropZTest on the TI-84 gives $z = 53.4$ with a P-value of 0.000.

Conclusion in context with linkage to the P-value: With this low of a P-value, $0.000 < 0.001$, there is strong evidence that African Americans are being stopped at a higher rate than what would be indicated by their population percentage.

(c) *Hypotheses:* H_0: Race (White, African American, Hispanic) and disposition (arrest/citation or not) are independent.

H_a: There is a relationship between race (White, African American, Hispanic) and disposition (arrest/citation or not).

Procedure: A chi-square test of independence.

Checks: Representative (random sample assumed) and all expected cells as shown below are ≥ 5.

OBS	White	African American	Hispanic
Citation	1058	481	237
None	652	942	306

EXP	White	African American	Hispanic
Citation	826.2	687.5	262.3
None	883.8	735.5	280.7

Mechanics: Software (such as χ^2-Test on the TI-84) gives the above expected values and $\chi^2 = 250.61$ with a *P*-value of 0.000.

Conclusion in context with linkage to the P-value: With this low of a *P*-value, 0.000 < 0.001, there is strong evidence of a relationship between race (White, African American, Hispanic) and disposition (arrest/citation or not).

6. (a) *Procedure:* Confidence interval for the slope of a regression line.

Checks: We assume a random sample of countries (or at least a representative sample). The scatterplot looks roughly linear. The residual plot shows no pattern. The distribution of residuals is very roughly normal.

Mechanics: From the printout, the standard error of the slope is $s_b = 0.0530$. With $df = 21 - 2 = 19$ and 0.025 in each tail, the critical *t*-values are ± 2.093. The 95% confidence interval of the true slope is $0.4404 \pm 2.093(0.0530) = 0.4404 \pm 0.1109$.

Conclusion in context: We are 95% confident that for every 1-unit increase in the Income ratio, the mean increase in the HSP Index is between 0.3295 and 0.5513.

(b) *Hypotheses:* H_0: $\beta = 0$ and H_a: $\beta > 0$

Procedure: Test of significance for the slope of the regression line.

Checks: The conditions for inference are the same as those in part (a).

Mechanics: The computer printout gives $t = 8.31$ with a *P*-value of $\frac{1}{2}(0.000) = 0.000$.

Conclusion in context with linkage to the P-value: With this small of a *P*-value, 0.000 < 0.05, there is strong evidence to reject H_0. In other words, there is strong evidence that the higher the income ratio (that is, the higher the income inequality in a country), the greater the HSP Index (that is, the greater the health and social problems of the country).

(c) There is strong evidence that income ratios (income inequality) have a positive linear association with HSP Indexes (health and social problems). In fact, according to the computer output, 78.41% of the variation in HSP Indexes can be explained by this linear model. However, we cannot conclude causation. We cannot say that if we work to lessen income inequality, we will reduce health and social problems in the country.

7. (a) *Hypotheses: H_0*: The population of local jails is distributed according to 34.4% convicted males, 5.0% convicted females, 53.4% unconvicted males, and 7.2% unconvicted females.

H_a: The population of local jails is not distributed according to 34.4% convicted males, 5.0% convicted females, 53.4% unconvicted males, and 7.2% unconvicted females.

Procedure: χ^2 goodness-of-fit test.

Checks: We are given a random sample. With a sample of size $81 + 5 + 110 + 4 = 200$, all expected cells are at least 5: 34.4% of 200 = 68.8, 5.0% of 200 = 10, 53.4% of 200 = 106.8, and 7.2% of 200 = 14.4.

Mechanics: A calculator gives $\chi^2 = \sum \dfrac{(\text{obs} - \text{exp})^2}{\text{exp}} = 12.27$. With $df = 4 - 1 = 3$, $P = 0.0065$.

Conclusion in context with linkage to the P-value: With this small of a *P*-value, $0.0065 < 0.05$, there is sufficient evidence to reject H_0. In other words, there is sufficient evidence that the population distribution of local jails has changed from the previous study's finding of 34.4% convicted males, 5.0% convicted females, 53.4% unconvicted males, and 7.2% unconvicted females.

(b) There are many possible ways of proportionately choosing the 500 inmates to study. We should stratify and randomly pick 44% of 500 = 220 inmates from small county jails and 56% of 500 = 280 inmates from large county jails. For example, we might use cluster sampling to pick 22 small county jails and 28 large county jails randomly. Then, in every chosen jail, number the inmates. Using a random integer generator, pick 10 unique numbers in the appropriate range corresponding to the number of inmates in each jail. The inmates with those numbers, together from all the 50 chosen jails, will form the sample to be studied.

8. (a) We have a margin of error of ± 0.03. The critical *z*-scores associated with 95% are ± 1.96. Assuming that the proportion of interest was unknown, we use $p = 0.5$ as a "worst case," which leads to $1.96\sqrt{\dfrac{0.5(1 - 0.5)}{n}} \le 0.03$, or $1.96\dfrac{0.5}{\sqrt{n}} \le 0.03, \sqrt{n} \ge 32.667$, $n \ge 1067.11$. So, the sample size should have been at least 1068.

(b) Both the ASD Case distribution and the Control distribution are bimodal. Their centers and spreads are roughly the same.

(c) Both the ASD Case distribution and the Control distribution are skewed left. Their centers and spreads are roughly the same.

(d) With a very large *P*-value, there is little or no evidence to reject the null hypothesis. In other words, there is little or no evidence of an association between autism and vaccines.

9. (a) There is a positive, linear, moderate to strong relationship between Temperature Anomaly and Time.

(b) *Parameters*: Let p_{Ind} represent the proportion of the population of Indians who express a great deal of worry about global warming. Let p_{Chi} represent the proportion of the population of Chinese who express a great deal of worry about global warming.

Hypotheses: H_0: $p_{\text{Ind}} = p_{\text{Chi}}$, H_a: $p_{\text{Ind}} \neq p_{\text{Chi}}$

Procedure: Hypothesis test for a difference between two proportions.

Checks: Independent random samples; $n_{\text{Ind}}\hat{p}_C = (2029)(0.417) = 846 \geq 10$, $n_{\text{Ind}}(1 - \hat{p}_C) = (2029)(0.583) = 1183 \geq 10$, $n_{\text{Chi}}\hat{p}_C = (2180)(0.417) = 909 \geq 10$, and $n_{\text{Chi}}(1 - \hat{p}_C) = (2180)(0.583) = 1271 \geq 10$, where $\hat{p}_C = \dfrac{x_1 + x_2}{n_1 + n_2} = \dfrac{1319 + 436}{2029 + 2180} = 0.417$; $n_{\text{Ind}} \leq 0.10 N_{\text{Ind}}$ and $n_{\text{Chi}} \leq 0.10 N_{\text{Chi}}$

Mechanics: $z = 29.59$, $P = 0.000$

Conclusion in context with linkage to the P-value: With this small of a P-value, $0.000 < 0.05$, there is sufficient evidence to reject H_0. In other words, there is sufficient evidence of a difference between the proportion of Indians who express a great deal of worry about global warming and the proportion of Chinese who express a great deal of worry about global warming.

(c) Phrases like "climate change deniers," "corporate lobbyists," and "a dying planet" are clearly nonneutral and leading. The calculated proportion of people who express a great deal of worry about global warming will most likely be greater than the true proportion.

Quiz 37 (Pages 393–399)

1. (a) **Think:** This sounds like a straightforward algebra problem in which you name a variable, set up an equation where two proportions (two ratios) are set equal, and solve for the unknown. So, let $N =$ the whole population size, and note that 108 springboks out of N were tagged, while 12 out of the sample of 80 springboks were seen to be tagged.

Answer: $\dfrac{108}{N} = \dfrac{12}{80}$ gives $N = 720$

(b) **Think:** That sure is a complex-looking definition of variance, but you're only asked to plug in. And note that the formula is for variance, and you're asked for the standard deviation.

Answer: $\text{var}(N) = \dfrac{108 \times 80 \times (108 - 12)(80 - 12)}{12^3} = 32{,}640$ and so, $\text{SD}(N) = \sqrt{32{,}640} = 180.67$

(c) **Think:** The sampling distribution is given to be approximately normal, so you can use a z-distribution. Even though this is not one of our standard confidence interval questions, it is still statistical inference, and, as always, you need to check conditions and give a conclusion in context, referring to the entire population.

Answer: Check conditions: Both captures are given to be random samples, $b = 12 \geq$ 10 and $c - b = 80 - 12 \geq 10$. A 90% z-interval has critical z-scores of ± 1.645, so the confidence interval is $720 \pm 1.645(180.67) = 720 \pm 297.2$. Thus, we are 90% confident that the total population of springboks in this wildlife preserve is between 422.8 and 1017.2.

(d) **Think:** This is not difficult, but careful reasoning is required! If some of the marked springboks are lost, then the proportion obtained in the recapture will be smaller than it probably would have been, and if you set $\frac{108}{N}$ equal to a smaller number, you'll get a larger N. Now your task is to fully and clearly explain this!

Answer: We are assuming that the proportion of marked individuals within the second sample is equal to the proportion of marked individuals in the whole population. If fewer of the marked springboks are available for recapture, the proportion of marked individuals in the second sample will be smaller than it should be. Therefore, we will think that the number of originally marked individuals is a smaller proportion of the population than it really is. Thus, we will think that the population is larger than it really is.

(e) **Think:** This seems to be the same as in (d), that is, fewer correctly marked springboks around.

Answer: Again, if we think that the proportion of marked individuals in the second sample is smaller than it really is, we will think that the number of originally marked individuals is a smaller proportion of the population than it really is. Thus, we will think that the population is larger than it really is.

SCORING

Part (a) is essentially correct for the correct proportion and calculation of N and is partially correct for the correct value of N with no work shown.

Part (b) is essentially correct for the correct calculation with work shown and is partially correct for a calculation of only variance with work shown.

Part (c) is essentially correct for a check of the ≥ 10 conditions, a correct calculation of the interval, and a correct statement in context. It is partially correct if missing the ≥ 10 check or if missing a statement in context.

Parts (d) and (e) together are essentially correct for the correct conclusions with reasonable explanations. They are partially correct for one correct conclusion with a reasonable explanation or for both correct conclusions but with weak explanations.

Count partially correct answers as one-half an essentially correct answer.

4 Complete Answer   Four essentially correct answers.

3 Substantial Answer   Three essentially correct answers.

2 Developing Answer   Two essentially correct answers.

1 Minimal Answer   One essentially correct answer.

Use a holistic approach to decide a score totaling between two numbers.

2. (a) **Think:** A complete answer considers shape, center, spread, and unusual features, and mentions context.

Answer:

Shape: Bimodal and roughly symmetric

Center: The center of the $W = \dfrac{\text{Max} + \text{Min}}{2}$ calculations (where Min and Max are the minimum and maximum weights, respectively, in a set of 12 containers) is around 64.00 ounces.

Spread: From 63.91 to 64.10 (or range of 0.19 or IQR of 0.07)

(b) **Think:** Outliers are values more than $1.5 \times$ IQR below the first quartile or $1.5 \times$ IQR above the third quartile. You are given the quartiles and then can look at the dotplot as to whether any values are more than $1.5 \times$ IQR from the quartiles.

Answer: IQR $= 64.04 - 63.97 = 0.07$ and $1.5(\text{IQR}) = 0.105$

$Q_1 - 0.105 = 63.865$ and $Q_3 + 0.105 = 64.145$

Since no values are below 63.865 or above 64.145, there are no outliers.

(c) **Think:** This is a two-sided hypothesis test because you are asked whether a value varies significantly from 64, not whether it is larger or smaller than 64.

Answer: $H_0: \dfrac{\text{Min} + \text{Max}}{2} = 64.0$ and $H_a: \dfrac{\text{Min} + \text{Max}}{2} \neq 64.0$

(d) **Think:** Remember that simulation can be used to determine what values of a test statistic are likely to occur by random chance alone, assuming the null hypothesis is true. Then looking at where the test statistic falls, you can estimate a *P*-value.

Answer: For this day's sample, $W = \dfrac{63.8 + 64.04}{2} = 63.92$, which is 0.08 away from 64.0. There are 4 values (2 above and 2 below) out of 100 this far away or farther from 64.0. With a *P*-value of 0.04, there is sufficient evidence at the 5% significance level $(0.04 < 0.05)$ to reject the null hypothesis. In other words, there is sufficient evidence to conclude that the machine needs adjustment.

Part (a) is essentially correct for correct statements about the shape, center, and spread and some mention of context. It is partially correct for three out of four of these parts correct.

Part (b) is essentially correct for correctly calculating $Q_1 - 1.5(IQR)$ and $Q_3 + 1.5(IQR)$ and for noting that there are no values outside these calculations. It is partially correct for a mistake in either calculating the cutoffs or in comparing values to the cutoffs.

Part (c) is essentially correct for two correct hypotheses and is partially correct if the only error is making the alternative hypothesis one-sided.

Part (d) is essentially correct for noting that we have a P-value of 0.4, linking this P-value to some significance level, and making a correct conclusion in context based on that linkage. It is partially correct if missing one of these parts.

Count partially correct answers as one-half an essentially correct answer.

4 Complete Answer   Four essentially correct answers.

3 Substantial Answer Three essentially correct answers.

2 Developing Answer   Two essentially correct answers.

1 Minimal Answer   One essentially correct answer.

Use a holistic approach to decide a score totaling between two numbers.

3. (a) **Think:** The conditions for regression inference are (1) the sample must be randomly selected; (2) the scatterplot should be approximately linear; (3) there should be no apparent pattern in the residual plot; and (4) the distribution of the residuals should be approximately normal.

   Answer: First, the scatterplot of Deaths vs. Cheese is roughly linear. Second, the residual plot shows no clear pattern. Third, the histogram of residuals appears roughly bell-shaped (unimodal, symmetric, and without clear skewness or outliers). We must assume a representative sample.

   (b) **Think:** Residual = observed − predicted; you're given the observed, and you can calculate the predicted from the regression line equation. Remember that the interpretation must be in context and discuss direction, not only that the residual is of a certain size.

   Answer: The predicted number of deaths from the regression line is

   $$-2977.3 + 113.13(30.1) = 427.9$$

   So, the residual = observed − predicted = 456 − 427.9 = 28.1. The actual number of deaths was 28.1 greater than what was predicted by the regression model for the year when the per capita consumption of cheese was 30.1 pounds.

(c) **Think:** Remember, this is a t-distribution with $df = n - 2$, and don't forget that confidence interval questions always involve conclusions in context.

Answer: With $df = 10 - 2 = 8$ and 0.025 in each tail, the critical t-values are ± 2.306.

$b \pm ts_b = 113.13 \pm 2.306(13.56) = 113.13 \pm 31.27$. We are 95% confident that for each additional pound of cheese consumption per capita, the average increase in deaths per year from bedsheet tangling is between 81.86 and 144.40.

(d) **Think:** Correlation does not imply causation! Most likely the two unrelated data sets follow the same time trend, resulting in correlation but no connection.

Answer: No cause-and-effect conclusion is appropriate. Evidence of an association is not evidence of a cause-and-effect relationship. In cases like this, most likely the two unrelated data sets happen to follow the same time trend, resulting in correlation but not connection.

(e) **Think:** The sum and thus the mean of the residuals is 0, and the SD is given by "S" in typical computer output. The residuals appear to have a roughly normal distribution, so the asked-for calculation is a straightforward normal probability!

Answer: The sum and thus the mean of the residuals is always 0. The standard deviation of the residuals is estimated with $s = 50.0007$. With a z-score of $\frac{75}{50} = 1.5$ and a roughly normal distribution, we have $P(z > 1.5) \approx 0.067$.

SCORING

Part (a) is essentially correct for noting the three conditions (roughly linear scatterplot, no clear pattern in the residual plot, and roughly normal histogram of residuals). Part (a) is partially correct for correctly noting two of the three conditions.

Part (b) is essentially correct for a correct calculation and a correct interpretation of the residual, including direction. Part (b) is partially correct for one of these two parts correct.

Parts (c) and (d) together are essentially correct for a correct calculation of the confidence interval, a correct interpretation in context, and a correct statement about causation. Parts (c) and (d) together are partially correct for two of the three steps correct.

Part (e) is essentially correct for noting that the distribution of residuals is roughly normal with mean 0 and standard deviation 50.0007 and then using this to calculate the probability correctly. Part (e) is partially correct for correctly noting the distribution of residuals but incorrectly calculating the probability, or for making a calculation based on a normal distribution with mean 0 but using an incorrect standard deviation.

Count partially correct answers as one-half an essentially correct answer.

4 Complete Answer   Four essentially correct answers.

3 Substantial Answer   Three essentially correct answers.

2 Developing Answer   Two essentially correct answers.

1 Minimal Answer   One essentially correct answer.

Use a holistic approach to decide a score totaling between two numbers.

4. (a) **Think:** This is a complex-looking graph right from the start! Take your time and look carefully at what the axes represent: values and their z-scores. You are only asked here for the range of CFU values, and the min and max CFU values can be easily estimated from the graph.

Answer: $33.5 - 28.5 = 5.0$ (billion)

(b) **Think:** There are 100 CFU values, so the fifth percentile is simply the value indicated by the fifth dot up from the minimum CFU on the graph.

Answer: Either counting 5 dots (out of 100) or going over to the data plot from a z-score of -1.645, the 5th percentile of the data is seen to be approximately 29.75 (billion).

(c) **Think:** Getting more complicated! You're not looking at the data here, only at that diagonal line. The 95th percentile in a normal distribution corresponds to a z-score of 1.645, which is fortunately marked on the graph.

Answer: Going over to the slant line from a z-score of 1.645, you can see that the 95th percentile of a normal distribution with the same mean and standard deviation is seen to be approximately 34.4 (billion).

(d) **Think:** What is the shape of a histogram? In the graph, you can see that lots of values are piled up at the upper end between 33 and 34, while at the lower end, the values are spread out from the minimum value toward the median. Remember that skewness is indicated when values are concentrated at one end and spread thinly on the other.

Answer: Note that there are relatively few CFU values between 28 and 32, that these lower values are spread out while the values between 32 and 33 are more concentrated, and that the values between 33 and 34 are very concentrated. Conclude that the distribution is skewed left (skewed toward the lower values).

TIP

Even if you have no idea how to do the first parts, a later part might be straightforward; in this case, the last part is simply a t-interval mean question.

(e) **Think:** To receive full credit for a confidence interval inference question, you must name the procedure, check conditions, find the interval, and give a conclusion in context about the whole population.

Answer:

Parameter: Let μ represent the mean CFU value (in billions) of the population of probiotic capsules of this brand.

Procedure: t-interval of a population mean.

Checks: Random sample (given), the sample size $n = 100$ is large enough for the CLT to apply, and $n = 100$ is less than 10% of all probiotic capsules of this brand.

Mechanics: With $\bar{x} = 32.48$, $s = 1.174$, and $n = 100$, a 95% t-interval is (32.247, 32.713).

Conclusion in context: We are 95% confident that the mean CFU value (in billions) of all probiotic capsules of this brand is between 32.247 and 32.713.

Parts (a), (b), and (c) together are essentially correct for the three *roughly* correct answers (no work need be shown) and are partially correct for two correct answers.

Part (d) is essentially correct for an answer of "skewed left" together with a clear explanation. It is partially correct for "skewed left" with a weak explanation.

Part (e1) is essentially correct if the correct procedure is named and the conditions of random sample and large sample size are confirmed. Part (e1) is partially correct for two out of three of these statements.

Part (e2) is essentially correct for a correct confidence interval and a correct conclusion in context. It is partially correct if a z-interval is calculated but everything else is correct or if the only part missing is context.

Count partially correct answers as one-half an essentially correct answer.

4 Complete Answer   Four essentially correct answers.

3 Substantial Answer   Three essentially correct answers.

2 Developing Answer   Two essentially correct answers.

1 Minimal Answer   One essentially correct answer.

Use a holistic approach to decide a score totaling between two numbers.

5. (a) **Think:** You're starting out with a straightforward binomial probability.

   Answer: The probability that Team A wins four straight games is $(0.5)^4 = 0.0625$. So, the probability that the World Series is over in four games is $0.0625 + 0.0625 = 0.125$.

   (b) **Think:** You're shown a helpful pattern, so say "thank you" and use it!

   Answer: $2\left[\binom{6}{3}(0.5)^3(0.5)^3\right](0.5) = 0.3125$, or $1 - (0.125 + 0.25 + 0.3125) = 0.3125$

   (c) **Think:** You calculated the relevant probabilities, and the number of games is 108.

   Answer: Multiplying each of the probabilities by 108 gives the expected number of occurrences:

	4	5	6	7
Expected Number of Years	13.5	27	33.75	33.75

   (d) **Think:** You're given observations and asked if a certain distribution is followed— a standard χ^2-GOF test! Remember to state the hypotheses, identify the test and check the assumptions, calculate the test statistic χ^2 and the P-value, and, linking to the P-value, give a conclusion in context.

   Answer:

   Hypotheses: H_0: The lengths of World Series follow a distribution of 4, 5, 6, and 7, with probabilities 0.125, 0.25, 0.3125, and 0.3125, respectively. H_a: The probability distribution is different from that of 0.125, 0.25, 0.3125, and 0.3125, respectively.

   Checks: This is a chi-square test of goodness-of-fit. We must assume that the 108 series are a representative sample. We note from part (c) that all expected cells are ≥ 5.

Mechanics: A calculator gives $\chi^2 = 7.437$ and $P = 0.0592$.

Conclusion in context with linkage to the P-value: With a *P*-value this large, 0.0592 > 0.05, there is not sufficient evidence to reject H_0. In other words, there is not sufficient evidence (at the 5% significance level) that the lengths of World Series do not follow a distribution of 4, 5, 6, and 7, with probabilities 0.125, 0.25, 0.3125, and 0.3125, respectively. Alternatively, with a *P*-value this small, 0.0592 < 0.10, there is sufficient evidence to reject H_0. In other words, there is sufficient evidence (at the 10% significance level) that the lengths of World Series do not follow a distribution of 4, 5, 6, and 7, with probabilities 0.125, 0.25, 0.3125, and 0.3125, respectively.

SCORING

Parts (a) and (b) together are **essentially correct** for correctly calculating 0.0625, 0.125, and 0.3125. They are **partially correct** for correctly calculating two of these three probabilities.

Part (c) is **essentially correct** for correctly calculating expected values using the probabilities calculated in (a) and (b), and **incorrect** otherwise.

Part (d1) is **essentially correct** for correctly stating the hypotheses, identifying the test, and noting that the expected cells are all ≥ 5. It is **partially correct** for two out of three of these statements.

Part (d2) is **essentially correct** for correctly giving the test statistic χ^2, giving the *P*-value, and giving a correct conclusion in context. It is **partially correct** for a mistake in one of these areas.

Count partially correct answers as one-half an essentially correct answer.

4 Complete Answer   Four essentially correct answers.

3 Substantial Answer   Three essentially correct answers.

2 Developing Answer   Two essentially correct answers.

1 Minimal Answer   One essentially correct answer.

Use a holistic approach to decide a score totaling between two numbers.

---

6. (a) **Think:** You are asked for a proportionate sample and are given the proportions and the total sample size.

    Answer: City: $(0.50)(2000) = 1000$ students

    Suburbs: $(0.30)(2000) = 600$ students

    Rural: $(0.20)(2000) = 400$ students

   (b) **Think:** You are given a formula and simply asked to plug in!

    Answer: $\bar{x}_{\text{overall}} = (0.5)(74.3) + (0.3)(80.4) + (0.2)(69.8) = 75.23$

   (c) **Think:** Remember that the standard error of the sampling distribution of $\bar{x}$ is $\frac{s}{\sqrt{n}}$.

    Answer: $SE\left(\bar{x}_{\text{city}}\right) = \frac{10.2}{\sqrt{1000}} = 0.3226$

    $SE\left(\bar{x}_{\text{suburb}}\right) = \frac{9.3}{\sqrt{600}} = 0.3797$

    $SE\left(\bar{x}_{\text{rural}}\right) = \frac{12.1}{\sqrt{400}} = 0.605$

(d) **Think:** Remember that when independent random variables are combined, it's the variances that add!

Answer:

$$SE(\bar{x}_{\text{overall}}) = \sqrt{[(0.5)(0.3226)]^2 + [(0.3)(0.3797)]^2 + [(0.2)(0.605)]^2} = 0.4105$$

(e) **Think:** You are given a statement of the procedure and are told to assume all conditions for inference are satisfied. You calculated $\bar{x}_{\text{overall}}$ in (b) and $SE(\bar{x}_{\text{overall}})$ in (d), so calculating the interval is straightforward. Don't forget a conclusion in context referring to the whole population.

Answer: $75.23 \pm 2.576(0.4105) = 75.23 \pm 1.06$, or $(74.17, 76.29)$. We are 99% confident that if all fifth graders took this mathematics exam, their mean score would be between 74.17 and 76.29.

SCORING

Parts (a) and (b) together are essentially correct for the three correct numbers of students and a correct calculation of the overall sample mean. Parts (a) and (b) together are partially correct for either the correct numbers of students and an incorrect mean calculation or for a correct mean calculation using an incorrect set of numbers of students.

Part (c) is essentially correct for three correct calculations of standard error. It is partially correct for just calculating variances or for dividing by n instead of $\sqrt{n}$. Do not penalize for arithmetic errors if correct formulas are shown.

Part (d) is essentially correct for a correct calculation of $SE(\bar{x})$ and is partially correct if the coefficients (0.5, 0.3, 0.2) are misplaced or missing.

Part (e) is essentially correct for a correct confidence interval based on the answers found in parts (b) and (d) together with a correct statement in context. It is partially correct if the only mistake is either using an incorrect z-score or not putting the statement in context.

Count partially correct answers as one-half an essentially correct answer.

4 Complete Answer   Four essentially correct answers.

3 Substantial Answer   Three essentially correct answers.

2 Developing Answer   Two essentially correct answers.

1 Minimal Answer   One essentially correct answer.

Use a holistic approach to decide a score totaling between two numbers.

7. (a) **Think:** Response bias occurs when the question can lead to misleading results because people don't want to be perceived as having unpopular views or looking bad to the interviewer. In this context, people might not want to admit to their dentist that they rush in brushing their teeth! Remember that when describing bias, you should always indicate in what direction the bias might distort the results.

Answer: The patients might not want to admit to the dentist that they don't brush their teeth for as long as they should. Thus, the resulting mean brushing time from the sample would be an overestimate of the true mean brushing time.

(b) **Think:** While the patients were selected at random, which is necessary, there is also the Normal/Large Sample condition, which says that either the population is approximately normal or the sample size is large enough ($n \geq 30$) for the CLT to kick in. When neither of these are given, then we can look at the sample data to see if it appears reasonable to say that they come from a roughly normal population. That is, do the sample data at least look unimodal and roughly symmetric? Clearly this is not true here!

Answer: We are not given that the population is roughly normal; the sample size, $n = 15$, is not large enough for the central limit theorem to apply; and the sample data are skewed left and so cannot be used to reasonably conclude that they come from a roughly normal population. Thus, it would not be appropriate to use a one-sample t-procedure to produce a confidence interval for the population mean.

(c) **Think:** The mean of the exponential of the brushing times is given to be $\bar{x} = 99.24$, and the standard deviation is given to be $s = 44.35$. With $df = 15 - 1 = 14$, the critical t-scores are $\pm \texttt{invT(0.975, 14)} = \pm 2.145$. The interval is $\bar{x} \pm t^* \left(\frac{s}{\sqrt{n}} \right)$.

Answer: $99.24 \pm 2.145 \left(\frac{44.35}{\sqrt{15}} \right) = 99.24 \pm 24.56$, or $(74.68, 123.8)$. We are 95 percent confident that the true mean of the population of exponential brushing times is between 74.68 and 123.8 $\left(10^{\text{minutes}} \right)$.

(d) **Think:** This is a simple calculation, and you are actually given an example to mimic.

Answer: The endpoints are $\log(74.68) = 1.87$ and $\log(123.8) = 2.09$, so the interval is 1.87 minutes to 2.09 minutes.

(e) **Think:** Looking at Graph 2, we see that $\log \mu$ has 50 percent of the data to each side, so it must represent the median of the distribution. Also, the distribution is skewed left, so the median is greater than the mean.

Answer: (i) The parameter $\log \mu$ is equal to the median of the population of brushing times.

(ii) The parameter $\log \mu$ is greater than the mean of the population of brushing times.

(f) **Think:** As you saw in (e), $\log \mu$ gives the median of the population of brushing times, so the logs that are being taken in (d) are giving us a confidence interval for the median!

Answer: The interval in (d) is a 95 percent confidence interval for the population median of the brushing times. So, we are 95 percent confident that the true median of the population of brushing times is between 1.87 minutes and 2.09 minutes.

SCORING

Section 1 is essentially correct for a correct example of response bias in (a), a correct statement that this bias would lead to an overestimate of the mean, and a correct explanation of why it would not be appropriate to use a one-sample *t*-procedure to produce a confidence interval for the population mean in (b). Section 1 is partially correct for two out of the three steps above.

Section 2 is essentially correct for calculating the interval in (c) and for putting the answer in proper context, and is partially correct for one out of these two steps correct.

Section 3 is essentially correct for the correct calculations in (d) and the proper conclusions for (i) and (ii) in (e), and is partially correct for two out of these three steps correct.

Section 4 is essentially correct for the correction conclusion about the median in context, and is partially correct for just missing context.

Count partially correct answers as one-half an essentially correct answer.

4 Complete Answer   Four essentially correct answers.

3 Substantial Answer   Three essentially correct answers.

2 Developing Answer   Two essentially correct answers.

1 Minimal Answer   One essentially correct answer.

Use a holistic approach to decide a score totaling between two numbers.

AP SCORING GUIDE

The Multiple-Choice and Free-Response sections are weighted equally.

There is no penalty for guessing in the Multiple-Choice section.

The Investigative Task counts for 25% of the Free-Response section. Each of the Free-Response questions has a possible four points.

To find your score use the following guide:

Multiple-Choice section (40 questions)

Number correct $\times$ 1.25 = _____

Free-Response section (5 Open-Ended questions plus an Investigative Task)

Question 1 _____ $\times$ 1.875 = _____
out of 4

Question 2 _____ $\times$ 1.875 = _____
out of 4

Question 3 _____ $\times$ 1.875 = _____
out of 4

Question 4 _____ $\times$ 1.875 = _____
out of 4

Question 5 _____ $\times$ 1.875 = _____
out of 4

Question 6 _____ $\times$ 3.125 = _____
out of 4

Total points from Multiple-Choice and Free-Response sections = _____

Conversion chart based on a recent AP exam

Total Points	AP Score
73–100	5
59–72	4
44–58	3
32–43	2
0–31	1

In the past, roughly 10% of students scored 5, 20% scored 4, 25% scored 3, 20% scored 2, and 25% scored 1 on the AP Statistics exam. Colleges generally require a score of at least 3 for a student to receive college credit.

BASIC USES OF THE TI-84

There are many more useful features than introduced below—see the guidebook that comes with the calculator. For reference, the following is a listing of the basic uses with which all students should be familiar. However, always remember that the calculator is only a tool, that it will be of minimal use in the multiple-choice section, and that "calculator talk" (calculator syntax) can be used in the free-response section as long as the inputs are labeled.

Plotting statistical data:

> **STAT PLOT** allows one to show scatterplots, histograms, modified boxplots, and regular boxplots of data stored in lists. Note the use of **TRACE** with the various plots.

Numerical statistical data:

> **1-Var Stats** gives the mean, standard deviation, and five-number summary of a list of data.

Binomial probabilities:

> **binompdf** (n, p, x) gives the probability of exactly x successes in n trials, where p is the probability of success on a single trial.

> **binomcdf** (n, p, x) gives the cumulative probability of x or fewer successes in n trials, where p is the probability of success on a single trial.

Geometric probabilities:

> **geometpdf** (p, x) gives the probability that the first success occurs on the x-th trial, where p is the probability of success on a single trial.

> **geometcdf** (p, x) gives the cumulative probability that the first success occurs on or before the x-th trial, where p is the probability of success on a single trial.

The normal distribution:

> **normalcdf** (lowerbound, upperbound, μ, σ) gives the probability that a score is between the two bounds for the designated mean μ and standard deviation σ. The defaults are $\mu = 0$ and $\sigma = 1$.

> **InvNorm** (area, μ, σ) gives the score associated with an area (probability) to the left of the score for the designated mean μ and standard deviation σ. The defaults are $\mu = 0$ and $\sigma = 1$.

The t-distribution:

> **tcdf** (lowerbound, upperbound, df) gives the probability a score is between the two bounds for the specified df (degrees of freedom).

> **invT** (area, df) gives the t-score associated with an area (probability) to the left of the score under the student t-probability function for the specified df (degrees of freedom).

The chi-square distribution:

> χ^2**cdf** (lowerbound, upperbound, df) gives the probability a score is between the two bounds for the specified df (degrees of freedom).

Linear regression and correlation:

LinReg (ax + b) fits the equation $y = ax + b$ to the data in lists L1 and L2 using a least-squares fit. When **DiagnosticOn** is set, the values for r^2 and r are also displayed.

Confidence intervals:

For proportions—

1-PropZInt gives a confidence interval for a proportion of successes.
2-PropZInt gives a confidence interval for the difference between the proportion of successes in two populations.

For means—

TInterval gives a confidence interval for a population mean (use the t-distribution because population variances are never really known).
2-SampTInt gives a confidence interval for the difference between two population means.

Hypothesis tests:

For proportions—

1-PropZTest
2-PropZTest compares the proportion of successes from two populations (making use of the pooled sample proportion).

For means—

T-Test
2-SampTTest

For chi-square test for association—

χ^2-**Test** gives the χ^2-value and P-value for the null hypothesis H_0: no association between row and column variables, and the alternative hypothesis H_a: the variables are related. The observed counts must first be entered into a matrix.

χ^2**GOF-Test** is a chi-square goodness-of-fit test to confirm whether sample data conform to a specified distribution. [Note: this is available on the new operating system for the TI-84+.]

For linear regression—

LinRegTTest calculates a linear regression and performs a t-test on the null hypothesis H_0: $\beta = 0$ (H_0: $\rho = 0$). The regression equation is stored in **RegEQ** (under **VARS Statistics EQ**) and the list of residuals is stored in **RESID** (under **LIST NAMES**).

Catalog help:

To activate Catalog Help, press APPS, choose CtlgHelp, and press ENTER. Then, for example, if you press 2nd, DISTR, arrow down to normalcdf, and press +, you are prompted to insert (lowerbound, upperbound, $[\mu, \sigma]$), that is, to insert the bounds and, optionally, the mean and SD.

BASIC USES OF THE TI-NSPIRE

The TI-Nspire has additional statistical features, more memory, and greater resolution than the TI-84+.

Entering data:

Add a **Lists & Spreadsheet** page. *Name* the list, and enter the data in the spreadsheet column. Data can then be graphed and/or used to create confidence intervals or run hypothesis tests.

Plotting statistical data:

Add a **Data & Statistics** page. At the bottom center of the screen, **Click to add variable** and select the list *name* to graph. The default graph is a dotplot. Use **Menu → Plot Type** to change to a Box Plot, Histogram, or Normal Probability Plot. If graphing bivariate data, click to add a variable and select second *name* of list to graph. Use Menu → Analyze → Regression → Show Linear(a+bx) to plot the regression line over the graph.

Probability functions:

All probability functions can be found in either of two places: under **Menu → Statistics → Distributions** or **Menu → Probability → Distributions**.

Normal distribution:

Normal Cdf (Lower Bound, Upper Bound, μ, σ) gives the probability that a score is between the two bounds for the designated mean and standard deviation.

Inverse Normal(Area, μ, σ) gives the value associated with an area (probability) to the left of the score for the designated mean and standard deviation.

t-distribution:

t Cdf (Lower Bound, Upper Bound, df) gives the probability a value is between the two bounds for the specified degrees of freedom.

Inverse t (Area, df) gives the *t*-score associated with an area (probability) to the left of the score for the designated degrees of freedom.

Chi-square distribution:

$\chi 2$ **Cdf** (Lower Bound, Upper Bound, df) gives the probability a score is between the two bounds for the specified degrees of freedom.

Binomial probabilities:

Binomial Pdf (*n*, *p*, *x*) gives the probability of exactly *x* successes in *n* trials, where *p* is the probability of success on a single trial.

Binomial Cdf (*n*, *p*, Lower Bound, Upper Bound) gives the cumulative probability of getting between a Lower and an Upper number of successes in *n* trials, where *p* is the probability of success on a single trial.

Confidence Intervals:

This can be accessed from either a Lists & Spreadsheet page (if using data) or a Calculator page (if using summary statistics or proportions).

Menu → Statistics → Confidence Intervals → desired interval

For proportions:

1-Prop z Interval (x, n, C-level) x has to be an integer value.

2-Prop z Interval (x1, n1, x2, n2, C-level) x1 and x2 have to be integer values.

For means:

t Interval can use data (list *name* from List & Spreadsheet) or summary statistics ($\bar{x}$, s, n) along with the level of confidence.

2-Sample t Interval can use data (list *names* from List & Spreadsheet) or two sample means, sample standard deviations, and sample sizes, along with the level of confidence.

For linear regression:

Linear Reg t Intervals (x list, y list, C-level) gives a confidence interval for a slope.

Hypothesis tests:

This can be accessed from either a Lists & Spreadsheet page (if using data) or a Calculator page (if using summary statistics or proportions).

Menu → Statistics → Stat Tests → desired test

For proportions:

1-Prop z Test(Po, x, n, alt. hyp) x has to be an integer value.

2-Prop z Test(x1, n1, x2, n2, alt. hyp) x1 and x2 have to be integer values.

For means:

t Test can use data or summary statistics.

2-Sample t Test can use data or summary statistics.

Chi-squared tests:

χ^2 GOF Test(observed list, expected list, df) gives the χ^2 statistic and the *P*-value.

$\chi2$ 2-way Test(name of matrix) for tests of association or homogeneity. It runs the hypothesis test and finds the test statistic and *P*-value. The observed counts must first be entered into a matrix.

Linear regression:

Linear Reg t Test(x list, y list, alt hyp) calculates the regression line and performs a *t*-test on the null hypothesis H_0: $\beta = 0$. Output gives both *s* and SE Slope, and it gives a list of residuals.

BASIC USES OF THE CASIO PRIZM

Casio calculators have a module completely devoted to Statistics. You can find this module on the Main Menu and can access it with number 2 on the key pad.

Plotting statistical data:

GRAPH allows one to show bar charts, pie charts, scatterplots, histograms, and boxplots of data in any of 26 stored lists. Note the use of **TRACE** with the various plots. With quantitative graphs, you can select **1-VAR** and access summary statistics. For scatterplots, selecting **DefG** will calculate predictions and **CALC** will find all manners of regressions.

Numerical statistical data:

CALC 1-VAR gives the mean, standard deviation, mode, and five-number summary of a list of data.

Probability Distributions are accessed in the $\boxed{\text{DIST}}$ menu. Here are some examples.

Binomial probabilities:

Bpd gives the probability of exactly x successes in n (Numtrial), where p is the probability of success on a single trial.

Bcd gives the cumulative probability of receiving between a Lower and an Upper number of successes, where p is the probability of success on a single trial.

Geometric probabilities:

Gpd gives the probability that the first success occurs on the x-th trial, where p is the probability of success on a single trial.

Gcd gives the cumulative probability of receiving your first success between a Lower and an Upper number of successes, where p is the probability of success on a single trial.

The normal distribution:

Ncd gives the probability that a score is between a Lower value and an Upper value with a given σ and μ. When using z-scores, the defaults of $\sigma = 1$ and $\mu = 0$ should be used. **InvN** gives the requested area that is either on the **Left** of the normal curve, on the **Right** of the normal curve, or in the **Central** area. When using z-scores, the defaults of $\sigma = 1$ and $\mu = 0$ should be used.

The t-distribution:

tcdf gives the probability a score is between the Lower and Upper bounds for the specified df (degrees of freedom).
Invt gives the t score associated with an area (probability) to the left of the score under the student t-probability function for the specified df (degrees of freedom).

The chi-square distribution:

Ccd gives the probability a score is between the Lower and Upper bounds for the specified df (degrees of freedom).

Linear regression and correlation:

From the $\boxed{\text{CALC}}$ menu, use the $\boxed{\text{X}}$ menu to fit the least-squares equation $\hat{y} = a + bx$ for data in any two lists. This command can also be accessed when viewing a scatterplot. Use the $\boxed{\text{SETUP}}$ command to choose where the residuals are stored.

Confidence Intervals are available in the INTR **menu:**

For proportions—

Z-1-PROP gives a confidence interval for a proportion of successes.

Z-2-PROP gives a confidence interval for the difference between the proportion of successes in two populations.

For means—

t-1-SAMPLE gives a confidence interval for a population mean.

t-2-SAMPLE gives a confidence interval for the difference between two population means.

Hypothesis Tests are available in the TEST **menu:**

For proportions—

Z-1-PROP is for a one-sample proportion test.

Z-2-PROP compares the proportion of successes from two populations (making use of the pooled sample proportion).

For means—

t-1-SAMPLE is for a one-sample mean test.

t-2-SAMPLE compares the difference between two population means.

For chi-square—

CHI-GOF is a chi-square goodness-of-fit test to confirm whether sample data conform to a specified distribution.

CHI-2WAY gives the χ^2-value and P-value for the null hypothesis H_o: no association between row and column variables, and the alternative hypothesis H_a: the variables are related. Data are entered in a matrix by pressing **F2** in the 2WAY menu.

Test or Interval for the slope of a regression equation—

t-REG in the INTR menu will calculate a confidence interval for slope.

t-REG in the TEST calculates a linear regression and performs a t-test on the null hypothesis H_0: $\beta = 0$ (H_0: $\rho = 0$). The regression equation can be saved into any **Y** variable of your choice with the **COPY** command, and the residuals will be placed in the **List** selected in the SETUP menu.

A 90-day free emulator trial of the Casio calculator is available at:

https://edu.casio.com/freetrial/en/freetrial_form.php

BASIC USES OF THE HP PRIME

The statistical apps:

Press `APPS INFO` and tap to open the app you want (Statistics 1Var, Statistics 2Var, or Inference). Press `NUM SETUP` for Numeric view, `SYMB SETUP` for Symbolic view, and `PLOT SETUP` for Plot view (or press `VIEW COPY` and select **Autoscale**).

Statistics 1Var App	
Numeric view	• `STATS` Enter data in list D1 or enter data in D1 and frequencies in D2 • Tap to calculate summary statistics (five-number summary and so on)
Symbolic view	• Set analysis H1 to use D1 as data or D1 as data and D2 as frequencies • Select a plot type (Histogram, Box and Whisker, Normal Probability, Dot, Stem and Leaf, Pie, and so on)
Plot view	• Press `VIEW COPY` and select **Autoscale** • View the graphical representation of your data; tap to move the cursor, pinch to zoom, drag to pan

Statistics 2Var App	
Numeric view	• Enter independent data in list C1 and dependent data in C2 • Tap `STATS` to calculate summary statistics (r, R^2, and so on)
Symbolic view	• Set analysis S1 to use C1 as independent data and C2 as dependent data • Select a fit type (Linear, Exponential, Logarithmic, and so on)
Plot view	• Press `VIEW COPY` and select **Autoscale** • View the scatterplot and fit; pinch to zoom, drag to pan • Tap `FIT` to toggle the fit on and off; the fit equation is saved to Symbolic view

Inference App	
Symbolic view	• Select a method (Hypothesis Test, Confidence Interval, χ^2 Test, Regression) • Select a type; options vary by Method; for example, if the Method is χ^2 Test, then the types are Goodness-of-fit and 2-way test • Select an alternative hypothesis for hypothesis tests or linear t-test
Numeric view	• Enter your sample statistics or import them from the Statistics 1Var or Statistics 2Var apps • Tap `CALC` to see the computed results in a table
Plot view	• View a graphical representation of your test results • For the linear t-test, use the U and D direction keys to cycle through the scatterplot, the residual plot, a histogram of the residuals, and the normal probability plot of the residuals

The probability density functions:

Press MEM B to open the Toolbox menu. Tap MATH to open the numerical Math menu. The fifth option is **Probability**. Under this option are Density, Cumulative Density, and Inverse Cumulative Density categories for the Normal, Student's T, Chi-Square, Binomial, Geometric, and other probability density functions.

NORMAL_CDF(μ, σ, x) returns the lower-tail probability of the normal probability density function for the value x, given the mean (μ) and standard deviation (σ) of a normal distribution. **NORMAL_CDF**(μ, σ, x_1, x_2) returns the area under the normal probability density function between the two x-values.

NORMALD_ICDF(μ, σ, p) returns the cumulative normal distribution x-value associated with the lower-tail probability p, given the mean (μ) and standard deviation (σ) of a normal distribution.

If μ and σ are omitted for the three commands above, results for the standard normal distribution will be returned.

STUDENT_CDF(n, x) returns the lower-tail probability of the Student-t probability density function at x, given n degrees of freedom. **STUDENT_CDF**(n, x_1, x_2) returns the area under the Student-t probability density function between the two x-values.

STUDENT_ICDF(n, p) returns the value x such that the Student-t lower-tail probability of x, with n degrees of freedom, is p.

The commands **CHISQUARE_CDF** and **CHISQUARE_ICDF** work in a manner similar to the above.

BINOMIAL(n, p, k) returns the probability of k successes out of n trials, each with a probability of success p.

BINOMIAL_CDF(n, p, k) returns the probability of k or fewer successes out of n trials, each with a probability of success p. **BINOMIAL_CDF**(n, p, k_1, k_2) returns the probability of at least k_1 but no more than k_2 successes out of n trials, each with a probability of success p.

BINOMIAL_ICDF(n, p, q) returns the number of successes, k, out of n trials, each with a probability of p, such that the probability of k or fewer successes is q.

The commands **GEOMETRIC_CDF** and **GEOMTRIC_ICDF** work in a manner similar to the above.

Help

Help on these and other apps, commands, and functions is always available. Press MEM B to open the Toolbox menus, and navigate to any command and press HELP USER to see help for that command. Tap EXAMPL and choose an example to paste it to the command line; press ENTER ≈ to see the result.

A 90-day free emulator trial of the HP Prime calculator is available at:

www.hp.com/go/primeapptrial

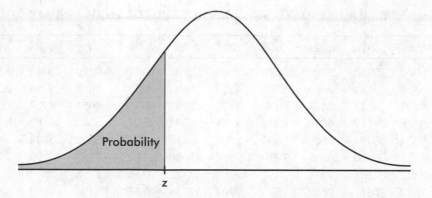

TABLE A: STANDARD NORMAL PROBABILITIES

z	.00	.01	.02	.03	.04	.05	.06	.07	.08	.09
−3.4	.0003	.0003	.0003	.0003	.0003	.0003	.0003	.0003	.0003	.0002
−3.3	.0005	.0005	.0005	.0004	.0004	.0004	.0004	.0004	.0004	.0003
−3.2	.0007	.0007	.0006	.0006	.0006	.0006	.0006	.0005	.0005	.0005
−3.1	.0010	.0009	.0009	.0009	.0008	.0008	.0008	.0008	.0007	.0007
−3.0	.0013	.0013	.0013	.0012	.0012	.0011	.0011	.0011	.0010	.0010
−2.9	.0019	.0018	.0018	.0017	.0016	.0016	.0015	.0015	.0014	.0014
−2.8	.0026	.0025	.0024	.0023	.0023	.0022	.0021	.0021	.0020	.0019
−2.7	.0035	.0034	.0033	.0032	.0031	.0030	.0029	.0028	.0027	.0026
−2.6	.0047	.0045	.0044	.0043	.0041	.0040	.0039	.0038	.0037	.0036
−2.5	.0062	.0060	.0059	.0057	.0055	.0054	.0052	.0051	.0049	.0048
−2.4	.0082	.0080	.0078	.0075	.0073	.0071	.0069	.0068	.0066	.0064
−2.3	.0107	.0104	.0102	.0099	.0096	.0094	.0091	.0089	.0087	.0084
−2.2	.0139	.0136	.0132	.0129	.0125	.0122	.0119	.0116	.0113	.0110
−2.1	.0179	.0174	.0170	.0166	.0162	.0158	.0154	.0150	.0146	.0143
−2.0	.0228	.0222	.0217	.0212	.0207	.0202	.0197	.0192	.0188	.0183
−1.9	.0287	.0281	.0274	.0268	.0262	.0256	.0250	.0244	.0239	.0233
−1.8	.0359	.0351	.0344	.0336	.0329	.0322	.0314	.0307	.0301	.0294
−1.7	.0446	.0436	.0427	.0418	.0409	.0401	.0392	.0384	.0375	.0367
−1.6	.0548	.0537	.0526	.0516	.0505	.0495	.0485	.0475	.0465	.0455
−1.5	.0668	.0655	.0643	.0630	.0618	.0606	.0594	.0582	.0571	.0559
−1.4	.0808	.0793	.0778	.0764	.0749	.0735	.0721	.0708	.0694	.0681
−1.3	.0968	.0951	.0934	.0918	.0901	.0885	.0869	.0853	.0838	.0823
−1.2	.1151	.1131	.1112	.1093	.1075	.1056	.1038	.1020	.1003	.0985
−1.1	.1357	.1335	.1314	.1292	.1271	.1251	.1230	.1210	.1190	.1170
−1.0	.1587	.1562	.1539	.1515	.1492	.1469	.1446	.1423	.1401	.1379
−0.9	.1841	.1814	.1788	.1762	.1736	.1711	.1685	.1660	.1635	.1611
−0.8	.2119	.2090	.2061	.2033	.2005	.1977	.1949	.1922	.1894	.1867
−0.7	.2420	.2389	.2358	.2327	.2296	.2266	.2236	.2206	.2177	.2148
−0.6	.2743	.2709	.2676	.2643	.2611	.2578	.2546	.2514	.2483	.2451
−0.5	.3085	.3050	.3015	.2981	.2946	.2912	.2877	.2843	.2810	.2776
−0.4	.3446	.3409	.3372	.3336	.3300	.3264	.3228	.3192	.3156	.3121
−0.3	.3821	.3783	.3745	.3707	.3669	.3632	.3594	.3557	.3520	.3483
−0.2	.4207	.4168	.4129	.4090	.4052	.4013	.3974	.3936	.3897	.3859
−0.1	.4602	.4562	.4522	.4483	.4443	.4404	.4364	.4325	.4286	.4247
−0.0	.5000	.4960	.4920	.4880	.4840	.4801	.4761	.4721	.4681	.4641

TABLE A: STANDARD NORMAL PROBABILITIES (CONTINUED)

z	.00	.01	.02	.03	.04	.05	.06	.07	.08	.09
0.0	.5000	.5040	.5080	.5120	.5160	.5199	.5239	.5279	.5319	.5359
0.1	.5398	.5438	.5478	.5517	.5557	.5596	.5636	.5675	.5714	.5753
0.2	.5793	.5832	.5871	.5910	.5948	.5987	.6026	.6064	.6103	.6141
0.3	.6179	.6217	.6255	.6293	.6331	.6368	.6406	.6643	.6480	.6517
0.4	.6554	.6591	.6628	.6664	.6700	.6736	.6772	.6808	.6844	.6879
0.5	.6915	.6950	.6985	.7019	.7054	.7088	.7123	.7157	.7190	.7224
0.6	.7257	.7291	.7324	.7357	.7389	.7422	.7454	.7486	.7517	.7549
0.7	.7580	.7611	.7642	.7673	.7704	.7734	.7764	.7794	.7823	.7852
0.8	.7881	.7910	.7939	.7967	.7995	.8023	.8051	.8078	.8106	.8133
0.9	.8159	.8186	.8212	.8238	.8264	.8289	.8315	.8340	.8365	.8389
1.0	.8413	.8438	.8461	.8485	.8508	.8531	.8554	.8577	.8599	.8621
1.1	.8643	.8665	.8686	.8708	.8729	.8749	.8770	.8790	.8810	.8830
1.2	.8849	.8869	.8888	.8907	.8925	.8944	.8962	.8980	.8997	.9015
1.3	.9032	.9049	.9066	.9082	.9099	.9115	.9131	.9147	.9162	.9177
1.4	.9192	.9207	.9222	.9236	.9251	.9265	.9279	.9292	.9306	.9319
1.5	.9332	.9345	.9357	.9370	.9382	.9394	.9406	.9418	.9429	.9441
1.6	.9452	.9463	.9474	.9484	.9495	.9505	.9515	.9525	.9535	.9545
1.7	.9554	.9564	.9573	.9582	.9591	.9599	.9608	.9616	.9625	.9633
1.8	.9641	.9649	.9656	.9664	.9671	.9678	.9686	.9693	.9699	.9706
1.9	.9713	.9719	.9726	.9732	.9738	.9744	.9750	.9756	.9761	.9767
2.0	.9772	.9778	.9783	.9788	.9793	.9798	.9803	.9808	.9812	.9817
2.1	.9821	.9826	.9830	.9834	.9838	.9842	.9846	.9850	.9854	.9857
2.2	.9861	.9864	.9868	.9871	.9875	.9878	.9881	.9884	.9887	.9890
2.3	.9893	.9896	.9898	.9901	.9904	.9906	.9909	.9911	.9913	.9916
2.4	.9918	.9920	.9922	.9925	.9927	.9929	.9931	.9932	.9934	.9936
2.5	.9938	.9940	.9941	.9943	.9945	.9946	.9948	.9949	.9951	.9952
2.6	.9953	.9955	.9956	.9957	.9959	.9960	.9961	.9962	.9963	.9964
2.7	.9965	.9966	.9967	.9968	.9969	.9970	.9971	.9972	.9973	.9974
2.8	.9974	.9975	.9976	.9977	.9977	.9978	.9979	.9979	.9980	.9981
2.9	.9981	.9982	.9982	.9983	.9984	.9984	.9985	.9985	.9986	.9986
3.0	.9987	.9987	.9987	.9988	.9988	.9989	.9989	.9989	.9990	.9990
3.1	.9990	.9991	.9991	.9991	.9992	.9992	.9992	.9992	.9993	.9993
3.2	.9993	.9993	.9994	.9994	.9994	.9994	.9994	.9995	.9995	.9995
3.3	.9995	.9995	.9995	.9996	.9996	.9996	.9996	.9996	.9996	.9997
3.4	.9997	.9997	.9997	.9997	.9997	.9997	.9997	.9997	.9997	.9998

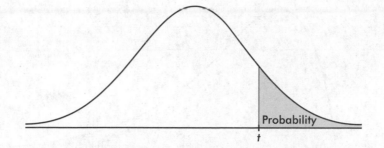

Probability

t

TABLE B: t-DISTRIBUTION CRITICAL VALUES

						Tail probability p						
df	.25	.20	.15	.10	.05	.025	.02	.01	.005	.0025	.001	.0005
1	1.000	1.376	1.963	3.078	6.314	12.71	15.89	31.82	63.66	127.3	318.3	636.6
2	.816	1.061	1.386	1.886	2.920	4.303	4.849	6.965	9.925	14.09	22.33	31.60
3	.765	.978	1.250	1.638	2.353	3.182	3.482	4.541	5.841	7.453	10.21	12.92
4	.741	.941	1.190	1.533	2.132	2.776	2.999	3.747	4.604	5.598	7.173	8.610
5	.727	.920	1.156	1.476	2.015	2.571	2.757	3.365	4.032	4.773	5.893	6.869
6	.718	.906	1.134	1.440	1.943	2.447	2.612	3.143	3.707	4.317	5.208	5.959
7	.711	.896	1.119	1.415	1.895	2.365	2.517	2.998	3.499	4.029	4.785	5.408
8	.706	.889	1.108	1.397	1.860	2.306	2.449	2.896	3.355	3.833	4.501	5.041
9	.703	.883	1.100	1.383	1.833	2.262	2.398	2.821	3.250	3.690	4.297	4.781
10	.700	.879	1.093	1.372	1.812	2.228	2.359	2.764	3.169	3.581	4.144	4.587
11	.697	.876	1.088	1.363	1.796	2.201	2.328	2.718	3.106	3.497	4.025	4.437
12	.695	.873	1.083	1.356	1.782	2.179	2.303	2.681	3.055	3.428	3.930	4.318
13	.694	.870	1.079	1.350	1.771	2.160	2.282	2.650	3.012	3.372	3.852	4.221
14	.692	.868	1.076	1.345	1.761	2.145	2.264	2.624	2.977	3.326	3.787	4.140
15	.691	.866	1.074	1.341	1.753	2.131	2.249	2.602	2.947	3.286	3.733	4.073
16	.690	.865	1.071	1.337	1.746	2.120	2.235	2.583	2.921	3.252	3.686	4.015
17	.689	.863	1.069	1.333	1.740	2.110	2.224	2.567	2.898	3.222	3.646	3.965
18	.688	.862	1.067	1.330	1.734	2.101	2.214	2.552	2.878	3.197	3.611	3.922
19	.688	.861	1.066	1.328	1.729	2.093	2.205	2.539	2.861	3.174	3.579	3.883
20	.687	.860	1.064	1.325	1.725	2.086	2.197	2.528	2.845	3.153	3.552	3.850
21	.686	.859	1.063	1.323	1.721	2.080	2.189	2.518	2.831	3.135	3.527	3.819
22	.686	.858	1.061	1.321	1.717	2.074	2.183	2.508	2.819	3.119	3.505	3.792
23	.685	.858	1.060	1.319	1.714	2.069	2.177	2.500	2.807	3.104	3.485	3.768
24	.685	.857	1.059	1.318	1.711	2.064	2.172	2.492	2.797	3.091	3.467	3.745
25	.684	.856	1.058	1.316	1.708	2.060	2.167	2.485	2.787	3.078	3.450	3.725
26	.684	.856	1.058	1.315	1.706	2.056	2.162	2.479	2.779	3.067	3.435	3.707
27	.684	.855	1.057	1.314	1.703	2.052	2.158	2.473	2.771	3.057	3.421	3.690
28	.683	.855	1.056	1.313	1.701	2.048	2.154	2.467	2.763	3.047	3.408	3.674
29	.683	.854	1.055	1.311	1.699	2.045	2.150	2.462	2.756	3.038	3.396	3.659
30	.683	.854	1.055	1.310	1.697	2.042	2.147	2.457	2.750	3.030	3.385	3.646
40	.681	.851	1.050	1.303	1.684	2.021	2.123	2.423	2.704	2.971	3.307	3.551
50	.679	.849	1.047	1.299	1.676	2.009	2.109	2.403	2.678	2.937	3.261	3.496
60	.679	.848	1.045	1.296	1.671	2.000	2.099	2.390	2.660	2.915	3.232	3.460
80	.678	.846	1.043	1.292	1.664	1.990	2.088	2.374	2.639	2.887	3.195	3.416
100	.677	.845	1.042	1.290	1.660	1.984	2.081	2.364	2.626	2.871	3.174	3.390
1000	.675	.842	1.037	1.282	1.646	1.962	2.056	2.330	2.581	2.813	3.098	3.300
∞	.674	.841	1.036	1.282	1.645	1.960	2.054	2.326	2.576	2.807	3.091	3.291
	50%	60%	70%	80%	90%	95%	96%	98%	99%	99.5%	99.8%	99.9%
						Confidence level C						

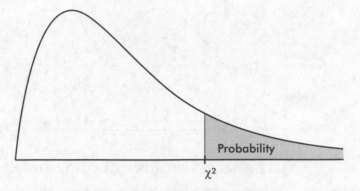

Probability
χ^2

TABLE C: χ^2 CRITICAL VALUES

	Tail probability p										
df	.25	.20	.15	.10	.05	.025	.02	.01	.005	.0025	.001
1	1.32	1.64	2.07	2.71	3.84	5.02	5.41	6.63	7.88	9.14	10.83
2	2.77	3.22	3.79	4.61	5.99	7.38	7.82	9.21	10.60	11.98	13.82
3	4.11	4.64	5.32	6.25	7.81	9.35	9.84	11.34	12.84	14.32	16.27
4	5.39	5.99	6.74	7.78	9.49	11.14	11.67	13.28	14.86	16.42	18.47
5	6.63	7.29	8.12	9.24	11.07	12.83	13.39	15.09	16.75	18.39	20.51
6	7.84	8.56	9.45	10.64	12.59	14.45	15.03	16.81	18.55	20.25	22.46
7	9.04	9.80	10.75	12.02	14.07	16.01	16.62	18.48	20.28	22.04	24.32
8	10.22	11.03	12.03	13.36	15.51	17.53	18.17	20.09	21.95	23.77	26.12
9	11.39	12.24	13.29	14.68	16.92	19.02	19.68	21.67	23.59	25.46	27.88
10	12.55	13.44	14.53	15.99	18.31	20.48	21.16	23.21	25.19	27.11	29.59
11	13.70	14.63	15.77	17.28	19.68	21.92	22.62	24.72	26.76	28.73	31.26
12	14.85	15.81	16.99	18.55	21.03	23.34	24.05	26.22	28.30	30.32	32.91
13	15.98	16.98	18.20	19.81	22.36	24.74	25.47	27.69	29.82	31.88	34.53
14	17.12	18.15	19.41	21.06	23.68	26.12	26.87	29.14	31.32	33.43	36.12
15	18.25	19.31	20.60	22.31	25.00	27.49	28.26	30.58	32.80	34.95	37.70
16	19.37	20.47	21.79	23.54	26.30	28.85	29.63	32.00	34.27	36.46	39.25
17	20.49	21.61	22.98	24.77	27.59	30.19	31.00	33.41	35.72	37.95	40.79
18	21.60	22.76	24.16	25.99	28.87	31.53	32.35	34.81	37.16	39.42	42.31
19	22.72	23.90	25.33	27.20	30.14	32.85	33.69	36.19	38.58	40.88	43.82
20	23.83	25.04	26.50	28.41	31.41	34.17	35.02	37.57	40.00	42.34	45.31
21	24.93	26.17	27.66	29.62	32.67	35.48	36.34	38.93	41.40	43.78	46.80
22	26.04	27.30	28.82	30.81	33.92	36.78	37.66	40.29	42.80	45.20	48.27
23	27.14	28.43	29.98	32.01	35.17	38.08	38.97	41.64	44.18	46.62	49.73
24	28.24	29.55	31.13	33.20	36.42	39.36	40.27	42.98	45.56	48.03	51.18
25	29.34	30.68	32.28	34.38	37.65	40.65	41.57	44.31	46.93	49.44	52.62
26	30.43	31.79	33.43	35.56	38.89	41.92	42.86	45.64	48.29	50.83	54.05
27	31.53	32.91	34.57	36.74	40.11	43.19	44.14	46.96	49.64	52.22	55.48
28	32.62	34.03	35.71	37.92	41.34	44.46	45.42	48.28	50.99	53.59	56.89
29	33.71	35.14	36.85	39.09	42.56	45.72	46.69	49.59	52.34	54.97	58.30
30	34.80	36.25	37.99	40.26	43.77	46.98	47.96	50.89	53.67	56.33	59.70
40	45.62	47.27	49.24	51.81	55.76	59.34	60.44	63.69	66.77	69.70	73.40
50	56.33	58.16	60.35	63.17	67.50	71.42	72.61	76.15	79.49	82.66	86.66
60	66.98	68.97	71.34	74.40	79.08	83.30	84.58	88.38	91.95	95.34	99.61
80	88.13	90.41	93.11	96.58	101.9	106.6	108.1	112.3	116.3	120.1	124.8
100	109.1	111.7	114.7	118.5	124.3	129.6	131.1	135.8	140.2	144.3	149.4

Index

Parentheses indicate problem numbers